Lecture Notes in Computer Science 5854

Wenyin Liu Xiangfeng Luo Fu Lee Wang
Jingsheng Lei (Eds.)

Web Information Systems and Mining

International Conference, WISM 2009
Shanghai, China, November 7-8, 2009
Proceedings

Springer

Volume Editors

Wenyin Liu
City University of Hong Kong, Department of Computer Science
83 Tat Chee Avenue, Kowloon Tong, Hong Kong, China
E-mail: csliuwy@cityu.edu.hk

Xiangfeng Luo
Shanghai University, School of Computer Engineering and Science
Digital Content Computing and Cognitive Informatics Group
Shanghai 200072, China
E-mail: luoxiangfeng@gmail.com

Fu Lee Wang
City University of Hong Kong, Department of Computer Science
83 Tat Chee Avenue, Kowloon Tong, Hong Kong, China
E-mail: flwang@cityu.edu.hk

Jingsheng Lei
Hainan University, College of Information Science and Technology
Haikou 570228, China
E-mail: jshlei@hainu.edu.cn

Library of Congress Control Number: 2009936960

CR Subject Classification (1998): H.3.5, I.2.6, H.2.8, C.2, J.1, K.4.4

LNCS Sublibrary: SL 3 – Information Systems and Application, incl. Internet/Web and HCI

ISSN 0302-9743

ISBN 978-3-642-05249-1 Springer Berlin Heidelberg New York

Typesetting: Camera-ready by author, data conversion by Scientific Publishing Services, Chennai, India
Printed on acid-free paper SPIN: 12781188 06/3180 5 4 3 2 1 0

Preface

The 2009 International Conference on Web Information Systems and Mining (WISM 2009) was held in Shanghai, China 7–8 November 2009. WISM 2009 received 598 submissions from 20 countries and regions. After rigorous reviews, 61 high-quality papers were selected for publication in this volume. The acceptance rate was 10.2%.

The aim of WISM 2009 was to bring together researchers working in many different areas of Web information systems and Web mining to foster exchange of new ideas and promote international collaborations. In addition to the large number of submitted papers and invited sessions, there were several internationally well-known keynote speeches.

On behalf of the Organizing Committee, we thank the Shanghai University of Electric Power for its sponsorship and logistics support. We also thank the members of the Organizing Committee and the Program Committee for their hard work. We are very grateful to the keynote speakers, invited session organizers, session chairs, reviewers, and student helpers. Last but not least, we thank all the authors and participants for their great contributions that made this conference possible.

November 2009

Wenyin Liu
Xiangfeng Luo
Fu Lee Wang
Jingsheng Lei

Organization

Organizing Committee

General Co-chairs

Jialin Cao	Shanghai University of Electric Power, China
Jingsheng Lei	Hainan University, China

Program Committee

Co-chairs

Wenyin Liu	City University of Hong Kong, Hong Kong
Xiangfeng Luo	Shanghai University, China

Local Arrangements

Chair

Hao Zhang	Shanghai University of Electric Power, China

Proceedings

Co-chairs

Fu Lee Wang	City University of Hong Kong, Hong Kong
Jun Yang	Shanghai University of Electric Power, China

Publicity

Chair

Tim Kovacs	University of Bristol, UK

Sponsorship

Chair

Zhiyu Zhou	Zhejiang Sci-Tech University, China

Program Committee

Ladjel Bellatreche	ENSMA - Poitiers University, France
Sourav Bhowmick	Nanyang Technological University, Singapore
Stephane Bressan	National University of Singapore, Singapore
Erik Buchmann	Universität Karlsruhe,Germany
Jinli Cao	La Trobe University, Australia
Jian Cao	Shanghai Jiao Tong University, China
Badrish Chandramouli	Microsoft Research, USA
Akmal Chaudhri	City University of London, UK
Qiming Chen	Hewlett-Packard Laboratories, USA
Lei Chen	The Hong Kong University of Science and Technology, China
Jinjun Chen	Swinburne University of Technology, Australia
Hong Cheng	The Chinese University of Hong Kong, China
Reynold Cheng	Hong Kong Polytechnic University, China
Bin Cui	Peking University, China
Alfredo Cuzzocrea	University of Calabria, Italy
Wanchun Dou	Nanjing University, China
Xiaoyong Du	Renmin University of China, China
Ling Feng	Tsinghua University, China
Cheng Fu	Nanyang Technological University, Singapore
Gabriel Fung	The University of Queensland, Australia
Byron Gao	University of Wisconsin, USA
Yunjun Gao	Zhejiang University, China
Bin Gao	Microsoft Research, China
Anandha	Gopalan Imperial College, UK
Stephane Grumbach	INRIA, France
Giovanna Guerrini	Università di Genova, Italy
Mohand-Said Hacid	Université Claude Bernard Lyon 1, France
Ming Hua	Simon Fraser University, Canada
Ela Hunt	University of Strathclyde, Glasgow, UK
Renato Iannella	National ICT Australia
Yan Jia	National University of Defence Technology, China
Yu-Kwong Ricky	Colorado State University, USA
Yoon Joon Lee	KAIST, Korea
Carson Leung	The University of Manitoba, Canada
Lily Li	CSIRO, Australia
Tao Li	Florida International University
Wenxin Liang	Dalian University of Technology, China
Chao Liu	Microsoft, USA
Qing Liu	CSIRO, Australia
Jie Liu	Chinese Academy of Sciences, China
JianXun Liu	Hunan University of Science and Technology, China
Peng Liu	PLA University of Science and Technology, China
Jiaheng Lu	University of California, Irvine, USA
Weiyi Meng	Binghamton University, USA

Wahab Mohd	Universiti Tun Hussein Malaysia, Malaysia
Miyuki Nakano	University of Tokyo, Japan
Wilfred Ng	The Hong Kong University of Science and Technology, China
Junfeng Pan	Google, USA
Zhiyong Peng	Wuhan University, China
Xuan-Hieu Phan	University of New South Wales (UNSW), Australia
Marc Plantevit	Montpellier 2, France
Tieyun Qian	Wuhan University, China
Kaijun Ren	National University of Defense Technology, China
Dou Shen	Microsoft, USA
Peter Stanchev	Kettering University, USA
Xiaoping Su	Chinese Academy of Sciences, China
Xiaoping Sun	Chinese Academy of Sciences, China
Jie Tang	Tsinghua University, China
Zhaohui Tang	Microsoft, USA
Yicheng Tu	University of South Florida, USA
Junhu Wang	Griffith University, Australia
Hua Wang	University of Southern Queensland, Australia
Guoren Wang	Northeastern University, USA
Lizhe Wang	Research Center Karlsruhe, Germany
Jianshu Weng	Singapore Management University, Singapore
Raymond Wong	The Hong Kong University of Science and Technology, China
Jemma Wu	CSIRO, Australia
Jitian Xiao Edith	Cowan University, Australia
Junyi Xie	Oracle Corp., USA
Wei Xiong	National University of Defense Technology, China
Hui Xiong	Rutgers University, USA
Jun Yan	University of Wollongong, Australia
Xiaochun Yang	Northeastern University, China
Jian Yang	Macquarie University, Australia
Jian Yin	Sun Yat-Sen University, China
Qing Zhang	CSIRO, Austrilia
Shichao Zhang	University of Technology, Australia
Yanchang Zhao	University of Technology, Australia
Sheng Zhong	State University of New York at Buffalo, USA
Aoying Zhou	East China Normal University, China
Xingquan Zhu	Florida Atlantic University, USA

Reviewers

Abbas Ayad
Bhowmick Sourav
Cai Hua Li
Cao Jinli
Cao Dong
Carmine Sellitto
Chandramouli Badrish
Chang Tieyuan
Chen Ling
Chen Yan
Chen Dong
Chen Qiming
Cheng Guo
Cheng Hong
Cheng Reynold
Cheng Hong
Choi Gi Heung
Cui Bin
Cui Jin-dong
Ding Juling
Dongqing Xie
Dou Wanchun
Feng Kunwu
Fu Cheng
Gao Yunjun
Gao Hui
Gao Ming
Gao Yunjun
Gopalan Anandha
Guerrini Giovanna
Guo Weijia
Guo Xiaoliang
Guofang Yu
Guo-Sheng Hao
He Frank
He yawen
Huang Shiguo
Huang Ruhua
Huang Yin
Huanhuan He
Ji Xiaopeng
Jia Yan
Jiang Tong
Jiang Yang
Jinfu wang
Jun Long
Jun Ma
Kai Luo
Kim Younghee
Kristensen Terje
Lamo Yngve
Lei Zhen
Leung Carson
Li Hongjiao
Li Min
Li Wei
Li Yingkui
Li Chong
Li Jing
Li Wenxiang
Li yingkui
Li Cong
Li Lily
Li Haiqing
Li Lily
Liang Hao
Liu Wei
Liu Juan
Liu Guohua
Liu Peng
Liu Yan
Liu Ying
Liu Hongyan
Liu Peng
Luo Xiangfeng
Ma Zhi
Man Junfeng
Mi Nian
Mingshun Yang
Nakano Miyuki
Ng Wilfred
Nie Jin
Niu Ling
OuYang Rong
Pan Shanliang
Pan Junfeng
Peng Hui
Peng Daogang
Peng Zhiyong
Peng Dewei
Peng Zhiyong
Plantevit Marc
Qian Tieyun
Qiu Qizhi
Qu Youtian
Qu Lili
Ren Zhikao
Ren Jiadong
Ren Kaijun
Ricky Yu-Kwong
Sazak B. Sami
Shang Helen
Shen Wenfeng
Shi Sanyuan
Shu Fengchun
Song Huazhu
Song Ling
Su Xiaoping
Sun Xiaoping
Tan Jian
Tang Ke-Ming
Tang Yan
Tu Yicheng
Wang Man
Wang Liang
Wang Fang
Wang Shengbao
Wang Lizhe
Wang Yude
Wang Zhen
Wang Xiaodong
Wang Yaxin
Wang Yuying
Wang Lizhe
Wei Li-sheng
Wei Ting
Wen Shan
Weng Jianshu
Wong Raymond
Wu Minghui
Wu xian
Wu Xuejun

Wu Jemma
Wu Ting-Nien
Wu Jemma
Xia Zhengmin
Xiancheng Zhou
Xiang Lei
Xie Zhengyu
Xin mingjun
Xiong Hui
Xu Cuihua
Xu Manwu
Xu Yang
Xue Mei
Yan Qiang
Yan Guanxiang
Yang LiXin
Yang Xiaochun
Yanmin Chen
Yao Zhong
Yin Wensheng
Yin Jian
Yoo Seong Joon
Yu Hua
Yu Zhengtao
Yuan Xiao-guang
Yuan Shizhong
Yue hongjiang
Zhang Xiwen
Zhang Cheng
Zhang Shuqing
Zhang Jianjun
Zhang Jiangang
Zhang Ning
Zhang Qinghua
Zhang Quan
Zhang Yuanyuan
Zhang Huai-Qing
Zhang Qing
Zhang Junsheng
Zhang Qing
Zhao Yuanping
Zhao Jie
Zhao Xiao-dong
Zhao Yanchang
Zheng jian
Zheng Cheng
Zheng Jianfeng
Zhong Sheng
Zhou Si
Zhou Zhiping
Zhou Zili
Zhou Jun
Zhu Hongtao
ZhuGe Bin
Zou Kaiqi

Table of Contents

Web Information Retrieval

Web Information Extraction

Web Information Classification

Web Mining

Semantic Web and Ontologies

Applications

XML and Semi-structured Data

Web Services

Intelligent Networked Systems

Information Security

E-Learning

E-commerce

Distributed Systems

The Research on Chinese Coreference Resolution Based on Maximum Entropy Model and Rules

Yihao Zhang [1], Jianyi Guo [1,2], Zhengtao Yu [1,2], Zhikun Zhang [1], and Xianming Yao [1]

[1] The School of Information Engineering and Automation, Kunming University of Science and Technology, Kunming 650051

[2] The Institute of Intelligent Information Processing, Computer Technology Application Key Laboratory of Yunnan Province, Kunming 650051

zhanghao8209@163.com, gjade86@hotmail.com, ztyu@bit.edu.cn, gemini413@sina.com, zyh0822@126.com

Abstract. Coreference resolution is an important research topic in natural language processing, including the coreference resolution of proper nouns, common nouns and pronouns. In this paper, a coreference resolution algorithm of the Chinese noun phrase and the pronoun is proposed that based on maximum entropy model and rules. The use of maximum entropy model can integrate effectively a variety of separate features, on this basis to use rules method to improve the recall rate of digestion, and then use filtering rules to remove "noise" to further improve the accuracy rate of digestion. Experiments show that the F value of the algorithm in a closed test and an open test can reach 85.2% and 76.2% respectively, which improve about 12.9 percentage points and 7.8 percentage points compare with the method of rules respectively.

Keywords: coreference resolution; named entity; maximum entropy model; rules.

1 Introduction

Coreference resolution is a key link in natural language processing [1]; it is also the core of the project tasks in many languages. It has great significance in Writing analysis, automatic abstracting, information extraction, machine translation and other natural language processing applications [7]. In the past 20 years, coreference resolution has received especial attention, and many research methods have been proposed, which mainly based on the syntactic level methods, based on statistical methods, based on statistical machine learning methods of corpus methods, and so on. So far, most methods adopt linguistics rules, while rules method has lots of shortcomings, such as the number of rule always limited, rule is difficult to cover all the natural language phenomenon; Even if being able to increase new rules unceasingly, but with the increasing number of rules, it is difficult to guarantee that the rules is overall, systemic and logic. At present, a relatively new method has been proposed that based on more powerful automatic analyzer and statistical learning theory, but the use of simple statistical method also has difficult in solving the problem with satisfaction.

W. Liu et al. (Eds.): WISM 2009, LNCS 5854, pp. 1–8, 2009.

Because the statistical method's knowledge granularity is coarse, failed to take into the difference in the senses of account attributes, can only guarantee that the statistical average sense is correct, can not guarantee the actual accuracy of each individual event's results. In this paper, a method of combining based on maximum entropy model and rules for Chinese coreference resolution is proposed. Experiments proved the digestion algorithm that is adopted in this article achieved good results.

2 The Coreference Resolution Processing

The coreference[2] also known as anaphora, it refers to a language unit (usually a word or phrase) in a chapter and the existence of special semantic association of the language unit before it, and its semantic interpretation depends on the former. The language units which are used to point to known as the anaphoric phrase, the language units which are pointed known as the first language (Antecedent). To determine the processing of the anaphoric phrase to refer to antecedent phrase is known as coreference resolution. In order to resolve the question, a method of based on maximum entropy model and rules is proposed in this paper.

The research processing to use of maximum entropy model methods and rules to solve Chinese coreference can be divided into the following steps, and the each step stands for a key link of the computer to processing the document, as shown in Fig.1.

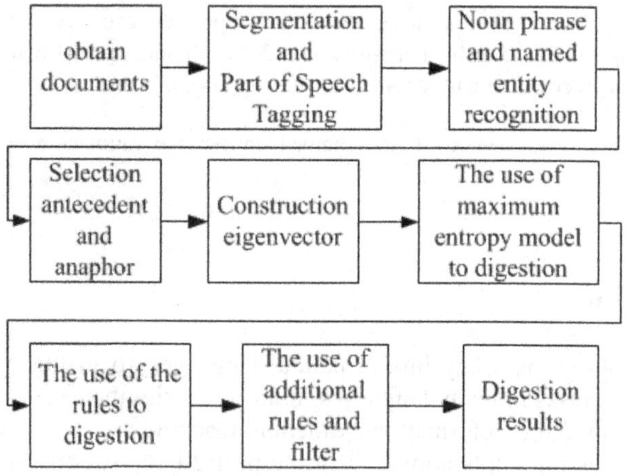

Fig. 1. The overall processing of coreference resolution

2.1 Document Preprocessing and the Selection of Antecedent and Anaphor

Pretreatment of documents include segmentation and part of speech tagging, and to identify the noun phrase and named entity. The process is: first of all, use of sub-word tools to segmentation and part of speech tagging. And then use generic training corpus model to identify named entity. On the choice of antecedent and anaphor, the typical way is to construct a candidate set at present, and then using certain of screening and priority strategies to deal with them. Initially, the antecedent candidate sets is identified as all noun phrases before it. But the possibility is almost zero that far away from the phrase becomes the antecedent, to consider such a large search space is obviously not necessary. However, in order to reduce the search space, simply defined antecedent candidate sets as the phrase before it, or the distance of the noun phrase collection less than or equal a sentence is not appropriate. This article make use of the digestion algorithm that are put forward by R.Mitkov[8] in 1997, the algorithm set separate window settings for two types of antecedent candidates: The phrase which describe the definition[2] is defined as ten sentences before anaphor. Pronoun is defined as three sentences before anaphor.

2.2 The Construction of Eigenvector

Based on the above selected collection of antecedent and anaphor, build the 14 dimensional eigenvector [3]. Assumed the collection of antecedent and anaphor that the antecedent in the front of the collection is for the P, the anaphor in the back of the collection is for the N. Select a representative 14 dimensional feature to construct the eigenvector. As follows:

(1) The single or plural of P: if P is a single, compared with S (Single); is a plural, compared with P (Plural), otherwise to determine U (Unknown).

(2) The single or plural of N: if N is a single, compared with S; is a plural, compared with P, otherwise to determine U.

(3) The gender of P: if P is male, compared with M (Male); is female, compared with F (Female), otherwise to determine U.

(4) The gender of N: if N is male, compared with M; is female, compared with F, otherwise to determine U.

(5) The type of P: if P is the name of a person, compared with H (Human Name); the name of a place, compared with P (Place Name); the name of a Organization, compared with O (Organization Name); the name of the time, compared with T (Time); pronoun, compared with D (Pronoun Name); the general noun phrase, compared with G (General Noun Phrase).

(6)The sentence structure of N: if syntactic structure of N is "noun (or pronoun) + verb" for the NPV (Noun Phrase + Verb), is "verb + noun (or pronoun)" for VNP (Verb + Noun Phrase), otherwise to determine U.

(7)The sentence structure of P and N: syntactic structure means" noun (or pronoun) + verb" or "verb + noun (or pronoun)", their syntactic structure the same as Y (Yes), different for the N (No).

(8) The distance of P and N: If P and N are in the same sentence, for 0; adjacent sentence for 1, spacing 1 for 2 (greater than or equal 3 for 2).

(9) The whole match information of P and N: If P and N are compared with the same, compared with True, otherwise False.

(10) The extraction abbreviated information of P and N: If N is P extraction (such as P for ABCDE, N for the ACE), compared with True, otherwise False.

(11) The substring abbreviated information of P and N: If N is P substring (such as P for ABCDE, N for the BCD), compared with True, otherwise False.

(12) The modified and number types of P: If the context of P is "numeral (+ quantifier) + noun" or "one (+ quantifier) + noun" or "quantifier + noun" type, while True, otherwise False.

(13) N is or not pronoun: if is the pronoun for T, otherwise F.

(14) The similarity of P, N: use of HowNet to calculate the similarity of the two terms.

3 The Coreference Based on Maximum Entropy Model and Rules

In this paper, maximum entropy model combine with the rules derived from decision tree algorithm are used to solve the document's coreference. First of all preprocessing documents and construct the eigenvectors, then select above characteristics to use of maximum entropy model for statistical digestion, reuse decision making rules to deal with the digestion, finally the use of filtering rules to filter the results of these digestion to be the ultimate resolution results.

3.1 The Coreference Based on Maximum Entropy Model

First of all, maximum entropy model are used in Chinese coreference. The thought of maximum entropy algorithm [4] is by calculating $P(y \mid x)$ to determine whether the digestion object point to the same entity, which X is the feature vector, to be received by comparing the characteristics of digestion objects (x_1, x_2), x_1 is the antecedent of candidate sets, x_2 is the anaphor of eigenvector, y is a two-valued attributes. When $y=1$, it represents the digestion object X point to the same entity; when $y=0$, it represents the digestion object X point to the different entity, $P(y \mid x)$ represents the probability that the digestion object x_1 and x_2 point to the same entity. This article selects the above 14 characteristics to structure the function $f_i(X, y)$;

I select GIS algorithm [5] to iterative calculation the probability $P(y \mid x)$ and the weight of characteristic function λ_i in the application.

$$P(y \mid x) = \frac{1}{z(x)} \exp\Big(\sum_i \lambda_i f_i(X, y)\Big)$$

Among, X is the eigenvector of the digestion object, y is a two-valued output, and $f_i(X, y)$ is the characteristic function which related to features. λ_i is the weight of characteristic function, represents the importance of the combination of each characteristic. $z(x)$ is a normalized factor, and its calculation formula is $z_\lambda(x) = \sum_y \exp \sum_i \lambda_i f_i(x, y)$.

In practice, the use of maximum entropy toolkit to train corpus and obtain the classifier model by training the corpus processed by computer and manual tagging, while the maximum entropy model toolkit developed by Zhang le, and then use the classifier model as a template to predict the data processed by computer and get the relationship between them, then obtain the final result.

3.2 The Coreference Based on Rules

Decision tree algorithm [6] is a commonly used method of supervised learning, is the most widely one of inductive inference algorithms. The algorithm aims at summarizing a general concept description from a range of known cases and counter examples. In this paper, the resolution rules is the training rules which are trained using C5.0 decision tree algorithm. The rules of the training are transforming into if-then rules to realize the digestion process based on rules. Then use rules by training decision tree algorithm as shown in Fig.2.

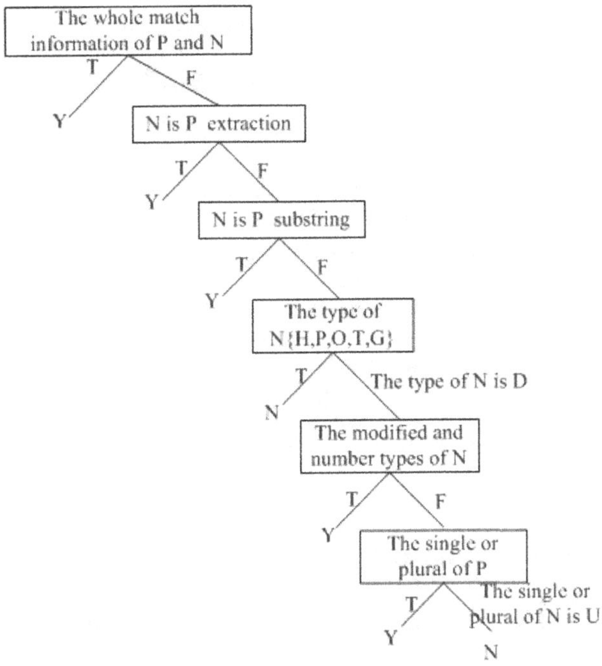

Fig. 2. Decision tree rules

Adopt decision tree algorithm to digest, which according to antecedent P and anaphor N to constitute eigenvector, to determine whether there were coreference relationship between them. P and N attributes vector $x \in X$ constitutes an example, which X represents the vector space of all vectors, and $y \in Y = \{Y, N\}$ represents the categories of the example, including "Y" indicated that it was cases, it indicate exist coreference relationship between P and N; while "N" indicated that counter examples, it indicate don't exist coreference relationship between them.

3.3 The Application of Rules and Filters

Based on maximum entropy model method is according to the statistics of training examples to make judgments, the result may be reflect the relationship between the properties to some extent, but sometimes may also appear error, which is inevitable use of statistical methods. The use of rules also exist digestion mistakes and recall rates, therefore additional rules and filtering is very necessary.

Additional Rules. These rules mainly targeted at some digestion of pronoun. Consider as follows:

Rule 1: If the antecedent of the digestion objects of is an entity that represents a person's names, anaphor is a pronoun "he" or "She", and the pronoun act as subject in the sentence. This situation can determine the coreference relationship between them.

Rule 2: If the antecedent of the digestion objects of is an entity that represents the names of organization, the names of place, or Proper noun. Anaphor is a pronoun "it", and the pronoun act as subject in the sentence. This situation can determine the coreference relationship between them.

The Filter of Results. Whether the use of statistical methods and rules, there will be some mistakes, in order to enhance the accuracy rate of the digestion results, filtering rules is also very necessary. Consider as the follows:

Rule 1: If the type of singular and plural is incompatible, considering the coreference relationship does not exist between them.

Rule 2: If the type of gender is incompatible, considering the coreference relationship does not exist between them.

Rule 3: If the anaphor is "he" or "she", C belongs to the candidate set, to cater for $C \in (P, O, T, G)$, considering the coreference relationship does not exist between them.

4 Experimental Results and Analysis

Depending on the different relations of the test set and training set, we measure the experiment result using closed and open measure. Because of the MUC-6 corpus of the coreference resolution has not been published publicly. I select the Lancaster Corpus of Chinese as the experimental data, as it is completely free and open to the public. I select LCMC_A and LCMC_C as the test corpus because of the two types' evaluation corpus of news reports and news evaluation has rich refer relationship. As the performance evaluation problem of coreference resolution, I have adopted the most commonly used evaluation criteria in information retrieval: accuracy rate, recall rate, and F values to measure. The experimental results as shown in table 1:

Accurate rate (P) = the number of correct digestion / the reference number tried to digest*100%.

Recall (R) rate= the number of correct digestion / the reference number of system identification*100%.

F value = 2 * P * R / (P + R) * 100%.

Table 1. The comparison of rules methods and the combination rules and statistic

model	type	total digestion	identification number	correct number	accurate rate	Recall rate	F value
Rules method	closed	10428	8173	6726	64.5%	82.3%	72.3%
	open	12030	9392	7326	60.9%	78%	68.4%
Maximum Entropy Model and rules	closed	10428	8803	8196	78.6%	93.1%	85.2%
	open	12030	9618	8252	68.6%	85.8%	76.2%

As can be seen from the experimental results, the combination of rule and statistic has its advantage compare to simple rule methods. The accurate rate, recall rate and F value has corresponding improve. The Accurate rate increased by about 14.1 percentage point in closed test and 7.7 percentage point in open test, The Recall rate increased by about 10.8 percentage point in closed test and 7.8 percentage point in open test, The F value increased by about 12.9 percentage point in closed test and 7.8 percentage point in open test, this show the methods based on Maximum Entropy Model and Rules has a greater advantage in Chinese coreference resolution.

5 Conclusions

In this paper, a coreference resolution algorithm that based on maximum entropy model and rule is proposed. Maximum entropy model can use of variety characteristics in the same framework, and can use of eigenvector in maximum extent possible that has been build up. The rules methods can make use of syntax and semantics to be reflected in the processing of digestion. The two methods have a certain complement each other, which overcome the shortcomings of using a single method to some extent. Then use rules to filter out the properties of the anti-conflict cases, also make up the shortcomings that rules ignore the relevance of the properties to a certain extent. The result shows that the above method obtains good results, but also has problem such as some digestion candidate set did not identify, the digestion of some pronouns also exist error. The next step is to improve the accuracy of identification of entities and discuss a variety of digestion rules.

Acknowledgments. This paper is supported by: The National Natural Science Foundation (60863011), the key project of Yunnan Natural Science Foundation of (2008CC023), Yunnan Province, middle-aged and young academic leaders Talents Fund Project (2007PY01-11), the key projects of Yunnan Provincial Education Department fund (07Z11139).

References

1. feng, W.H.: Computational Models and Technologies in Anaphora Resolution. Journal of Chinese Information Processing 16(6), 9–17 (2002)
2. feng, W.H.: On Anaphora Resolution within Chinese Text. Applied Linguistics 11(4), 113–119 (2004)
3. Wei, Q., Kun, G.Y., qian, Z.Y.: English Noun Phrase Coreference Resolution Based on Maximum Entropy Model. Journal of Computer Research and Development 40(9), 1337–1343 (2003)
4. juan, L.Z., ning, H.C.: An Improved Maximum Entropy Language Model and Its Application. Journal of Software 10(3), 257–263 (1999)
5. Ning, P., Erhong, Y.: Study on Pronoun Clearance based on Statistical Mode and Rule. Journal of Chinese Information Processing 22(2), 24–27 (2008)
6. Jun, L., Ting, L., Bing, Q.: Decision Trees-based Chinese Noun Phrase Coreference Resolution. Harbin Institute of Technology, Harbin (2004)
7. Qiang, W.Z., Xin, Z.Y.: Research on Chinese Coreference Resolution and Its Related Technologies. Beijing University of Posts and Telecommunications, Beijing (2006)
8. Sha, L.S., Jun, L.Z., Wang, C.H.: Research on Resolution with in Text. Computer Science 34(7), 138–141 (2007)

Performance Improvement in Automatic Question Answering System Based on Dependency Term

Jianxing Shi, Xiaojie Yuan, Shitao Yu, Hua Ning, and Chenying Wang

Department of Computer Science and Technology, Nankai University
Tianjin 300071, P.R. China
{shijianxing,yuanxiaojie,yushitao,ninghua,
wangchenying}@dbis.nankai.edu.cn

Abstract. Automatic Question Answering (QA) system has become quite popular in recent years, especially since the QA tracks appeared at Text REtrieval Conference (TREC). However, using only lexical information, the keyword-based information retrieval cannot fully describe the characteristics of natural language, thus the system performance cannot make people satisfied. It is proposed in this paper a definition of dependency term, based on the dependency grammar, employing the natural language dependency structure, as the improvement of the term, to support the typical information retrieval models. It is in fact a solution for a special application in XML information retrieval (XML IR) field. Experiments show that: dependency-term-based information retrieval model effectively describes the characteristics of natural language questions, and improves the performance of automatic question answering system.

Keywords: Question and Answer Web Forum, XML Information Retrieval, Automatic Question Answering System, Dependency Grammar, Dependency Term.

1 Introduction and Related Work

With the development of Internet, there are various Internet services, one of which is question answering (QA) system [1]. Unlike search engines which return a few relevant documents, QA systems accept user's natural-language questions, and provide one or several exact answers to each question. It is more preferable, so has become quite popular in recent years, especially since the QA tracks appeared at Text REtrieval Conference (TREC)[1].

There are mainly two kinds of QA systems. One of them is open-domain QA system [2]. It uses the Web data sets, with broad knowledge, while its answers are not in very high quality. The other kind is Frequently Asked Question (FAQ) answering system [3]. FAQ answering module retrieves answers from existing question/answer pairs (Q/A pairs) in the database which was previously posted. The corresponding correct answer is returned immediately to the user as the final answer [4]. The data

[1] http://trec.nist.gov/

W. Liu et al. (Eds.): WISM 2009, LNCS 5854, pp. 9–18, 2009.

sets are usually manually marked with high quality answers. However, the system performance is usually limited by the small-scale of the data.

There is another important type of Internet service - QnA Web forum. It is not a QA system actually, but a Web platform where users can ask or answer a question, vote for a best answer, such as Yahoo! Answers[2], Live QnA[3] and Baidu Knows[4], etc.

Moreover, QnA forums provide quite suitable data sets for building QA systems. Using the question matching strategy of FAQ answering system, based on the large-scale of threads data in QnA forums, we can implement a high-quality QA system.

Within the whole procedure, question similarity calculation is the key step which mainly determines the answer quality. To address this problem, the general methods are using term to describe the basic feature of the document by employing keyword retrieval, then calculating the similarity using the typical model of information retrieval such as vector space model (VSM). The original definition of term relies only on lexical level matching to rank passages. The underlying assumption is that each question term is considered an independent token. However, this simplification does not hold in many cases because dependency relation exists between words [5].

In 1959, the French linguist, L. Tesiniere proposed the dependency grammar to express natural language. According to the dependency grammar, the dependency relationships normally form a tree that connects all the words in a sentence, called dependency tree. To store the dependency tree, XML technology has its own special advantages. A dependency tree can be represented as an XML document with losslessness of information. Thus, the similarity measuring between questions is an issue of XML document similarity calculation, is the field of XML IR/keyword [6].

XML similarity measuring takes both the structure and the content into account. There are many related researches in this filed [7-9]. The current study focus on two aspects: one is based on tree edit distance method (TED) [10]; the other is approximation method which is impacted by the TF*IDF technology of typical information retrieval [11].

We approach this problem by using the TF*IDF technology to achieve similarity measuring between questions. Based on the dependency grammar, the dependency term (DT) is defined instead of term, taking into account the XML node structure and content information. Then, the application-specific XML information retrieval (XML IR) solution is proposed, including the DT generation, the similarity and weight calculation. And the dependency term is not limited to be used in FAQ answering systems; it is applicable to general QA systems. The architecture of QA system based on DT is showed in Fig. 1.

More specifically, the main contributions of this work are summarized as follows:

- Provide a feasibility solution for QA system, using the large quantities of QnA threads data;
- Introduce syntactic information into the process of information retrieval, based on the dependency grammar;
- Design an XML presentation strategy for the question, thus the problem of similarity measuring between questions can be treated as an issue of XML IR;

---

[2] http://answers.yahoo.com/

[3] http://qna.live.com/

[4] http://zhidao.baidu.com/

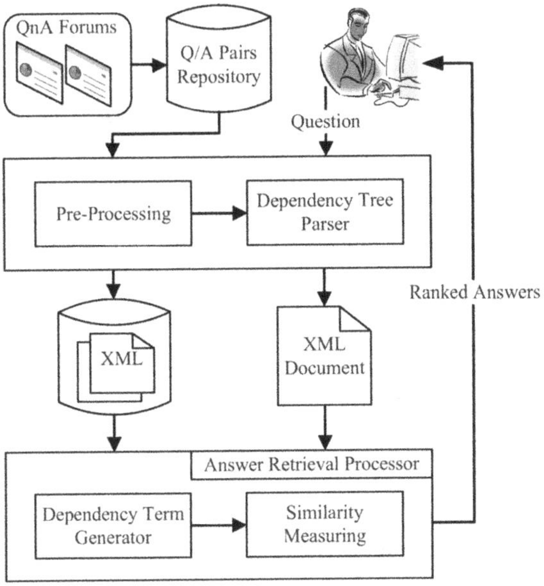

Fig. 1. The Architecture of QA system

- Propose a definition of dependency term, to enhance the efficiency of term. The performance of the QA system based on it is improved. It is in fact a solution for a special application in XML IR field.

2 Dependency Tree

2.1 Definition of Dependency Tree

A dependency relationship is an asymmetric binary relationship between a word called head (or governor, parent), and another word called modifier (or dependent, daughter). Dependency grammars represent sentence structures as a set of dependency relationships.

Definition 1 (Dependency Tree). *The dependency relationships form a tree that connects all the words in a sentence, called dependency tree. A word in the sentence may have several modifiers, but each word could modify at most one word. The root of the dependency tree does not modify any word. It is also called the head (mostly predicate verb) of the sentence.*

The dependency grammar is comprised of several grammatical categories; some examples are listed as follows:

- A sentence has a predicate verb as the core;
- The predicate verb can be modified by subject;
- The predicate verb can be modified by object directly or indirectly;
- The subject and object (noun) can be modified by attributes;
- The predicate verb can be modified by adverbs.

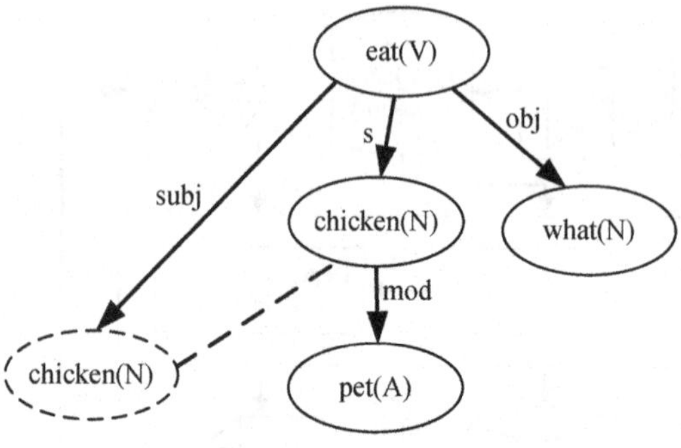

Fig. 2. Dependency tree for the sample question generated by Minipar

We employ Minipar [12], a fast and robust dependency parser, to accomplish dependency parsing. For example, the parsed result of question "What do pet chicken eat?" is shown in Fig. 2.

2.2 Representation of Dependency Tree

It is known that, the dependency parsed result of natural language is in the form of semi-structured tree. So plain text is unable to meet the needs of its representation. XML technology has its own special advantages on this issue:

1. Firstly, XML has become a standard format for data exchange on the Web. The hierarchical structure of the dependency tree can be accurately represented as an XML document, with losslessness of information;
2. Secondly, XML storage and access technology is becoming more and more mature. Especially most mainstream commercial databases have supported for the storage and access of XML data type, ensuring the operation of the XML data quickly and conveniently.

We propose a unified XML representation of dependency tree. The standard format definition of an XML document is its XML Schema model. The XML Schema of dependency tree is described as:

```
<?xml version="1.0" encoding="utf-8"?>
<?xs:schema
xmlns:xs="http://www.w3.org/2001/XMLSchema">
 <xs:element name="word">
  <xs:complexType>
   <xs:sequence minOccurs="0">
    <xs:element ref="word" maxOccurs="unbounded" />
   </xs:sequence>
   <xs:attribute name="pos" type="xs:string" />
   <xs:attribute name="stem" type="xs:string" />
   <xs:attribute name="rel" type="xs:string" />
  </xs:complexType>
 </xs:element>
</xs:schema>
```

As indicated above, a complete dependency tree can be expressed as an XML document. Each XML document element node corresponds to a word containing three attributes (pos, stem, and dependency relation). The value of the attributes can be exacted from Minipar's output results.

2.3 Generation of Dependency Tree

We can develop a dependency tree generator using Minipar. The output result is an XML document with the XML Schema introduced above. The XML document can clearly and accurately represent all dependency relationships between head and modifier with losslessness of information.

Take the question "What do pet chicken eat?" for example, the XML result is:

```
<?xml version="1.0" encoding="utf-8"?>
<word pos="V" stem="eat" rel="">
  <word pos="N" stem="chicken" rel="subj">
    <word pos="A" stem="pet" rel="mod" />
  </word>
  <word pos="N" stem="what" rel="obj" />
</word>
```

We can see that all the words in a sentence are no longer independent, but act as a certain element of the sentence. Among the elements, the predicate verb, subject and object are obvious more important than others. So words in a sentence should have different weights. It is appropriate to assign different value as weight for each word.

3 Dependency Term

XML similarity includes two aspects: structure similarity and content similarity. In the dependency tree, XML structure reflects the dependency relation, and XML content refer to words and POS information.

After analyzing a large number of XML IR methods, taking into account the computational complexity of edit-distance algorithm, we adopt TF*IDF method of XML IR/keyword model. Therefore, the definition of dependency term is proposed next.

In the typical model of information retrieval, term is a very important concept, based on which a lot of information retrieval models are implemented. In this paper, we introduce the syntactic information into the process of information retrieval. To enhance the expression capabilities of term without changing of retrieval models themselves, the term combined with dependency structure information is defined as dependency term.

3.1 Definition of Dependency Term

Definition 2 (Dependency Term, DT). *It refers to the term integrated with dependency relation of natural language processing (NLP), to serve as term with the same functionality in the information retrieval model. DT records not only the word stem, but also the dependency relation and weight information. Thus it is defined in the form of triple as:*

DT = <*stem, relation, weight*>

The stem element is equal to the stem of term; the relation element is the modifying relation between the *DT* and its XML parent node, and the value is ε (empty) if the *DT* refers to XML root node (mostly predicate verb). The stem and relation elements are combined together as an identity. In other words, only when both of them are equal, the two *DTs* are considered as the same.

3.2 Weight of Dependency Term

Weight represents the status of the *DT* in a sentence, is usually associated with the XML node depth: the depth of the root node is defined as 0; for other nodes, the more close a certain node is to the root, the smaller its depth is, and the higher its weight is. The definition of node weight is discussed in [7]. It should show certain discrimination, but couldn't converge too fast. It can be defined:

$$weight(n) = \frac{1}{(1 + level(n))^k} \tag{1}$$

where *n* represents an XML node, *weight(n)* is the weight of the node, *level(n)* is the depth of the node, *k* is used as an constant and $k \geq 1$. In our system, *k* is 1.

3.3 Generation of Dependency Term

This section discusses the generation process of the *DT*. According to the definition of dependency term, the three element values of *DT* can be extracted or computed by dependency tree in terms of XML document. The process in fact is the dependency tree recursion, filling the triple of each *DT*.

For example, the generation result of question "What do pet chicken eat?" is as follows:

```
<eat, ε , 1>
<chicken, subj, 1/2>
<pet, mod, 1/3>
<what, obj, 1/2>
```

In addition, Stop words pruning is useful here. If the stem of a *DT* is a stop word, the *DT* should be removed. We construct a stop word list which is slightly different from the normal one to do this pruning.

3.4 Use of Dependency Term

In the information retrieval model, the term is replaced by dependency term. It is necessary to re-calculate their statistical features: *tf* (Term Frequency), *idf* (Inverse Document Frequency).We impose three constraints when using *DT*:

- When matching dependency terms. The two *DTs* are considered the same when both stem and relation equals at the same time. Otherwise, the statistical weights should be measured separately;

- When calculating the statistics characteristics (such as *tf*) relevant to the count of *DT*. It approximates to the measuring process based on term. However, the cumulative value of each *DT* is not 1, but the value of weight attribute;
- When calculating the statistics characteristics (such as *idf*) irrelevant to number of *DT*. It is exactly the same with the measuring process based on term.

After calculating the weight of *DT*, the cosine similarity between questions can be computed by:

$$SC(q,d) = \frac{\sum_{i=1}^{m} \omega_{qi} \bullet \omega_{di}}{\sqrt{\sum_{i=1}^{m} w_{qi}^2} \sqrt{\sum_{i=1}^{m} w_{di}^2}} \quad (2)$$

where q is user's questions and d is the questions in a Q/A pair database. Each question is represented as a vector. ω_{ij} is the weight of each dimension corresponds to a dependency term.

4 Experiments

4.1 Experimental Setup

Dependency term is an improvement on term. We use the VSM as the retrieval model based on both of them to implement two similar QA systems, by calculating the similarity between a user's question and each question in the FAQ database. We also use the similar-question-search service of QnA forums for comparison.

To conduct the experiment, we use the Q/A pairs got from QnA forums for our QA system. Statistics of these sets are given in Table 1.

Table 1. Statistics of data sets used in our experiments

Dataset	#
Q/A pairs from Yahoo! Answers	18000
Q/A pairs from MSN QnA	9000
Mixed	27000

We repeat five independent experiments over the three datasets. For each experiment, 99% of Q/A pairs are randomly selected from the data set as the training set to build the QA system; and the remaining 1% as the test set.

To evaluate the performance of QA system, the mean reciprocal rank (MRR) formula provided by TREC is used as the evaluation criterion. The score for an individual question is the reciprocal of the rank at which the first correct answer was returned or 0 if no correct response is returned. The score is then the mean over the set of questions in the test. For every question, our system returns top 20 answers ranked by their score, and each answer's accuracy is determined manually.

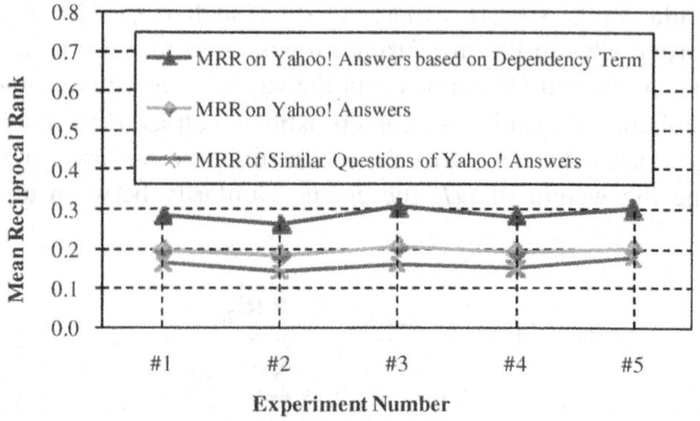

(a) Performance improvement result of QA system base on term using the data set from Yahoo! Answers forum (Repeating 5 times independent experiments, each of them have 17820 Q/A pairs as the training set, and 180 questions as test set)

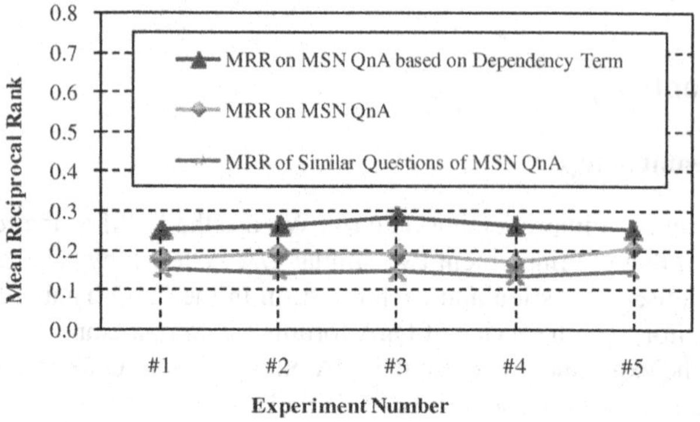

(b) Performance improvement result of QA system base on term using the data set from MSN QnA forum (Repeating 5 times independent experiments, each of them have 8910 Q/A pairs as the training set, and 90 questions as test set)

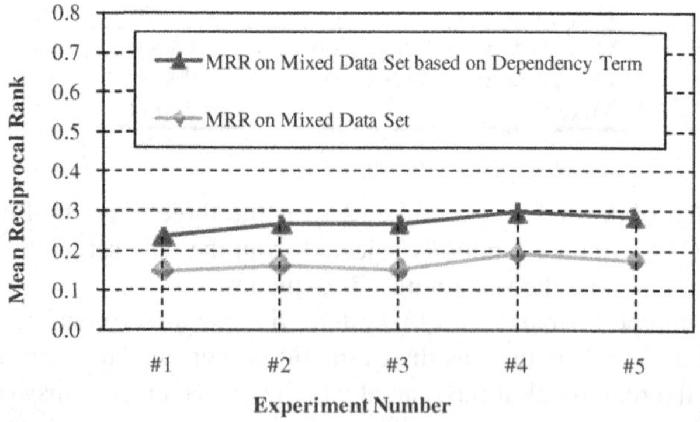

(c) Performance improvement result of QA system base on term using the mixed data set (Repeating 5 times independent experiments, each of them have 26730 Q/A pairs as the training set, and 270 questions as test set)

Fig. 3. Performance improvement result of QA system base on term using different data sets

4.2 Results Analysis

The experimental results are shown in Fig. 3, divided into (a), (b), (c) three parts, corresponding to the three data sets. We can see that, both of the QA systems get higher MRR score than the similar-question-search service in QnA forums. And our system based on DT shows better performance than the one based on term.

Moreover, the DT has strong abilities in adapting to the real data sets from QnA forums, and improves the system performance stably. The main reasons are as follows:

1. DT reduces the statistical proportion of the secondary element in a sentence. All the words are treated as the same when using term. While predicate verb, subject and object are in fact considered as the core of the whole sentence when using DT, and other modifiers have been given lower weights;
2. It also reduces the error probability of term matching. The DT matching takes both stem and dependency relation into account.

5 Conclusion

As mentioned above, using the question matching strategy of FAQ answering system, with the large number of threads data in QnA forum, a high-quality QA system can be implemented. Based on the dependency grammar, the dependency tree is represented as an XML document. The similarity measuring between questions is an issue of XML document similarity calculation, is the field of XML IR/keyword. In order to achieve this XML information retrieval tasks, the definition of dependency term is proposed. It takes both the XML node structure and content information into account, as the improvement of the term in typical information retrieval.

Experiments show that: dependency term can effectively express the characteristics of natural language questions, and improves the performance of automatic question answering system.

Acknowledgment. This work is supported by National 863 Plans Projects of China (No.2009AA01Z152) and Microsoft Research Asia Internet Services in Academic Research Fund (No.FY07-RES-OPP-116). The authors are very grateful to the anonymous reviewers for their comments.

References

1. Voorhees, E.M.: Overview of the TREC 2003 question answering track. In: Proceedings of the Twelfth Text REtrieval Conference (TREC 2003), pp. 54–68 (2004)
2. Moldovan, D., Pasca, M., Harabagiu, S., Surdeanu, M.: Performance issues and error analysis in an open-domain Question Answering system. In: Proceedings of the 40th Annual Meeting on Association for Computational Linguistics, Philadelphia, Pennsylvania, July 07-12 (2002)
3. Burke, R., Hammond, K., Kulyukin, V., Lytinen, S., Tomuro, N., Schoenberg, S.: Question Answering from Frequently Asked Question Files. AI Magazine 18(2), 57–66 (1997)

4. Song, W., Feng, M., Gu, N., Wenyin, L.: Question Similarity Calculation for FAQ Answering. In: Proceedings of the Third International Conference on Semantics, Knowledge and Grid, October 29-31, pp. 298–301 (2007)
5. Cui, H., Sun, R., Li, K., Kan, M.-Y., Chua, T.-S.: Question Answering Passage Retrieval Using Dependency Relations. In: Proc. of SIGIR 2005, Salvador, Brazil (2005)
6. Anh, V.N., Moffat, A.: Compression and an IR Approach to XML Retrieval. In: Proceedings of the Workshop of the Initiative for the Evaluation of XML Retrieval (INEX 2002), Germany, pp. 99–104 (2002)
7. Bertino, E., Guerrini, G., Mesiti, M.: A Matching Algorithm for Measuring the Structural Similarity between an XML Document and a DTD and Its Applications. Information Systems 29(1) (2004)
8. Tekli, J., Chbeir, R., Yetongnon, K.: Semantic and Structure based XML Similarity: An Integrated Approach. In: Proceedings of the 13th Interventional Conference on Management of Data (COMAD 2006), New Delhi, India, pp. 32–43 (2006)
9. Nierman, A., Jagadish, H.V.: Evaluating structural similarity in XML documents. In: Proceedings of the 5th ACM SIGMOD International Workshop on the Web and Databases (WebDB), pp. 61–66 (2002)
10. Zhang, K., Shasha, D.: Simple Fast Algorithms for the Editing Distance between Trees and Related Problems. SIAM Journal of Computing 18(6), 1245–1262 (1989)
11. Lingbo, K., Shiwei, T., Dongqing, Y., Tengjiao, W., Jun, G.: Query Techniques for XML Data. Journal of Software 18(6), 1400–1418 (2007) (in Chinese)
12. Lin, D.: Dependency-based Evaluation of MINIPAR. In: Workshop on the Evaluation of Parsing Systems, Granada, Spain (1998)

An Expert System Based Approach to Modeling and Selecting Requirement Engineering Techniques

Yan Tang and Kunwu Feng

Department of Computer Science,
The University of Texas at Dallas 800 West Campbell Road, Richardson, TX 75080-3021
{yxt063000,kxf041000}@utdallas.edu

Abstract. The importance of requirements engineering (RE) has been raised numerous times in literatures. To choose suitable RE techniques for a particular project in a given situation is a challenging task, requiring substantial expertise and efforts. To help solving this problem, an expert system based approach is proposed. This expert system uses the knowledge from domain experts to model the causal factors of RE techniques. It can select suitable RE techniques for a software project. A web-based questionnaire is created in the first place to collect the expertise available in the community. The information collected by the questionnaire is analyzed and transformed into a new dataset for constructing a Bayesian Belief Network (BBN). The resulting BBN integrated with a GUI forms an expert system for RE techniques modeling and selection. Empirical study validates the transformed dataset and shows that the expert system outperforms other predictors in selecting suitable RE techniques in different RE phases.

Keywords: Requirements Engineering Techniques, Expert System, Bayesian Belief Network, Data Transformation, Probability Mass function, Classifiers.

1 Introduction

Defining software requirements is recognized as critical to successful software projects. The importance of requirements engineering (RE) has been raised numerous times in literatures. For example, empirical evidence shown the benefits of RE are discussed in [10]; Many literatures show that using appropriate RE techniques has a positive impact on software quality [12]. The RE techniques are quite different from organization to organization, and from project to project. Influence factors of selecting RE techniques include characteristics of the projects, technical maturity of the organization, the involvement of engineering and managerial disciplines, organizational culture, etc. In general, software projects can be classified into two categories: Single product or a Software Product Line Engineering (SLPE) product [13]. The requirement engineering process consists of six phases: Requirement *Elicitation*, *Specification*, *Analysis*, *Negotiation*, *Validation*, and *Management*. RE techniques can be classified into to three different groups: Agile [2], Traditional [20]and Product Line [13]. There are different RE techniques under each group and each RE phase.

W. Liu et al. (Eds.): WISM 2009, LNCS 5854, pp. 19–30, 2009.

Selecting suitable techniques for a particular project is a challenging task, requiring substantial expertise and efforts. To help solve this problem, a BBN based expert system is proposed. The BBN is built via state-of-the-art questionnaire design methods, information retrieval techniques and accurate machine learning algorithms.

The BBN is a probabilistic graphical model that represents a set of variables and their probabilistic dependencies. BBN is selected to be the expert system's engine because it exhibits a powerful ability to model the causal relationships between variables involving complex variant factors. It provides a consistent, theoretically solid mechanism for processing uncertain information and solves both discriminative problems (classification) and regression problems (configuration problems and prediction). The BBN is a fundamental tool in Modeling, Prediction, Risk Analysis, Diagnosis and Decision Support, having numerous applications in a wide range of areas like Medical, Education, Finance, Telecommunication, Environment and Information Technology. A formal description of the BBN is presented in the Appendix.

This is an interdisciplinary study on data mining and software engineering. To our best knowledge, this study is the first to build an expert system based on BBN for RE techniques modeling and selection. Our study is the first to collect APLE-RE (Agile and Product Line Engineering Requirements Engineering) expertise using an online questionnaire. The causal factors of RE techniques are identified and modeled using the knowledge collected by the questionnaire. This is also the first study to identify and address issues about questionnaire data in the automated construction of expert systems. Although from a SE questionnaire's perspective, the collected data sample is large (50 respondents), from an expert system's perspective, it is still considered small. The questionnaire allows the respondent to select multiple answers for some questions, resulting in a raw data with multiple values in one data entry. It is challenging to learn a BBN from this raw data. A new data conversion algorithm is proposed to solve the small sample-size issue and multiple-value challenge. The expert system construction process is reusable for building similar systems in other areas using datasets collected by questionnaires. This could ultimately assist people in making better decisions in many problem domains.

2 Methodology

This section describes the overall methodology of this study shown in Figure 1. Our study consists of two phases: Data Generation and Expert System Construction.

In the Data Generation Phase, the first task is to collect data using an online questionnaire, which is designed and implemented to collect expert's knowledge of RE techniques selection. A broad range of project scenarios are pre-defined in the first place to cover different project domains. The second task is the data transformation. Respondents' answers are extracted from the questionnaire's backend database to form a raw data. As mentioned in Section 1, the raw data is not suitable for BBN learning. A probability mass function (PMF) based algorithm is proposed to convert the raw data into single value data and to increase its size. Finally, the generated data is spited into two sets, one for training the BBN and the other for testing the BBN.

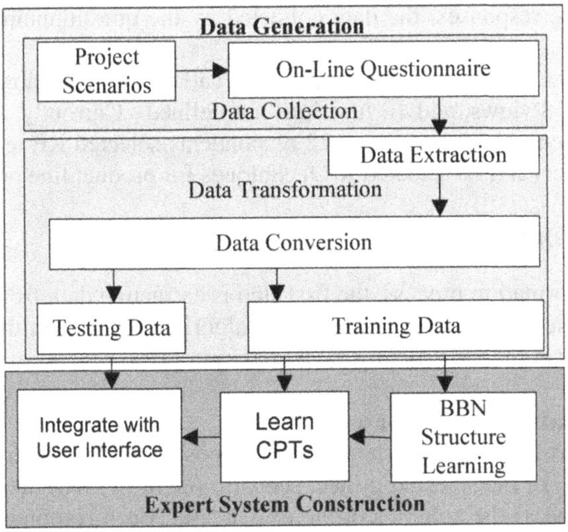

Fig. 1. Overview of This Study

The second phase is Expert System Construction. A state-of-the-art BBN structure learning algorithm (BBN-SLA) identifies the BBN structure using the training dataset. Then the parameters of the BBN are learned. The learned BBN is integrated with user interface to form a new expert system for RE techniques modeling and selection.

3 Data Generation

3.1 Collect Data Using Questionnaire

The on-line questionnaire [3] is composed of two parts. Part I collects information about the respondent to build a project profile. This part includes following questions about the project characteristics the respondent is working on / worked on: project *type*, *domain*, *size*, *geographic distribution*, *duration*, *demo frequency* and *non-functional requirements*. We established a project scenario pool containing 342 project scenarios across 19 application domains. Based on respondent's background collected in Part I, Part II selects one matching project scenario from the pool and asks the respondent to select RE techniques in different RE phases for the given scenario. Part II also asks several RE related questions. We collected 95 RE techniques in great level of detail. These techniques are categorized according to six phases of the RE process and are listed as options for the questions in Part II.

Once the questionnaire is implemented and tested, it is disseminated to the experts in the community. Researchers and practitioners involved with program committees and/or participants at related conferences and workshops (e.g., ICSE 2008, APLE 2006, RWASE 2007, SPLC 2007) are invited to respond. A list containing over 200 candidates with great experience in RE from both academia and industry is constructed. An invitation letter is sent out to each candidate. The invitation email is also sent to three major groups in the SE community. By carefully targeting the audience and reviewing the responses, the data collected by the questionnaire is very reliable and practical.

The user's answers are stored in the database called APLE Database. In APLE Database, 20 tables, 8 views, and 10 functions are defined. Currently, the questionnaire collected 50 responses. Among them, 22 respondents selected RE techniques for single product projects and 28 selected RE techniques for product line projects.

3.2 Transform Data

In the data transformation process, the first step is extracting data from the database to create a raw dataset. Then we propose a novel algorithm to convert the raw data into a suitable new dataset for constructing the expert system.

3.2.1 Extract Data from Database

Questionnaire Part II is directly related to the proposed project scenario. Therefore, all the characteristics of the selected project scenario (eight in total) and ten questions in Part II are selected as the columns of the raw dataset. Each respondent's project scenario and answers to questions in questionnaire Part II are extracted from the database and put into the raw dataset. The raw dataset is available at [1]. It contains 18 columns; nine of them contain multiple values in the data entry. These nine columns correspond to RE techniques in six RE phases and other three multi-answer-multi-choice questions. For example, one question in Part II is: which RE techniques will you use in the requirement elicitation phase? 17 RE techniques for requirement elicitation are listed as the question's options. If the fifth respondent selected four RE techniques, then the fifth row and jth column (j is the column number of the above question) of the raw dataset contains four values.

As mentioned in Section 3.1, the respondents' answers naturally form two distinct raw datasets. For this study, the raw dataset of single product projects is used to construct the expert system for RE techniques modeling and selection. The total number of rows in the raw dataset is 22, corresponding to the number of respondents answering the questionnaire for single product projects.

3.2.2 Convert Multi-to-Single Value

BBN Structure Learning Algorithms (BBN-SLAs) require all the dataset entries to be a single value and often require large-size training dataset for good learning accuracy [18]. After carefully reviewing and analyzing the raw data, we propose a probability mass function (PMF) based algorithm (Figure 2) to convert the raw data. PMF gives the probability that a discrete random variable is equal to an exact value. Suppose X: $S \rightarrow$ R is a discrete random variable defined on a sample space S. Then the probability mass function f_X: R $\rightarrow$ [0, 1] for X is defined as:

$$f_X(x) = \begin{cases} \Pr(X = x), x \in S \\ 0, x \in R \setminus S \end{cases} \quad (1)$$

Throughout the paper, the PMF of a random variable X is calculated by counting the frequency of each value in the sample space of X.

```
Input:
rawData: the raw dataset after the data extraction
fold:  number of times to multiply the raw data
numRow = the number of rows in rawData

output:
data: The  new data

1. f_qi = PMF of each multi-value column qi of rawData
For i=1:fold
 For k=1:numRow

   2.Scan through rawData, copy the single values.
   for each multi-value questions  qi, if the kth  row has
   n answers EA_qi, then sample n answers a_qi  using f_qi

   3. If a_qi ∩ EA_qi ≠ φ
    Let CommonAnswer = a_qi ∩ EA_qi
            3.1 If | CommonAnswer | == 1
                  data(k,qi) = CommonAnswer;
             3.2  Else
                  Choose one answer an_qi
                  from CommonAnswer following f_qi ;
                  data(k,qi) = an_qi ;
             End If

    4. Else
           4.1 If ( |EA_qi| ==1 )
                  data(k,qi) = EA_qi ;

         4.2 Else
                  Choose one answer
                  an_qi from EA_qi following f_qi ;
                  data(k,qi) = an_qi ;
         End If
  End For
End For
```

Fig. 2. The Data Conversion Algorithm

This data conversion algorithm is self-explanatory. In step 1, each multi-value column is considered as a random variable and its sample space is a distinct set of all its values. The PMF of each multi-value column is obtained. In step 2, the algorithm scans through the raw dataset; the single value in the data entry is treated as invariant and is kept the same. When the algorithm finds a raw data entry (k,qi) with n multiple values, it samples n values using the PMF of the column qi and intersects the sampled values with the raw data entry (k,qi). Step 3 deals with cases when the intersection is not null. Step 4 deals with cases when the intersection is null. These two steps always try to replace the original multi-values in the data entry with a single value. This single value not only belongs to the original multi-values, but also follows the probability distribution of the multi-values column. The inner loop generates one dataset containing only single values with the same number of rows and columns as the raw dataset. The outer loop iterates *fold* times to generate a large dataset having *numRow*fold* rows containing only single values.

The proposed algorithm solves the challenge imposed by the raw data as mentioned in Section 1. It makes full use of the raw data to generate a consistent large-size dataset suitable for BBN learning. The PMF acts like a board of experts offering suggestion, so it is used to consolidate and enhance individual respondent's choice. The conversion process replaces the multi-value by a single value not only belonging to the respondent's original answers but also consistent with the preference of all experts. The conversion algorithm is purely based on the nature (probability distribution) of the raw data and limits the introduction of bias and noise into the generated data. The algorithm also keeps respondents' answer order intact. In addition, since some respondents selected more options than others, the raw data is unbalanced. The conversion algorithm could even out the unbalance by generating the same number of rows for each respondent.

Many BBN learning algorithms requires a large training dataset size [18]. In this study, we generated a new dataset of 18 columns and 4400 rows – 200 times the size of the raw dataset. Each column is a node of the BBN. The distinct set of all values of each column becomes the values of the node in the BBN. Each RE phase is a node in the BBN and all the RE techniques in that phase are values of the node. The new dataset is divided into a training dataset with 3300 rows and a testing dataset with 1100 rows.

4 Expert System Construction

Figure 3 shows the process for constructing the expert system. The process of learning the BBN from data contains two steps: 1.Structure Learning—identifying the BBN's structure and 2. Parameters Learning—identifying the BBN's parameters or Conditional Probability Tables (CPTs).

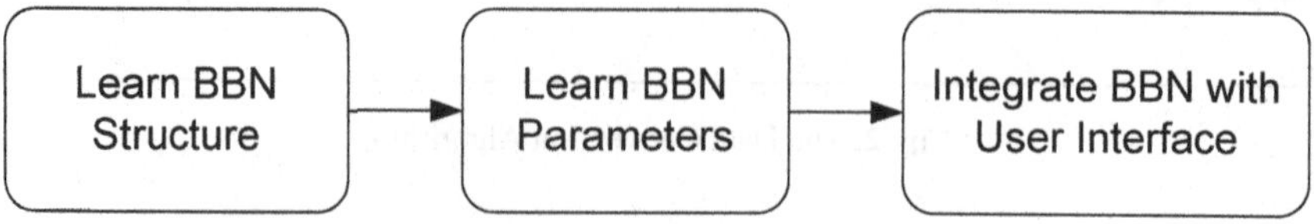

Fig. 3. The Expert system Construction Process

4.1 Learning BBN from Data

There are more than 40 BBN Structure Learning Algorithms (BBN-SLAs) dating back to 1968. It is important to use those algorithms that have good learning accuracy. In our independent empirical study [17], we evaluated seven state-of-the-art BBN-SLAs: Three-Phase Dependency Analysis (TPDA)[5], Optimal Reinsertion (OR)[11], Greedy Equivalent Search (GES) [6], Max-Min-Hill-Climbing (MMHC) [18], CB [15], PC-GBPS [16] and REC[19]. We choose MMHC for learning the network structure because it achieves the best learning accuracy among all algorithms on synthetic datasets. In addition, MMHC algorithm is relatively robust against the presence of noise in the dataset. It also has good predictive accuracy on six real-world datasets.

Input:
D: Dataset
ε: threshold for the Conditional Independence test

MMHC (*D*, ε)
Phase I:
1. For all variables X_i, $PC(X_i) = MMPC(X_i, D, \varepsilon)$;

Phase II:
2. Start from an empty graph; perform greedy hill-climbing with operators: *add_edge*, *delete_edge* and *reverse_edge*. Only try operator *add_edge* $Y \rightarrow X_i$ if $Y \in PC(X_i)$

3. Return the highest scoring DAG found

Fig. 4. The MMHC Algorithm

Figure 4 outlines the MMHC algorithm. Phase I of MMHC identifies the candidate set $PC(X_i)$ (possible parents and children) for each node X_i by calling the procedure *Max-Min-Parent-Children* (*MMPC*). *MMPC* uses *Min-Max* heuristic to select the node that remains highly associated with X despite best efforts to make the node independent of X. Phase II performs a Bayesian-scoring greedy hill-climbing search starting from the empty graph to add, delete and orient the edges. Threshold for the Conditional Independence test is set to be 0.05 as suggested in [18].

After learning the structure of the BBN, the next step is learning the parameters - the CPTs of the BBN. This is achieved by estimating Pr_{ijk}, the probability that node X_i is in a state k given that its parents are in a configuration j. In this study, the parameters are calculated using sequential Bayesian parameter updating with Dirichlet priors [9].

Once the complete BBN is built from the training dataset, we export the BBN into XMLBIF format and integrate the BBN with a BBN visualization and inference tool JavaBayes [4]. As a result, a fully functional expert system is constructed for modeling and selecting RE techniques for single product project.

4.2 Results

The learned BBN has 18 nodes and 42 arcs. It models the causal factors of RE techniques in six RE phases as well as dependency relationships between all other nodes. It identifies following key causal factors of RE techniques—*Project Domain, Project Type, Project Geographic distributions, Agility of the project* and *Project Duration.*

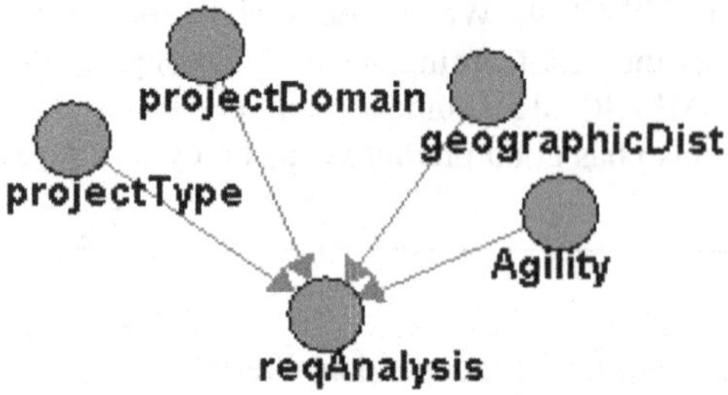

Fig. 5. Causal Factors for RE Techniques in Requirement Analysis Phase

Due to the space limitation, part of the BBN is shown instead of the complete graph. Figure 5 shows the four causal factors identified for RE techniques in *Requirement Analysis* phase (node *reqAnalysis*). A total of 15 RE techniques are listed as values of the node *reqAnalysis.* Similarly, causal factors are identified for RE techniques in other RE phases.

5 Empirical Study

We show that the learned BBN models the causal factors of the RE techniques. However, there are still some unanswered questions like: 1. What is the quality of the generated dataset? 2. How accurate is the expert system in selecting RE techniques? 3. Is the BBN better than other predictors? This section answers these questions.

5.1 Probability Distribution Study

To evaluate the quality of the generated dataset, we compare the PMFs of the columns of original raw dataset with those of the generated dataset (Figure 6). Each column is considered as a random variable and its sample space is the distinct set of all its values.

Figure 6 compares PMFs of six columns (RE techniques in six RE phases) of both datasets. In general, PMFs of the generated dataset are very similar to the original PMFs of the raw dataset. This indicates that the original probability distributions are well preserved in the generated dataset by the data conversion algorithm. Therefore, the generated dataset has trustable quality to replace the raw dataset for BBN learning.

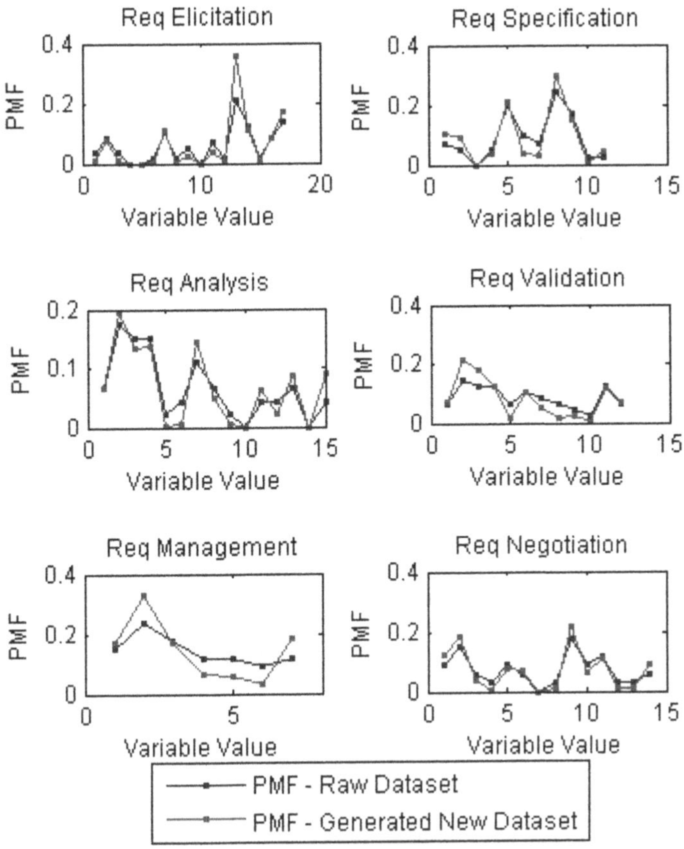

Fig. 6. PMFs of RE Techniques in Six RE Phases

5.2 Classification Accuracy Study

One objective of this paper is to build an expert system that selects appropriate RE techniques for any future projects. The selection is achieved by inference on the BBN and choosing the RE technique that has the maximum posterior probability in each RE phase given evidences. We use classification accuracy [8] to evaluate the expert system's ability of selecting correct RE techniques. We compare the expert system's classification accuracy on RE techniques with that of two baseline classifiers: Naïve Bayes and J48 [14] (a decision tree based classifier). All the classifiers are learned using the training dataset with 3300 rows. The testing dataset with 1100 rows is used to obtain the classification accuracy. The results are listed in table 1. The number of classes for each class variable is also listed. The highest accuracy is marked bold.

It is clearly shown that the expert system is more accurate in selecting RE techniques in most RE phases (except *Requirement Management*) using the separate testing dataset. The expert system performs better than two baseline classifiers mainly because the BBN identifies the causal factors of RE techniques and use the values of

Table 1. Classification Accuracy of the Expert System (E.S), Naïve Bayes(NB) and J48 classifier

Class Variable	#Classes	E.S	NB	J48
RE Techniques in Req Elicitation:	17	**62.27**	44.43	56.14
RE Techniques in Req Analysis:	15	**67.63**	58.86	66.59
RE Techniques in Req Management:	7	82.45	74.20	**84.32**
RE Techniques in Req Negotiation:	14	**82.29**	73.52	81.02
RE Techniques in Req Specification:	11	**70.45**	56.59	70.11
RE Techniques in Req Validation:	12	**67.09**	52.61	66.36

these causal factors to effectively select the RE technique in each RE phase. Naïve Bayes has the lowest classification accuracy because it assumes independence between all the variables, omitting the causal factors of RE techniques. The results confirm the usage of BBN as the core of the expert system. On the generated data, the BBN is a more accurate predictor than Naïve Bayes and J48.

6 Conclusions

In this study, a BBN based expert system is built for modeling and selecting RE techniques. A new web-based questionnaire is developed to collect expertise available in the RE community. For building the expert system, a novel data conversion algorithm is proposed to transform the raw dataset. This process can be re-applied to situations where the raw data is small in size and contains multiple values in one entry. This study builds a new BBN based expert system using state-of-the-art machine learning algorithm. In the empirical study, the data generated by the data conversion algorithm is analyzed and validated. The expert system outperforms other base line predictors in selecting RE techniques. The future work could include the following tasks: 1. The collection of more responses from RE community to enhance the expert system 2. Applying the same process to build an expert system for product line projects.

References

[1] Agile Development in Product Line Engineering Website: `http://www.utdallas.edu/~kxf041000/research/aple.html`

[2] Manifesto for Agile Software Development, `http://www.agilemanifesto.org`

[3] Agile Product Line Engineering Questionnaire, `http://129.110.92.41/Default.aspx`

[4] Bayesian Networks in Java, `http://www.cs.cmu.edu/~javabayes`

[5] Cheng, J., Greiner, R., Kelly, J., Bell, D.A., Liu, W.: Learning Bayesian networks from data: An information-theory based approach. Artificial Intelligence 137, 43–90 (2002)
[6] Chickering, D.M.: Optimal structure identification with greedy search. Journal of Machine Learning Research, 507–554 (2002)
[7] Feng, K., Lempert, M., Tang, Y., Tian, K., Cooper, K., Franch, X.: Developing a Survey to Collect Expertise in Agile Product Line Requirements Engineering. In: Inaugural International Research-in-Progress Workshop on Agile Software Engineering, Washington DC (2007)
[8] Friedman, N., Geiger, D., Goldszmidt, M.: Bayesian network classifiers. Machine Learning 29, 131–163 (1997)
[9] Heckerman, D.E., Geiger, D., Chickering, D.M.: Learning Bayesian networks: The combination of knowledge and statistical data. Machine Learning, 197–243 (1995)
[10] Macaulay, L.A.: Requirements Engineering. In: Applied Computing. Springer, Heidelberg (1996)
[11] Moore, A., Wong, W.: Optimal reinsertion: A new search operator for accelerated and more accurate Bayesian network structure learning. In: Twentieth International Conference on Machine Learning (2003)
[12] Nuseibeh, B., Easterbrook, S.: Requirements engineering: a roadmap. In: Finkelstein, A. (ed.) The Future of Software Engineering. ACM Press, New York (2000)
[13] Pohl, K., Böckle, G., van der Linden, F.J.: Software Product Line Engineering: Foundations. Principles and Techniques. Springer, Heidelberg (2005)
[14] Quinlan, R.: C4.5: Programs for Machine Learning. Morgan Kaufmann, San Francisco (1992)
[15] Singh, M., Valtorta, M.: An algorithm for the construction of Bayesian network structures from data. In: 9th Conference on Uncertainty in Artificial Intelligence, pp. 259–265 (1993)
[16] Spirtes, P., Meek, C.: Learning Bayesian networks with discrete variables from data. In: Proceedings from First Annual Conference on Knowledge Discovery and Data Mining, pp. 294–299 (1995)
[17] Tang, Y., Cooper, K., Cangussu, C.: A Survey and Analysis of Bayesian Belief Network Structure Learning Algorithms: Accuracy and Sensitivity to Noise, Technical Report. Department of Computer Science. University of Texas at Dallas (2009)
[18] Tsamardinos, I., Brown, L.E., Aliferis, C.F.: The max-min hill-climbing Bayesian network structure learning algorithm. Machine Learning (2006)
[19] Xie, X., Geng, Z.: A Recursive Method for Structural Learning of Directed Acyclic Graphs. Journal of Machine Learning Research 9, 459–483 (2008)
[20] Yadav, S.B., Bravoco, R.R., Chatfield, A.T., RA-Jukumar, T.M.: Comparison of analysis techniques for information requirement determination. ACM 31(9), 1090–1097 (1988)

Appendix

A BBN is a probabilistic graphical model that represents a set of variables and their probabilistic independencies [9]. A Bayesian-Belief Network (BBN) has two components:

1. A Directed Acyclic Graph (DAG) whose nodes represent random variables and whose arcs represent the dependencies between the variables. If there is an Arc from variable A to variable B, then A is called a *parent* of B, and B is a *child* of A;

2. The Conditional Probability Tables (CPTs). Each variable has a CPT that defines the prior probability of the variable given the value of its parents. Each variable is assumed to be conditionally independent of its non-descendents given its parents.

Formally, a BBN is denoted as B = <G, θ>. G = <V, A>, V={X1, X2, X3,...Xn} is the set of variables presented by the BBN. A is the set of Arcs in the DAG. θ is a set of parameters. Each parameter in θ is the CPT of Variable X_i denoted as $\theta_i = P(X_i \mid Pa(Xi))$ where Pa(Xi) are parents of variable Xi. The BBN *B* defines a unique joint probability distribution over *V*:

$$P(X_1, X_2, X_3, \ldots X_n) = \prod_{i=1}^{n} P(X_i \mid Pa(X_i))$$

If Xi has no parents, its CPT is said to be unconditional, otherwise it is conditional. If the variable represented by a node is observed, then the node is said to be an evidence node, otherwise the node is said to be hidden or latent.

Web Access Latency Reduction Using CRF-Based Predictive Caching

Yong Zhen Guo, Kotagiri Ramamohanarao, and Laurence A.F. Park

Department of Computer Science and Software Engineering
University of Melbourne, Australia
{yzguo,rao,lapark}@csse.unimelb.edu.au

Abstract. Reducing the Web access latency perceived by a Web user has become a problem of interest. Web prefetching and caching are two effective techniques that can be used together to reduce the access latency problem on the Internet. Because the success of Web prefetching mainly relies on the prediction accuracy of prediction methods, in this paper we employ a powerful sequential learning model, Conditional Random Fields (CRFs), to improve the Web page prediction accuracy for Web prefetching. We also propose a predictive caching scheme by incorporating CRF-based Web prefetching and caching together to reduce the perceived waiting time of Web users further. We show in our experiments that by using CRF-based Web predictive caching, we can achieve higher cache hit ratio and thus reduce more access latency with less extra transmission cost when compared with the predictive caching methods based on the well known Markov Chain models.

Keywords: Web Page Prediction, Conditional Random Fields, Web Predictive Caching.

1 Introduction

The World Wide Web is a vast information source in which many people turn to for daily news and general knowledge. The easy and convenient access to information in remote locations has attracted more and more people to use the Internet, which also results in a rapid increase in the Network traffic. The popularization and convenience of wireless connections has turned a large amount of Web users to use handheld devices (such as mobile phones or PDAs) to surf the Internet, although the transmission speed of these wireless connections are usually slow. These low bandwidth Web users may spend lengthy times waiting for requested Web pages to be transferred to them from Internet, which leads to intolerable access delays.

The perceived access latency of Web users comes from several aspects. First, Web servers need to process user requests. When a Web server is overloaded, the latency caused by its processing time is noticeable. Second, Web clients need to spend time on parsing the data received (*i.e.*, interpreting a segment of script or running a Java program) and displaying the contents to Web users. Third,

W. Liu et al. (Eds.): WISM 2009, LNCS 5854, pp. 31–44, 2009.

the transmission of a file from a Web server to a user will consume a period of time. Due to the rapid development of computer hardware, the processing power of both Web servers and clients have improved dramatically, therefore, the access latency caused by the processing time of Web servers and clients is negligible. Because the network bandwidth (especially the wireless network) is limited, the transmission of a large amount of data across a long distance over narrow bandwidth will take a relatively long time, thus the transmission time from Web servers to users results in most of the perceived latency of Web users.

Researchers have proposed many techniques to decrease the Internet access latency, among which *caching* and *prefetching* are two main techniques. In Web prefetching if most prefetched Web pages are not visited by a Web user in his subsequent accesses (implying that the prefetching method has predicted the user's actions poorly), the limited network bandwidth and server resources will not be used efficiently, increasing the access delay to the user-requested pages and worsening the access latency problem. Consequently, the success of Web prefetching relies mainly on the Web page prediction accuracy. In this paper, we introduce the use of a powerful prediction algorithm, Conditional Random Fields (CRFs) [1], to improve the prediction accuracy of Web prefetching. Furthermore, although a prefetched Web page is not the page a user wants to visit currently, it might be requested by the user in his subsequent access, we can save these prefetched pages in the user's cache to reduce the waiting time further. In this way caching and prefetching can be combined together as predictive caching to improve the cache hit ratio and provide better performance in reducing Web access latency. In this paper we propose the CRF-based predictive caching of this kind for Web access latency reduction by combining CRF-based prefetching and caching together. We show in our experiments the merits of CRF-based Web predictive caching in reducing Web access latency over the one based on the popular Markov Chain models [2].

The rest of this paper is organized as follows: In Section 2 we briefly describe the Web page prediction method based on Conditional Random Fields. In Section 3 we illustrate how we combine CRF-based prefetching and caching together as predictive caching for Web access latency reduction. In Section 4 we compare the performance of CRF and Markov Chain-based predictive caching in improving cache hit ratio and reducing transmission cost. In this section we also show the impact of CRF-based predictive caching in reducing access latency by simulations. Finally, we conclude this paper in Section 5.

2 Conditional Random Field-Based Web Page Prediction

The prediction algorithm employed to estimate the probability of each Web page being requested by a Web user in the immediate future is very important for Web prefetching. In this section we will first briefly review the basic principle of Conditional Random Fields (CRFs) and then explain how to use the CRF-based Web page prediction in Web prefetching.

2.1 Conditional Random Fields

A Conditional Random Field [1] is an undirected graphical model, it defines a conditional probability distribution of a label sequence $Y = (y_1, y_2, \cdots, y_n)$, given an observation sequence $X = (x_1, x_2, \cdots, x_n)$. Although theoretically the structure of a Conditional Random Field can be an arbitrary undirected graph that obeys the Markov property, for the tasks of labeling the most common graphical structure is an undirected linear chain of first-order among label sequence Y. A linear chain CRF has the form as below:

$$P_\theta(Y|X) = \frac{1}{Z(X)} \exp\left[\sum_{t=1}^{T}\left(\sum_i \lambda_i f_i(y_{t-1}, y_t, X, t) + \sum_j \mu_j s_j(y_t, X, t)\right)\right] \tag{1}$$

Where $f_i\,(y_{t-1}, y_t, X, t)$ is a transition feature function between the states (labels) at position $t-1$ and t, while $s_j\,(y_t, X, t)$ is a state feature function of the state at position t. $Z(X)$ is a global normalization factor over all possible label sequences. The parameters $\theta = (\lambda_i, \mu_j)$ can be estimated by maximizing the log-likelihood of the training data by using approach such as L-BFGS [5]. In terms of the size of training data, the CRF training using L-BFGS usually requires many iterations, each of which calculates the log-likelihood and its derivative [3]. The larger number of sequences a dataset has, the more iterations it needs for the L-BFGS algorithm to converge. The time complexity of such an iteration is $O(L^2NT)$, where L is the number of labels, N is the number of sequences (*e.g.*, Web page sessions) and T is the average length of sequences. Therefore, the time complexity of a CRF training is quadratic with respect to the number of unique labels. When there are a large number of unique labels, the CRF training can become very slow or intractable even with efficient training algorithm like L-BFGS. However, in the next section we will describe how to reduce the training complexity of CRFs. After the parameters are trained, the Viterbi [4] algorithm can be used to label the testing data and perform the prediction.

Because CRFs directly model the conditional distribution $P(Y|X)$, they do not need to model the visible observation sequences X, which results in the relaxation of unwarranted independence assumptions over observation sequences. Actually, CRFs can model long-range dependencies between observation elements. Moreover, due to the conditional nature, CRFs are able to model arbitrary features of observation sequences, regardless of the relationships between them. Because of these advantages, CRFs can yield more accurate prediction than other popular prediction methods such as Markov Chain models.

2.2 Grouped ECOC-CRFs for Web Page Prediction

In the Web page prefetching scenario, each unique Web page of a Website is regarded as a unique label. Then we can treat each previous user session of the Website as an observation sequence, and in a user session we can consider each pageview's subsequent pageview as its label. In this way, we can make use of all the previous uesr sessions to obtain observation sequences and the corresponding label sequences which can be used to train a CRF model for making prediction on this Website. It has been shown in [8] that CRFs can be used efficiently in Web page prediction on Websites whose number of unique Web pages is small.

In a Website where there are a large number of unique Web pages, the direct use of CRF-based prefetching is infeasible due to the inherent high complexity of CRF training (quadratic to the number of unique labels). In this case, we can reduce the overall training complexity of CRFs for Web prefetching by using Error Correcting Output Coding (ECOC) [6] method to decompose a multi-label CRF training into a set of binary-label sub-CRF trainings, and then combine the output of all the sub-CRFs to obtain the possible original result [9]. Since all the sub-CRFs are binary, they can be trained very fast and efficiently.

Moreover, although by using ECOC method we can achieve faster training time for CRF-based Web prefetching, we also drastically reduced the class information given to each sub-CRF, which will decrease the overall prediction accuracy slightly when compared to a multi-label CRF training. We can improve the prediction accuracy of ECOC-CRFs by using the grouping technique [10], in which the columns of the ECOC code matrix are divided into several groups and each group is used to train a sub-CRF. By doing this, each sub-CRF can obtain more refined class information and the prediction accuracy can be enhanced further. In this way, we can utilize grouped ECOC-CRFs to obtain satisfactory prediction accuracy for Web page prefetching on large-size websites.

3 CRF-Based Predictive Caching

Caching techniques save the copies of the Web pages that a user has visited in his local cache, in the future if this user wants to visit these pages again his browser can display these pages immediately from the cache. Every time when a user requests a Web page, if this page is in his cache, then there is a cache hit; otherwise there is a cache miss, and this page will be put into the user's cache. The cache hit ratio is then defined as *cache hits / (cache hits + cache misses)*. The more requested Web pages can be found in cache, the higher the cache hit ratio is, and consequently more perceived access waiting time can be saved. However, practically in a surfing process a Web user usually tends to jump from the current Web page to a new Web page that was not accessed before, seldom going back to the Web pages he has previously visited. In this case, most cached Web pages will not be used again during this user's surfing on the Internet. Therefore, using caching solely might not easily achieve a satisfactory performance in decreasing Web access latency.

Not like caching, prefetching techniques usually prefetch Web pages to a user before this user makes request to them. With a high-accuracy prefetching, most Web pages requesed by the Web user can be downloaded in advance. Furthermore, although some of these prefetched Web pages might be currently unwanted, they may be requested by the user in his subsequent access, and these

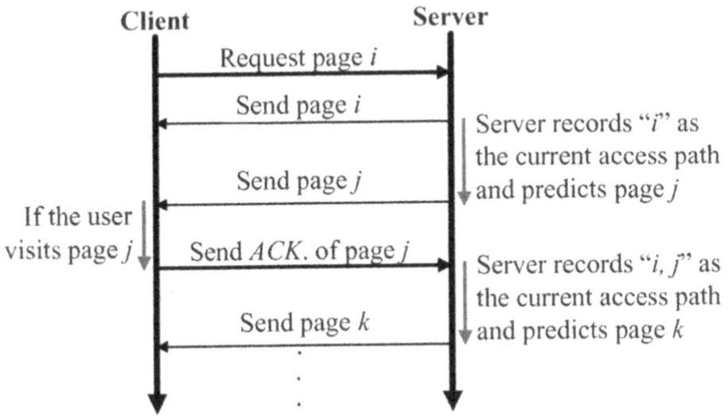

Fig. 1. Data exchange between client side and server side in a practical predictive caching system. *ACK* is the acknowledgement package.

prematurely prefetched Web pages can be stored in the user's cache to enhance the cache hit ratio. By doing so, caching and prefetching can be combined together as predictive caching to reduce more Web access latency, where cache is more like a buffer for prefetched Web pages. In this paper, we incorporate CRF-based prefetching and caching togegher in this way as CRF-based predictive caching to reduce the access latency of Web users.

Moreover, because Web page prediction needs to utilize the historical access information of previous users stored on Web servers, prefetching is usually made by the server side. In order to predict a possible Web page correctly for a Web user, the Web server needs to keep track of this user's current access path. However, in a predictive caching system, if the requested Web page can be found in the client's cache, the client will not make a request to the Web server for this Web page, in which case the server can not be aware of the user's current behaviour and thus can not make a correct prediction for the user. Therefore, in practice it is necessary for the client to send an acknowledgement package to the server informing the server about the user's access to a cached Web page. This allows the server to maintain this user's current access path exactly. Since the acknowledgement package only contains the ID information of a Web page, its size is very small and the caused network traffic of it can nearly be neglected. An example process of this is illustrated in Figure 1.

4 Experimental Evaluations

In this section we conducted experiments to evaluate the effect of predictive caching based on CRFs and Markov Chain models in reducing the Web access latency. In our experiments we implemented the first, second and third order Markov Chains (referred as 1st-MC, 2nd-MC and 3rd-MC respectively in the future) for Web page prefetching. For the implementation of CRF training and decoding, we use the CRF++ toolkit [13]. We use three different feature templates of CRF++ to generate the feature functions that will be used in the

CRF training in our experiments. In the first template (referred to as CRF1), we define the current and previous one observation and their combination as the unigram features; the second and the third template (CRF2 and CRF3) are defined similarly.

Since the cache hit ratio will directly influence the perceived waiting time that can be saved, we first examine how the predictive caching based on CRFs and Markov Chains can improve the cache hit ratio on two datasets: the publicly accessible *msnbc* dataset [11] and the larger *CSSE* dataset [12].

4.1 Experiments on the *msnbc* Dataset

The first experiment about cache hit ratio is carried out on the *msnbc* dataset [11], which is obtained from the Web logs of *www.msnbc.com.* In this dataset, all the user visits are recorded in session format at the level of Web page categories such as *health*, *sports* and so on. There are 17 different page categories of this kind which can also be treated as 17 unique Web pages. After preprocessing we randomly selected 35,000 distinct sessions with length more than 8 and less than 100 from this dataset and divided them into two subsets: 30,000 sessions for training and 5,000 for testing.

In this experiment the training dataset is used to train the CRF and Markov Chain prediction models and the calculation of the average cache hit ratio is based on the testing dataset. All the user sessions in the testing dataset are assumed to be from different users, each of whom has an empty cache or buffer of a fixed size n (for simplicity we assume all Web pages are of the same size 1). Then for every pageview of each user session in the testing dataset a procedure is conducted: First, if this pageview is found existing in its user's cache, then there is a cache hit; otherwise, a cache miss occurs and this pageview is put into its user's cache. Second, the prediction models obtained from the training stage are employed to predict the most possible next Web page for this pageview and prefetch this page to its user's cache. Because prefetching only happens after a Web user requests the first Web page, there is always a miss for the first Web page of every user session. Every time when the user's cache is full, the "LRU" cache replacement strategy is adopted to remove the least recently used Web page from the cache. This procedure continues from the first pageview to the last pageview of a session, then the cache hit ratio for this session (or user) can be calculated. In this way, we can obtain the cache hit ratio of all sessions (or users) in the testing dataset and calculate the average value of them.

The average cache hit ratio of the predictive caching based on CRFs and Markov Chains at different cache sizes on the *msnbc* dataset are shown in Figure 2, in which the result of the pure caching method without prefetching (referred as LRU) is also shown as a baseline. From this figure we can see that while the cache size increases, the cache hit ratios of all the methods with or without prefetching increase correspondingly. The reason of this is straightforward: with a bigger cache size more Web pages can be cached, thus more Web pages a user requests can be found in the cache. When compare the cache hit ratios of these different methods at a same cache size, we can find that all prefetching methods perform much better than

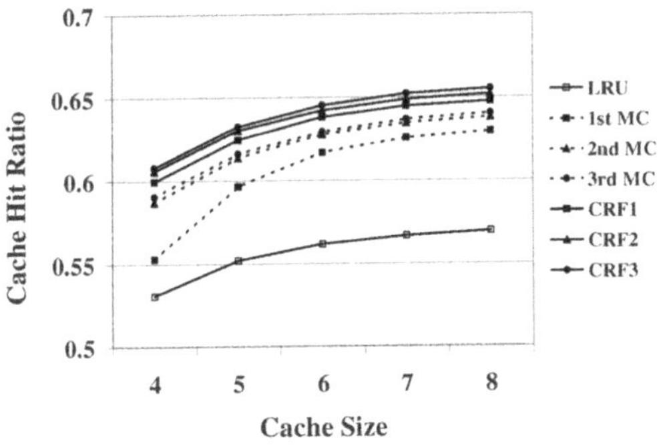

Fig. 2. Cache hit ratios at different cache sizes on the *msnbc* dataset

LRU (the method without prefetching), this is because at every visit the prefetching method will bring in the possible "next" pages that might be requested by the user immediately, which in turn improve the cache hit ratio. Furthermore, it is noticeable that CRF-based predictive caching can outperform Markov Chain-based predictive caching significantly, this is because CRFs are more accurate models than Markov Chains in predicting Web user behaviours.

4.2 Experiments on the *CSSE* Dataset

The same experiment is carried out on the *CSSE* dataset [12] from the Web log of the Department of Computers Science and Software Engineering (*CSSE*), the University of Melbourne, which contains 3,829 unique Web pages. After preprocessing, we randomly select 2,723 user sessions as the training dataset and 544 user sessions as the testing dataset.

Because there are many labels (3,829 unique pages) in this dataset, the training of a multi-label CRF here is infeasible due to the high complexity of CRF training. Therefore, we use grouped ECOC-CRFs for the CRF training in this dataset. We made use of the Search Coding [7] method to design a $3,829 \times 24$ ECOC code matrix whose minimum Hamming distance d is 8, and then grouped every 8 columns of this code matrix into one group, which means we have 3 groups and need to train 3 sub-CRFs.

In this experiment we also show the cache hit ratio of an *Optimal* scheme as a comparison. The *Optimal* scheme uses an ideal prefetching method whose prediction accuracy is 100% correct to every page except the first one of each user session (we assume prefetching only happens after a Web user requests the first Web page, and the first Web page of a user session is always unpredictable). Moreover, the *Optimal* scheme uses an optimal cache replacement strategy to replace Web pages. The optimal cache replacement is a theoretical page replacement algorithm, which works as follows: when a page needs to be swapped in, the cache system discards the page whose next use will occur farthest in the future. For example, a page that is not going to be used for the next 10 visits will be discarded over a page that will be used within the next 3 visits. Although the

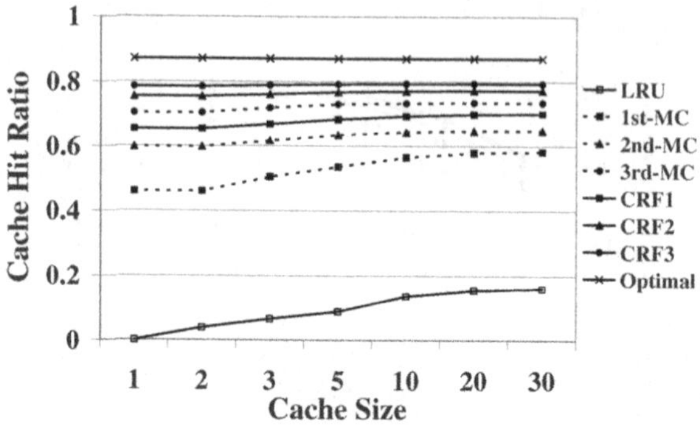

Fig. 3. Cache hit ratios at different cache sizes on the *CSSE* dataset

optimal cache replacement strategy is the best cache replacement strategy, it is not realizable in practice since it is impossible to predict how far in the future a page will be needed. However, because in our experiment from each user session we can know exactly what page sequence the user will request subsequently at each visit in advance, we can simulate the optimal replacement strategy. The *Optimal* scheme can yield a higher cache hit ratio than any other methods. The closer the cache hit ratio a predictive caching method with "LRU" cache replacement strategy can obtain to that of the *Optimal* scheme, the better the method.

The cache hit ratios of different methods at different cache sizes on the *CSSE* dataset can be found in Figure 3. In this figure we can see that the *Optimal* scheme has the highest cache hit ratio, while the other predictive caching methods with relatively higher prediction accuracy (such as CRF2 and CRF3) can obtain a closer cache hit ratio to it, which indicates these methods can reduce waiting time more for Web users. We can also notice that *LRU* (the method without prefetching) yields the lowest cache hit ratio (lower than 20% when at the cache size of 30), which shows the effect of prefetching in improving the cache hit ratio. Moreover, from this figure we can see that the cache hit ratios of all methods except *Optimal* increase while cache size enlarges, which again indicates that increasing cache size can help to increase cache hit ratio. The reason that the cache hit ratio of *Optimal* remains constant while cache size enlarges is due to its 100% correct prediction accuracy: for a user every requested Web page (except the first Web page) has already been prefetched into his cache, no matter what the cache size is. Finally, it is obvious in this plot that the increase rate of the cache hit ratio of the predictive caching methods with low prediction accuracy (such as 1st-MC and 2nd-MC) is higher than the predictive caching methods with high prediction accuracy (such as CRF2 and CRF3). Therefore, caching is more important when the predictive caching method's prediction accuracy is low.

In Figure 3 we showed the effect of predictive caching on increasing the cache hit ratio of Web pages requested by Web users, in Figure 4 we will examine the impact of predictive caching on the hit ratio of predicted Web pages (referred as

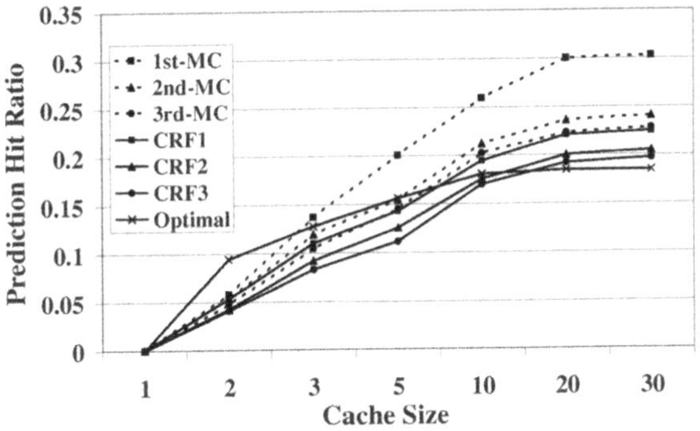

Fig. 4. Prediction hit ratios at different cache sizes on the *CSSE* dataset

prediction hit ratio, if the predicted Web page can be found in the cache, then there is a hit). The higher the prediction hit ratio is, more predicted Web pages do not need to be transmitted and thus can save more bandwidth. In Figure 4 we can see that while the cache size increases the prediction hit ratio of all predictive caching methods increase significantly. It is noticeable that the general trend in this plot shows that the prediction hit ratio of low-accuracy predictive caching methods (such as 1st-MC) is higher than that of high-accuracy predictive caching methods (such as CRF3). This is because generally a Web user will jump from a Web page to a new page he does not visit before, and the low-accuracy predictive caching methods have less chance to predict the user's real intention correctly and tend to predict more randomly than the high-accuracy predictive caching methods.

We also evaluated the transmission cost of each method in Figure 5. Every time when the Web server transmits a page (whether it is requested by the user explicitly or prefetched by the prefetching system) to a user, a transmission cost will occur. Transmission cost can be used to indicate the extra network traffic incurred by prefetching. The lower a predictive caching method's caused transmission cost, the better this method. From Figure 5 we can see that the transmission cost of all methods decrease as the cache size increases due to more Web pages can be saved with a larger cache. The transmission cost of predictive caching methods (except *Optimal*) are much higher than that of the method without prefetching (*LRU*), the reason is that every time these predictive caching methods need to prefetch extra pages into the cache, and if the exact page the user accesses is not in the cache after prefetching, the user's browser still needs to connect to the Web server to download this page. Because of the 100% prediction accuracy and the optimal cache replacement strategy, the transmission cost of *Optimal* is lower than *LRU*. By comparing the predictive caching methods, we can notice that the methods with higher prediction accuracy can produce lower transmission cost, this is because higher prediction accuracy results in higher cache hit ratio and thus more Web pages a user requests can be found in his cache. Among all the predictive caching methods, CRF3 produces the closest transmission cost to the *Optimal* scheme, denoting the merits of CRF-based predictive caching.

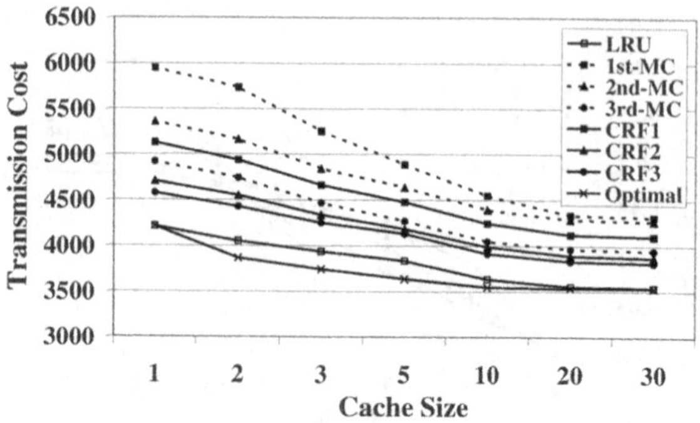

Fig. 5. Transmission cost at different cache sizes on the *CSSE* dataset

4.3 Simulations on Access Latency Reduced by CRF-Based Predictive Caching

In order to evaluate how well CRF-based predictive caching can reduce a Web user's perceived access latency, we run two more simulations regarding the average waiting time Web users spent on each Web page on the *CSSE* dataset.

In these simulations, each Web page of all user sessions in the testing dataset will be assigned a random transmission time (the time needed to transmit this Web page from the Web server to the user) and a random reading time (the time the user spent on reading this page). The calculation of the waiting time on every Web page of a session can be described by an example as follows: If a user session in the testing dataset is *"A B"*, which means the user's exact browsing path is from Web page A to page B, then the user's waiting time on page A is page A's transmission time because page A is the first page the user requests, his browser needs to download page A from the very beginning. Then there are two cases that need to be considered: (1) Page B is prefetched correctly as the next requested page by a prefetching method, and (2) the prefetching method fails to predict the user's behaviour correctly and prefetchs pages other than B for the user. In the first case, if page B's transmission time is less than the reading time the user spent on page A, then there is zero waiting time for page B since while the user is reading page A, page B has already been prefetched to his computer; Otherwise, if page B's transmission time is longer than page A's reading time, then the user's perceived waiting time on page B is calculated as page B's transmission time minus page A's reading time. In the second case, because page B is not correctly prefetched while the user is reading page A, his browser needs to download page B when he requests it, thus the user's perceived waiting time on page B is page B's transmission time. The calculation of the waiting time on each Web page of this example is shown in Figure 6. In addition, the influence of transmitting a prefetched Web page on the transmission time of the Web pages that a user explicitly requests is also considered in these simulations. For example, if a requested Web page and a prefetched Web page are transmitting simultaneously, the transmission time of the requested Web page will double because the transmission of the prefetched

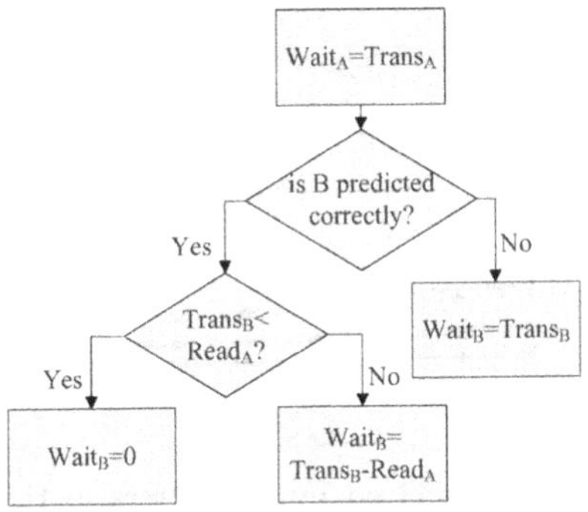

Fig. 6. Calculation of the waiting time on each Web page of the example session *"A B"*. $Wait_A$, $Trans_A$ and $Read_A$ mean the waiting time, transmission time, and reading time of Web page A respectively.

Web page takes up half of the user's bandwidth (assuming the user's bandwidth is constant). After the waiting time on every Web page is calculated, we sum them and divide this value by the total number of Web pages in the testing dataset to get the average waiting time on each Web page.

We generated a random transmission time that follows the Poisson distribution and a random reading time that obeys the Exponential distribution for each Web page of every session in the testing dataset. In the first simulation, we set the *mean transmission time* of all Web pages to 20 seconds and the *mean reading time* of all Web pages to 60 seconds to represent the case where the mean transmission time of Web pages is shorter than the mean reading time. In the second simulation, we set the *mean transmission time* to 60 seconds and the *mean reading time* to 20 seconds to represent the case where the mean transmission time of Web pages is longer than the mean reading time. Then we calculated the average waiting time of different predictive caching methods based on CRFs and Markov Chains in these two simulations, where the results of *LRU* and *Optimal* are depicted as well. We run each simulation 10 times and calculate the average value as the final result. The results of these two simulations are shown in Figure 7 and Figure 8 respectively.

From the results of these two simulations we find that with a bigger cache size, the average waiting time of all methods decrease drastically (especially in Simulation 2), thus caching is very useful in reducing a Web user's waiting time. Of all the methods, the *Optimal* scheme yields the shortest average waiting time. When compared at a certain cache size, the average waiting time of the predictive caching methods with high prediction accuracy (such as CRF3) is much shorter than the methods with low prediction accuracy (such as 1st-MC), and the higher prediction accuracy a predictive caching method has, the closer the performance it can achieve to the *Optimal* scheme. Moreover, in Figure 8 we also observe that when the cache size is smaller than 5, the average waiting time of the predictive caching method based on the 1st-order Markov Chain is longer than the non-prefetching method *LRU*, this is due to the low prediction accuracy

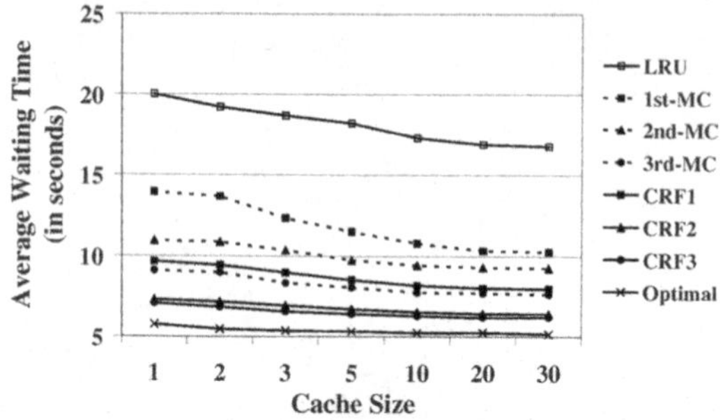

Fig. 7. Average waiting time perceived by Web users in the first simulation (*mean transmission time*=20 seconds and *mean reading time*=60 seconds) on the *CSSE* dataset

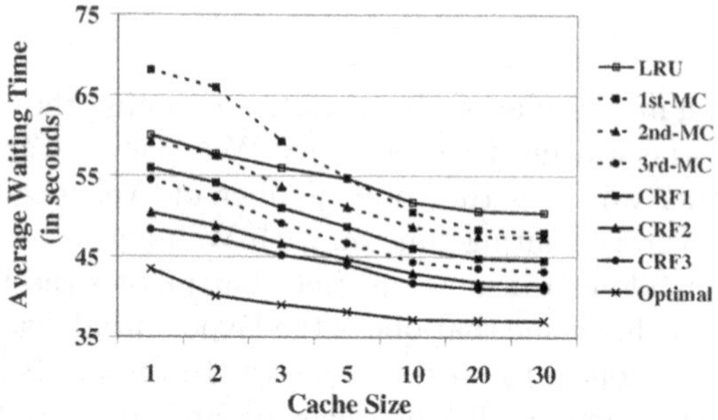

Fig. 8. Average waiting time perceived by Web users in the second simulation (*mean transmission time*=60 seconds and *mean reading time*=20 seconds) on the *CSSE* dataset

of the 1st-order Markov Chain and in Simulation 2 the mean transmission time of Web pages is much longer than the mean reading time. However, with a cache size bigger than 5 the 1st-MC can perform better than *LRU*.

Finally, in Figure 9 we depict the results concerning the percentage of perceived access latency reduction that can be achieved in these two simulations and the corresponding extra transmission cost for each predictive caching method when compared with *LRU* at the cache size of 30. From Figure 9 we can see that the percentage of reduced waiting time is proportional to the prediction accuracy of predictive caching methods, while the extra transmission cost presents an opposite trend. Predictive caching methods with high accuracy prediction can reduce substantial waiting time while only incur slight extra transmission cost. For example, CRF3 has the highest prediction accuracy and thus has the best performance here: by using CRF3 the percentage of perceived waiting time reduction can achieve 63.0% and 18.9% in Simulation 1 and Simulation 2 respectively, while the extra transmission cost is only about 7.3%, which is less than all other Markov Chain-based predictive caching methods. Therefore, CRF-based predictive caching can reduce Web access latency more efficiently than the one based on Markov Chains.

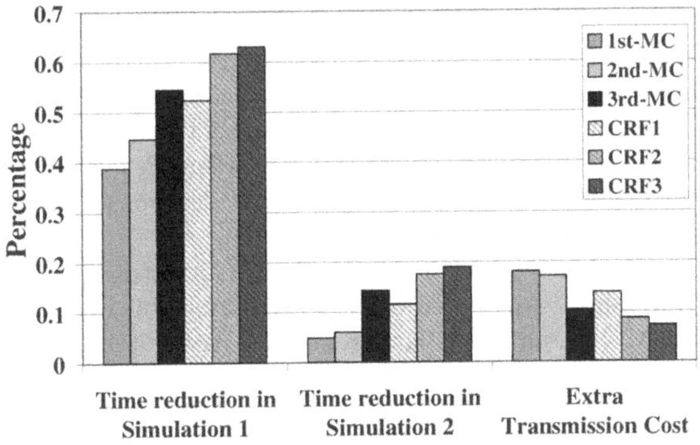

Fig. 9. Reduction of perceived waiting time VS. Extra transmission cost when compared with *LRU* at the cache size of 30 on the *CSSE* dataset

5 Conclusion

Low-bandwidth Web users usually suffer a lot from waiting for the Web pages they request being tranferred to them. In this paper, we propose the use of Web predictive caching for Web access latency reduction by combining Web prefetching and caching together. Since the performance of predictive caching relies crucially on the Web page prediction algorithm used in prefetching, in this paper we utilize a powerful prediction algorithm based on Conditional Random Fields in prefetching to save more waiting time for Web users. We examine the effect of predictive caching in improving the cache hit ratio and decreasing the transmisstion cost, and show the advantages of CRF-based predictive caching over the one based on Markov Chains in reducing a Web user's access latency. Our simulation results show that by using CRF-based Web predictive caching (CRF3), the average access latency of Web users can be dramatically decreased by up to 63% with a slight extra transmission cost of 7.3% when compared with *LRU*.

References

1. Lafferty, J., McCallum, A., Pereir, F.: Conditional Random Fields: Probabilistic Models for Segmenting and Labeling Sequence Data. In: Proceedings of the 18th International Conference on Machine Learning, pp. 282–289 (2001)
2. Markov, A.: Extension of the Limit Theorems of Probability Theory to a Sum of Variables Connected in a Chain. In: Dynamic Probabilistic Systems, vol. 1, Markov Chains (1971)
3. Wallach, H.: Efficient Training of Conditional Random Fields. Master thesis, Division of Informatics, University of Edinburgh (2002)
4. Viterbi, A.J.: Error Bounds for Convolutional Codes and an Asymptotically Optimum Decoding Algorithm. IEEE Trans. Info. Theory, 260–269 (1967)
5. Nocedal, J.: Updating Quasi-Newton Matrices with Limited Storage. In: Mathematics of Computation, pp. 773–782 (1980)

6. Dietterich, T., Bakiri, G.: Solving Multiclass Learning Problems via Error-Correcting Output Codes. J. Arti. Inte. Rese (1995)
7. Jiang, Y.H., Zhao, Q.L., Yang, X.J.: A Search Coding Method and Its Application in Supervised Classification. J. Software (2005) (in Chinese)
8. Guo, Y.Z., Ramamohanarao, K., Park, L.: Web Page Prediction Based on Conditional Random Fields. In: Proceedings of the 18th European Conference on Artificial Intelligence (ECAI 2008), pp. 251–255 (2008)
9. Guo, Y.Z., Ramamohanarao, K., Park, L.: Error Correcting Output Coding-Based Conditional Random Fields for Web Page Prediction. In: Proceedings of the 2008 IEEE/WIC/ACM International Conference on Web Intelligence (WI 2008), pp. 743–746 (2008)
10. Guo, Y.Z., Ramamohanarao, K., Park, L.: Grouped ECOC Conditional Random Fields for Prediction of Web User Behavior. In: Proceedings of the 13th Pacific-Asia Conference on Knowledge Discovery and Data Mining, pp. 757–763 (2009)
11. UCI KDD Archive, `http://kdd.ics.uci.edu/databases/msnbc/msnbc.html`
12. Department of Computer Science and Software Engineering, University of Melbourne, `http://csse.unimelb.edu.au`
13. CRF++, Yet another CRF toolkit, `http://crfpp.sourceforge.net`

A Property Restriction Based Knowledge Merging Method

Haiyan Che, Wei Chen, Tie Feng, and Jiachen Zhang

College of Computer Science and Technology, Jilin University, Changchun, 130012, China
{chehy,chw,fengtie,zhangjc}@jlu.edu.cn

Abstract. Merging new instance knowledge extracted from the Web according to certain domain ontology into the knowledge base (KB for short) is essential for the knowledge management and should be processed carefully, since this may introduce redundant or contradictory knowledge, and the quality of the knowledge in the KB, which is very important for a knowledge-based system to provide users high quality services, will suffer from such "bad" knowledge. Advocates a property restriction based knowledge merging method, it can identify the equivalent instances, redundant or contradictory knowledge according to the property restrictions defined in the domain ontology and can consolidate the knowledge about equivalent instances and discard the redundancy and conflict to keep the KB compact and consistent. This knowledge merging method has been used in a semantic-based search engine project: *CRAB* and the effect is satisfactory.

Keywords: knowledge merging, property restriction based, equivalent instances recognition, Semantic Web, ontology.

1 Introduction

Merging new instance knowledge to get an updated and consolidated KB is very important for a system that provides users knowledge-based services. However, due to the fact of information explosion of current Web, there will undoubtedly be a lot of "bad" knowledge coming from various sources. Here, the "bad" knowledge means the redundant or contradictory knowledge. So appropriate method is necessary to detect and discard such "bad" knowledge to ensure the good quality of the KB.

Identifying the equivalent instances which refer to the same real-world entity is the core task, and only basing on this can we perform correct knowledge consolidation and efficient detection of contradict and redundancy. It is thus necessary for the efficient usage of Semantic Web data. There are already research works concerning about equivalent instances recognition and knowledge consolidation (see the related works section). But most of them do not give a systematic method to deal with the redundant or contradictory knowledge.

The property restriction based knowledge merging method advocated in this paper provides an equivalent instances identification method, which can identify the equivalence by comparing their *key* property values. According to the recognized

W. Liu et al. (Eds.): WISM 2009, LNCS 5854, pp. 45–52, 2009.

equivalences, the merging method can consolidate the knowledge about equivalent instances, reject the redundant knowledge and choose from the contradictory knowledge.

The rest of this paper is organized as follows: first, we introduce some related works in section 2. Then, we give a brief introduction to our ontology definition method in section 3, which is the prerequisite for understanding our knowledge merging method described in section 4. In section 5, we describe a trivial experiment to validate the effectiveness of the method, and last, we conclude our work and point out some future work in section 6.

2 Related Works

The work of [1] advocates an object consolidation method which identifies the equivalent instances by analyzing the *inverse functional* properties and merges the equivalent instances by rewriting their URIs. The problems are: 1) the *inverse functional* property is not reliable for judging equivalence and more worse, some organizations do not obey the semantics of such *inverse functional* properties in their usage; 2) it only considers those properties whose values are simple data type values, ignoring the fact that the values may also be instances which need to be judged equivalence; 3) after comparison it does not deal with the redundant or contradictory knowledge, which will lead to a rapid expanding and inconsistent KB.

The strategy of [2] is similar to ours, which identifies the equivalence between anonymous instances by comparing the key property value or clusters of properties. But it does not describe about how to select the key properties and how to consolidate the factual knowledge about the equivalent instances.

The L2R system [3] implements a logical method for reference reconciliation, based on rules of reconciliation that are automatically generated from the axioms of the schema. However, this method infers the equivalence result from all the identifiers in the data source for one time and thus, it does not fit for the application that merges new knowledge into the KB incrementally.

3 Ontology Definition Method

Before introducing our knowledge merging method, we will first give a brief review of the domain ontology, which is constructed according to our aggregated knowledge concept (AKC for short) based ontology definition method [4]. This domain ontology can not only characterize the complex n-ary relations but also emphasize the usage of property restrictions. The dataset to be merged are RDF documents about the instance knowledge of this ontology.

The AKC based ontology definition method treats the n-ary relation knowledge as a kind of aggregated knowledge and characterizes it by AKC in the domain ontology. AKC is such a kind of concept that its only function is to organize all the related information in an n-ary relation together and characterize each dimension of the n-ary relation as its properties. So the property values of an AKC instance are the true meaningful content of the aggregated structure and the AKC instances are blank nodes in RDF in nature. They are assigned special URIs, which can only indicate the instances are AKC instances but can not identify them (two AKC instances with different URIs may denote the same group of aggregated knowledge).

Considering the possibility that there will also be aggregated structures inside an AKC, the AKCs are further categorized into Outer-AKCs and Inner-AKCs.

The Outer-AKC is the outmost AKC in an aggregated knowledge structure, i.e. the Outer-AKC instances can only be the property values of non-AKCs' instances. The Outer-AKC instance denotes the whole group of related knowledge in an *n*-ary relation, so it may be referred to from outside of a particular graph and should be identified uniquely. In the domain ontology, we specify key properties (the range types of key properties can only be named entities or simple data types) for each Outer-AKC and define a *cardinality* restriction with the cardinality of 1 on each of these properties, requiring that each Outer-AKC instance must have exactly one value for each key property. So we can get the set of key properties of an Outer-AKC according to the ontology and identify the Outer-AKC instances by the combination of the values of their key properties.

The Inner-AKC is the internal AKC in an aggregated knowledge structure and is used to characterize the aggregated knowledge structure of the property values of the AKCs. So they are subordinate to the outmost Outer-AKC instances and need not be identified.

To depict an *n*-ary relation, we define an Outer-AKC to represent the whole aggregated knowledge of the *n*-ary relation and define each dimension of the *n*-ary relation as a property of the Outer-AKC. If some property value of the Outer-AKC is also aggregated knowledge, then define an Inner-AKC to represent the range type of the property.

The application domain we are interested is the financial stock field of China and the knowledge in this field are all about stocks and companies. So for simplicity, the ontology concepts we concern can fall into two categories: named entities (for example the *Stock* and the *Company*,etc), whose instances can be identified by their names, and AKCs, which are used to characterize the aggregated knowledge structure of *n*-ary relations and depict certain properties of the stocks or companies.

Take the following paragraph about the annual financial report as an example:

"本报讯岳阳兴长今日公布2006年年度报告,报告称该公司实现销售收入18.49亿元,同比增长12.01%;净利润4377万元,同比增长45.49%. (Yueyang Xingchang released the annual financial report of 2006 today. The report said the company's sales income was 1849 million Yuan, increased by 12.01%; net profit was 43.77 million Yuan, increased by 45.49%.)" –Paragraph 1

This paragraph describes an *n*-ary relation among multiple individuals and values: the data and the trend values over the same period of previous year (Tosppy for short) of several financial indicators of the Yueyang Xingchang Co.ltd in 2006. The domain ontology related to Paragraph 1 is shown in Fig.1. The ellipses in dotted lines represent the AKCs, among which the *Annual financial report entity* is an Outer-AKC, the *Net profit entity* and *Total income entity* are Inner-AKCs The properties with an asterisk (*) are key properties.

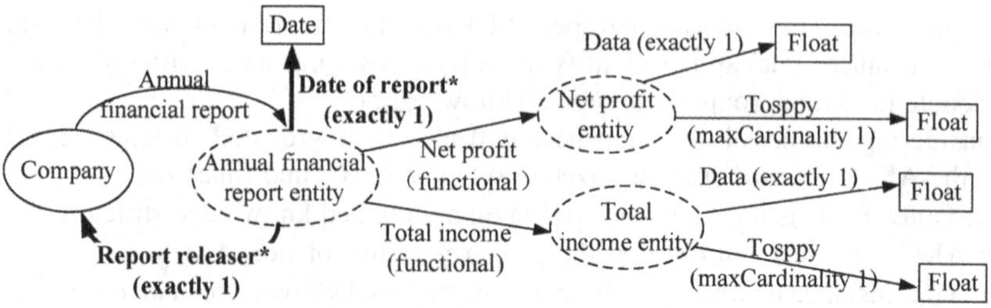

Fig. 1. Ontology definition of the Annual financial report entity

For a knowledge triple (or a statement) *t* :< *s; p; o*>, we use *S* (*t*), *P* (*t*) and *O* (*t*) to denote the subject, the predicate and the object of *t* respectively.

4 Knowledge Merging

The knowledge acquisition (KA for short) system of the *CRAB* [5] extracts instance knowledge from the domain related Web documents and gets a corpus of RDF documents. Then, the statements in these RDF documents should be merged into the KB to populate the domain ontology.

4.1 Equivalent Instances Recognition

The equivalent instances refer to the instances that denote the same real world entity but with different URIs. We see all the resources in the data set to be RDF nodes and define different methods for different RDF nodes. For simple RDF literals (such as Date values) and named entities (such as the *Stock* instances), we can compare them directly by their values or names. But for AKC instances, their URIs can not be used to identify them. Since every Inner-AKC instance is only subordinate to some Outer-AKC instance, we need not judge the equivalence between them. For Outer-AKC instances, we compare their key property values to judge their equivalence.

Definition 4.1 (*Identicalness*): Two RDF nodes are identical, if they are two RDF literals with the same type and values; or if they are two named entity instances of the same concept and with the same names; or if they are instances of the same Inner-AKC and all of their property values are identical respectively.

Definition 4.2 (Equivalence): Two Outer-AKC instances with different URIs are equivalent if they are instances of the same Outer-AKC and all of their key property values are identical respectively.

For example, *Annual financial report entity inst 1* and *Annual financial report entity inst 2* are equivalent Outer-AKC instances in Fig. 2.

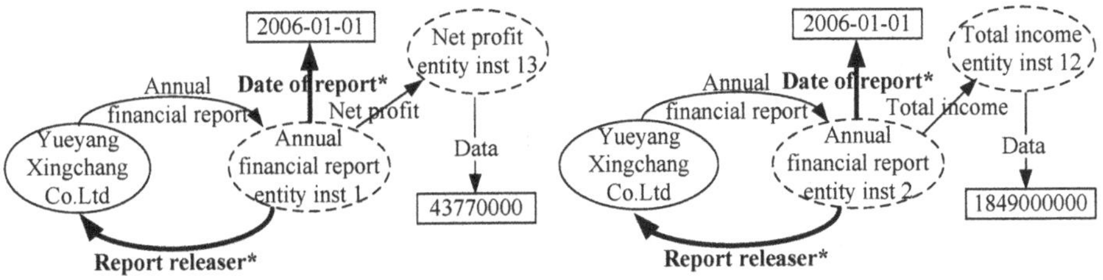

Fig. 2. Example of the equivalent instances

4.2 Redundant Knowledge Recognition

Definition 4.3 (Redundant knowledge): For two statements *stat*<*s*, *p*, *o*> and *stat*<*s*, *p*, *o*>, If *s* and *s* are equivalent or identical, and *o* and *o* are equivalent or identical, then *stat* and *stat* are redundant knowledge.

Since the instance equivalence is taken into account, this definition can depict the semantically redundant knowledge. For example, the statements in Fig.2.: <*Annual financial report entity inst 1*, *Date of report*, "2006-01-01"> and <*Annual financial report entity inst 2*, *Date of report*, "2006-01-01"> are semantically redundant knowledge.

4.3 Contradictory Knowledge Recognition

Definition 4.4 (*Contradictory knowledge*): For two statements *stat*<*s*, *p*, *o*> and *stat*<*s*, *p*, *o*>, If *s* and *s* are equivalent or identical, *p* is a *functional* property or there is a *cardinality* restriction with the cardinality of 1 on *p*, *o* is not identical nor equivalent to *o*, then *stat* and *stat* are contradictory knowledge; for two sets of statements *statSet* <*s*, *p*, *o*> (*i*=1..*l*) and *statSet* <*s*, *p*, *o*> (*j*= 1..*m*), if *s* and *s* are equivalent or identical, *C* is the concept that *s* and *s* belong to, and *C* has a *maxCardinality* restriction on *p* with the cardinality of *n* ($n \geq 1$), and the instances in $\{o_{1i}\} \cup \{o_{2j}\}$ are not identical nor equivalent to each other, and $l+m>n$, then *statSet* and *statSet* are contradictory knowledge sets.

To avoid incurring conflict, we should select more reliable ones from the contradictory statements to add to the KB. We can make a choice basing on the meta information (the origin of the Web page, Last-Modified date, etc) of the corresponding Web pages, from which the statements are extracted.

4.4 Knowledge Merging Algorithms

When merging new instance knowledge into the KB, equivalent instances should be identified first, and then if the new knowledge is redundant to some in the KB, do not add it; if the new knowledge is contradictory to some in the KB, choose the better one and add it into the KB. Algorithm *merge* depicts the merging process.

Algorithm 1. *merge*

INPUT: *newModel*, set of newly extracted instance knowledge to be merged;
INPUT: *totalKBModel*, set of instance knowledge in the current KB;
OUTPUT: set of instance knowledge of the KB after merging the new statements in *newModel*;

```
1: let newStatList be the set of RDF statements in newModel;
2: let oAKCInsList be the set of Outer-AKC instances in newModel;
3: for each oAKCIns_i ∈ oAKCInsList do
4:    let subStatList be the set of statements in newModel whose subject is oAKCIns_i;
5:    let objStatList be the set of statements in newModel whose object is oAKCIns_i;
6:    if oldIns ∈ totalKBModel, such that oldIns is equivalent to oAKCIns_i then
7:       mergeEquivIns( oAKCIns_i, oldIns, subStatList, objStatList, newStatList, totalKBModel);
8:    else
9:       recursively add each statement in subStatList to totalKBModel and remove them from newStatList;
10:      add all the statements in objStatList to totalKBModel and remove them from newStatList;
11: simpleMerge (newStatList, totalKBModel);
12: return totalKBModel;
```

After performing the step 2~10 in *merge*, the statements in *newStatList* do not relative to any AKC instances, and they are merged into the KB by *simpleMerge*. If a new instance is equivalent to an old one in KB, the statements about the new instance should be consolidated with those about the old one. Algorithm 3 depicts this process.

For a statement *stat*, recursively add/delete/remove *stat* means: add/delete/remove *stat* itself if *O* (*stat*) is a RDF literal or named entity instance; add/delete/remove *stat* and recursively add/delete/remove every statement whose subject is *O* (*stat*) if *O* (*stat*) is an AKC instance or anonymous instance.

Algorithm 2. *simpleMerge*

INPUT: *newStatList*, set of statements to be added into the KB;
INPUT: *totalKBModel*, set of instance knowledge in the current KB;
OUTPUT: set of instance knowledge of the KB after merging the statements in *newStatList*;

```
1: for each state_i ∈ newStatList do
2:    if state_i is not redundant to any knowledge in totalKBModel then
3:       if state_i is contradictory to some statements( marked as statSet) in totalKBModel then
4:          selectedSet ← choose(state_i ∪ statSet, n);
5:          if state_i ∈ selectedSet then
6:             delete { statSet-selectedSet/state_i } from totalKBModel;
7:             add state_i to totalKBModel;
8:       else
9:          add state_i to totalKBModel;
```

The *n* in step 4 of algorithm 2 and in step 6, 16 of algorithm 3 is 1 if the related property is *functional* or there is a cardinality restriction with the cardinality of 1 on the related property, or is the cardinality if there is a *maxCardinality* restriction on the related property.

Algorithm 3. *mergeEquivIns*

INPUT: *newIns*, *oldIns*, two equivalent instances, *newIns* is the new instance to be added and *oldIns* is the old one in the KB;
INPUT: *subStatList* and *objStatList*, sets of statements whose subject is *newIns* and whose object is *newIns*;
INPUT: *newStatList*, set of new statements to be added into the KB;
INPUT: *totalKBModel*, set of instance knowledge in the current KB;
OUTPUT: set of instance knowledge of the KB after merging the new knowledge about *newIns*;

```
remove all the statements from newStatList and subStatList whose subject is
  newIns and the predicate is the key property of newIns ;
for each stat_i ∈ objStatList do
  remove stat_i from newStatList;
  if totalKBModel does not contain the statement: <S(stat_i), P(stat_i), oldIns>
  then
    if stat_i is contradictory to some statements( marked as statSet) in
    totalKBModel then
      selectedSet ← choose(state_i ∪ statSet, n);
      if state_i ∈ selectedSet then
        delete {statSet − selectedSet/state_i}from totalKBModel;
        add <S(stat_i), P(stat_i), oldIns> to totalKBModel;
    else
      add <S(stat_i), P(stat_i), oldIns> to totalKBModel;
for each stat_i ∈ subStatList do
  if stat_old ∈ totalKBModel, such that stat_i is redundant to stat_old then
    recursively remove stat_i from newStatList;
  else if stat_i is contradictory to some statements( marked as statSet) in
  totalKBModel then
    selectedSet ← choose(state_i ∪ statSet, n);
    if state_i ∈ selectedSet then
      recursively delete {statSet − selectedSet/state_i}from totalKBModel;
      recursively add <oldIns, P(state_i), O(state_i)> to totalKBModel;
      remove the statements added in last step from newStatList;
      remove stat_i from newStatList;
    else
      recursively remove stat_i from newStatList;
  else
    recursively add <oldIns, P(state_i), O(state_i)> to totalKBModel;
    remove the statements added in last step from newStatList;
    remove stat_i from newStatList;
```

5 Experiment and Discussion

For an experiment we downloaded 219 Chinese Web news about the annual financial reports of different companies from http://finance.sina.com.cn/ as our Web pages set. Then we ran the KA system of *CRAB* to extract the instance knowledge automatically from these documents and got 219 rdf files. The resulting rdf files describe instance knowledge about 170 companies and there are totally 1119 AKC instances. The ratio of company instances to AKC instances is 1:6.58, which means that averagely 6~7 AKC instances are constructed to represent the annual financial status of a company.

Among the 1119 AKC instances there are 850 Inner-AKCs' instances, such as the instance of *Net profit entity* and *Total income entity*, etc; and 269 Outer-AKCs' instances, all of which are instances of the concept *Annual financial report entity*. Among all these 269 *Annual financial report entity* instances, 63 have equivalent instances and the number of the equivalent instances is 162. The ratio of *Annual financial report entity* instances to equivalent instances is 4.26:1, which means that averagely for every 4~5 *Annual financial report entity* instances there is an equivalent instance. So identifying the equivalent instances and consolidating the knowledge about them is necessary for the knowledge merging.

At last, we run the program based on the algorithm *merge* to merge all these 219 rdf files into an empty KB one by one. The result is that all these 162 equivalent instances are identified and the knowledge about them is consolidated; 98 conflicts are detected, and the updated KB is consistent checked by the Pellet reasoner.

The experimental result is satisfactory and demonstrates the feasibility of this approach.

6 Conclusion

Knowledge merging is necessary for knowledge management in a knowledge acquisition system. The property restriction based knowledge merging method advocated in this paper can identify the equivalent, identical, redundant and contradictory knowledge according to the property restrictions defined in the domain ontology, and basing on this it can consolidate the instance knowledge about the equivalent instances, and get rid of the redundant or contradictory knowledge to ensure the KB compact and consistent. The usage of this method in *CRAB* proved its applicability. However, in order to detect more complex conflicts we should take advantage of reasoning rules, which is our further research direction.

References

1. Aidan, H., Andreas, H., Stefan, D.: Performing Object Consolidation on the Semantic Web Data Graph. In: WWW 2007 Workshop on Entity-Centric Approaches to Information and Knowledge Management on the Web, Banff, Canada (2007), `http://ftp.informatik.rwth-aachen.de/Publications/CEUR-WS/Vol-249/submission_135.pdf`
2. Brickley D.: Rdfweb notebook: aggregation strategies (2002), `http://rdfweb.org/2001/01/design/smush.html`
3. Sais, F., Pernelle, N., Rousset, M.C.: L2R: a logical method for reference reconciliation. In: 22nd AAAI Conference on Artificial Intelligence, Vancouver, BC, Canada, pp. 329–334 (2007)
4. Che, H., Sun, J., Bai, X., Shi, L.: Application of aggregated knowledge concept in automatic knowledge acquisition from Chinese web pages. In: 2008 International Symposium on Information Processing, Moscow, Russia, pp. 439–443 (2008)
5. Che, H., Sun, J., Jing, T., Bai, X.: A prototype of semantic-based intelligent search engine for Chinese documents. In: 4th International Conference on Fuzzy System and Knowledge Discovery, Haikou, Wuhan, China, pp. 663–667 (2007)

A Multi-view Approach for Relation Extraction

Junsheng Zhou[1,2], Qian Xu[3], Jiajun Chen[3], and Weiguang Qu[1,2]

[1] Department of Computer Science and technology, Nanjing Normal University, Nanjing, China
[2] Jiangsu Research Center of Information Security & Confidential Engineering, Nanjing, China
[3] State Key Laboratory for Novel Software Technology, Nanjing University, Nanjing, China
{Zhoujs,Xuq,Chenjj}@nlp.nju.edu.cn,
Wgqu@njnu.edu.cn

Abstract. Relation extraction is an important problem in information extraction. In this paper, we explore a multi-view strategy for relation extracting task. Motivated by the fact, as in work of Jiang and Zhai's [1], that combining different feature subspaces into a single view does not generate much improvement, we propose a two-stage multi-view learning approach. First, we learn two different classifiers from two different views of relation instances: sequence representation and syntactic parse tree representation, respectively. Then, a meta-learner is trained using the meta data constructed along with other contextual information to achieve a strong predictive performance, as the final classification model. The experimental results conducted on ACE 2005 corpus show that the multi-view approach outperforms each single-view one for relation extraction task.

1 Introduction

Relation extraction is the task of finding tuples of entities for which there exists textual evidence that supports a predefined type of relationship. It is an important problem in information extraction. Many works on relation extraction have been presented in the literature, such as rule-based, feature-based and kernel-based. Recent studies have shown that feature-based and kernel-based methods are mainstream approaches to this problem.

The feature-based methods transform the relation context into features. These methods employ a large amount of diverse linguistic features, such as lexical, syntactic and semantic features [2][3]. As an attractive alternative to feature-based methods, kernel methods retain the original representation of objects and use the object in algorithms only via computing a kernel function between a pair of objects. A variety of kernel functions have been proposed to compute the similarity over the syntactic parse trees or dependency parse tree of relation instances [5, 6, 7, 8, 9, 10]. Of various kernels proposed, convolution kernels give state-of-art performance [8, 10]. Apart from their computational efficiency, convolution kernels also implicitly correspond to some feature space. Therefore, the feature-based and kernel-based methods both depend on an appropriately defined set of features.

W. Liu et al. (Eds.): WISM 2009, LNCS 5854, pp. 53–62, 2009.

Jiang and Zhai then systematically explored a large space of features and evaluated the effectiveness of different feature subspaces corresponding to sequence, syntactic parse tree and dependency parse tree [1]. Their experiments showed that using only the basic unit features within each feature subspace can already achieve state-of-art performance, while over-inclusion of complex features might hurt the performance. Motivated by their finding in that paper, we try to investigate the effectiveness of multi-view learning for relation extraction. In the problem of relation extraction, the datasets are naturally comprised of multiple views, e.g. sequence feature subspace, syntactic parse tree feature subspace, and dependency parse tree feature subspace. But simply combining the three feature subspaces in a single view does not generate much improvement. The reason is probably that the sequence, syntactic and dependency relations have much overlap for the task of relation extraction.

In this paper, we propose a multi-view approach to relation extraction to avoid the problem of features overlapping caused by combining different feature subspaces in a single view. As an empirical study, we first learn two different classifiers from two different views: sequences and syntactic parse trees, respectively. Then, we apply a meta-learning approach to combining the two classifiers to produce the final class label. Finally, we use the three trained classifiers to predicate the test data. The experimental results conducted on ACE 2005 corpus show that the multi-view approach outperforms each single-view one for relation extraction task.

2 Related Work

Rule-based methods for relation extraction describe various relation patterns with a number of linguistic rules. Miller et al. [11] used statistical parsing models to extract relational facts from text. They integrate various tasks such as POS tagging, NE tagging, template extraction and relation extraction into a generative model. Their results essentially depend on the entire full parse tree.

The feature-based methods achieve promising performance and competitive efficiency by transforming a relation example into a set of syntactic and semantic features, such as lexical knowledge, entity-related information, syntactic parse trees and deep semantic information [2, 3]. However, the feature space didn't be defined in a unified representative form.

Kernel function implicitly calculates the dot-product of feature vectors of objects in high-dimensional feature spaces. Zelenko et al. [4] first developed a kernel over parse trees for relation extraction. The kernel matches nodes from roots to leaf nodes recursively layer by layer in a topdown manner. Culotta et al. [5] generalized it to estimate similarity between dependency trees. Bunescu et al. [6] proposed another dependency tree kernel for relation extraction. Their kernel simply counts the number of common word classes at each position in the shortest paths between two entities in dependency trees. Zhang et al. [8, 9] described a convolution tree kernel (CTK) to investigate various structured information for relation extraction. Zhou et al. [10] pointed out that the convolution tree kernel is context-free. They expanded it by dynamically including necessary predicate-linked path information and extending the standard CTK to context sensitive CTK. The convolution tree kernels achieved state-of-the-art performance. Since convolution kernels correspond to some explicit large feature spaces, the feature selection problem still remains.

Jiang and Zhai [1] then systematically explored a large space of features for relation extraction. They proposed and defined a unified graphic representation of features for relation extraction. With this framework, they explored three different representations of sentences—sequences, syntactic parse trees, and dependency trees—which lead to three feature subspaces. Their experiments showed that using only the basic unit features within each feature subspace can already achieve state-of-art performance, while combining the three subspaces in a single view generate slight improvement.

In this paper, we will study to use multi-view strategy to relation extraction. Our aim is to make different feature views complement each other, while avoiding the features overlapping between different feature subspaces.

3 Motivation

Our approach was inspired by a promising strategy, i.e. multi-view learning. Multi-view learning describes the problem of learning from multiple independent sets of features of the presented data. This framework has been successfully applied to many real world applications such as web pages classification [2] and face recognition [12].

Recently, Jiang and Zhai [1] systematically explored and evaluated the effectiveness of different feature subspaces corresponding to sequence, syntactic parse tree and dependency parse tree for relation extraction. Their experiments showed that each feature subspace can achieve state-of-art performance, while simply combining all feature subspaces in a single view can only generate a slight improvement than a single feature subspace. The reason is probably that the sequence, syntactic and dependency relations have much overlap for the task of relation extraction.

It is interesting to investigate the effectiveness of different classifiers learned from different feature views for relation extraction problem. We conducted some experiments on ACE 2005 corpus. First, we trained two different SVM classifiers on training data, from sequence feature subspace and syntactic parse tree feature subspace, respectively. Then, we apply the two classifiers to the evaluation data. The experimental results show that there exits a large difference between the classification results produced by the two classifiers learned from the different feature views. There are 1942 positive examples in the evaluation data in total. The number of examples correctly recognized positive by both the two classifiers is 923, while the number of examples that are correctly recognized by the classifier learned from syntactic parse tree view and not correctly recognized by the classifier learned from sequence view is 424. On the other hand, the number of examples that are correctly recognized by the classifier learned from sequence view and not correctly recognized by the classifier learned from syntactic parse tree view is 414. Furthermore, we investigate the types of errors produced by the two different classifiers from the two feature views, and find that the classifier learned from sequence view has good effect on instances that the relations are local, while the classifier learned from syntactic parse tree view is more reliable on recognizing relation instances that the two entity mentions are separated by multiple words. The facts show that there exists complementarity between the two feature views: the sequence view and the syntactic parse tree view. So, if we can appropriately combine the classification results from the classifiers learned from the different feature views, a better performance for relation extraction can be achieved.

4 The Unified Feature Representation of Relation Instances

The problem of relation extraction is to assign an appropriate relation type to an occurrence of two entity pairs in a given context. It can be represented as follows:

$$R \rightarrow \left(C_{pre}, e_1, C_{mid}, e_2, C_{pos}\right)$$

where e_1 and e_2 denote the two entity mentions, and C_{pre}, C_{mid},and C_{post} denote the contexts before, between and after the entity mention pairs.

Similar to the method proposed in [1], we also represent relation instances as graph, with nodes denoting tokens or syntactic categories, and edges denoting various types of relations between the nodes. Formally, the directed graph representing relation instances can be defined as: $G=(V, E, A, B)$, where V is the set of nodes in the graph, E is the set of directed edges in the graph, and A, B are functions that assign labels to the nodes. For each node $v \in V$, $A(v)$ is a set of attributes associated with node v. Another function $B:V \rightarrow \{0, 1, 2, 3\}$ is introduced to distinguish argument nodes from non-argument nodes. For each node $v \in V$, $B(v)$ indicates how node v is related to e_1-or e_2. We refer to $B(v)$ as the argument tag of v. To enhance the distinguishability of function B for the words in sentences, we introduce 6 argument tags for sequence representation, more than the number of argument tag in [1]. 0 indicates that v does not cover any argument, 1 or 2 indicates that v covers head word of e_1-or e_2, respectively, 3 or 4 indicates that v covers other word, besides head word, of e_1-or e_2, respectively, and 5 indicates that v covers both arguments. The nodes representing syntactic categories in a syntactic parse tree can possibly be assigned 5.

Consider the relation instance "President Bush returned to the United States yesterday." with $e1$ = "President Bush s" and $e2$ = "the United States". The head word of $e1$ is "Bush", and the head word of $e2$ is "United States". The graphs of sequence representation and syntactic parse tree representation of this relation instance are shown as Figures 1 and 2. Some examples of subgraph denoting various types of features are also shown in Figures 1 and 2.

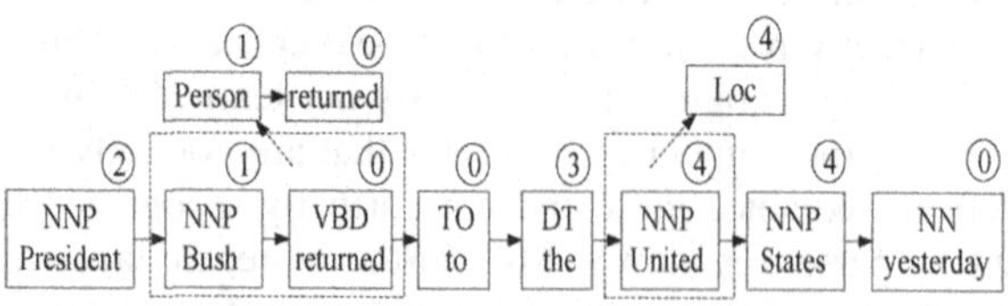

Fig. 1. An example of sequence representation

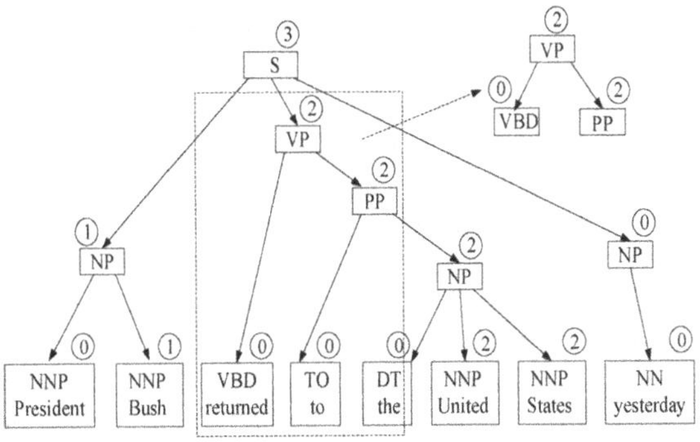

Fig. 2. An example of syntactic parse tree representation

When generating feature subspace, we also start with conjunctive entity features, which are conjunctions of two entity attributes. Then we add three levels of small unit features in order. First, we add unigram features G_{uni} = ({v}, ∅ , A_{uni}, B). Next, we add bigram features G_{bi} =({u,v},{(u,v)}, A_{uni}, B). The third level of attributes we add are trigram features G_{tri}=({u,v,w}, {(u,v),(u,w)}, *Auni, B*) or G_{tri}=({u,v,w}, {(u,v),(v,w)}, *Auni, B*). The trigram features consist of two connected edges and three nodes, where each node is also labeled with a single attribute.

5 The Multi-view Approach

Multi-view learning [2, 13] can exploit multiple redundant views to effectively learn from unlabeled data by mutually training a set of classifiers defined in each view. The theoretical foundations of multi-view learning are based on the assumptions that the views are independent [2]. However, research has shown that in real-world domains, the ideal assumption of multiple strictly independent views is not fully satisfied. As pointed out by Muslea et al. [14], in real-world problems, one seldom encounters problems with independent views. Multi-view learning can be advantageous when compared to learning with only a single view [15], especially when the weaknesses of one view complement the strengths of the other.

Due to the aforementioned motivation, we propose a supervised multi-view approach for relation extraction. Our so-called Multi-view relation extraction strategy learns from multiple views (feature set) of a relation instance, and then information acquired by view learners are integrated to construct a final classification model. In detail, the multi-view relation extraction approach, as depicted in Fig. 3, consists of two sequential stages:

(1) Multiple Views Construction Stage: The relation instances are naturally comprised of three different views: sequences, syntactic parse trees, and dependency parse trees. As showed in [1] , among the best performances achieved in every feature subspace, syntactic parse tree is the most effective feature space, while sequence and dependency tree are similar. On the other hand, combining sequence subspace and dependency tree subspace does not generate any performance improvement. These facts suggest that there is likely much overlap between sequence information and

dependency information. So, We learn two different classifiers from the two views: sequences and syntactic parse trees, respectively.
(2) View Combination Stage: In this stage, we aim to combine the two classifiers to produce the final class label. The most popular are those such as bagging [16], boosting [18] and stacking [17]. A meta learner method, such as the one used in stacking, is designed to learn which base classifiers are the reliable ones, using another learning algorithm. In the multi-view learning framework, multiple view learners may results in various performances, thus it is hard to guarantee the comparable performances of the view learners. Therefore, the meta-learning method is more suitable to the multi-view learning framework. In meta-learning schemes, a learning algorithm is usually used to construct the function that combines the predictions of the individual learners. This combination process contains two steps. Firstly, a meta training data set is generated. Each instance of the meta training data consists of the class label predictions, made by the individual learner on a specific training example, along with the original class label for that example. Secondly, a meta-learner is trained using the meta data constructed to achieve a strong predictive performance, as the final classification model.

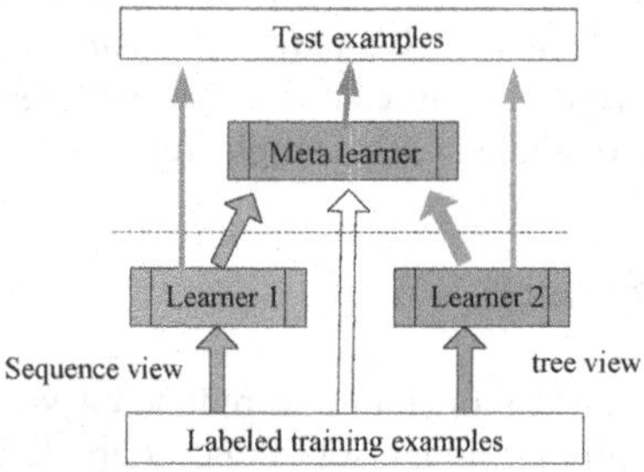

Fig. 3. An illustration of the multi-view procedure for relation extraction

When generating the training data for the meta learner, we utilize not only the class label predictions made by the individual learner, but also other contextual information, to help to make final decision. After making investigation into the effectiveness of two different classifiers learned from sequence view and syntactic parse tree view respectively, we find that the classification effect of different classifier is largely related to the positional structures between two entities. By examining the relation instances in ACE corpus, we define four positional structures between two entities, as depicted in Figure 4. The nested positional structure indicates entity $e2$ is included in entity $e1$, while the superposition positional structure indicates that entity $e1$ overlaps with entity $e2$, the adjacent positional structure indicates that entity $e2$ is adjacent to entity $e1$, with no words between the two entities, and the separated positional structure indicates that there are one or more words between the two entities. For the separated positional structures, we also provide distance information specifying the number of words between entity $e1$ and entity $e2$. The values of distance are set to 0 for other three types of positional structures.

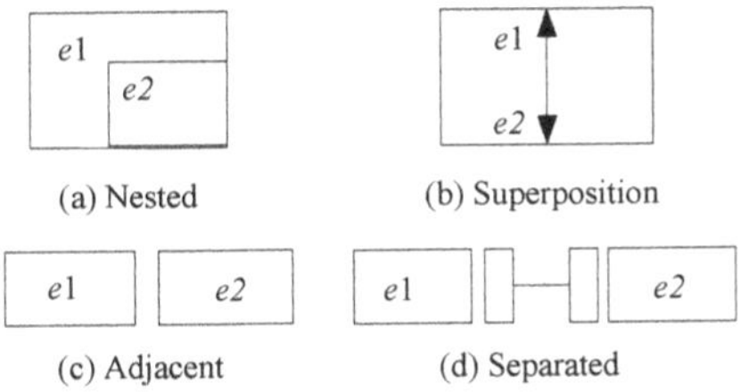

Fig. 4. Four positional structures: Nested structure, superposition structure, adjacent structure and separated structure

Thus, the multi-view approach returns multiple view learners, along with a meta learner which knows how to combine these multiple learners to achieve a strong predictive performance.

6 Experiments

This section will evaluate the effectiveness of our multi-view approach for relation extraction through experiments.

6.1 Experimental Setting

We use the official ACE 2005 corpus provided by LDC to train and evaluate our Multi-view relation extraction approach.

The ACE 2005 corpus contains documents from six sources: newswire, broadcast conversation, broadcast news, telephone, usenet and weblog. The ACE 2005 RDR task supported three languages, which are Arabic, Chinese and English. We use English portion of the training and evaluation corpus in our experiments. This corpus defines 7 types of relations: Physical, Personal-Social, Artifact, Org-Affiliation, Gen-Affiliation, Part-Whole and METONYMY.

We parse the sentences in the corpus using Charniak parser and we discarded sentences that could not be parsed by Charniak parser. The candidate relation instances were generated by iterating all pairs of entity mentions that occur in the same sentence and we find the negative samples are 10 times more than the positive samples. The training data contains 599 documents, and we obtained 97019 candidate relation instances in total, among which 8687 instances were positive. The evaluation data contains 155 documents, and we obtained 19895 candidate relation instances in total, among which 1963 instances were positive. As in most existing work, instead of using the entire sentence, we used only the sequence of tokens that are inside the minimum complete subtree covering the two entities. Presumably, tokens outside of this subtree are not so relevant to the task. In our graphic representation of relation instances, the attribute set for a token node includes the token itself, its POS tag, and entity type, entity subtype and entity mention type. The attribute set for a syntactic category node includes only the syntactic tag.

We used SVM as view learners and C4.5 decision trees as meta learners. When using SVM learner, we adopted one vs. others strategy for the multi-class classification problem. We implemented the C4.5 decision trees algorithm using Weka [19]. The C4.5 decision tree learner was used due to its de facto standard for empirical comparisons.

As commonly used, we conduct our evaluations on terms of precision of relation extraction (P), recall of relation extraction (R) and F-measure (F) that is weighted combination of P and R.

6.2 Experimental Results

To verify the effectiveness of our approach, we conduct the following experiments. First, we use all training data to train three different SVM classifiers. Among them, the first two are learned from sequence feature representation and syntactic parse trees feature representation, respectively. For each feature subspace, we use unigram features, bigram features and trigram features, plus conjunctive features. The third classifier is learned by combining sequence feature subspace and syntactic parse trees feature subspace in a single view. We apply the three trained classifier to extracting relation instances in evaluation data. The experiment results are showed in Table 1.

Our experimental results in Table 1 are all lower than the results conducted in ACE 2003 corpus in [1]. The first reason is the difference between the two corpora. The second reason may be due to the fact that their experiments are carried out on explicit relations, while the types of relation instances in ACE 2005 corpus are not divided into two classes: explicit and implicit. However, the fact conforming to the results of their is that syntactic parse tree is the more effective feature space than sequence, and combining syntactic parse tree and sequence feature subspaces achieves better performance than each single feature subspace, and the improvement is slight.

Table 1. The performance of SVM classifiers learned from different feature representation, with all training data

	Precision	Recall	F1
Seq	0.725	0.562	0.633
Syn	0.741	0.567	0.643
Seq+Syn	0.713	0.596	0.649

Next, we apply multi-view approach to relation extraction, using k-fold cross validation method. We partition the training data into 4 subsets of equal size. We train the classifiers 4 times, each time leaving out one of the subsets from training, but using only the omitted subset to train the meta learner for multi-view approach. When training the meta learner, we use j48 algorithm in Weka, and set the parameter confidence factor to 0.55. The experiment results are showed in Table 2. The first line and second line in Table 2 show results produced by classifiers learned from sequence feature subspace and syntactic parse tree feature subspace respectively, when using only 3 subsets out of 4 subsets of training data. The experimental results from Table 1 and Table 2 show that our multi-view approach gives better performance than each single view approach with different feature subspace or combining feature subspaces. The reason why multi-view approach achieves better performance than single-view one is probably that the strength of sequence view or syntactic parse tree view may complement to the other. For many short-range relation instances in the ACE corpus,

Table 2. Comparison among the two single-view classifiers and the multi-view approach

	Precision	Recall	F1
Seq	0.716	0.564	0.631
Syn	0.695	0.592	0.639
multi-view	0.741	0.567	**0.661**

sequence information may be even more reliable than syntactic information, while syntactic information may has advantage over sequence information when handling the long-range relation instances. This reason seemed to be proved by the fact that multi-view approach obtains a good tradeoff between the recall and the precision so that it gives a better F1 value.

7 Conclusions and Future Work

In this paper, we explore a multi-view strategy for relation extracting task. Motivated by the fact that combining different feature subspaces into a single view does not generate much improvement, we propose a effective two-stage multi-view learning approach. First, we learn two different classifiers from two different views: sequences and syntactic parse trees, respectively. Then, a meta-learner is trained using the meta data constructed along with other contextual information, such as positional structures, to achieve a strong predictive performance, as the final classification model. The experimental results demonstrate the performance of multi-view approach.

Our approach can also handle semi-supervised learning in a straightforward manner. In our future work, we will study how to enable multi-view strategy for relation extraction task bootstrapping from a small set of labeled training data via a large set of unlabeled data. In addition, there exists view disagreement between different views in our multi-view approach, due to unavoidable parsing errors. In the future, we will attempt to solve the problem of view disagreement by treating each view as corrupted by a structured noise process and detecting view disagreement by exploiting the joint view statistics.

Acknowledgement

This work is supported by the National Natural Science Foundation of China under Grant No.60673043, 60773173; and the Natural Science Foundation of Jiangsu Higher Education Institutions of China under Grant No. 07KJB520057.

References

1. Jiang, J., Zhai, C.: A Systematic Exploration of the Feature Space for Relation Extraction. In: Proceedings of NAACL/HLT, pp. 113–120. Association for Computational Linguistics, New York (2007)
2. Zhou, G., Su, J., Zhang, J., Zhang, M.: Exploring various knowledge in relation extraction. In: Proceedings of ACL, pp. 427–434. Association for Computational Linguistics, Michigan (2005)

3. Zhao, S., Grishman, R.: Extracting relations with integrated information using kernel methods. In: Proceedings of ACL, pp. 419–426. Association for Computational Linguistics, Michigan (2005)
4. Zelenko, D., Aone, C., Richardella, A.: Kernel methods for relation extraction. Journal of Machine Learning Research 3, 1083–1106 (2003)
5. Culotta, A., Sorensen, J.: Dependency tree kernels for relation extraction. In: Proceedings of ACL, pp. 423–429. Association for Computational Linguistics, Spain (2004)
6. Bunescu, R.C., Mooney, R.J.: A shortest path dependency kenrel for relation extraction. In: Proceedings of HLT/EMNLP, pp. 724–731. Association for Computational Linguistics, Canada (2005)
7. Bunescu, R.C., Mooney, R.J.: Subsequence kernels for relation extraction. In: Proceedings of NIPS, pp. 171–178. MIT Press, Cambridge (2005)
8. Zhang, M., Zhang, J., Su, J.: Exploring syntactic features for relation extraction using a convolution tree kernel. In: Proceedings of HLT/NAACL, pp. 288–295. Association for Computational Linguistics, New York (2006)
9. Zhang, M., Zhang, J., Su, J., Zhou, G.: A composite kernel to extract relations between entities with both flat and structured features. In: Proceedings of ACL, pp. 825–832. Association for Computational Linguistics, Sydney (2006)
10. Zhou, G., Zhang, M., Ji, D., Zhu, Q.: Tree Kernel-based Relation Extraction with Context-Sensitive Structured Parse Tree Information. In: Proceedings of EMNLP/CoNLL, pp. 728–736. Association for Computational Linguistics, Prague (2007)
11. Miller, S., Fox, H., Ramshaw, L., Weischedel, R.: A novel use of statistical parsing to extract information from text. In: Proceedings of NAACL, pp. 226–233 (2000)
12. de Sa, V.R., Ballard, D.: Category learning through multi-modality sensing. Neural Comput. 10(5), 1097–1117 (1998)
13. Ando, R.K., Zhang, T.: Two-view feature generation model for semi-supervised learning. In: Proceedings of ICML, pp. 25–32. Morgan Kaufmann Publishers, San Francisco (2007)
14. IA Muslea: Active learning with multiple views. Ph.D. thesis, Department of Computer Science, University of Southern California (2002)
15. Christoudias, C.M., Saenko, K., Morency, L.-P., Darrell, T.: Co-adaptation of audio-visual speech and gesture classifiers. In: Proceedings of ICMI, pp. 84–91. ACM, Alberta (2006)
16. Breiman, L.: Bagging predictors. Machine Learning 24(2), 123–140 (1996)
17. Wolpert, D.H.: Stacked generalization. Neural Networks 5(2), 241–259 (1992)
18. Freund, Y.: RE Schapire: Experiments with a new boosting algorithm. In: Proceedings of ICML, pp. 148–156 (1996)
19. Witten, I.H., Frank, E.: Data mining: practical machine learning tools and techniques with Java implementations. Morgan Kaufmann, San Francisco (2000)

Self-similarity Clustering Event Detection Based on Triggers Guidance*

Xianfei Zhang, Bicheng Li, and Yuxuan Tian

Zhengzhou Information Science and Technology Institute
No. 837, P.O. BOX 1001, Zhengzhou, Henan, 450002, China
zhangxianfei2003@126.com,
lbclm@263.net,
showyuxuan@sina.com

Abstract. Traditional method of Event Detection and Characterization (EDC) regards event detection task as classification problem. It makes words as samples to train classifier, which can lead to positive and negative samples of classifier imbalance. Meanwhile, there is data sparseness problem of this method when the corpus is small. This paper doesn't classify event using word as samples, but cluster event in judging event types. It adopts self-similarity to convergence the value of K in K-means algorithm by the guidance of event triggers, and optimizes clustering algorithm. Then, combining with named entity and its comparative position information, the new method further make sure the pinpoint type of event. The new method avoids depending on template of event in tradition methods, and its result of event detection can well be used in automatic text summarization, text retrieval, and topic detection and tracking.

Keywords: Event Detection and Characterization; self-similarity; triggers; text summarization; topic detection and tracking.

1 Introduction

Event Detection and Characterization (EDC) comes from Automatic Content Extraction (ACE) meeting, which research objectives are viewed as the detection and characterization of Entities, Relations and events from news corpora. Nowadays, there are three primary ACE annotation tasks corresponding to the three research objectives: Entity Detection and Tracking (EDT), Relation Detection and Characterization (RDC), and Event Detection and Characterization (EDC) [1]. The definition of event in this paper comes from ACE meeting, which is composed by triggers and arguments, for example: ZhangSan was born in ShanDong QingDao in 1981. "Born" is this event's trigger, "ZhangSan", "1981" and "Shandong Qingdao" are arguments which make up of this event.

---

* Supported by the national high-teach research and development plan of China under grant (863). NO. 2007AA01Z439.

W. Liu et al. (Eds.): WISM 2009, LNCS 5854, pp. 63–70, 2009.

There are two kinds of methods for EDC: Pattern Matching Method (PMM) and Machine Learning Method (MLM). PMM adopts pattern matching method to match fixed model of event with detected sentence [2]. The representation of PMM is FSA system, which aims at region event detection by Surdeanu and harabagiu [3]. This method has high precision, but only for fixed region, and it is not easy to transplant. MLM looks event detection as classification problem, which has two major steps: event type detection and event arguments detection. In which event types is decided by event model defined in advance. In ACE, annotators identify and characterize five types of events and tag the textual mention or anchor for each event, and categorize it by type and subtype. EDC most adopts classification method at present. In 2002, hai Leong Chieu and Heww Tou Ng first adopted maximum entropy approach for event argument detection [4]. David Ahn carried out event types and argument detection combining with Megam and Timbl's machine learning methods [5]. For Chinese Event Detection and Characterization, the representation method is based on event trigger expansion and binary classification by Zhao Yanyan in 2008 [6]. However, because classification method puts every word as sample of classifier, it can import large number of counterexamples, which can lead to an imperfect result of detection. Furthermore, multi-classification of event types and multi-classifier constructed by arguments of each event lie some data sparseness problem when corpora are small.

Aiming at these problems, this paper avoids classification by word, adopts clustering ideology in event type judgment, and puts forward self-similarity clustering event detection method based on triggers guidance. The new method first uses triggers to fix original centroid of K-means clustering method, and fixes on the value of K with similarity strategy, which solves the problem of original centroid selecting in clustering method and realizes text event type judgment. Then, combining with named entities such as name, placement, organization and time information and its position information, the new method makes sure event types more and completes event detection. Results of experiment show that recall of event detection of new method is obviously improved because of breaking the limitation of event types in classification methods, which result is well used in text summarization, classification, retrieval and topic detection and tracking.

2 K-means Clustering Algorithm

For d dimensions data set, $X = \{x_i | x_i \in R^d, i = 1,2,\cdots,N\}$ is clustered to K categories $\omega_1, \omega_2, \cdots, \omega_K$, in which centroids are $c_1, c_2, \cdots, c_K$, where $c_i = (1/n_i)\sum_{x\in\omega_i} x$, and n_i is the number of data point of category of ω_i. The goal function of clustering is $J = \sum_{i=1}^{k}\sum_{j=1}^{n_i} d_{ij}(x_j, c_i)$, where $d_{ij}(x_j, c_i)$ is the distance of x_j and c_i.

The clustering steps of K-means algorithm can be described as follows:

(1) Choose K beginning centroids $c_1, c_2, \cdots, c_K$ from data set X stochastically.

(2) Cluster X by $c_1, c_2, \cdots, c_K$. The principle of clustering is that if $d_{ij}(x_i, c_j) = \min_{m=1,2,\cdots,k} d_{im}(x_i, c_m)$ where $j = 1,2,\cdots,K$ and $i = 1,2,\cdots,N$, x_i is put to category of ω_j.

(3) Recalculate centroid of $c_1^*, c_2^*, \cdots, c_K^*$ by $c_i = (1/n_i)\sum_{x \in \omega_i} x$.

(4) If for all $i \in \{1, 2, \cdots, K\}$, the formula of $c_i^* = c_i$ can come into existence, computation finishes and $c_1^*, c_2^*, \cdots, c_K^*$ are the final categories. Otherwise, put $c_i = c_i^*$ and go to step (2).

Considering there is possibility of infinite loop in step (4), the method usually gives a threshold th. When for all c_i, the formula $\left| c_i^* - c_i \right| < th$ comes into existence, the algorithm finishes.

Event detection based on K-means clustering algorithm needs to resolve the following problems:

- Confirming the number of category K. We don't know the number of event before clustering, so the value of K needs to be confirmed.
- Choosing beginning centroids. K-means is clustered by determinate centroids. Selecting beginning centroids determines event content of result of clustering. So, how to choose beginning centroids is very important.

3 Self-Similarity-Based Max-Min Clustering

To resolve the problem of K-means algorithm, we adopt self-similarity-based-on max-min principle [7] [8], compute the value of K by self-similarity strategy and complete clustering event detection based on triggers guidance.

Assumed an event is a text block, so we can compute cosine-similarity using distance of two text blocks. For text block i, its corresponding feature vector is $x_i = (a_1, a_2, \cdots, a_n)$, text block j and feature vector $x_j = (b_1, b_2, \cdots, b_n)$, similarity of two events can be described as follows:

$$sim(x_i, x_j) = (\sum_{m=1}^{n} a_m b_m) / sqrt(\sum_{m=1}^{n} a_m^2) \times sqrt(\sum_{m=1}^{n} b_m^2) \tag{1}$$

Assumed samples set is $X = \{x_1, x_2, \cdots, x_N\}$, where N is the number of sample. Calculating all the similarity of two events, $sim_{ij} = sim(x_i, x_j)$ when $i = j$ and $sim_{ij} = 1$.

Defined whole average similarity (WAS):

$$\bar{s} = \sum_{i=1}^{N-1} \sum_{j=i+1}^{N} sim_{ij} \Big/ N(N-1)/2 \quad , \ i \neq j \tag{2}$$

Defined max-min average similarity (MMAS):

$$\bar{s}_{minmax} = \{\max(sim_{ij}) + \min(sim_{ij})\}/2 \tag{3}$$

Where $i = 1, 2, \cdots, N-1$, $j = i+1, i+2, \cdots, N$.

Defined average self-similarity of category ω_k*:*

$$\bar{s}_{kk} = \frac{s_{kk}}{n_k(n_k - 1)/2} \quad , \ n_k = |\, \omega_k \,| \tag{4}$$

Where

$$s_{kk} = \sum_{i \in \omega_k} \sum_{\substack{j \in \omega_k \\ j=i}} sim_{ij} \tag{5}$$

For the samples i and j of the same category ω_k, s_{kk} is self-similarity of ω_k. If $i \in \omega_{k_1}$ and $j \in \omega_{k_2}$, $s_{k_1k_2} = \sum_{i \in \omega_{k_1}} \sum_{j \in \omega_{k_2}} sim_{ij}$ is mutual-similarity of ω_{k_1} and ω_{k_2}.

Defined threshold of whole average self-similarity:

$$\overline{s}_{Gth} = \max(\overline{s}, \overline{s}_{minmax}) \tag{6}$$

Assumed choosing k centroids, average self-similarity of each category are $\overline{s}_{11}, \overline{s}_{22}, \cdots, \overline{s}_{kk}$. When one centroid is added, average self-similarity of each category becomes $\overline{s}_{11}^{'}, \overline{s}_{22}^{'}, \cdots, \overline{s}_{kk}^{'}, \overline{s}_{k+1k+1}$.

Defined $\overline{s}_{th}$ is the threshold of local average self-similarity, which is dynamic in each clustering.

$$\overline{s}_{th} = (\overline{s}_{11} + \overline{s}_{22} + \cdots + \overline{s}_{kk} + \overline{s}_{11}^{'} + \overline{s}_{22}^{'} + \cdots + \overline{s}_{kk}^{'})/2k \tag{7}$$

If the following instance appears, go on for the next centroid.

$$\frac{(\overline{s}_{11}^{'} + \overline{s}_{22}^{'} + \cdots + \overline{s}_{kk}^{'})}{(|\overline{s}_{11}^{'} - \overline{s}_{th}| + |\overline{s}_{22}^{'} - \overline{s}_{th}| + \cdots + |\overline{s}_{kk}^{'} - \overline{s}_{th}|)} \geq \frac{(\overline{s}_{11} + \overline{s}_{22} + \cdots + \overline{s}_{kk})}{(|\overline{s}_{11} - \overline{s}_{th}| + |\overline{s}_{22} - \overline{s}_{th}| + \cdots + |\overline{s}_{kk} - \overline{s}_{th}|)} \quad \overline{s}_{k+1k+1} \geq \overline{s}_{Gth} \tag{8}$$

The steps of self-similarity-based-on max-min clustering algorithm:

(1) Put triggers as beginning centroids of clustering.

Put triggers of samples $s_1 s_2$ as beginning centroid, which self-similarity of event is smallest where triggers are involved in. Other samples are put on another category which has bigger similarity, and then compute $\overline{s}_{11}\overline{s}_{22}$. In this case, the value of K is 2. If there are more than one smallest similarity samples, compute every average self-similarity $\overline{s}_{11}^{'}\overline{s}_{22}^{'}$ and choose one of it according to the following principle:

① compute every mutual-similarity $s_{k_1k_2}$ and $s_{k_1k_2}^{'}$, choose samples of smaller value as beginning centroids.

② Instead of computing mutual-similarity, choose beginning centroid according to average self-similarity. Compare $(\overline{s}_{11} + \overline{s}_{22})/(|\overline{s}_{11} - \overline{s}_{Gth}| + |\overline{s}_{22} - \overline{s}_{Gth}|)$ with $(\overline{s}_{11}^{'} + \overline{s}_{22}^{'})/(|\overline{s}_{11}^{'} - \overline{s}_{Gth}| + |\overline{s}_{22}^{'} - \overline{s}_{Gth}|)$, and choose samples of bigger value as beginning centroids. If the value is equal to each other, choose samples which have the smaller value of $|\overline{s}_{11} - \overline{s}_{22}|$ and $|\overline{s}_{11}^{'} - \overline{s}_{22}^{'}|$.

The advantage of this strategy is not only ensuring average self-similarity of each category is not too small; the discrepancy of two average self-similarities is also not too big after clustering.

(2) Choose the next centroid using the same strategy. Compute average self-similarity of each category until break inequality (8), finish clustering and confirm the value of K.

4 Self-similarity Clustering Event Detection Based on Triggers Guidance

The main ideology of the new method is discharge limitation of event type defined in advance in classification method, cluster event in original text corpora in guidance of triggers, and detect event of text corpora combining with named entities and position information. Flow chart of the new method is as follows.

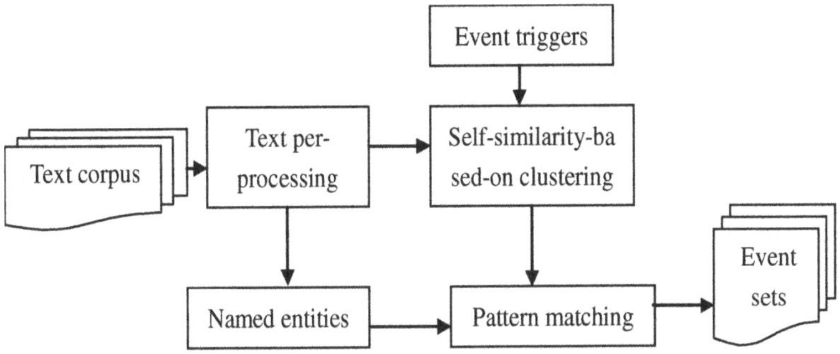

Fig. 1. Flow chart of self-similarity clustering event detection based on triggers guidance

The steps of self-similarity clustering event detection based on triggers guidance can be described as follows according to Fig 1.

(1) Pre-process corpora, which mainly include Chinese word segment and word taggability and so on, and finish pre-processing of natural text.

(2) Find triggers in corpora by the result of pre-processing, and match with tagged triggers after eliminating thesaurus.

(3) In the guidance of the matched triggers, cluster text block according to corresponding triggers using self-similarity K-means clustering algorithm and confirm event categories primarily.

(4) Extract named entities such as name, toponym, organization and time information by the result of text pre-processing, and meanwhile, memorize the comparative position information of entities.

(5) Match named entities with the result of clustering in step (3). The result of matching is the event of detecting using the new method.

5 Experiment and Performance Analysis

The experiment data comes from Internet, which contains 500 news corpora after extracting text contents. We tag these corpora, which involve named entity, such as name, toponym, organization and event triggers, and also polarity, tense, genericity and modality and so on, and make it very propitious to test. This paper also adopts F_1-measure, Precision (P) and Recall (R), which are standard measure in ACE meeting, to evaluate performance of algorithms. The definition of measures is as follows.

$$F = \frac{2PR}{P+R} \qquad (9)$$

$$P = \frac{\text{correct event number detected}}{\text{sum event number detected}} \qquad (10)$$

$$R = \frac{\text{correct event number detected}}{\text{sum event number tagged}} \qquad (11)$$

5.1 Detect Result of Different Clustering Algorithm

This experiment compares the result of three clustering algorithms; include traditional K-means (algorithm1), self-similarity K-means (algorithm 2) and self-similarity K-means based on triggers guidance (algorithm 3). In algorithm 2, its beginning centroid is not event trigger, but the result of computing smallest similarity every time. The result of event detection by three clustering algorithm is as follows.

Table 1. Result of event detection by three algorithms

	R(Recall)	*P(Precision)*	*F1-measure*
algorithm 1	38.91%	50.36%	43.91%
algorithm 2	56.18%	61.24%	58.61%
algorithm 3	61.31%	65.47%	63.32%

Because the category number of K-means is transcendental, and its beginning centroid is choose stochastically, so its recall is the lowest in three algorithms and its precision is not well too. Algorithm 2 adopts self-similarity strategy in choosing beginning centroid and confirms the value of K by self-constringency, so its performance is better than traditional K-means algorithm. However, because it only depends on self-similarity strategy, if the first centroid is error, it can lead to error continuance in the following clustering. So, its performance is not well enough too. In the guidance of event triggers for choosing beginning centroid, this paper's method can avoid the shortcoming of error expanding and its performance of event detection is the best in three algorithms.

5.2 Detection Result Comparison

This experiment compares the new method with Ahn's event type and argument detection method based on machine learning [5] (ML method) and event detection method based on triggers and duality-classification [6] (DC method). For creditability of the result comparison, this paper calculates methods of machine learning and duality-classification in the same corpora. Comparison result is as follows.

Table 2. Comparison results with classification methods

	R(Recall)	*P(Precision)*	*F1-measure*
ML method	43.18%	57.93%	49.48%
DC method	53.27%	68.87%	65.47%
the new method	61.31%	65.47%	63.32%

Machine learning method first limits event types of text corpora, and then imports plenty of counterexamples by using word as classifier sample, which is the main reason of its bad result of event detection. Although event detection method based on trigger and duality-classification add trigger to help event type construction and can improve detection precision by reducing training counterexamples, this method is also limited in defined types in advance, which leads to a low recall and can not detect undefined event type. The new method of this paper breaks through limitation of event types and clusters event based on triggers in text corpora, which can improve recall obviously. However, because clustering method is unsupervised method, it can lead to reduplicated event in detection result and make precision lower. So, some manicure strategy can be considered in the future research to improve precision.

6 Conclusion and Future Work

This paper does not adopt traditional event detection way of classifying event by limited knowledge first and fixing on category by event template. It concentrates on self-similarity clustering algorithm based on triggers guidance to detect text event. The new method advances performance of event detection. However, Event Detection and Characterization is on its primary status. Text event can not be detected very well only by limited triggers guidance. It needs a brand-new model to detect event very well, in the condition of making the full use of triggers, named entities information and semantic model.

References

1. ACE (Automatic Content Extraction) Chinese Annotation Guidelines for Events, National Institute of Standards and Technology (2005)
2. Surdeanu, M., Harabagiu, S., Williams, J., et al.: Using Predicate-Argument Structures for Information Extraction. In: Proceedings of ACL, pp. 8–15 (2003)
3. Surdeanu, M., Harabagiu, S.: Infrastructure for open-domain information extraction. In: Proceedings of the Human Language Technology Conference, pp. 325–330 (2002)
4. Chieu, H.L., Ng, H.T.: A Maximum entropy Approach to Information Extraction from Semi-Structured and Free Text. In: Proceedings of the 18th National Conference on Artificial Intelligence, pp. 786–791 (2002)

5. Ahn, D.: The Stages of Event Extraction. In: Proceedings of the Workshop on Annotations and Reasoning about Time and Events, pp. 1–8 (2006)
6. Yanyan, Z., bing, Q., Wanxiang, C., Ting, L.: Technology Research on Chinese Event Extraction. Journal of Chinese Information 22(1), 3–8 (2008)
7. Ding, C., He, X.: Cluster Merging and Splitting in Hierarchical Clustering Algorithms. In: Proceedings of the 2002 IEEE International Conference on Data Mining, Maebashi City, Japan, pp. 139–146. Maebashi TERRSA (2002)
8. Ding, C., He, X., Zha, H., et al.: A Min-Max Cut Algorithm for Graph Partitioning and Data Clustering. In: Proceedings of the IEEE International Conference on Data Mining, San Jose, California, USA, pp. 107–114 (2001)

Analysis and Interpretation of Semantic HTML Tables

Wensheng Yin, Feifei Guo, Fan Xu, and Xiuguo Chen

School of Mechanical Science and Engineering,
Huazhong University of Science and Technology,
Wuhan 430074, China
wsyin@mail.hust.edu.cn, scarlett_5008@yahoo.com.cn,
xufan3344@qq.com, cxshen@qq.com

Abstract. Table is an effective manifestation of structural knowledge, on which the semantic analysis is a very important part in semantic document analysis. To interpret the structure and the semantic relations of the HTML documents, definitions of normalized table and tabular coordinate system are proposed according to database relation theory. This paper classifies cells into normalized cells and visual cells, indicates that row or column and its combined cell are the primary semantic expression forms of table and nested tables are the further expansion of a certain table cell. Finally, a table analyzing algorithm is given based on tabular coordinate system. Practice shows that the algorithm is simple, fast and having certain practical significance.

Keywords: HTML; table; tabular coordinate system; visual cell; row or column combined cell; nested table.

1 Introduction

Table is a structured format for data description, widely used in various documents due to it can describe concrete instances of one or more class objects effectively. With the development of Internet technology and its application, tables constructed by Hyper Text Markup Language (HTML) etc. are used frequently in web documents. For table cells have certain semantic relations, in order to extract the semantic relations in tables we should parse the tables first.

Language like HTML uses a particular series of tags: <TABLE>, <TBODY>, <TR>, <TH>, <TD> (hereafter, table tags) as the framework to construct tables. The purpose of table analyzing is to convey documents containing these table tags into Document Object Model (DOM), establish the logical relations between different cells and finally integrate a relatively complete semantic description relied on table cells. The flexible table construction and complexity of semantic relations make the cognition of tables a big challenge, so it is very meaningful to parse the tables.

2 Related Works

Several researchers have reported their work on extracting information from the web table [1-8]. In [1][2], Jung et al. used a machine learning algorithm C4.5 to separate

W. Liu et al. (Eds.): WISM 2009, LNCS 5854, pp. 71–79, 2009.

the table head from the body according to rows and columns' characteristics [1], and complemented the head extraction method with supplementary heuristic rules [2]. They also proposed heuristic rules for extracting table semantic information based on SCE (semantic-core element) and table structures [2]. Yoshida et al. proposed a method to extract attributes and their values from HTML tables based on Hidden Markov Models technique (HMMs) relying on distributions of words [3]. Tanaka et al. proposed a semi-automatic method interpreting the table structure by humans to extract relations from the tables [4]. Liu et al. designed algorithms that parse HTML tables to a content tree [5]. Kim et al. segmented HTML tables into attribute and value areas by checking visual and semantic coherency [6]. Li et al. capture the semantic hierarchies of HTML tables using a heuristic rule based on the notion of eigenvalue in formatting information, and semi-automatically integrate HTML tables [7]. Chen et al. used heuristic rules and cell similarities to identify tables [8].

In this paper, we propose an approach to interpret the structure and the semantic relations of the HTML tables based on the definition of normalized table, tabular coordinate system and column or row combined cell according to database relation theory.

The rest of the paper is organized as follows: In Section 3, we give the definition of table structure. Section 4 describes the semantic relation of cell. Section 5 presents the table analysis and interpretation algorithm. Then we make an analysis of the results in Section 6. Finally we conclude in Section 7.

3 Definition of Table Structure

Tables can be divided into two groups according to their expressing purposes: (1) tables based on semantic to describe the relations of object and its attribute and value; (2) tables (non-semantic tables) to construct the layouts, mainly used for the layouts of words or images without or with less semantic relations among them. This paper just concentrates the former group of tables.

Here HTML tables are generally referred to the tables constructed by TABLE, TBODY, TR, TH, TD etc. table tags in accordance with SGML (Standard Generalized Markup Language).

3.1 Tabular Coordinate System

Definition 1: Given a set of domain $D_1, D_2, \cdots, D_n$, their Cartesian Product is defined as $D_1 \times D_2 \times \cdots \times D_n = \{d_1, d_2, \cdots, d_n) \mid d_i \in D_i, i = 1,2,\cdots,n\}$, evenly putting the name of every domain D_i and Cartesian Product into some 2D rectangular grids in the horizontal and vertical directions, and making every grid in a row and a column have equal height and width respectively, thus, such table is called normalized table.

Definition 2: For normalized table, horizontal-right direction is defined as x positive direction, vertical-down direction as y positive direction, and left endpoint as origin, one single grid as the unit of coordinate, and then the tabular coordinate system is derived.

3.2 Cell

Definition 3: In a normalized table, one grid is called a normalized cell.

If the top-left coordinate of the normalized cell is (*x*,*y*), then its 4 endpoints are (*x*,*y*), (*x*,*y*+1), (*x*+1,*y*+1), (*x*+1,*y*) respectively and it is abbreviated to *R*(*x*,*y*).

Definition 4: The inseparable grid in the table, that can be showed on the web-browser and described in HTML documents, is called visual cell.

HTML documents record left endpoint's information in this row for each cell row by row and column by column. Obviously, different from normalized cell, visual cell could be the combination of multiple normalized cells but must be rectangle. If the start row, end row, start column, end column of a visual cell U are r_s, r_e, c_s, c_e, this visual cell marked as $U(r_s, r_e, c_s, c_e)$.

As all the cells on the web pages are visual ones, we often call them cells for short.

3.3 Table Matrix

Definition 5: If down-right coordinate of table *T* is (n_r, n_c), the matrix recording information of all the normalized cells in *T* is called table matrix of *T*, denoted by $M(n_r, n_c)$.

Each element $M(i, j)$ of *M* records the information of visual cell at (i, j), thus, when designing algorithm we could search the whole table simply by traversing matrix *M*.

3.4 Combined Cell

Combined cell is an important form for people to design and understand tables expressing semantic relations of table. As seen from the form, combined cell $C = \{U_1, U_2, \cdots, U_n\}$, where $U_1, U_2, \cdots, U_n$ are visual cells. *C* may have different geometrical shapes; however, only several common forms are discussed in the following.

Definition 6: For $C = \{U_1, U_2, \cdots, U_n\}$, $\forall U_i(r_{si}, r_{ei}, c_{si}, c_{ei})$, $i = 1,2,\cdots,n$, if at least one of c_{si} or c_{ei} falls between $[c_{\min}, c_{\max}]$, *C* is called a column combined cell between $[c_{\min}, c_{\max}]$.

Definition 7: For $C = \{U_1, U_2, \cdots, U_n\}$, $\forall U_i(r_{si}, r_{ei}, c_{si}, c_{ei})$, $i = 1,2,\cdots,n$, if c_{si} and c_{ei} all fall between $[c_{\min}, c_{\max}]$, and $\exists U_j$, $c_{sj} = c_{\min}, c_{rj} = c_{\max}$, *C* is called a normalized column combined cell.

Row combined cell and normalized row combined cell can be defined in the same way.

3.5 Nested Table

Definition 8: In HTML documents, *T1* is the nested table of *T2*, if a table tags block of table *T1* is embedded in the table tags block of table *T2*.

Generally, nested table is completely contained in a cell, which is a further expansion of this cell content. Moreover, a cell can have multiple nested tables.

Nested table defined in HTML documents is a hierarchy structure of table, and can be implemented through recursive program. Therefore we will not discuss it in the following.

4 Semantic Relation of Cell

A semantic based table is the expression of semantic relation. This expression requires people to obey some recognized rules to maintain a clear logic relation among cells of the table. The purpose of table's semantic relations analysis is to determine the main semantic relations among the cells.

4.1 Semantic Relation of Normalized Cell

According to relation theory, the table defined in definition 1 is a description to certain objects. Domain D_i depicts attribute of the objects and each element of the Cartesian Product describes the attribute value of the specific objects.

Definition 9: Each normalized cell used to describe the domain name of Domain D_i in the first row of normalized table *T* is called attribute cell of table *T*, while any normalized cell in other rows is the value of domain D_i of the specific objects, which is called value cell of *T*.

Definition 10: The table with the domain names and values of domain D_i formatted in column direction is called column-wise table, and called row-wise table when formatted in row direction.

It is clear that a normalized table is a column-wise table, as shown in Fig. 1. The row-wise table has the similar characters like column-wise table and will not be discussed in the following paper.

System 486	System 586	System 686
486DX2-66 CPU	120 MHZ AMD586	200 MHZ Pentium Pro
8 MB RAM	16 MB RAM	16 MB RAM
500 MB HD	1 GB HD	1.4 GB HD
14.4 Modem	28.8 Modem	28.8 Modem
desktop case	minitower case	tower case
DOS/Win 3.1	Windows 95	Windows NT 4.0

Fig. 1. Column normalized semantic table

The semantics of normalized table is as follows: each value cell is the cell that has the value of the attribute corresponding to current objects and the semantic relations among them are expressed by the predicate relations defined by the table.

4.2 Basic Assumptions

As to the semantic relations of semantic table, we give the following assumptions:

Assumption 1: A table describes the attributes and values of all the instances in the set belonging to one or more class objects.

Assumption 2: In the same row or column, intervals between the attribute cell and value cell will not appear.

Assumption 3: The attribute cell is always in the top or left of the value cell.

4.3 Combined Cell Table

In general, row or column combined cell table (hereinafter collectively referred to as combined cell table) has the most evident semantic relations. For example, Fig. 2 shows a normalized column combined cell, which describes some related attributes of computer components.

For a normalized table, a column combined cell has only one attribute cell. But for non-normalized tables, column combined cell may have more than one attribute cell. According to assumption 2 and 3, we can obtain rule 1:

Rule 1: The attribute cells of column combined cell table must be located in the top of the column. If there are multiple attribute cells, they must be conjoint and the widths of the cells are different.

For example, all the columns in Fig. 2(a) and 2(b) constitute a column combined cell, each cell in the first row and second row is attribute cell and the widths of two adjacent ones are different.

4.4 Scope of Attribute

Definition 11: All the value cells of attribute *A* in the table *T* are called the scope of *A* in *T*.

From the assumption 2 and 3, we can derive that scope of the attribute of normalized table is all the cells in its located column except itself. According to the properties of column combined cell table, we make the following assumption:

Assumption 4: the widths of value cells with the same attribute are identical in normalized column combined cell table and all equivalent to the width of attribute cell.

Therefore the scope of an attribute contains all the value cells in this attribute column.

Definition 12: if the attribute cell U_A of attribute *A* is below the attribute cell U_B of attribute *B*, it is called that U_A is subordinate to U_B, which can be denoted as

$U_A \subset U_B$. If U_A and U_B are adjacent, it is called that U_B is the parent attribute of U_A and U_A is the sub-attribute of U_B.

For example, each cell in the second row subordinates to corresponding one of the first row in Fig. 2.

PC Component		
ProductID	ProductName	Price
Memory		
M27_512	PC2700 512MB	$70
M27_256	PC2700 256MB	$40
Processor		
P4_340	Pentium 4 3.40E GHz	$260
P4_280	Pentium 4 2.80E GHz	$140
A64_320	Athlon 64 3200+	$160

(a)

Rate(%)	Regular	Float
Regular Fixed Deposit		
1 Year	5.25	5.25
2 Years	5.35	5.35
3 Years	5.35	5.35
Fixed Deposit		
3 Months	4.4	4.4
6 Months	4.95	4.95
9 Months	5.05	5.05
1 Year	5.15	5.15
2 Years	5.25	5.25
3 Years	5.25	5.25

(b)

Fig. 2. Normalized column composite cell

4.5 Content of Value Cell

Definition 13: Description of all the semantic relations of value cell is called content of the cell.

According to the scope of attribute, rule 2 is derived as follows:

Rule 2: if U_A is a sub-attributes of U_B, the content of its one value cell contains the values of both attribute *A* and attribute *B*.

For example, the value of cell $U(3,4,0,1)$ in Fig. 2(a) is "M27_512", which is the value of attribute "ProductID". At the same time, the "ProductID" has the parent attribute "PC Component", so $U(3,4,0,1)$ has another value "Memory". If the table in Fig. 2(a) describes computer components, the content of the value cell is a component of PC with Component=Memory and ProductID=M27_512.

4.6 General Combined Cell

In practice, most tables are the normalized column or row combined cell tables. However, some un-normalized combined cell tables with various forms will be used too. For example, Fig.3 contravene assumption 2 in the column direction and the affiliation in U(1,2,1,3) (the value is "AB1") is unclear in Fig. 4.

Tour Code			DP91.AX01AB	
Valid			1999.04.01-2000.03.31	
Class/Extension			Economic Class	Extension
Adult	P	Single Room	35,450	2,510
	R	Double Room	32,500	1,430
		Extra Room	30,550	720
Child	I	Occupation	25,800	1,430
	C	Extra Bed	23,850	720
	E	No Occupation	22,900	360

Fig. 3. A table contravening to the assumption 2

A		B
A1	AB1	B1
A2	AB2	B2

Fig. 4. A table with unclear affiliation

5 Table Analysis Algorithm

5.1 Table Analysis Steps

To provide data for table analysis program, a HTML file is read and decoded at first, and then a DOM tree is formed. Analytical algorithm of table is as follows:

1) Search the block composed of tags such as TABLE, TR, TH and TD in DOM tree, and then analyze the data in the block. Even though there are still TEBLE tags in the block, it should not be expanded.

2) Cycle in the tag block, and then calculate the number of rows n_r and columns n_c of the table.

3) Construct table matrix $M(n_r, n_c)$ and record the relationship between normalized cells and visual cells.

4) Traverse table row by row or column by column, and establish the father-son relationship of every cell.

5) Call this analysis process recursively if there are TABLE tags in cell, and turn it into one part of the current cell.

6) Acquire the content value of each cell.

5.2 Cell Analysis Algorithm

The cell analysis algorithm is as follows:

Algorithm 1:

```
Search Node of TABLE type
Create Table associate with Node
Get Table.Row and Table.Col according to Node
Make TableMatrix[Table.Row][Table.Col]
For i=0 to Table.Row
For j=0 to Table.Col
{
    Get Unit from TableMatrix[i][j]
    Build the parent-son relation between Unit
                    and the above grid
    Print the content of Unit and its patent grid
}
```

6 Result Analysis

A table semantic analysis system called TableToSS has been developed. More than thirty tables shown in Fig. 1-4 or collected from Internet randomly have been analyzed and interpreted by TableToSS. The main conclusions are as follows:

- Because the semantic relationships of tables shown in Fig.1, Fig.2(a) are comparatively distinct and the father-son relationship between attributes is the one-to-many relationship, they can be identified completely.
- In fact, there are two rows of attribute cell in Fig. 2(b), but one row is omitted, so the system recognizes the value cell as attribute cell.
- Fig. 3 doesn't accord with assumption 2. Because the table shown in Fig.3 is combined by two tables actually, it must be split based on assumption 2.
- In Fig. 4, a semantic mistake will occur because of the unclear affiliation.
- The analysis of nested table is comparatively easy, and its analytical accuracy mainly depends on that of non-nested table. The analysis of nested table can be carried out by the recursive algorithm of table analysis.
- If a semantic table is composed of more than one TABLE tag blocks, we must integrate the blocks by the structure and semantic analysis.
- The mistakes taken in the analysis on the Fig.1-4 can be modified, but due to the complicated and changing combination of cell, it's very difficult to determine the attributes. The semantic ambiguity problem can not be solved only depending on the structural analysis, and in many cases analysis of the value of cells is needed.

7 Conclusion

Table is the manifestation of single or multiple relations in two-dimensional plane. When manifesting a single relationship, the widths of each cell are the same. However, when manifesting multiple relationships, changes will happen inevitably in the width. Moreover, certain subordinate relations exist in cells with different widths. Since most tables are constructed in this approach, the algorithm proposed in this paper has some practicality. In order to expand table analytical capability of the algorithm, further analyses are needed about the combinatorial and semantic relationships engaged in the process of people making tables among cells.

References

1. Jung, S.W., Kwon, H.C.: Hybrid Approach to Extracting Information from Web-Tables. In: Matsumoto, Y., Sproat, R.W., Wong, K.-F., Zhang, M. (eds.) ICCPOL 2006. LNCS (LNAI), vol. 4285, pp. 109–119. Springer, Heidelberg (2006)
2. Jung, S.W., Kwon, H.C.: A Machine Learning Based Approach for Separating Head from Body in Web-Tables. In: Gelbukh, A. (ed.) CICLing 2006. LNCS, vol. 3878, pp. 524–535. Springer, Heidelberg (2006)
3. Yoshida, M., Torisawa, K., Tsujii, J.: Extracting attributes and their values from web pages. In: Proceedings of the ACL Student Research Workshop, Japan, pp. 72–77 (2002)
4. Tanaka, M., Ishida, T.: Ontology extraction from tables on the web. In: Proceedings of the 2005 Symposium on Application and the Internet (SAINT 2006), pp. 284–290. IEEE, Los Alamitos (2006)
5. Jiexue, L., Zhuoyun, A., Park, H.H., et al.: An XML approach to semantically extract data from HTML tables. In: Andersen, K.V., Debenham, J., Wagner, R. (eds.) DEXA 2005. LNCS, vol. 3588, pp. 696–705. Springer, Heidelberg (2005)
6. Kim, Y.S., Lee, K.H.: Extracting logical structures from HTML tables. Computer Standards & Interfaces, 296–308 (2007)
7. Li, S., Peng, Z., Liu, M.: Extraction and Integration Information in HTML Tables. In: Proc. CIT 2004, IEEE Computer Society digital library, pp. 315–320. IEEE, Wuhan (2004)
8. Chen, H., Tsai, S., Tsai, J.: Mining Tables from Large Scale HTML Texts. In: Proceedings of the 18th International Conference on Computational Linguistics, pp. 166–172. Association for Computational Linguistics, New Jersey (2000)

An Improved Feature Selection for Categorization Based on Mutual Information

Haifeng Liu, Zhan Su, Zeqing Yao, and Shousheng Liu

Institute of Sciences, PLA University of Science and Technology
210007 Nanjing, China
liuhaifeng19620717@sina.com, suz123456789@163.com,
yzqnj@yahoo.com.cn, ssliunuaa@sina.com

Abstract. The feature reduction is one of the core techniques in text categorization. But there is no consideration of text position factor to the differentiation of labeling text capability in the method of weighting basing on multi-information (MI) in features. So in this paper, we put forward an improved feature selection method that based on MI. By adding the amending parameters in different positions, we have increased the using efficiency about the character information. The result of experiment shows that this method has improved the accuracy of the text classification.

Keywords: multi-information; feature selection; text categorization; feature reduction.

1 Introduction

In modern society , the rapid growth of text message of which the major formal is electronic makes text mining data technology gain more and more attention, and as the main elements of text pre-processing steps, automatic text categorization technology research has become one of the hot areas of text mining. The efficiency of text information indexing based on content is decided by the development level of text automatic classification techniques.

How to deal with the high-dimensional eigenvector is one of the main bottlenecks of text classification. To count tens of thousands of dimensions makes common computing platform endurable, at the same time, the efficiency of some classification machine which runs well in low-dimension space has dropped drastically. Thus, the text feature reduction research become to the mainly apartment of text categorization, and the efficiency of feature reduction greatly influent the effect of text categorization.

For its semantic integrity, easy to understand, easy to be used, the method of feature reduction based on selection is widely applied. At present, the common feature selection methods are feature frequency [1], multi-information [2], information gaining [3], expected cross entropy [4], χ^2 statistic [5], text evidence weight [6],etc. The research on the feature evaluation function has made some fruits which focus on the

W. Liu et al. (Eds.): WISM 2009, LNCS 5854, pp. 80–87, 2009.

relativity of feature and text form different knowledge domain, but there are also many deficiencies in different levels. The common problem is that all of this functions give weight mainly by text frequency, feature frequency, etc., without considering the differences of text indexing capabilities of features regarding to their position in text. In this paper, we draw a new weight empowering method by analyzing the features and problems of the feature selection methods which based on multi-information. Experiments show that this empowering algorithm acquires satisfying results.

2 Characters, Existed Problems and Improving Idea of Feature Selection Method Based on Multi-information

As a feature selection method which based on statistics and information theory, the multi-information selection method reflects the relativity degree of feature and text. The results show that the multi-information method is less effective in several commonly used feature reduction methods [7]. Analyzing the features and shortcomings of this model, and finding an effective idea to improve this method is meaningful in theory and application prospects.

2.1 Characteristics and Problems of Multi-information (MI)

MI is the measurement of relativity with feature and each category of the text. First of all, this method calculates MI with feature and each category. Secondly, selects the weighted average as the MI value of the feature, and the MI value with feature option and text categories is calculated as follow:

$$MI(w\,,\ c_k) = P(c_k)\log\frac{p(w,c_k)}{p(w)p(c_k)}$$

$$= P(c_k)[\log p(w,c_k) - \log p(w) - \log p(c_k)] \qquad (1)$$

MI with the text set and is calculated as follow:

$$MI(w) = \sum_{k=1}^{l} P(c_k)\log\frac{p(w,c_k)}{p(w)p(c_k)} \qquad (2)$$

Among them, $p(w,c_k)$ is the rate that number of text in which feature w is in category c_k divide number of text in training set, $p(w)$ is the rate that number of text that have the feature w in training set divide number of text in training set; $p(w\,|\,c_k)$ is the probability that the number of text that have the feature w. $p(c_k)$ is the probability that the text belong to c_k in the text set, and l is the category number of text.

The more multi-information of w, the stronger indexing capability this feature has and the more power to determine text category.

However, the flaw of this feature selection method is obvious.

For the first time, the role of low-frequency words is magnified in this model. It can be seen in analysis (2) that $p(c_k)$ is certain number for category c_k and the difference between frequency $p(w,c_k)$ relating with c_k is unobvious, so the number of $p(w)$ greatly impacts $MI(w)$. In another words, if $-\log p(w)$ is larger, $MI(w)$ is larger. Seen in this light, $MI(w)$ primarily reflects the impact degree in feature selection caused by low-frequency words. Compared with other commonly used feature selection methods such as TF-IDF, the advantage of MI method is that it absorbs the text indexing information which contained in the low-frequency words, but at the same time, it is also the shortcoming of this method that it is too sensitive to low-frequency words! Actually, the category topic words with important information in text often are considered as low-frequency words, because they occur in less frequency in text, but occur in more frequency in some texts within one category due to the used feature information. At the same time, the role of other low-frequency words is enlarged which actually play a small role in the model of $MI(w)$, so these low-frequency words with small contribution in the feature selection should be removed. But the formula (2) could not solve this problem, and the flaw is especially obvious when the distribution of training set is uneven.

Secondly, there is no consideration with the factor of feature text position to the differentiation of labeling text capability in this model. From the MI calculating formula (2), we find that this method mainly take the relevant information of feature and category into account basing on the frequency, and statistic the distribution of feature in the whole text set, without considering feature's location in the text.

In fact, the expression capability of features are different regarding to their position in text, that compared with the feature in the text, the indexing efforts of the title are much greater! Therefore, this factor should be reflected in feature empowering.

2.2 The Improved Idea in Feature Selection Method Based on MI

To solve the first problem, we can take this method into account that moderately reduces the weight of low-frequency words by amending decision-making functions. In the second problem, we can give different weight to different feature based on the location, and increase the location influence to weight. Based on this thinking way, we improve the MI model in two aspects as follow.

3 The Improvement of Feature Weight Empowering Based on MI

To solve the problem of low-frequency words in MI, we will improve the model:

$$MI(w) = \sum_{k=1}^{l} P(c_k) \left| \log \frac{p(w,c_k)}{p(w)p(c_k)} \right| \tag{3}$$

It aims at avoiding the information lack that caused by numerical offset of positive and negative numbers.

Then, the occurrence frequency deviation of feature in different category should be considered and adjusting factor should be inducted to adjust the situation in which the uneven distribution of text in different category of text collection will cause that multi-information especially in low-frequency words related does not reflect expression capability of feature to text category properly.So, if $n_w(j)$ is the frequency of feature w in Class c_j ,and if $C=\{c_1,c_2,\cdots,c_l\}$ is the set composed by all classes in text set, so

$$\lambda_{wj}=\frac{\max\limits_{1\le k\le l}\{n_w(k)\}}{n_w(j)} \tag{4}$$

reflects change trend of w along with change of numbers of text in class c_j .

We make the further improvement of (3):

$$M\tilde{I}(t_i)=\sum_{i=1}^{l}\lambda_{wj}\times P(c_k)\,|\log\frac{p(w,c_k)}{p(w)p(c_k)}| \tag{5}$$

Meanwhile, it should be cared that there are probably some low-frequency uncommon words in the feature set when using this method. So uncommon words should be filtered firstly by conducting appropriate methods when this feature selection method is used, such as constructing dictionary or interfering mutually.

4 Feature Weighting Method Based on Position

The text indexing capabilities of features are different regarding to their position in text, even in the same word, there is a big deviation in text expressing capability due to the different position of a feature in text. "The words in document title are most useful for the overall document searching hits rate, followed the sub-title, chapter name or abstract in document, and finally the words in document "Prof. Zhang Qiyu said . Document title is the summary of whole document from the author, and the abstract is the important part of document, which reflects the main content of whole document in short; the keywords , subtitles and so on compose the key content. Combined with the weighting of word frequency, treating the feature position in text as one of key factors to determine its weight will use the information in the feature more sufficiently

In general, for the expressing capability in document content, the descendible text expressing capability order of feature in document title, abstract, keywords, subtitle, first sentence in first paragraph, last sentence in first paragraph, last paragraph is as below:

Title>Abstract>Keyword>Sub-title>First Sentence in First Paragraph>Last Sentence in First Paragraph>Last Paragraph>Other

Analyzing and experimenting 1800 text article from economic, education, literature and psychology for the overall information distribution position: Title(bt), First Sentence in First Paragraph (ds1), Last Sentence in First Paragraph (dw1), First Sentence in Second Paragraph (ds2), Last Sentence in Second Paragraph (dw2), First

Sentence in Third Paragraph (ds3), Last Sentence in Third Paragraph (dw3), First Paragraph (sd), Last Paragraph (wd), Others(qt) (remained text part excluding sd, wd, ds2, ds2, dw2, ds3, dw3), the following topic expressing capability ascending order is derived by Article [9] in detailed statistic analysis:

bt > Sd > Ds1 > Dw1 > Qt> Wd > Ds2 > Dw2 > Ds3 > Dw3;

and proposed position weighting solution is :

$$Bt:Sd:Ds1:Dw1:Qt:Wd:Ds2:Dw2:Ds3:Dw3=5:5:4:4:2:2:2:2:2:2 \quad (6)$$

Derived by this experiment, the document title and first paragraph are most powerful in topic expressing capability, followed by First Sentence in First Paragraph and Last Sentence in First Paragraph, finally remained text parts such as First Sentence in Second Paragraph (ds2), Last Sentence in Second Paragraph (dw2), First Sentence in Third Paragraph (ds3), Last Sentence in Third Paragraph (dw3).

So it is reasonable that text processing is divided into three levels:

First of all, the title is a refined summary for the text content with less words, which is composed by feature words almost except for few form word, and reflect main content of text;

Secondly, first paragraph reflects the writing intention of the author, which plays a significant role in text topic indication.

We consider the two aspects as the first level;

Secondly, First Sentence in First Paragraph, Last Sentence in First Paragraph have stronger text expressing capability. For the Chinese text writing, the writing style of "pass, connect, transfer, merge", "straightforward" is often employed for pointing out the topic in first paragraph.

So compared with other parts in text, those sentences have more text topic expressing capability, which will be treated as second level.

At last, although First Sentence in Second Paragraph (ds2), Last Sentence in Second Paragraph (dw2), First Sentence in Third Paragraph (ds3), Last Sentence in Third Paragraph (dw3), First Paragraph (sd), Last Paragraph (wd), Others(qt) have some difference, those parts still be combined as the third level due to the trivial difference.

From this analysis, we draw a text feature weighting solution based on position:

1) Divide the texts in text set into the above three levels which are treated as 3 sets S1, S2, S3. The texts in those 3 sets are composed of 3 "pseudo-text collection";

2) Every text d_j is indicated by 3 feature vector: first vector V_{1j} is composed by the feature weight of text d_j in first level; respectively, the second and the third feature V_{2j}, V_{3j} is composed by the feature weight in second and third level; the weight empowering is calculated by the formula (5):

3) The number of those 3 "pseudo-text collection" dimension should be limited for the sake of reducing the computing time and storage. It's common to select some highest features' weight as the pseudo-text vector:

$$V_{kj} = (w_{k1j}, w_{k2j}, \cdots, w_{ktj}); k = 1,2,3 \quad (7)$$

Among them, w_{klj} is the weight of feature l in the level k related to text d_j, $1 \le l \le t$.

4) According the impacting degree of weight on different position of feature reflected by formula (2-9), the vector of d_j is derived as below:

$$d_j = (2.5w_{11j}, 2.5w_{12j}, \cdots, 2.5w_{1tj}, 2w_{21j}, 2w_{22j}, \cdots, 2w_{2tj}, w_{31j}, w_{32j}, \cdots, w_{3tj})^T \quad (8)$$

Finally, test similarity degree of the text and training text by calculating cosine:

$$Sim(d_i, d_j) = \frac{\sum_{k=1}^{n} w_{ik} w_{jk}}{\sqrt{\sum_{k=1}^{n} w^2_{ik}} \sqrt{\sum_{k=1}^{n} w^2_{ik}}} \quad (9)$$

5 Text Categorization Experiment and Result Analysis

This article does the experiment for categorization effect using the method mentioned. The test data in which there is 3720's article are downloaded from Baidu, Sina, which are composed by economic(612 piece), political(567 piece), sport(813 piece), computer(770 piece), literature(523 piece) and art(435 piece). In this experiment using cross-3 test the 3720's text are divided into 3 pieces averagely, 2 piece for training set, 1 piece for test set; Every piece will be test set for 3 times round test, the results in average will be the test result. After breaking the words using the work-breaking software provided by nature language processing lab in Northeast University, filtering banned words using banned words table, deleting the rarely frequent used words manually, and getting rid of Empty Words, particle, personal pronouns, ultra-high-frequency words, 5217 features is derived to the candidate feature collection, select features into 3 "pseudo-text sets" using formula(5), in every "pseudo-text set " there are 40 features, which have the highest multi-information value, composed to feature set for weighting the text feature, the number of text feature dimension is 120, take the KNN classifier as employed classifier, the parameter k=30; the precision rate, recall rate, F1 test value will be employed as evaluation function:

Precision rate = number of correct classified text/ number of actual classified text

Recall rate = number of correct classified text/ number of whole text

$$F_1 = \frac{PercisionRate \times \mathrm{RecallRate} \times 2}{PercisionRate + \mathrm{RecallRate}}$$

We add F_1 test value from formula (5) in which the position factor is excluded as F_{1old} for comparing classification effeteness from the improved model, the related 3 group improved data will be marked new for the difference.

The statistic classification result is as below:

Table 1. Text categorization test result statistics

Category	Economic	Art	Sport	Computer
$PercisionRate_{new}$	0.921	0.875	0.934	0.906
$RecallRate_{new}$	0.893	0.882	0.945	0.879
F_{1new}	0.907	0.878	0.939	0.892
F_{1old}	0.812	0.803	0.846	0.802

Category	Literature	Politic	Average
$PercisionRate_{new}$	0.868	0.925	0.905
$RecallRate_{new}$	0.857	0.896	0.892
F_{1new}	0.862	0.910	0.898
F_{1old}	0.825	0.849	0.823

From the above table, the text classification of economic, sport, computer and political is better than art and literature, probably there are more common words and less articles in art and literature. It's a good effeteness that average F_1 test value is 89.8% which is improved to 9.1% than non-position factor test value.

6 Summary

The efficiency of text feature selection algorithm has a direct impact to precision rate of text categorization. The relevance between feature and text, feature and text categorization, features themselves are very complicated so exploring more tangible feature reduction method and reducing the theory lack in all model are more practice ways to improve the text categorization precision based on all existed feature reduction method. The feature reduction problem is one of mainly aspects of text information processing. The problems that need to be resolved in text categorization domain include drawing feature reduction solution regarding to different distribution of training sample in training set and improving the robustness of classifier in the environment of large sample set and uneven sample distribution. Those are our continuous research topics in future.

References

1. De Villiers, G., Linford Vogt, P., De Wit, P.: Business Logistics Management. Oxford University Press, Oxford (2002)
2. Sheng, Y., Jun, G.: Feature selection based on mutual information and redundancy-synergy coefficient. Journal of Zhejiang University Science A 5(11), 1382–1391 (2004)
3. Huan, L., Lei, Y.: Toward Integrating Feature Selection Algorithms for classification and Clustering. IEEE Transactions on Knowledge and Data Engineering 17(5), 491–502 (2005)
4. Qian, Z., Ming-sheng, Z., Wen, H.: Study on Feature Selection in Chinese Text Categorization. Journal of Chinese Information Processing 18(3), 17–23 (2004)

5. Hai-feng, L., Yuan-yuan, W., Ze-qing, Y., et al.: A Research of Text Categorization Model Based on Feature Clustering. Journal of the China society for scientific and technical information 27(2), 224–228 (2008)
6. Wenqian, S., Houkuan, H., Haibin, Z., et al.: A novel feature selection algorithm for text categorization. Expert Systems with Applications 33(1), 1–5 (2007)
7. Guo-ju, S., jie, Z.: An Evaluation of Feature Selection Methods for Text Categorization. Journal of Harbin University of Science and Technology 10(1), 76–78 (2005)
8. Qi-yu, Z.: Basic of information and philology. Wu Han University publishing company, Wuhan (1997)
9. Han-qing, H., Cheng-zhi, Z., Hong, Z.: Research On the Weighting of Indexing Sources for Web Concept Mining. Journal of the China society for scientific and technical information 24(1), 87–92 (2005)

A Feature Selection Method Based on Fisher's Discriminant Ratio for Text Sentiment Classification

Suge Wang[1,3], Deyu Li[2,3], Yingjie Wei[4], and Hongxia Li[1]

[1] School of Mathematics Science, Shanxi University, Taiyuan 030006, China
[2] School of Computer & Information Technology, Shanxi University, Taiyuan 030006, China
[3] Key Laboratory of Computational Intelligence and Chinese Information Processing of Ministry of Education, Shanxi University, Taiyuan 030006, China
[4] Science Press, Beijing 100717, China
wsg@sxu.edu.cn

Abstract. With the rapid growth of e-commerce, product reviews on the Web have become an important information source for customers' decision making when they intend to buy some product. As the reviews are often too many for customers to go through, how to automatically classify them into different sentiment orientation categories (i.e. positive/negative) has become a research problem. In this paper, based on Fisher's discriminant ratio, an effective feature selection method is proposed for product review text sentiment classification. In order to validate the validity of the proposed method, we compared it with other methods respectively based on information gain and mutual information while support vector machine is adopted as the classifier. In this paper, 6 subexperiments are conducted by combining different feature selection methods with 2 kinds of candidate feature sets. Under 1006 review documents of cars, the experimental results indicate that the Fisher's discriminant ratio based on word frequency estimation has the best performance with F value 83.3% while the candidate features are the words which appear in both positive and negative texts.

1 Introduction

Owing to its openness, virtualization and sharing criterion, the Internet has been rapidly becoming a platform for people to express their opinion, attitude, feeling and emotion. With the rapid development of e-commerce, product reviews on the Web have become a very important information source for customers' decision making when they plan to buy some products online or offline. As the reviews are often too many for customers to go through, so an automatically text sentiment classifier may be very helpful to a customer to rapidly know the review orientations (e.g. positive/negative) about some product. Text sentiment classification is aim to automatically judge what sentiment orientation, the positive ('thumbs up') or negative ('thumbs down'), a review text is with by mining and analyzing the subjective information in the text, such as standpoint, view, attitude, mood and so on. Text sentiment classification can be widely applied to many fields such as public opinion analysis,

W. Liu et al. (Eds.): WISM 2009, LNCS 5854, pp. 88–97, 2009.

product online tracking, question answer system and text summarization. Whereas, the review texts on the Web, such as on BBS, Blogs or forum websites are often with non-structured or semi-structured form. Different from structured data, feature selection is a crucial problem to the non-structured or semi-structured data classification.

Although there has been a recent surge of interest in text sentiment classification, the state-of-the-art techniques for text sentiment classification are much less mature than those for text topic classification. This is partially attributed to the fact that topics are represented objectively and explicitly with keywords while the topics are expressed with subtlety. However, the text sentiments are hidden in a large of subjective information in the text, such as standpoint, view, attitude, mood and so on. Therefore, the text sentiment classification requires deeper analyzing and understanding of textual statement information and thus is more challenging. In recent years, by employing some machine learning techniques i.e. Naive Bayes (NB), maximum entropy classification (ME), and support vector machine (SVM)), many researches have been conducted on English text sentiment classification[1-7] and on Chinese text sentiment classification[8-11].

One major particularity or difficulty of the text sentiment classification problem is the high dimension of the features used to describe texts, which raises big hurdles in applying many sophisticated learning algorithms to text sentiment classification. The aim of feature selection methods is to obtain a reduction of the original feature set by removing some features that are considered irrelevant for text sentiment classification to yield improved classification accuracy and decrease the running time of learning algorithms[11]. Tan et al. (2008) presented an empirical study of sentiment categorization on Chinese documents[12]. In their work four feature selection methods, Information Gain (IG), MI(Mutual Information), CHI and DF(Document Frequency) were adopted for feature selection. The experimental results indicate that IG performs the best for sentimental terms selection and SVM exhibits the best performance for Chinese text sentiment classification. Yi et al. (2003) developed and tested two feature term selection algorithms based on a mixture language model and likelihood ratio[13]. Wang et al. (2007) presented a hybrid method for feature selection based on the category distinguishing capability of feature words and IG[10].

In this paper, from the contribute views of distinguishing text sort, a kind of effective feature selection method is proposed for text sentiment classification. By using two kinds of probability distribution methods, i.e., Boolean value and word frequently, four kinds of feature selecting method is proposed. By using support vector machine (SVM), the experiment results indicate this approach is the efficacy for review text sentiment classification.

The remainder of this paper is organized as follows: Section 2 presents proposed feature selection methods. Section 3 presents experiments used to evaluate the effectiveness of the proposed approach and discussion of the results. Section 4 concludes with closing remarks and future directions.

2 Feature Selection Method Based on Fisher's Discriminant Ratio

Fisher linear discriminant is one of an efficient approaches for dimension reduction in statistical pattern recognition[14]. Its main idea can be briefly described as follows.

Suppose that there are two kinds of sample points in a d-dimension data space. We hope to find a line in the original space such that the projective points on the line of the sample points can be separated as much as possible by some point on the line. In other words, the bigger the square of the difference between the means of two kinds projected sample points is and at the same time the smaller the within-class scatters are, the better the expected line is. More formally, construct the following function, so-called Fisher's discriminant ratio.

$$J_F(w) = \frac{(\overline{m_1} - \overline{m_2})}{\overline{S_1}^2 + \overline{S_2}^2} \tag{1}$$

where w is the direction vector of the expected line, $\overline{m_i}$ and $\overline{S_i}^2$ $(i = 1, 2)$ are the mean and the within-class scatter of the i th class respectively. So the above idea is to find a w such that $J_F(w)$ achieve its maximum.

2.1 Improved Fisher's Discriminant Ratio

In fact, Fisher's discriminant ratio can be improved for evaluating the category distinguishing capability of a feature by replacing w with the dimension of the feature. For this purpose, Fisher's discriminant ratio is reconstructed as follows.

$$F(t_k) = \frac{(E(t_k \mid P) - E(t_k \mid N))^2}{D(t_k \mid P) + D(t_k \mid N)} \tag{2}$$

where $E(t_k \mid P)$ and $E(t_k \mid N)$ are the conditional mean of the feature t_k with respect to the categories P and N respectively, $D(t_k \mid P)$ and $D(t_k \mid N)$ are the conditional variances of the feature t_k with respect to the categories P and N respectively.

It is obvious that $(E(t_k \mid P) - E(t_k \mid N))^2$ is the between-class scatter degree and $D(t_k \mid P) + D(t_k \mid N)$ is the sum of within-class scatter degrees.

In applications, the probabilities involved in Formula (2) can be estimated in different ways. In this paper, two kinds of methods respectively based on Boolean value and frequency are adopted for probability estimation.

(1) Fisher's discriminant ratio based on Boolean value

Let $d_{P,i} (i = 1, 2, ..., m)$ and $d_{N,j} (j = 1, 2, ..., n)$ denote the ith positive text and the jth negative text respectively. Random variables $d_{P,i}(t_k)$ and $d_{N,j}(t_k)$ are defined as follows.

$$d_{P,i}(t_k) = \begin{cases} 1 & \text{if } t_k \text{ occurs in } d_{P,i} \\ 0 & \text{otherwise} \end{cases}$$

$$d_{N,i}(t_k) = \begin{cases} 1 & \text{if } t_k \text{ occurs in } d_{N,i} \\ 0 & \text{otherwise} \end{cases}$$

Let $m_1(n_1)$ and $m_0(n_0)$ be the numbers of positive (negative) texts within feature t_k and without feature t_k respectively.

Obviously, random variables $d_{P,i}(t_k)$ and $d_{N,j}(t_k)$ are depicted by the following distributions respectively.

$$P(d_{P,i}(t_k) = l) = m_i / m, l = 1,0;$$

$$P(d_{N,j}(t_k) = l) = n_i / n, l = 1,0 \text{ ,.}$$

It should be note that $d_{P,i}(t_k)$ ($i = 1,2,...,m$) are independent with the same distribution, and so is $d_{N,j}(t_k)$ ($j = 1,2,...,n$). $d_{P,i}(t_k)$ ($i = 1,2,...,m$) can be regarded as a sample with size m. Then the conditional means and the conditional variances of the feature t_k with respect to the categories P and N in Formula (2) can be estimated by using Formulas (3)-(6).

$$\begin{aligned} E(t_k \mid P) &= E(\frac{1}{m}\sum_{i=1}^{m} d_{P,i}(t_k)) \\ &= \frac{1}{m}\sum_{i=1}^{m} E(d_{P,i}(t_k)) = \frac{m_1}{m} \end{aligned} \quad (3)$$

$$E(t_k \mid N) = E(\frac{1}{n}\sum_{j=1}^{n} d_{N,j}(t_k)) = \frac{n_1}{n} \quad (4)$$

$$D(t_k \mid P) = \frac{1}{m}\sum_{i=1}^{m} (d_{P,i}(t_k) - \frac{m_1}{m})^2 \quad (5)$$

$$D(t_k \mid N) = \frac{1}{n}\sum_{j=1}^{n} (d_{N,j}(t_k) - \frac{n_1}{n})^2 \quad (6)$$

Hence, we have that

$$F(t_k)=\frac{(E(t_k \mid P)-E(t_k \mid N))^2}{D(t_k \mid P)+D(t_k \mid N)}$$
$$=\frac{(m_1 n - m n_1)^2}{mn^2\sum_{i=1}^{m}(d_{P,i}(t_k)-\frac{m_1}{m})^2+nm^2\sum_{j=1}^{n}(d_{N,j}(t_k)-\frac{n_1}{n})^2} \quad (7)$$

Fisher's discriminant ratio $F(t_k)$ based on Boolean value is subsequently denoted by $F_B(t_k)$.

(2) The Fisher's discriminant ratio based on frequency

It is obvious that Fisher's discriminant ratio based on Boolean value dose not consider the appearing frequency of a feature in a certain text, but only consider whether or not it appears. In order to examine the influence of the frequence on the significance of a feature another probability estimation method based on frequency is adopted in Formula (2).

Let $v_{P,i}$ and $v_{N,j}$ be the word tokens of texts $d_{P,i}$ and $d_{N,j}$, $w_{P,i}(t_k)$ and $w_{N,j}(t_k)$ be the frequencies of t_k appearing in $d_{P,i}$ and $d_{N,j}$ respectively.

Then the conditional means and the conditional variances of the feature t_k with respect to the categories P and N in Formula (1) can be estimated by using Formulas (8)-(11).

$$E(t_k \mid P)=\frac{1}{m}\sum_{i=1}^{m}\frac{w_{P,i}(t_k)}{v_{P,i}} \quad (8)$$

$$E(t_k \mid N)=\frac{1}{n}\sum_{j=1}^{n}\frac{w_{N,j}(t_k)}{v_{N,j}} \quad (9)$$

$$D(t_k \mid P)=\frac{1}{m}\sum_{i=1}^{m}(\frac{w_{P,i}(t_k)}{v_{P,i}}-E(t_k \mid P))^2 \quad (10)$$

$$D(t_k \mid N)=\frac{1}{n}\sum_{j=1}^{m}(\frac{w_{N,j}(t_k)}{v_{N,j}}-E(t_k \mid N))^2 \quad (11)$$

Hence, we have that

$$F(t_k)=\frac{(E(t_k \mid P)-E(t_k \mid N))^2}{D(t_k \mid P)+D(t_k \mid N)}$$

$$= \frac{(\frac{1}{m}\sum_{i=1}^{m}\frac{w_{P,i}(t_k)}{v_{P,i}} - \frac{1}{n}\sum_{j=1}^{n}\frac{w_{N,j}(t_k)}{v_{N,j}})^2}{\frac{1}{m}\sum_{i=1}^{m}(\frac{w_{P,j}(t_k)}{v_{P,j}} - \frac{1}{m}\sum_{i=1}^{m}\frac{w_{P,i}(t_k)}{v_{P,i}})^2 + \frac{1}{n}\sum_{j=1}^{n}(\frac{w_{N,j}(t_k)}{v_{N,j}} - \frac{1}{n}\sum_{j=1}^{n}\frac{w_{N,j}(t_k)}{v_{N,j}})^2}$$

$$= \frac{mn(n\sum_{i=1}^{m}\frac{w_{P,i}(t_k)}{v_{P,i}} - m\sum_{j=1}^{n}\frac{w_{N,j}(t_k)}{v_{N,j}})^2}{n^2\sum_{i=1}^{m}(m\frac{w_{P,i}(t_k)}{v_{P,i}} - \sum_{i=1}^{m}\frac{w_{P,i}(t_k)}{v_{P,i}})^2 + m^2\sum_{j=1}^{n}(n\frac{w_{N,j}(t_k)}{v_{N,j}} - \sum_{j=1}^{n}\frac{w_{N,j}(t_k)}{v_{N,j}})^2} \qquad (12)$$

Fisher's discriminant ratio $F(t_k)$ based on frequency is subsequently denoted by $F_F(t_k)$.

2.2 Process of Feature Selection

(1) Candidate feature sets

In order to compare the classification effects of features from the different regions. We design two kinds of word sets as the candidate feature sets. One of them denoted by U consists of all words in the text set. Another candidate feature set I contains all words which appear in both positive and negative texts.

(2) Features used in the classification model

The idea of Fisher's discriminant ratio implies that it can be used as a significance measure of features for classification problems. The larger the value of Fisher's discriminant ratio of a feature is, the stronger the classification capability of the feature is. So we can compute the value of Fisher's discriminant ratio for every feature and rank them in the descending order. And then choose the best features with a certain number.

3 The Process of Text Sentiment Classification

The whole experiment process is divided into training and testing parts.

(1) Segment preprocess of review texts.

(2) By using the methods introduced in Section 2, obtain the candidate feature sets, and then select the features for the classification model.

(3) Express each text in the form of feature weight vector. Here, feature weights are computed by using TFIDF[2].

(4) Train the support vector classification machine by using training data, and then obtain a classifier.

(5) Test the performance of the classifier by using testing data.

Up to now, it is verified that support vector machine possesses the best performance for the text sentiment classification problem[1,10,11]. Therefore we adopt support vector machine to construct the classifier in the presented paper.

4 Experiment Result

4.1 Experiment Corpus and Evaluation Measures

(1) We collected 1006 Chinese review texts about 11 kinds of car trademarks published on http://www.xche.com.cn/baocar/ from January 2006 to March 2007. In the corpus there are 578 positive reviews and 428 negative reviews. The total reviews contain 1000 thousands words.
(2) To evaluate the effectiveness of the proposed feature selection methods, three kinds of classical evaluation measures generally used in text classification, Precision, Recall and F1 value are adopted in this paper. By PP(PN), RP(RN) and FP(FN) we denote Precision, Recall and F1 value of positive(negative) review texts respectively. By F we denote F1 value of all review texts.

4.2 Experiment Result and Analysis

For convenience, a kind of simple symbol, Candidate Feature Set + Feature Selection Method, shows which candidate feature set and which feature selection method were adopted in an experiment. For example, $U + F_B(t_k)$ stands for that the candidate feature set U and the feature significance measure $F_B(t_k)$ are adopted in an experiment.

Experiment 1: In order to exam the influence of feature dimension on classification performance 500, 1000 and 2000 features are selected from the candidate feature set I by using the methods in Section 2. The experiment result is shown in Table 1.

Table 1. Classification effects of the different feature dimensions by using $I + F_F(t_k)$ approach

Measure \ Dimension	500	1000	2000
PP	82.65	82.90	80.39
RP	85.57	89.39	89.39
FP	84.08	86.00	84.60
NP	79.81	84.20	82.80
NR	75.53	75.06	70.12
NF	77.61	79.30	75.79
F	81.3	83.3	81.2

Table 1 shows that almost all evaluation measures are best under 1000 feature dimension. In other words, a larger amount of features need not imply a good classification result. So all the feature dimensions in the succedent experiments are 1000.

Experiment 2: The aim of this experiment is to compare the effects of 6 kinds of feature selection approaches with 1000 dimension for text sentiment classification. Here MI and IG stand for Mutual Information and Information Gain respectively. The text sentiment classification experiment results are shown in Table2.

Table 2. Classification effects based on 6 kinds of feature selection approaches with 1000 dimension

Approach \ Measure	PP(%)	RP(%)	FP(%)	PN(%)	RN(%)	FN(%)	FT(%)
$U+MI$	71.68	49.39	58.48	53.25	74.35	62.06	62.84
$U+IG$	81.19	88.52	84.61	82.58	72.00	76.72	81.50
$U+F_B(t_k)$	81.75	90.61	85.86	85.25	72.24	77.95	82.80
$U+F_F(t_k)$	76.73	84.35	80.32	78.85	65.41	70.15	76.30
$I+F_B(t_k)$	81.67	85.04	83.32	79.25	74.12	76.60	80.40
$I+F_F(t_k)$	82.90	89.39	86.00	84.20	75.06	79.29	83.30

From Table 2 it can be seen that:

(1) Among the 3 kinds of feature significance measures Mutual Information, Information Gain and Fisher's discriminant ratio for feature selection the performance of MI is worst.

(2) For IG and Fisher's discriminant ratio, only in the third instance $U+F_F(t_k)$ the performance of Fisher's discriminant ratio is worse than that of IG.

(3) The performances of two kinds of variations of Fisher's discriminant ratio rely on the candidate feature set to some extent. For Fisher's discriminant ratio based on Boolean value the performance of it is better when U is adopted as the candidate set.

(4) Among all the 6 kinds of approaches, the performance of Fisher's discriminant ratio based on frequency is best when the candidate feature set is I.

Remarks: (1) The computing time cost of the feature selection process depends upon the size of the candidate feature set, it is advisable to design a smaller candidate feature set such as I in this paper. (2) For text sentiment classification problem many kinds of feature selection methods such as MI, IG, CHI and DF are compared in some literatures [11]. Among these methods IG is validated to be best in the past research works. However, the experiments in this paper shown that Fisher's discriminant ratio based on frequency is a better choice than IG for feature selection.

5 Conclusions

Text sentiment classification can be widely applied to text filtering, online tracking opinions, analysis of public-opinion poll, and chat systems. However, compared with traditional subject classification, there are more factors that need to be considered in text sentiment classification. Especially, the feature selection for text sentiment classification is more difficult. In this paper, under the granularity level of words, a new feature selection method based on Fisher's discriminant ratio is proposed. In order to validate the validity of the proposed method, we compared it with other methods respectively based on information gain and mutual information while support vector machine is adopted as the classifier. In this paper, 6 subexperiments are conducted by combining different feature selection methods with 2 kinds of candidate feature sets. The experiment results show that $I + F_F(t_k)$ obtain the best classification effectiveness, its F value achieves 83.3%. Our further research works will focus on establishing a sentiment knowledge base based on vocabulary, syntactic, semantic and ontology.

Acknowledgements

This work was supported by the National Natural Science Foundation (No.60875040), the Foundation of Doctoral Program Research of Ministry of Education of China (No.200801080006), Key project of Science and Technology Research of the Ministry of Education of China (No.207018), Natural Science Foundation of Shanxi Province (No.2007011042), the Open-End Fund of Key Laboratory of Shanxi Province, Shanxi Foundation of Tackling Key Problem in Science and Technology (No.051129), Science and Technology Development Foundation of Colleges in Shanxi Province (No. 200611002).

References

[1] Pang, B., Lee, L., Vaithyanathan, S.: Thumbs up? Sentiment Classification Using Machine Learning Techniques. In: EMNLP 2002 (2002)

[2] Turney, P.D., Littman, M.L.: Unsupervised Learning of Semantic Orientation from A Hundred-billion-word Corpus. Technical Report EGB-1094, National Research Council Canada (2002)

[3] Turney, P.D., Littman, M.L.: Measuring Praise and Criticism: Inference of Semantic Orientation from Association. ACM Transactions on Information Systems (TOIS) 21(4), 315–346 (2003)

[4] Tony, M., Nigel, C.: Sentiment Analysis Using Support Vector Machines with Diverse Information Sources. In: Lin, D., Wu, D. (eds.) Proceedings of EMNLP 2004, Barcelona, Spain, July 2004, pp. 412–418 (2004)

[5] Michael, G.: Sentiment Classification on Customer Feedback Data: Noisy Data, Large Feature Vectors, and the Role of Linguistic Analysis. In: Proceedings the 20th international conference on computational linguistics (2004)

[6] Kennedy, A., Inkpen, D.: Sentiment Classification of Movie and Product Reviews Using Contextual Valence Shifters. In: Workshop on the analysis of Informal and Formal Information Exchange during negotiations (2005)

[7] Chaovalit, P., Zhou, L.: Movie Review Mining: A Comparison Between Supervised and Unsupervised Classification Approaches. In: IEEE proceedings of the 38th Hawaii International Conference on System Sciences, Big Island, Hawaii, pp. 1–9 (2005)

[8] Ye, Q., Lin, B., Li, Y.: Sentiment Classification for Chinese Reviews: A Comparison Between SVM and Semantic Approaches. In: The 4th International Inference on Machine Learning and Cybernetics ICMLC 2005 (June 2005)
[9] Ye, Q., Zhang, Z., Rob, L.: Sentiment Classification of Online Reviews to Travel Destinations by Supervised Machine Learning Approaches. Expert System with Application (in press)
[10] Wang, S., Wei, Y., Li, D., Zhang, W., Li, W.: A Hybrid Method of Feature Selection for Chinese Text Sentiment Classification. In: Proceedings of the 4th International Conference on Fuzzy Systems and Knowledge Discovery, pp. 435–439. IEEE Computer Society, Los Alamitos (2007)
[11] Ahmed, A., Chen, H., Salem, A.: Sentiment Analysis in Multiple Languages: Feature Selection for Opinion Classification in Web Forums. ACM Transactions on Information Systems 26(3) (2008)
[12] Tan, S., Zhang, J.: An Empirical Study of Sen-timent Analysis for Chinese Documents. Expert Systems with Applications 34, 2622–2629 (2008)
[13] Yi, J., Nasukawa, T., Bunescu, R., Niblack, W.: Sentiment Analyzer: Extracting Sentiments About A Given Topic Using Natural Language Processing Techniques. In: Proceedings of the 3rd IEEE International Conference on Data Mining, pp. 427–434 (2003)
[14] Webb, A.R.: Statistical Pattern Recognition, 2nd edn. John Wiley & Sons, Inc., Chichester
[15] Yang, Y., Liu, X.: A Re-examination of Text Categorization Methods. In: SIGIR, pp. 42–49 (1999)
[16] Yang, Y., Pedersen, J.O.: A Comparative Study on Feature Selection in Text Categorization. In: ICML, pp. 412–420 (1997)

Mining Preferred Traversal Paths with HITS

Jieh-Shan Yeh, Ying-Lin Lin, and Yu-Cheng Chen

Department of Computer Science and Information Management,
Providence University, Taichung 433, Taiwan
{jsyeh,g9671001,g9772002}@pu.edu.tw

Abstract. Web usage mining can discover useful information hidden in web logs data. However, many previous algorithms do not consider the structure of web pages, but regard all web pages with the same importance. This paper utilizes HITS values and PNT preferences as measures to mine users' preferred traversal paths. Wë structure mining uses HITS (hypertext induced topic selection) to rank web pages. PNT (preferred navigation tree) is an algorithm that finds users' preferred navigation paths. This paper introduces the *Preferred Navigation Tree with HITS* (PNTH) algorithm, which is an extension of PNT. This algorithm uses the concept of PNT and takes into account the relationships among web pages using HITS algorithm. This algorithm is suitable for E-commerce applications such as improving web site design and web server performance.

Keywords: Preferred navigation patterns; Preference; Web usage mining; Web structure mining; HITS algorithm.

1 Introduction

With the rapid growth of the web, web servers continue to gather huge amounts of data. This data contains rich information that reveals web user behaviors. Web mining can be used to discover useful information hidden in this data. Web mining is classified into three types: web structure mining (WSM), web content mining (WCM), and web usage mining (WUM) [7]. WSM analyzes web structures to find relationships between web pages; WCM gleans meaningful information from web contents; WUM discovers users' past behavior and predicts web pages that users will want to visit in the future.

Many researchers have studied WUM, and proposed several algorithms to discover trends and regularities in web users' navigation patterns. WUM findings are very helpful in improving web site design and web performance. There are many researches mining patterns from web logs [3], [4]. Several studies use sequential pattern mining techniques [1], [2], which are primarily based on association rule mining (ARM). Some studies take the concept of weights into ARM [8], [9], [10], they thought that different items should have different weights according to their importance. Chen et al. [5] introduced the concept of using maximal forward references to break down user sessions into transactions to easily mine navigation patterns.

W. Liu et al. (Eds.): WISM 2009, LNCS 5854, pp. 98–107, 2009.

In addition to WUM, WSM is an important approach to web mining. The number of in-degree (links to a page) and out-degree (links from a page) occurrences are valuable pieces of information because popular pages are often linked by other pages. Therefore, an important page contains a high number of out-degree links. However, many algorithms assess user navigation patterns based on according to the access frequency. The paths visited most frequently are the preferred navigation paths. Previous algorithms do not consider the structure of web pages, but regard all web pages with the same importance.

This paper considers the structure of web pages when mining user navigation patterns. This approach uses the HITS algorithm [6] to discover the relationships among web pages and give each page its own weight. This paper proposes a new measure that considers the structure of web pages in measuring user interest and preference of a page, replacing the *preference* variable in Xing [12]. Moreover, this study proposes the PNTH structure, which is an extended algorithm of PNT, to mine preferred navigation paths.

The rest of this paper is organized as follows. Section 2 provides an overview of the HITS algorithm. Section 3 describes the modified *preference* variable. Section 4 describes the algorithm for mining preferred navigation paths. Section 5 provides an example to illustrate the feasibility of the proposed algorithm. Section 6 summarizes the conclusions.

2 HITS Algorithm

HITS is primarily used in web structure mining. HITS ranks web pages by analyzing the in-degree and out-degree of a webpage. In this algorithm, web pages pointed to by many other pages are called *authorities,* while web pages that point to many other pages are called *hubs* [6]. Authorities and hubs have mutual effects on each other. The relationship between authorities and hubs could be called a *mutually reinforcing relationship*: a good hub is a page that points to many good authorities; a good authority page is a page pointed to by many good hubs. Authorities and hubs can be depicted as a bipartite graph. Fig. 1 illustrates authorities and hubs.

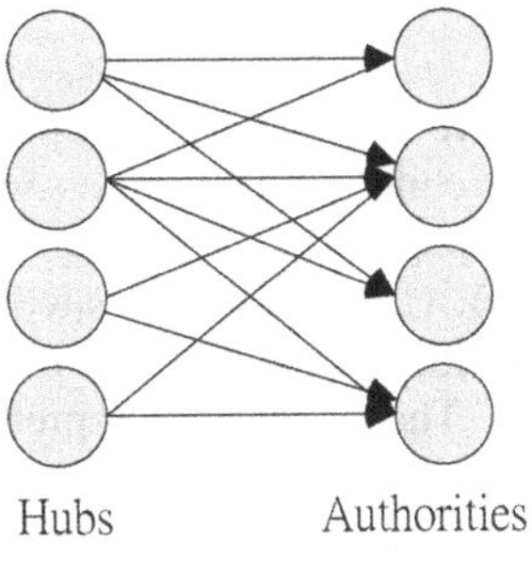

Fig. 1. Authorities and Hubs

Authorities and hubs are assigned respective values. Let $auth_p$ and hub_p represent the authority and hub values of page *p*, respectively. The values of authorities and hubs can be obtained from an iterative algorithm that maintains and updates numerical values for each page. This algorithm performs the following equations:

$$auth_p \leftarrow \sum_{q:(q,p)\in E} hub_q \quad . \tag{1}$$

$$hub_p \leftarrow \sum_{q:(p,q)\in E} auth_q \tag{2}$$

where $(p,q)\in E$ denotes that there is a link from page *p* to page *q*. Pages with larger $auth_p$ and hub_p are regarded as better authorities and hubs, respectively.

In addition to assigning the weights to web pages, HITS algorithm can be used to assign the weights to transaction data. Sun [8] regarded the "transactions" as hubs and "items" as authorities.

3 Preference with HITS

The concept of preference is that if there are many different options to leave a page, the options that are selected most frequently and the next page viewed reveal user interest and preference. The proposed measure for judging the degree of user interest in a webpage includes four factors: *authority*, *hub*, *selection preference*, and *time preference*. This new measure replaces the original preference in Xing [12].

The previous section describes *authority* and *hub*. These two factors are used to rank the web pages via the HITS algorithm [6]. *Selection preference* and *time preference* are elements of the original preference in [12]. Let *SP* denotes selection preference. *SP* is defined as

$$SP_k = C_k \Big/ \left(\left(\sum_{i=1}^{n} C_i \right) / n \right) . \tag{3}$$

where C_k is the times that users browse from a parent page to the *k-th* child page. C_i is the times that users visit the *i-th* child page from the same parent page. *n* is the number of child pages that the parent page offers.

- *Example.* 3 child pages {B, C, D} can be visited after browsing page A. The times of visiting Page A is 25, and the times of visiting Page {B, C, D} are {10, 10, 5}, respectively. The selection preference of Page B is $SP_B = 10/[(10+10+5)/3] = 1.2$. The sum of the times of all child pages is equal to the times of the parent page (Page A).

Let *TP* represent time preference. The definition of *TP* is shown as

$$TP_k = T_k / \left(\left(\sum_{i=1}^{n} T_i \right) / n \right). \tag{4}$$

where T_k is the browsing time of the *k-th* child page a user visits from a parent page. T_i is the browsing time of the *i-th* child page a user visits from the same parent page. *n* is the number of child pages that the parent page offers.

- *Example.* 3 child pages {B, C, D} are visited after browsing page A. The browsing time on Page A is 25, and the browsing time on Page {B, C, D} are {34, 78, 20}, respectively. The time preference of Page B is $TP_B = 34/[(34+78+20)/3] = 0.77$.

If the *k-th* child page is visited from the parent page *p*. Based on the four factors stated above, the original preference is modified as

$$PH_k = hub_p \times auth_k \times SP_k \times TP_k \tag{5}$$

4 Algorithm of PNTH for Mining the Preferred Paths

This section describes the algorithm for mining the preferred navigation path with the modified preference term.

4.1 Structure of PNTH

The structure of PNTH is as follows:

```
TYPE Nodeptr=^Nodetype
    Nodetype=RECORD
             URL: string;
             Count: integer;
             Average Time:  real;
           Auth: real;
             Hub: real;
             Preference: real;
             Child, Younger, Ref: Nodeptr
           END;
```

URL denotes the address of the current web page. Count is the number of times a web page has been visited through the same route. Time is the time that the user spends viewing a page. Auth is the authority value of a webpage. Hub is the hub value of a web page. Preference is defined as Eq. 5. Child and younger are pointers that point to child and brother nodes, respectively. Ref is also a pointer that points to the parent node.

4.2 Construction of PNTH

There are two steps involved in constructing PNTH: Step1 calculates the *authorities* and *hubs* of the web pages of the website, and Step 2 actually constructs the PNTH.

- **Algorithm1.** Calculate authorities and hubs of web pages

Input: a site graph G=(V,E), the number of iteration *num_it*.
Output: Hub(p) and Auth(p) for each page p.
(1) *Auth(i)* = 1 for each page *i*;
(2) **For** (*loop* =0; *loop<num_it*; *loop*++) **do**
(3) Auth′(*i*) = 0 for each page *i*;
(4) **For**(*k*=0; *k<num_page*; *k*++) **do**
(5) $Hub(k) = \sum_{p \in O(p)} Auth(p)$, where *O(p)*denotes the set of all pages pointed to by page *p*
(6) **EndFor**
(7) **For**(*k*=0; *k<num_page*; *k*++) **do**
(8) $Auth'(k) = \sum_{p \in I(p)} Hub(p)$, where *I(p* denotes the set of all pages that point to *p*
(9) **EndFor**
(10) *Auth(i)* = *Auth′(i)* for each page *i*
(11) normalize *auth*
(12) **EndFor**

- **Algorithm2.** Construction of PNTH

Input: User access session database $D = \{(i, S_i) | S_i = (S_{i1}, S_{i2}, \ldots, S_{in})$, *where* S_{ij} *is a webpage, n is the size of the session* $S_i\}$.
Output: PNTH.
(1) Create a root node.
Its Count is the number of user session in *Ds*.
(2) **For**(*i*=1; *i*<=*num_session*; *i*++) **do**
(3) Let current pointer P ← root node.
(4) **For** (*j*=1; *j*<= size_sessionS_i; *j*++) **do**
(5) **If** ((P → Child)==Null)
Create child of P;
(P → Child) → URL= S_{ij}.URL;
(P → Child) → Count= 1;
(P → Child) → AverageTime= S_{ij}.Time;
(P → Child) → Auth= S_{ij}.Auth;
(P → Child) → Hub= S_{ij}.Hub;
(P → Child) → Ref= P;
(6) **Else**
(7) **If**((P → child) → URL==S_{ij}.URL)
(P → child) → Count ++;
(P → Child) → AverageTime+= S_{ij}.Time;
(8) **Else**
(9) P=P → Child;
(10) **While**(P!=NULL) **do**

```
(11)        If ((P → Younger) → URL== Sij.URL)
              Break;
            EndIf
          EndWhile
(12)      If ((P → Younger) →URL== Sij.URL)
            (P →Younger) → Count ++;
            (P →Younger) → AverageTime+= Sij.Time;
          Else
            Create Younger of P;
            (P → Younger) → URL= Sij.URL;
            (P → Younger) → Count= 1;
            (P →Younger) → AverageTime= Sij.Time;
            (P → Younger) → Auth= Sij.Auth;
            (P → Younger) → Hub= Sij.Hub;
            (P → Younger) → Ref= P → Ref;
(13)      EndIf
(14)    EndIf
(15)  EndIf
(16)  Let P ← current node.
(17)  EndFor
(18) EndFor
(19) Calculate the average browsing time for each node .
(20) Store the web page preference PHk gained by Eq.(5) for each
     page.
(21) END
```

The pointer *ref* links to the parent page to fetch the hub of the parent page when the algorithm calculates the web page preferences.

4.3 Mining Preferred Navigation Path from PNTH

- **Algorithm3.** Mining preferred paths from a PNTH

```
Input: a PNTH, Preference threshold (Pthreshold), Support
       threshold (Sthreshold).
Output: User Preferred Paths.
Let R be pointer that points to the root node of the PNTH and
let O be an auxiliary stack.
(1) O = NULL, P ← R
(2) If (P == NULL)
      goto (4) EndIf
(3) visit node P;
     If (P.Preference>=Pthreshold and P.Count>=Sthreshold)
       mark P as a preferred node
     Else
       mark P as a non-preferred node
     EndIf
     push P to O
     P ← Child(P)
     Goto(2)
(4) If (O == NULL)
      End
     Else
```

```
pop node x from O
If( x is a preferred node)
 Create a new path in candidate preferred paths set
  Output x into path
  Do
   Next node y in O
   Output y into path
  Until (y is a non-preferred node or y = NULL)
  P ← Younger(P)
  goto(2)
 EndIf
EndIf
```

(5) Match all the sequences in the candidate preferred paths set. The sequence that is not included in any other sequences of the set is the preferred navigation path.

This algorithm is proposed by Xing [12]. It uses a Depth First Traversal (DFT) strategy to mine the preferred navigation patterns. The difference between the proposed algorithm and Xing's is that the current study considers the web structure, but Xing's only considers the selection preference and time preference to reveal users interest and preference.

5 Example

This section provides an example for mining users' preferred navigation path with the proposed algorithm. Fig. 2 shows the structure of the sample web site.

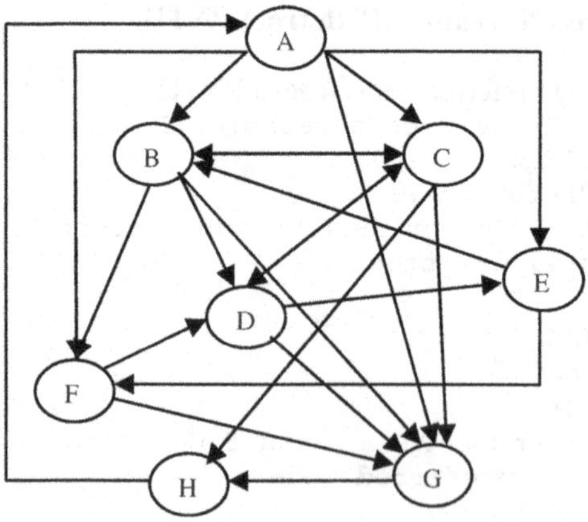

Fig. 2. The structure of the sample web site

Table 1 illustrates the browsing sessions that used in [11]. The elements in each session are represented as (URL, Time duration), where URL is the visited web page. Time duration denotes the time spent viewing the selected web page.

Table 1. Browsing sessions

UID	Browsing sessions
1	(A,23), (B,68), (D,98), (E,130)
2	(A,45), (B,89), (D,102)
3	(A,27), (B,56), (F,86)
4	(A,32), (B,87), (D,45), (G,115), (H,118)
5	(A,30), (C,65), (H,78)
6	(A,12), (C,34), (G,32)
7	(A,78), (C,89), (G,110), (H,123)
8	(A,10), (B,24), (G,45), (H,34)

Fig. 3 shows the structure of PNTH node according to Algorithm. The PNTH constructed by Algorithm 2 appears in Fig. 4.

URL	Count	Ref
Auth	Hub	Younger
Average Time	Preference	Child

Fig. 3. The structure of PNTH node

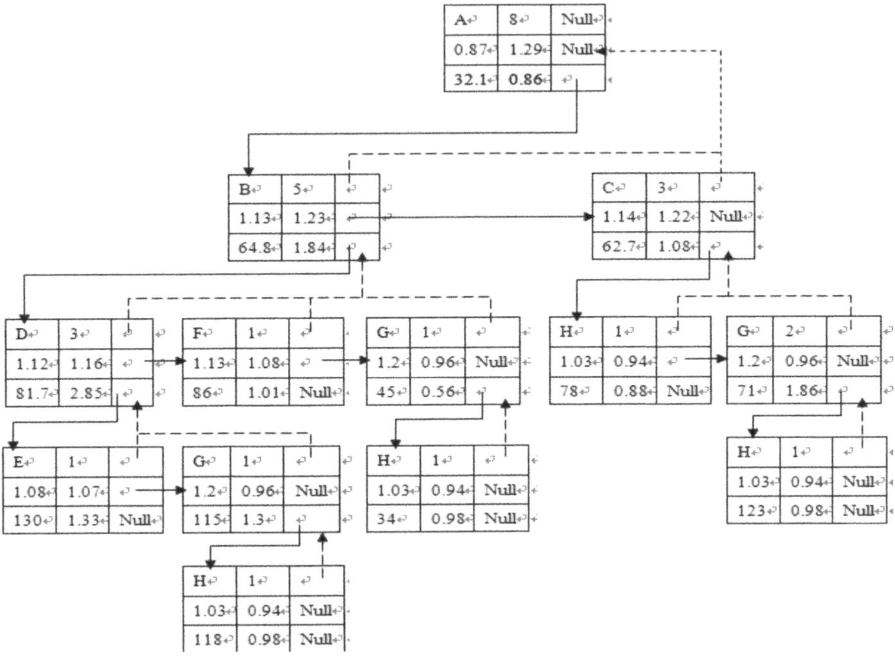

Fig. 4. The constructed PNTH

Assume that the *support* threshold and *preference* threshold are set at 2 and 1, respectively. Thus, all mined preferred navigation paths are {B, D} and {C, G}. The results in Xing's were {A, B, D}. The reasons that cause the difference are:

(1) Although page A is the most frequently visited page, the authority of page A is low. It means that the content of page A is not popular. Users just use page A to link other pages.

(2) According to the example, Xing's algorithm cannot find the preferred navigation pattern {C, G} because it does not consider the structure of web pages.

6 Experimental Evaluation

All the experiments were conducted on a 2.40GHz Core 2 Duo P8600 with 2GB of RAM, running Microsoft Windows Vista. All code was compiled using C.

We compare the accuracy of the proposed algorithm with that of Xing based on the information resource shown in Table 2.

Table 2. The web log data of Providence University

Web URLs	161
Experimental period	2009/1/1-2009/1/5
Records	3027096
Records after data cleaning	86733

We use the records of 2009/1/1 to be the training set to mine the preferred navigation patterns and the remainder to be the testing set. The result is shown in Fig. 5 (a).

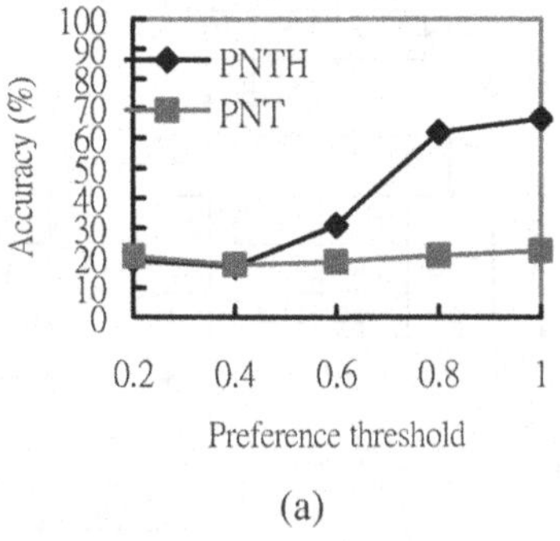

(a)

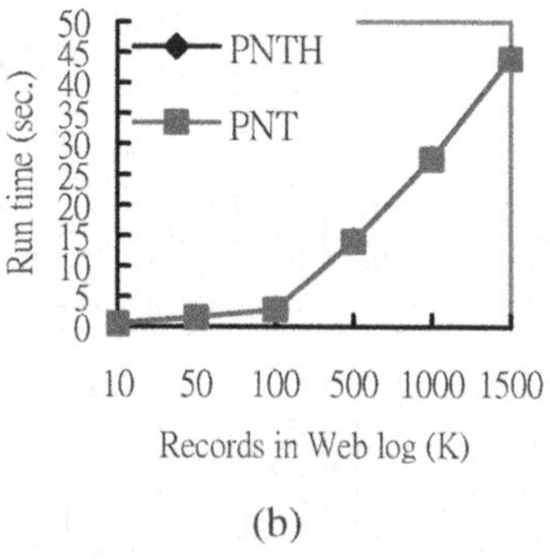

(b)

Fig. 5. (a) Accuracy comparison; (b) Run time comparison

According to Fig. 5 (a), PNTH algorithm is more accurate than PNT algorithm. It should be noticed that the lower preference threshold, the lower accuracy is, because the number of mined preferred paths increases as the threshold reduction. Then, we compare the runtime of the PNTH with PNT shown as Fig. 5 (b). Although the proposed algorithm has to implement the HITS algorithm, its run time is close to the PNT algorithm. The reason is that the HITS algorithm just needs to be implemented once. Moreover, authorities and hubs will converge about just three or four iterations, so it doesn't take a lot of time.

7 Conclusion

This paper proposes a new measure, modified-preference, to replace the original preference when mining the preferred navigation path. The modified-preference considers the web structure. Based on this measure, this study develops an algorithm to mine the preferred navigation paths from PNTH. PNTH can discover more accurate navigation path than PNT.

References

1. Agrawal, R., Srikant, R.: Fast algorithms for mining association rules. In: 20th International Conference Very Large Data Bases (VLDB 1994), pp. 487–499 (1994)
2. Agrawal, R., Srikant, R.: Mining sequential patterns. In: 11th International Conference on Data Engineering, pp. 3–14 (1995)
3. Cooley, R., Mobasher, B., Srivastava, J.: Web mining: Information and pattern discovery on the World Wide Web. In: 9th IEEE International Conference on Tools with Artificial Intelligence, pp. 558–567 (1997)
4. Cooley, R., Mobasher, B., Srivastava, J.: Data preparation for mining world wide web browsing patterns. Journal of Knowledge and Information System 1, 5–32 (1999)
5. Chen, M.S., Park, J.S., Yu, P.S.: Efficient data mining for path traversal patterns. IEEE Transactions on Knowledge and Data Engineering 10, 209–221 (1998)
6. Kleinberg, J.M.: Authoritative sources in a hyperlinked environment. Journal of the ACM 46, 604–632 (1999)
7. Kosala, R., Blockeel, H.: Web Mining Research: A Survey. ACM SIGKDD Explorations Newsletter 2, 1–15 (2000)
8. Sun, K., Bai, F.: Mining weighted association rules without preassigned weights. IEEE Transactions on Knowledge and Data Engineering 20, 489–495 (2008)
9. Tao, F., Mutagh, F., Farid, M.: Weighted association rule mining using weighted support and significance framework. In: 9th ACM SIGKDD International Conference on Knowledge Discovery and Data Mining (KDD 2003), pp. 661–666 (2003)
10. Wang, W., Yang, J., Yu, P.S.: Efficient mining of weighted association rules (WAR). In: 6th ACM SIGKDD International Conference on Knowledge Discovery and Data Mining (KDD 2000), pp. 270–274 (2000)
11. Wu, R., Tang, W., Zhao, R.: Web Mining of Preferred Traversal Patterns in Fuzzy Environments. In: Ślęzak, D., Yao, J., Peters, J.F., Ziarko, W.P., Hu, X. (eds.) RSFDGrC 2005. LNCS (LNAI), vol. 3642, pp. 456–465. Springer, Heidelberg (2005)
12. Xing, D., Shen, J.: Efficient data mining for web navigation patterns. Information and Software Technology 46, 55–63 (2004)

Link Analysis on Government Website Based-on Factor

Changling Li and Xinjin Fu

Science and Technology Information Research Institute,
Shandong University of Technology
Zibo, P.R. China
lichl@sdut.edu.cn, fuxj_1984@hotmail.com

Abstract. 15 provincial government websites were chosen as a sample, and their link data were searched. To use Factor to analyze the data, and gave the evaluation results. Correlated analyzed the results with the government efficiency scores, finding the two were correlated. Also correlated analyzed the link indexes with the government efficiency scores and the most obviously correlated indexes were total link and external link. So the two link indexes are available for evaluation.

Keywords: link analysis; government website evaluation; factor.

1 Introduction

Government website is a window to communicate with outside, just like an appearance. Evaluation can help government improve its website construction and its e-government efficiency.

E-government evaluation has already become an important part in the USA, UK, Canada, Japan, Korea and etc. Some international organizations and consulting institutions also have deep researches. For example, Department of Economic and Social Affairs of United Nations used "E-government Readiness Index" and "E-government Participation Index" to investigate 191 countries' e-government processes in 2003[1]. European Commission made "eEurope2002" in 2000, and made "eEurope2005" in 2002, to evaluate the effect on policy [2]. In 2003, Accenture used "overall maturity" to evaluate 22 countries' e-governments quantitatively [3].

In China, there are also many researches on e-government evaluation. For instance, China Computer World Research posted Appraisal Report on City E-governments in China which measured and compared 36 chief cities' e-governments in 2002[4]. Website Economy Research Center of Peking University posted Research on Prefecture-level City E-government in China in Oct.2003, evaluating these cities by using the method of combining content and technology [5]. Besides, many researchers mainly focused on two aspects, review on method and cases study. For example, Hao Yuanyuan introduced evaluation methods for e-government based on websites [6].Yu Fang built an evaluating index system to assess the website of e-government of Hefei [7].Cao Ping used fuzzy evaluation to build an government website evaluation model[8].

W. Liu et al. (Eds.): WISM 2009, LNCS 5854, pp. 108–115, 2009.

Different researchers have different approaches. For link analysis is more object, we use it to evaluate the government website to see its web impact and construction level. Evaluation the government website also can help to evaluate the e-government and improve its efficiency.

2 Index Description and Data Acquisition

2.1 Index Description

There are 5 indexes:

① Total link(X_1) is the amount of website pages which link the target website. The more links it has, the higher degree of information shown, and the better organization.

② Ex-link(X_2) is the pages linking the target website, except the ones inside of itself. The number of this index can reflect the degree of utilization efficiency and impact of this website.

③ Total pages(X_3) is the number of all pages in one website. The more these pages, the higher scale and more information the website has. In general, this index value is acquired by the search engine. So it also can show the usage degree by public.

④ Total WIF(X_4) can be acquired by total links divided by whole website pages.

⑤ Ex-WIF(X_5) can be acquired by Ex-link divided by whole website pages.

2.2 Data Acquisition

We chose top 15 government websites as a sample from "Evaluation on Provincial E-government Efficiency in 2008" [9]. And we got their "Web Index" and "Score".

The links can be searched from the search engine AltaVista. And the WIFs can be computed based on the above. We take the Beijing government website as an example to show the search logics.

Total link: link:www.beijing.gov.cn;

Ex-link: link:www.beijing.gov.cn –host:www.beijing.gov.cn;

Total pages: host:www.beijing.gov.cn.

So we got the link data of the 15 websites. The original link data are shown in Table 1.

Table 1. Link data of government websites

R	Province	Domain Name	X_1	X_2	X_3	X_4	X_5
1	Beijing	www.beijing.gov.cn	556000	397000	103000	5.40	3.85
2	Shanghai	www.shanghai.gov.cn	140000	131000	379000	0.37	0.35
3	Zhejiang	www.zj.gov.cn	42800	42300	20400	2.10	2.07
4	Hainan	www.hainan.gov.cn	25000	22600	108000	0.23	0.21
5	Shaanxi	www.shaanxi.gov.cn	118000	103000	22600	5.22	4.56
6	Guangdong	www.gd.gov.cn	404000	370000	42300	9.55	8.75

Table 1. (*Continued*)

R	Province	Domain Name	X_1	X_2	X_3	X_4	X_5
7	Anhui	www.ah.gov.cn	104000	99600	50900	2.04	1.96
8	Sichuan	www.sichuan.gov.cn	1410	1260	36100	0.04	0.03
9	Fujian	www.fujian.gov.cn	103000	43200	72600	1.42	0.60
10	Hunan	www.hunan.gov.cn	99000	83500	264000	0.38	0.32
11	Tianjin	www.tj.gov.cn	155000	109000	49400	3.14	2.21
12	Jiangsu	www.jiangsu.gov.cn	59800	31600	55100	1.09	0.57
13	Heilongjiang	www.hlj.gov.cn	155000	82500	56500	2.74	1.46
14	Yunnan	www.yn.gov.cn	62700	62300	120000	0.52	0.52
15	Jilin	www.jl.gov.cn	54800	48700	57300	0.96	0.85

Note: time of link data collection is 2009.05.06 10:10-10:40.

3 Link Analysis on Government Website Based-on Factor

3.1 Data Standardization

Indexes' orders of magnitude are different, so we use range converter technique to standardize the data. All standardized data meet 0≤yij≤1, and the maximum value is 1, the minimum value is 0. All indexes are positive, so the formula is:

$$y_{ij} = \frac{x_{ij} - \min x_{ij}}{\max x_{ij} - \min x_{ij}} (1 \le i \le m, 1 \le j \le n) \tag{1}$$

The standardized data are shown in Table 2.

Table 2. Standardized Data

R	Province	X_1	X_2	X_3	X_4	X_5
1	Beijing	1	1	0.23	0.56	0.44
2	Shanghai	0.25	0.33	1	0.03	0.04
3	Zhejiang	0.07	0.10	0	0.22	0.23
4	Hainan	0.04	0.05	0.24	0.02	0.02
5	Shaanxi	0.21	0.26	0.01	0.54	0.52
6	Guangdong	0.73	0.93	0.06	1	1
7	Anhui	0.18	0.25	0.09	0.21	0.22

Table 2. (*Continued*)

R	Province	X_1	X_2	X_3	X_4	X_5
8	Sichuan	0	0	0.04	0	0
9	Fujian	0.18	0.11	0.15	0.15	0.06
10	Hunan	0.18	0.21	0.68	0.04	0.03
11	Tianjin	0.28	0.27	0.08	0.33	0.25
12	Jiangsu	0.11	0.08	0.10	0.11	0.06
13	Heilongjiang	0.28	0.21	0.10	0.28	0.16
14	Yunnan	0.11	0.15	0.28	0.05	0.06
15	Jilin	0.10	0.12	0.10	0.10	0.09

3.2 Factor Analysis

At first, we should know whether the data are suit to do the Factor analysis, so we use Bartlett test of sphericity as Table 3 shown. The observed quantity is 133.609, and the corresponding probability p is 0, less than the given significance level $\alpha=0.05$, so the data can do Factor analysis.

Table 3. Bartlett test of sphericity

Bartlett's Test of Sphericity	Approx. Chi-Square	133.609
	df	10
	Sig.	.000

Secondly, extraction method is principal component analysis. Fig. 1 is the scree plot, and its X-Axis is component number, Y-Axis is Eigenvalue. The first component's Eigenvalue is high, so it explains original data the most. After the third component, their Eigenvalues are lower, so they explain a little.

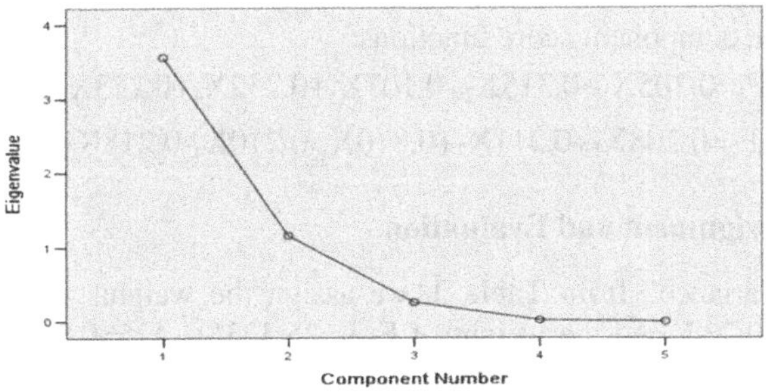

Fig. 1. Scree plot

Choose components which Eigenvalue is more than 1, then output component matrix in descending order of the first component shown as Table 4.

Table 4. Component Matrix

	Component	
	F_1	F_2
X_2	0.982	0.112
X_1	0.949	0.113
X_4	0.905	-0.381
X_5	0.876	-0.386
X_3	-0.030	0.973
Total	3.455	1.267
% of Variance	69.093	25.333
Cumulative %	69.093	94.425

From Table 4, we can know the first component's Eigenvalue is 3.455, mainly interpret total link, ex-link, total WIF, ex-WIF; the second component's Eigenvalue is 1.267, mainly interpret total pages. The two interpret 94.425% of original data total variance, the Factor analysis is ideal.

Thirdly, use Regression to reckon and output the component score coefficient, shown as Table 5.

Table 5. Component score coefficient matrix

	Component	
	F_1	F_2
X_1	0.305	0.208
X_2	0.315	0.211
X_3	0.107	0.810
X_4	0.232	-0.210
X_5	0.223	-0.218

So, we get the component score functions:

$$F_1=0.305X_1+0.315X_2+0.107X_3+0.232X_4+0.223X_5 \quad (2)$$

$$F_2=0.208X_1+0.211X_2+0.810X_3-0.210X_4-0.218X_5 \quad (3)$$

3.3 Weight Assignment and Evaluation

Using "% of Variance" from Table 3, we assign the weights. The weight of F_1 is 69.093/94.425=0.732, and the weight of F_2 is 25.333/94.425=0.268. So link analysis score F should be:

$$F=0.732F_1+0.268F_2 \quad (4)$$

According to the model, we compute every website link analysis score F shown as Table 6. R2 means the link analysis rank. "Web Index" and "Score" are from "Evaluation on Provincial E-government Efficiency in 2008".

Table 6. Link analysis results

R	Province	Web Index	Score	F	R2
1	Beijing	0.88	78.12	0.744	1
2	Shanghai	0.85	72.60	0.467	3
3	Zhejiang	0.76	65.07	0.100	13
4	Hainan	0.72	64.96	0.104	12
5	Shaanxi	0.68	62.90	0.250	5
6	Guangdong	0.88	59.39	0.706	2
7	Anhui	0.78	56.72	0.195	8
8	Sichuan	0.70	55.25	0.013	15
9	Fujian	0.77	54.78	0.148	10
10	Hunan	0.65	53.62	0.317	4
11	Tianjin	0.84	52.89	0.242	6
12	Jiangsu	0.80	51.98	0.099	14
13	Heilongjiang	0.65	50.67	0.215	7
14	Yunnan	0.67	50.28	0.169	9
15	Jilin	0.69	48.08	0.112	11

From Table 6, we can know that Beijing government website is at the top, Guangdong government website is the second, and Sichuan government website is the final. Meanwhile, we can know which component of a government website is prior directly.

3.4 Correlated Analysis

We make link analysis score F do correlated analysis with "Web Index" and "Score". Their Pearson correlations are shown in Table 7.

Table 7. Pearson Correlation for F

	Web Index	Score
F	0.636*	0.588*
*. Correlation is significant at the 0.05 level (2-tailed).		

F is correlated with "Web Index", the Pearson correlation is 0.636; it is also correlated with "Score", their Pearson correlation is 0.588. Thus, link analysis for government website can be a reference of e-government evaluation.

Table 8. Pearson Correlation for Indexes

	F	Web Index	Score
X_1	0.931**	0.639*	0.533*
X_2	0.962**	0.658**	0.536*
X_3	0.304	0.077	0.328
X_4	0.720**	0.493	0.274
X_5	0.691**	0.473	0.268

*. Correlation is significant at the 0.05 level (2-tailed).
**. Correlation is significant at the 0.01 level (2-tailed).

Table 8 shows the Pearson correlations of link indexes with F, Web Index and Score. As for F, the link analysis score, ex-link is the most obvious correlated index, their Pearson correlation is 0.962; the following are total link, total WIF and ex-WIF; total pages is not correlated. Ex-link and total link are correlated with Web Index. And the two are correlated with Score as well. So, links are the most obviously correlated with e-government efficiency.

4 Conclusion

We link analyzed 15 government websites based on Factor analysis. Beijing government website does the best, while Sichuan government website is the last one. According to the results, we can also do comparative research to find out which website is better and which component of a website is prior. So they can perfect themselves.

From the correlated analysis results, we knew the link analysis results are correlated with e-government efficiency score, so link analysis can help do e-government evaluation. Total link and ex-link have good performance in evaluation, so the two can be used in government website evaluation.

Commercial search engine, dynamic web data and etc. are still the inevitable issues in following study. Thus, we need to do more case studies and develop research tools to make up present researches, and improve link analysis theories and methods.

References

1. Department of Economic and Social Affairs of United Nations: World Public Sector Report 2003: E-government at the Crossroads, `http://unpan1.un.org/intradoc/groups/public/documents/UN/UNPAN012733.pdf`
2. European Commission: eEurope 2002: creating an EU framework for the exploitation of public sector information. Office for Official Publications of the European Communities, Luxembourg (2002)
3. Accenture: E-government Leadership: Engaging the Customer, `http://www.taxadmin.org/fta/meet/03am_pres/finnegan.pdf`

4. Sunding: The Truth of Chinese Government Websites, http://www2.ccw.com.cn/02/0234/a/0234a16_2.asp
5. Weiying, Z., He, L.: Research on Prefecture-level City E-government in China. Economy and Science Publisher, Beijing (2003)
6. Yuanyuan, H., Yijun, L.: Review of Research Methods of Websites-based E-government Evaluation. Information Science, 148–152 (2006)
7. Fang, Y., Yanyan, W.: Research on the Government Website Evaluation Based on AHP. Sci.-Tech. Information Development and Economy, 226–228 (2007)
8. Ping, C., Jian, Z.: Evaluation of the Government Websites Performance Based on Fuzzy Comprehensive Evaluation Methods. Journal of Beijing Institute of Technology (Social Sciences Edition), 90–92 (2008)
9. Evaluation on Provincial E-government Efficiency in 2008, http://www.cstc.org.cn/2008wzpg/jixiaopinggu/wenjian/paimingbaogao/shengji.htm

WAPS: An Audio Program Surveillance System for Large Scale Web Data Stream

Jie Gao, Yanqing Sun, Hongbin Suo, Qingwei Zhao, and Yonghong Yan

ThinkIT Speech Lab, Institute of Acoustics,
Chinese Academy of Sciences, Beijing 100190, P.R. China
{jgao,ysun,hsuo,qwzhao,yonghong.yang}@hccl.ioa.ac.cn

Abstract. We address the problem of effectively monitoring audio programs on the web. The paper tries to present how to construct such an audio program surveillance system using several state-of-the-art speech technologies. A real-world system WAPS (**W**eb **A**udio **P**rogram **S**urveillance) is used as an example. WAPS is described in details in terms of the challenges it faces, it system architecture and its component modules. Objective evaluation of the whole WAPS is also given. Experiments show that WAPS shows satisfying performance on both artificially created data and real web data.

1 Introduction

Ever-increasing connectivity bandwidth of the web, together with falling storage costs are resulting in increasing amount of audio data produced and exchanged. Consequently, it has draw much research interest that how to effectively monitor, organize and retrieve this audio data [1, 2, 3, 4, 5]. Since information retrieval has became one of the key applications on web [1, 2, 3, 4, 5]. Much previous work concerning the audio data on the web focuses on organization and retrieval of audio. And their goals are to build search engines of audio contents like Google, but whose indexing objects are audio data instead of the textual web pages. Many prototype systems have been built, for example, SpeechSpot [3], Everyzing [4] and Google Audio Indexing [5].

This paper addresses the task of building a highly precise web audio program surveillance system. Specifically, given a predefined set of textual keywords (in Chinese) and a web data stream, any audio program[1] in the data stream containing any of the keywords has to be detected and reported. This web audio surveillance task is in essence similar to the task of keyword spotting in continuous speech on telephony switch network [6]. However the task addressed in this work still has its own features since it is targeted for the large scale, heterogonous web audio stream. First, the audio stream data is much complicated than the telephony speech. The telephony speech typically consists of a two-way conversation. But audio on the web contains much non-speech audio (music etc.) and speech of foreign languages. These non-speech

[1] Audio programs include both pure audio program and sound track of video program. And each audio program is uniquely identified by its source URL.

W. Liu et al. (Eds.): WISM 2009, LNCS 5854, pp. 116–128, 2009.

audio and foreign speeches actually pose a challenge on the precision aspect of the surveillance system because these data may all cause false alarms, especially when they account for a large portion of whole audio stream. Secondly, the surveillance system focus on detection results of a whole audio program (a specific URL in the web stream data) this also provides chances to make use of program-level information in the system development.

This paper introduces our surveillance system WAPS (Web Audio Program Surveillance) developed recently. It is developed by ThinkIT speech lab, Chinese Academy of Sciences and will be deployed in daily use for the web surveillance in a certain security department in China. And the major contribution of the work is that it presents how to integrate many state-of-the-art speech technologies into a real-world surveillance application for audio on the web. This paper is organized as follows: Section 2 gives a brief overview of requirements for the WAPS and analyzes characteristics of the audio data on the web. Section 3 details development of the WAPS system, i.e. its overall architecture and working of each component blocks. Section 5 gives out experimental results to objectively evaluate the WAPS system. Finally, Section 6 concludes this paper and gives directions for future work.

2 Task Description

As mentioned above, the task is to spot Chinese keyword from audio programs on the web and construct a surveillance system. The requirements and challenges faced by our systems are:

- **High processing ability:** The raw web data stream maybe as high as 700M bps or more, WAPS must effectively extract and process the audio data in the data stream.
- **Live feature of audio data**: All audio data are extracted online from the web and they must process lively.
- **Heterogeneity of the audio data**: The content of the audio data varies greatly. It may contain many non-speech audio, such as music and pop songs. In addition, the speech part alone may be of different languages. Since our goal is to do Chinese keyword spotting, these foreign languages may pose challenges on the precision.
- **High Precision**: Because the amount of audio on the web is vast, the users of WAPS system emphasize on the precision more than on the recall rate of the keywords. It is expect that the highest precision achievable may be around ---

First a set of real-world audio data is collected from web data streams. The following presents some statistics on these data. Human annotation is performed to classify all audio programs into three categories: audio programs which do not contain speech (Non-Speech), audio programs contains speech which is not Chinese (Non-Chinese) and audio programs of Chinese speech (Chinese). The distribution of these audio is shown in Table 1.

Table 1. Statistics of audio dataset collected from the web

Corpus	#of Programs	Duration(sec.)
Non-Speech	863(72.8%)	90,228
Non-Chinese	198(16.7%)	40,617
Chinese	125(10.5%)	19,929
All	1186(100%)	150,774

Totally 1186 programs are collected. We can easily find that 72.8% of the audio stream collected contains no speech at all, which are pure music or pop songs. 16.7% numbers of programs are foreign languages. Only about 10% of the programs contain Chinese. This analysis indicates that a proper audio pre-processing module is desirable to filter out these non-Chinese speech audio, which can cause great degradation of the precision.

3 Global System Description

The overall WAPs system can be represented as a pipeline of several processing blocks, as shown in Fig.1.

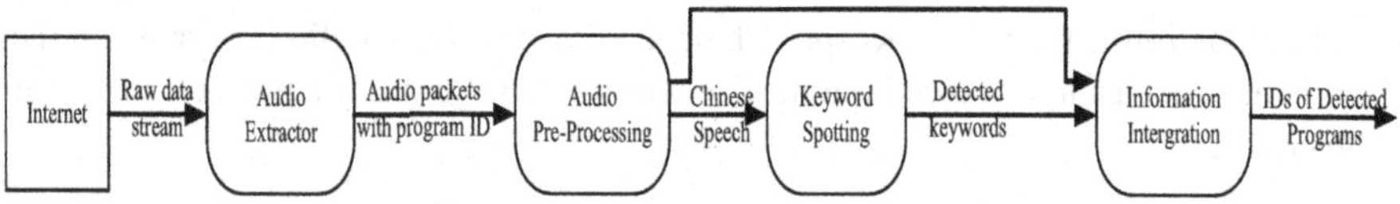

Fig. 1. System Architecture of the whole WAPS

At system operation, **Audio Streaming Extracting (ASE)** block searches for audio in the raw web data stream, converts into the linear PCM format for following processing and marks the data with a unique program ID, which indicates its original source URL. It then feeds the audio into the Audio Pre-Processing block in the format of audio packages of fixed duration (say, 3 seconds), together with its corresponding program ID. The **Audio Pre-Processing (APP)** block receives pure audio stream input from ASE. It first discriminates between speech and non-speech, then it further divides speech part into Chinese speech and non-Chinese speech. To make the discrimination procedures more reliable, longer audio packets are needed. Therefore, a packet cache strategy is used to concatenate audio packets of the same program to a longer packet. The **Keyword Spotting (KWS)** block plays the key role in the WAPS system. It detects the pre-defined Chinese keywords from the Chinese speech. Detected results will be output with a soft confidence score. The **Information Integration and Decision** block receives the detected keywords by the **KWS** block and use the confidence scores of keywords and information from the **APP** block to make the final decision. Finally, keywords decided as putative hits will be output with corresponding program IDs.

The overall system works in a pipeline operation mode. Each block first fulfills its own task and propagates the results to the next block. Next, a brief description of the related techniques used in main blocks is presented.

3.1 Audio Stream Extracting (ASE)

The block diagram of the Audio Stream Extracting (ASE)[2] module is shown in Fig.2. As illustrated in Fig.2, ASE block plays three roles: 1) it fetches pure audio stream or audio track of the video stream from raw web data flow. 2)Further more, the fetched audio are all converted to a uncompressed audio format (typically, 8k sample rate 16 bits encoded linear PCM). 3)For the convenience of online processing, the audio data is organized in packets of fixed length of 3 seconds so that the following operation can be performed without waiting for all the data of an audio program. Each packet is labeled with an identifier (ID), which indicates it original program source (source URL). An empirical observation for the raw web data stream of 700Mbps, an uncompressed audio stream of 7Mbps must be processed lively

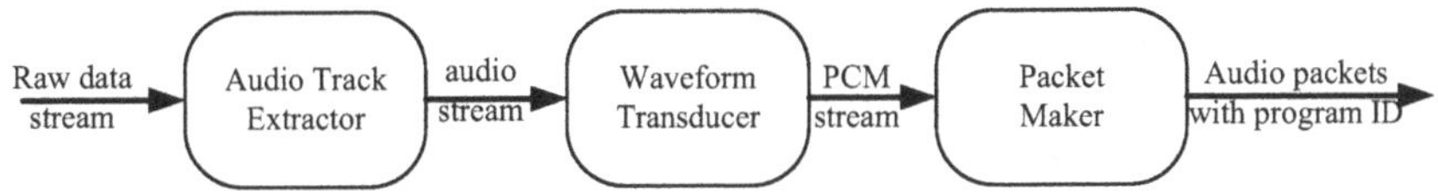

Fig. 2. Block diagram of the Audio Stream Extracting

3.2 Audio Pre-Processing (APP)

As shown in Section 2, the major function of the APP block is to filter out the non-speech audio and non-Chinese speech from the complex audio stream. It is further divided into three sub-modules: Audio Packet Cache, Audio Classification and Language Identification, as shown in Fig. 3.

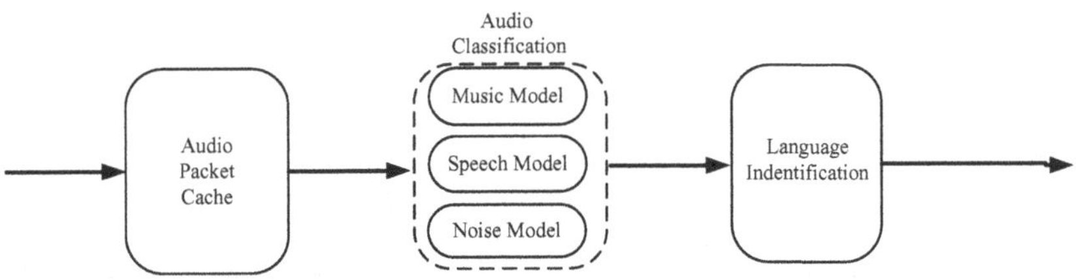

Fig. 3. Block diagram of the Audio Pre-Processing

3.2.1. Audio Packet Cache

Audio stream are passed in the form of audio packets of fixed length of 3 seconds. Classification on audio segment of such duration may no not an easier task. To make

---

[2] It is implemented by a third party as a cooperator of ThinkIT Speech lab, and only intergraded into WAPS as a component.

the discrimination procedures more reliable, longer audio packets are needed. Therefore, a packet cache strategy is used to concatenate audio packets of the same program to a longer packet. Considering unlimited amount of incoming audio packets and limited of the cache, it is expected only data of the *latest n* incoming programs is kept. Therefore, the cache is organized in a queue structure. The queue is also augmented with a lookup table of the ID of the program so that the incoming data packet can be easily concatenated with data with the same ID.

3.2.2. Audio Classification

For audio classification, three classes of audio are assumed, i.e. *Speech*, *Music*, *Noise*. A GMM classifier is used for the audio classification. A Gaussian mixture model (GMM) is a weighted sum of *M* component densities and is given as :

$$p(\vec{x}|\lambda) = \sum_{i=i}^{M} p_i b_i(\vec{x}) \tag{1}$$

where $\vec{x}$ is a d-dimensional random vector, $b_i(t)$ is the component density and p_i is the mixture weight. Each component density is a d-variate Gaussian function. Assume each audio packet is parameterized as a set of *T* observations:$[\vec{x}_1, \vec{x}_2 \cdots, \vec{x}_T]$ Each of three class is represented as a GMM, λ_k , $k \in \{1, 2, 3\}$ as shown in Fig 2.

The objective of the audio classification module is to find the model with the maximum *a posterior* for a given audio packet..

$$C^* = \arg \max_{1 \leq k \leq 3} Pr(\lambda_k|X) = \arg\max \frac{Pr(X|\lambda_k)Pr(\lambda_k)}{Pr(X)} \tag{2}$$

Note that $Pr(X)$ is the same for all audio classes; $Pr(\lambda_k)$ is the prior probability of class k, which can be set empirically; is the likelihood on *k*, which can be evaluated as following:

$$Pr(X|\lambda_k) = \quad {}_{t=1}^{T} p(\vec{x}_t|\lambda_k) \tag{3}$$

3.2.3. Language Identification

To filter out non-Chinese speech, our LID system adopts a state-of-the-art scheme for language identification, i.e. a GMM-supervector(GSV) system [7, 8]. This system is based on GMMs and means of GMMs are further classified by support vector machines (SVM).

As shown in Fig. 4, speech is first parameterized at the feature extraction module. Then a GMM-supervector is generated for the speech segment from a GMM called Universal Background Model (UBM). The means of components of the UBM are adapted using MAP-adaptation. And each audio segment is represented by means of all these Gaussian components. All mean vectors of all GMM mixture components are then concatenated to form the supervector. Finally, The SVMs are used to classify mean super-vectors using linear kernel. Please refer to [8] for details.

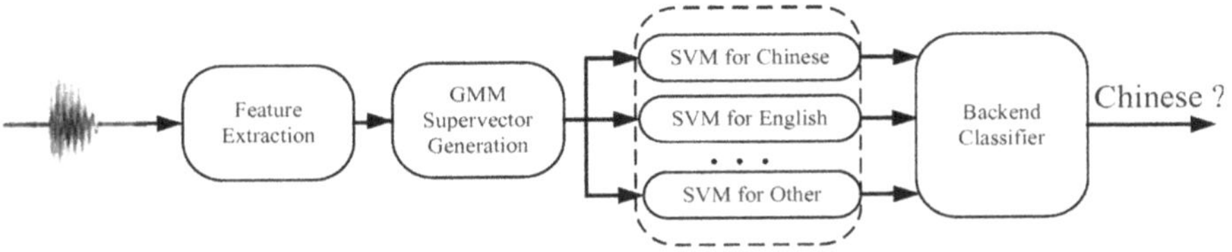

Fig. 4. Block diagram of Language Identification

3.3 Keyword Spotting (KWS)

The block diagram of the Keyword spotting (KWS) is shown in Fig. 5. It is an automatic speech recognition (ASR)- based keyword spotting system. It consists of several cascading modules. Firstly, the input of Chinese speech from APP goes through the ASR system. Lattices are generated by the ASR systems, which contains multiple hypothesis of the recognition results. Then confusion networks are generated from these lattices with the algorithm as described in [9]. Finally, a real-time keyword detector can be configured to spot the keywords online.

Our speech recognition recognizer is real-time recognizer using a statically expanded search space [10]. All knowledge sources except language models are statically integrated to form a memory-efficient state network. To compact the search networks dummy nodes are introduce to cluster the cross-word contexts and a so-called forward-backward merging is used to merge both the prefix and the suffix of the network. Many other techniques are also used to speed up the search procedure, such as word identity forwarding, LM look- ahead, LM cache and layer-dependent beam pruning.

The working of the real-time detector is briefly described as follows. Given a query $Q = q_1, q_2, q_3 \cdots q_N$ in its subword representation, where N is the number of the subwords and a confusion network of a utterance in a vector of its confusion sets $S = s_1, s_2, s_3, \cdots s_M$ The keyword spotting detection procedure amounts to transforming the query (keyword) into partial confusion network using basic operations including substitution, deletion and insertion in the viewpoint of a minimum edit distance (MED). Hence, the keyword spotting can be also be formulated as exploiting paths in the conception search space as shown in Fig. 6. A valid path is also given in the figure. A dynamic matching is finally implemented with a token passing scheme. Please refer to [11] for details.

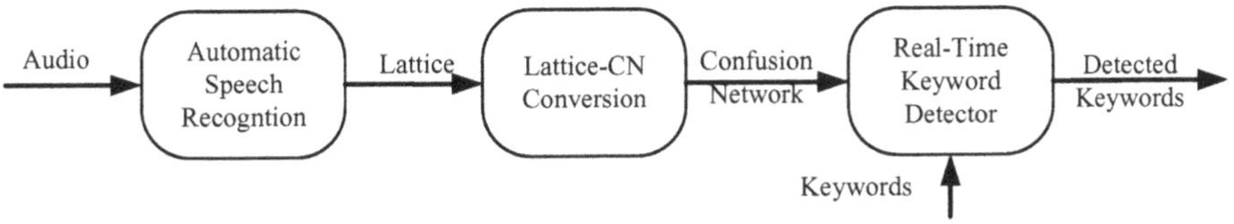

Fig. 5. Block diagram of keyword spotting [11]

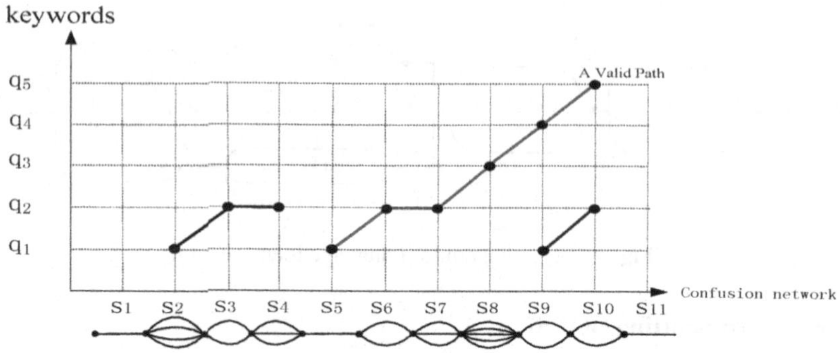

Fig. 6. Search space for keyword spotting [11]

3.4 Information Integration and Decision (IID)

The Information Integration and Decision (IID) collects all the possible information about the audio programs and uses it to make decision on the detected keywords. Currently, a simple scheme is used. First, it collects the classification results of packets of a particular program (specified by a ID *id*) from the APP module. Then the probability that this audio program is a Chinese program (not music or foreign speech) is estimated. The estimation is based on maximum likelihood estimation (MLE).

$$Pr_{chn}(id) = \frac{C_{chn}(id)}{C(id)} \tag{4}$$

where $C(id)$ and $C_{chn}(id)$ is the number of audio packets for program specified by *id* and these classified as Chinese speech by the APP. Finally, this estimated probability is integrated into the final decision of detected keywords by modifying the confidence score of a detected keyword $P(Q)$;

$$P(Q) = \begin{array}{ll} -\infty & \text{if } Pr_{chn}(id) \geq \tau \; and \, C(id) \geq C \\ P(Q) & \text{else} \end{array} \tag{5}$$

where $C$and $\tau$are empirically set constant. C assures that the number of packet is large enough so that $Pr_{chn}(id)$ is a relatively reliable estimation.

4 Experiments

4.1 Experimental Setup

The whole system is evaluated on three data set.

- **ASet**: Artificially created data sets to evaluate each module alone. It consists of three parts: Chinese telephony speech (*ChnSpeech*), foreign telephony speech (*ForSpeech*) and music and songs (*NonSpeech*). Details are shown in Table 2. The

ASet is carefully design to make assure that the proportion of each part in the audio is similar to that in Table 1.

- **WSet**: Real world data collected from the web in October, 2008. 250 hours in total, without any annotation. It is only used as a test set for the final WAPS system. An analysis of a subset of 40 hours is analyzed in Section 2.

Table 2. *Description of the Aset*

Corpus	Dur.(h)	Data Src& Style	Trans.
ChnSpeech	1.0	CTS data	Yes
ForChinese	3.2	CTS data	No
NonSpeech	7.5	Music& Songs	No

The system development experiments will be all performed on *ASet* and experiment on the final WAPS on *WSet* will be given for reference.

4.2 Evaluation of APP

4.2.1 Audio Packet Cache

The audio packet cache module itself only cache the audio packets. The evaluation measure used is its memory consumption. Table 3 shows how the memory consumption (Mem.) varies with the duration of cached packets (Dur.) when a fixed number of ID of 100 is tracked.

Table 3. Memory consumption of Audio Packet Cache

# of Packets	Dur.(sec)	# of ID tracked	Mem.(MB)
2	9	100	9.6
3	12	100	14.4
4	15	100	19.2
5	18	100	24.0

4.2.2 Audio Classification

Three GMMs are trained for speech, music and noise class respectively. However, what we are really interested in is whether an audio segment is speech or non-speech (music and noise). Therefore, the Audio Classification module is evaluated as binary classifier for Speech or Non-speech (music and noise) classes. And precision and Recall of each audio class is used.

The recall rate and the precision rate are shown in Fig. 7. It is notable that the audio classification in our case is a relative easy task that the recall and the precision rate for both classes can be higher than 95%. But still there are margins for improvements. Longer packet size tends to lending to better classification performance.

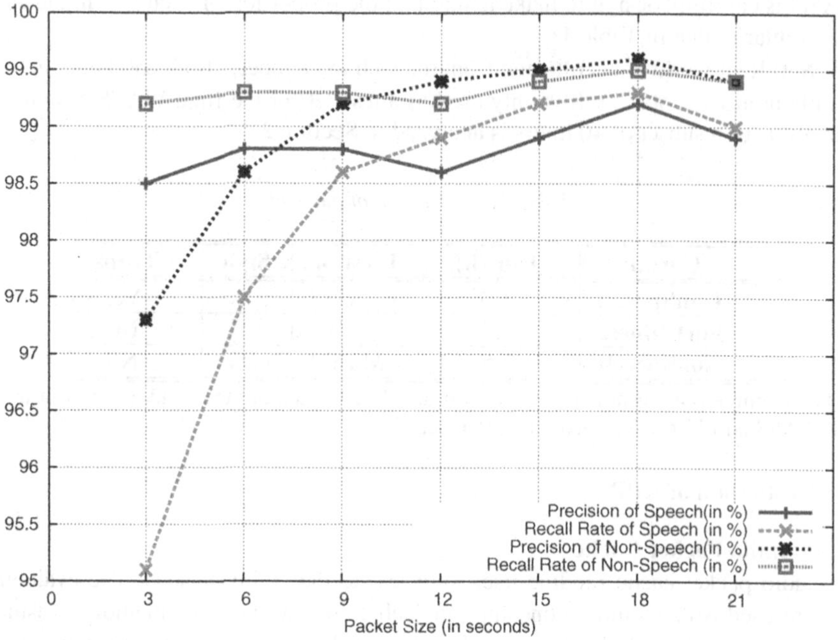

Fig. 7. Performance of the Audio Classification module with respect to the size of cached packet

4.2.3 LID

A GMM with 32 mixtures is trained for the UBM, and 19 languages are token into consideration. Although there are many metrics for evaluate the LID module. We simply evaluate the results as binary classification problem, i.e. whether the audio segment is a Chinese speech or non-Chinese speech. The LID is evaluated on the *ChnSpeech* and *ForSpeech* of *ASet*, excluding *NonSpeech*. And the results are shown in Fig. 11.

We can find that when no packet caching scheme is used (packet size =3s). The precision of Chinese speech is as low as 60% or less. We believe that this is caused by imbalanced amounts of Chinese speech and Non-Chinese speech. Because there are much more non-Chinese speech than Chinese speech. A few misclassification of Non-Chinese speech packets may cause great degradation in precision of Chinese Speech. We can also find how the packet cache module greatly improves the performance of LID module. Especially, the precision of Chinese part, the precision is improved from less than 60% to about 95%. This is similar to our observation in results of NIST LID evaluations [8].

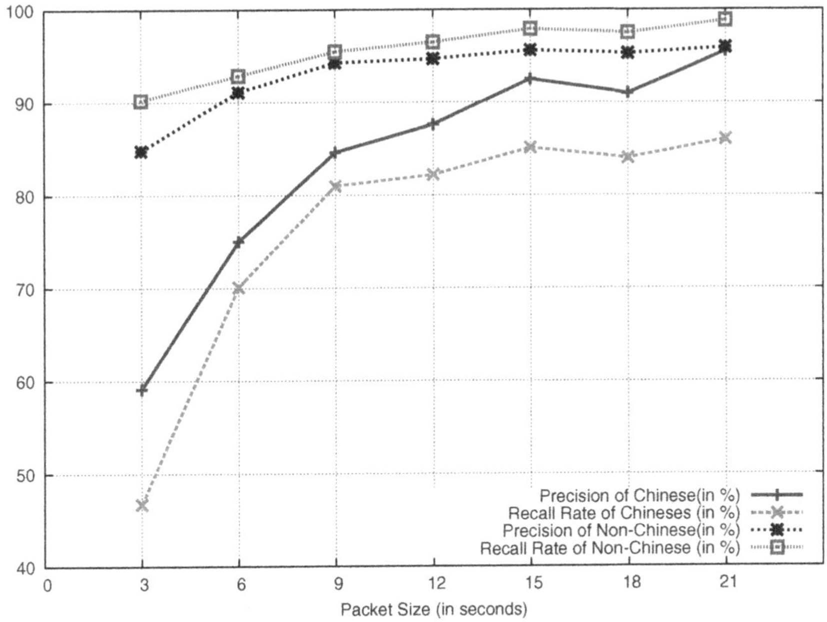

Fig. 8. Performance of the LID module with respect to the size of cached packet

4.2.4 Whole APP

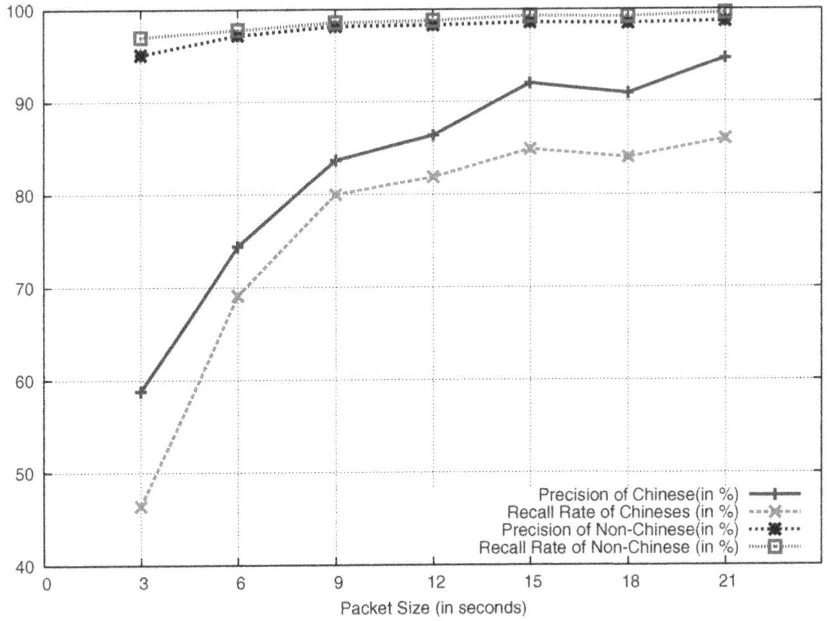

Fig. 9. Performance of the LID module with respect to the size cached packet

Finally, the performance of the whole APP module is shown in Fig. 9. We can see how the audio cache scheme is important and greatly improves the performance of whole APP module. Finally, a packet cache size of 15 seconds is chosen empirically and a precision of 93% is achieved for the Chinese speech part.

4.3 Evaluation of KWS

As described in Section 2.4, our keyword spotting system is an ASR based system. Because state-of-the-art ASR system is inherently computationally expensive, measures must be taken to make the ASR decoding fast enough to fulfill the requirement of live processing. First, a syllable recognition system is used instead of a word-based system. Second, a set of search beams for the Viterbi search [10] is tried for fast decoding. The performance of our KWS system under different real time factors (RT) is shown in Fig. 10. The evaluation is only performed on *ChnSpeech* of *ASet*. Finally, the setting of 0.14RT is used as balance between performance and speed.

4.4 Whole WAPS

Finally, the performance of the WAPS is evaluated on the whole *ASet*. The evaluation measure is the recall rate and precision of all keywords. First, the performance on the

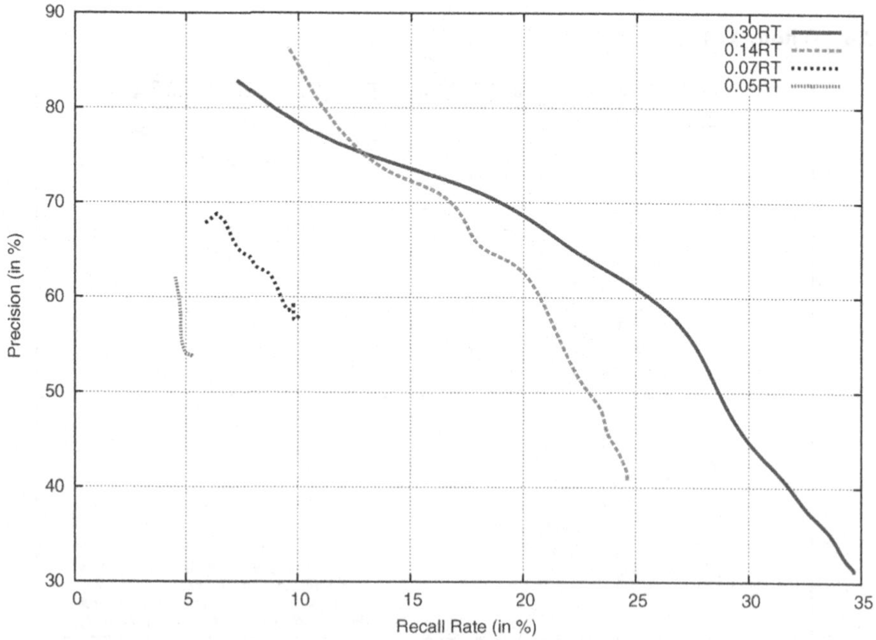

Fig. 10. Performance of the KWS module under different RT factors on *ChnSpeech*

ChnSpeech subset (*Pure Speech*) is first listed, which can be thought as a kind of upper limits of the performance on the whole *ASet*. When the APP is not used (*APP Not Used*), the best precision achievable is only about 40%. There is a great degradation compared to the performance on *ChnSpeech*. However, when the APP is adopted (*APP Used*), the performance is greatly improved to about 85%compared to the case of *APP Not Used*. It is very close on that on the pure speech subset (*Pure Speech*). When **IID** is incorporated, the performance is further improvement slightly.

Finally, a decision threshold corresponding to 80% of precision is used. Open set test is performed on the *WSet*. Because the transcription of *WSet* is unknown. We only check for the precision of detected keywords. A set of experiments show that the precision roughly varies between 80% and 85%, which is a satisfying result.

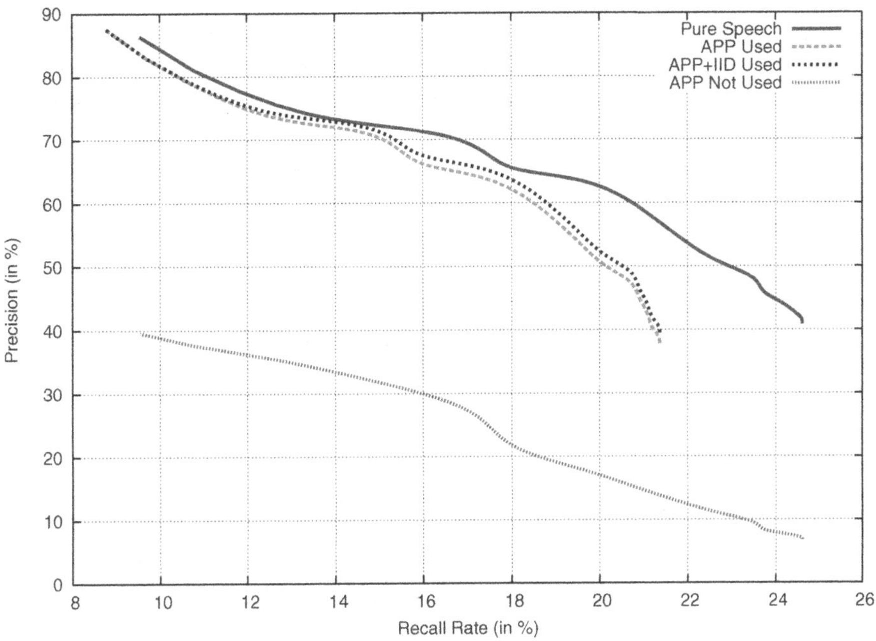

Fig. 11. Performance of Whole WAPs

5 Conclusion and Future Work

In this paper, we present our effort in constructing an audio program surveillance system WAPS. We first give out the challenges faced by WAPS by empirically analyzing the audio data on the web. Then the construction of WAPS system is described in details, illustrating how a few of state-of-art speech technologies are integrated effectively for the system development. Finally, the WAPS system is evaluated objectively.

If more meta data about the audio programs are available. The whole surveillance system may be further integrated with other mode. Moreover, the confidence annotation of the detected keyword merely relies on the acoustic information and the decoding process, therefore we will further tries high-level information in the final decision module. We also plan to do more human annotation on the real web data and further experiment with it.

Acknowledgement

The construction of the whole WAPS system is a demanding work, and only possible with collaboration of many people and many groups within ThinkIT speech lab. We would like to thank everyone in the ThinkIT speech lab who has helps us in the system development. This work is partially supported by The National High Technology Research and Development Program of China (863 program, 2006AA010102), National Science & Technology Pillar Program (2008BAI50B00), MOST (973 program, 2004CB318106), National Natural Science Foundation of China (10874203, 60875014, 60535030).

References

[1] Koumpis, K., Renals, S.: Content-based access to spoken audio. IEEE Signal Processing Magazine 22(5), 61–69 (2005)
[2] Chelba, C., Hazen, T., Saraclar, M.: Retrieval and browsing of spoken content. IEEE Signal Processing Magazine 25(3), 39–49 (2008)
[3] Manuel, J., Thong, V., Moreno, P., et al.: Speechbot: An Experimental Speech-based Search Engine for Multimedia Content on the Web. IEEE Trans. on Mutimedias 3(4), 88–96 (2002)
[4] http://www.everyzing.com/
[5] Christopher, A., Michiel, B., Ari, B., Ciprian, C., Anastassia, D.: An audio indexing system for election video material. In: ICASSP 2009, vol. 77(2), pp. 596–599 (2003)
[6] Rose, R.: Keyword detection in conversational speech utterances using hidden Markov model based continuous speech recognition. Computer speech & language (Print) 9(4), 309–333 (1995)
[7] Campbell, W., Sturim, D., Reynolds, D.: Support vector machines using GMM supervectors for speaker verification. IEEE Signal Processing Letters 13(5), 308–311 (2006)
[8] Suo, H., Li, M., Xiao, X., Zhang, X., Wang, X., Lv, P., Yan, Y.: IOA ThinkIT Speech Laboratory System Description for NIST LRE 2007. In: Workshop of NIST LRE 2007 (2007)
[9] Hhkkan-Tur, D., Riccardi, G.: A General Algorithm For Word Graph Matrix Decomposition. In: ICASSP 2003, vol. 77(2), pp. 596–599 (2003)
[10] Shao, J., Li, T., Zhang, Q., Zhao, Q., Yan, Y.: A One-Pass Real- Time Decoder Using Memory-Efficient State Network. IEICE Transactions on Information and Systems 91(3), 529 (2008)
[11] Gao, J., Shao, J., Zhang, Q., Zhao, Q., Yan, Y.: Spoken Term Detection Using Dynamic Match Subword Confusion Network. In: Fourth International Conference on Natural Computation. ICNC 2008, vol. 4 (2008)

Hot Topic Detection on BBS Using Aging Theory*

Donghui Zheng and Fang Li

UDS-SJTU Joint Research Lab for Language Technology
Dept. of Computer Science and Engineering
Shanghai Jiaotong University, Shanghai, China
zhdhui@gmail.com, fli@sjtu.edu.cn

Abstract. BBS(Bulletin Board Systems) is one of the most common places for threaded discussion. It becomes more and more popular among web users, especially in China. Everyday a huge amount of new discussions are generated on BBS. It is too difficult to find hot topics. To solve this issue, we propose a novel approach to detect hot topics on BBS for any period of time. Our solution consists of three steps. First of all, candidate topics are extracted using the clustering method. Secondly, based on the extracted topics, aging theory is employed to valuate the hotness of topics. Both two steps above are carried out incrementally over time. Finally, topics are ranked and hot topics are detected. Experiments performed on practical BBS data show that our method is quite effective.

Keywords: Hot topic detection, Aging theory, BBS.

1 Introduction

As ReadWriteWeb[1] indicates, there are nearly 2 hundred million of BBS users in early 2008 in China. The total number of daily page views across BBS has reached over 1.6 billion, with 10 million posts published every day. However, too many topics are discussed on BBS. How to identify hot topics on BBS becomes more and more important.

Here are some basic concepts for BBS:

Post: A post is a user submitted message. The first post starts the thread; this may be called the *entry*. Posts that follow in the thread are meant to continue the discussion about the *entry*, or respond to other replies. A post mainly contains four parts: the author, the title, the time-stamp and the content.

Thread: A thread is a collection of posts. Each post belongs to and only belongs to one thread. A thread can contain any number of posts, including multiple posts from the same member, even if they are one after the other.

Topic: A topic on BBS is formed with one or more threads.

* Supported by the National Science Foundation of China under Grant No.60873134.
[1] http://www.readwriteweb.com/

W. Liu et al. (Eds.): WISM 2009, LNCS 5854, pp. 129–138, 2009.

Nowadays, to get hot topics on BBS, the most popular method is to see Top-N topics based on the number of page views and replies each day, such as "Today's TOP HOT" in SMTH[2] BBS . However, this method may not be a suitable solution to user's concern due to its limits. The users usually want to have an outline of discussions to see "what's hot now" or "what's hot today".

Hot Topic Detection algorithm proposed in this paper is a good practice of topic detection and tracking [1](TDT) on BBS. We find some common features on BBS: Firstly, language used in BBS is more informal, colloquial and with many abbreviations. Even those experienced users could not know all these abbreviations. Secondly, it is very common using emotional signs like "!!!", "^_^", "re". Some posts include nothing but those signs. Thirdly, a single thread may reflect different topics. A survey thread like "how are you going on these days", may trigger other topics during the discussion. In addition, unlike traditional news reports, the information of users' participation is helpful to find hot topics on BBS.

The remainder of this paper is organized as follows: In Section 2, we discuss some related work on hot topic detection. In Section 3, we first define the hot topic, and introduce the aging theory. Then we describe our approach for hot topic detection on BBS using aging theory. In Section 4, we discuss the results of the experiments run. Finally, in section 5, we present our conclusions and some future research directions.

2 Related Work

Topic Detection and Tracking (TDT) has been researched for a very long time [1] [2] [3], however, former researches on hot topic detection are mainly based on those traditional, formal news reports [5] [6] [7]. If these methods are applied to BBS directly, it will surely lead to bad results [4]. With the rapid development of social networking recent years, researchers get to focus on topic detection towards social media such as BBS, BLOG, etc [4].

In this paper, we try to find hot topics on BBS, which shows a high threaded discussion level with strong interactivity and many informal terms. In 2005, Lan You elt [8] did a similar research to find hot topics on BBS. In his method, threads were first clustered into topics based on their lexical similarity. Then a BPNN(Back-Propagation Neural Network) based classification algorithm was used to judge the hotness of topics according to their popularity, quality as well as thread distribution over time. Although this method can find hot topics in a certain period of time, it would be better if we enhance the influence of hot topics at the observation point.

In addition, BBS is a social network, its interaction often impacts the generation of hot topic. Hot topic, is always accompanied by the abundance of users' participation. Researchers who study the social networking have already found some interesting results [9] [10]. There is a simple but effective way to analyze the relationship of posts and measure the importance of them: *out-degree* and

[2] http://www.newsmth.net/

in-degree. It's very helpful to improve the topic model [11] using this way. *out-degree* means how many posts this post replies to and *in-degree* means how many posts reply to this post.

3 Hot Topic Detection on BBS Using Aging Theory

Although daily posts involve a lot of topics, more than 60% of the posts just focus on a few topics. Therefore, it is very meaningful to find hot topics on BBS.

3.1 Definition of Hot Topic

To find out hot topics, we need to define the hot topics. There are four distinct characteristics of "hot topic":

Massive Posts: Only an attractive topic can assemble lots of users' discussion, which, in turn, becomes a prerequisite for a hot topic. This factor is comprised in our energy calculation process. Each post contains certain nutrition which can be transformed to the energy, therefore, topics with more posts could gain more energy.

High Quality Posts: Compared with those junk posts (like "I agree", "good"), a hot topic always has more posts of high quality. The relationship among posts could help us to identify which posts have high quality.

High Cohesion: Since scattered content has less attraction, the content of a hot topic is usually compact and centralized. We use the threshold of Single-Pass clustering to strictly control the number of threads to form the topic.

Bursting: For a hot topic, it often gathers a large number of posts in a short period of time, and then gets to a stable state until slowly disappear, which implies a life cycle of the topic.

3.2 Aging Theory

We regard the Rise and Fall of the topic as a life cycle. Chen [12] is the first person who models the topic using aging theory in 2003. He divides the life cycle into four stages: birth, growth, decay and death. To track life cycles of topics, aging theory uses the concept of energy function. The value of energy function shows how active a topic is. The energy of a topic increases when the topic becomes popular, and diminishes with the time. If we use the concept of nutrition, each post can be seen as a food to the topic, which contain certain nutrients, and these nutrients can be transformed to the energy value using energy function.

Time line can be equally divided into time slots(we take 3 hours as a time slot in our experiment),and then we employ three functions of aging theory to update the energy value of the topic at the end of each time slot.

getNutrition(): Calculate how much nutrition a topic can get from posts.

energyFunction(): This is a monotonically increasing function. It can transform the nutritional value into energy value. $energyFunction^{-1}()$ transforms the energy values into the nutritional value of the topic.

energyDecay(): In the end of each time slot, we perform the attenuation of energy to the topic. If the energy value declines below β, this topic will be moved to removed list, and not used for following steps. β is a decay factor, which can be obtained from the training data.

3.3 Hot Topic Detection on BBS

After given a observation point, we start to analyze the discussion data several time slots before the point. Our method will perform in three steps: (1) Candidate Topic Discovery. We use incremental Single-Pass clustering method to get the candidate topic list in each time slot. (2) Topic energy calculation. In this step, it will give each topic an energy value at the end of each time slot. Both two steps above are carried out incrementally.(3) Hot topic ranking. We rank the topics according to their energy value and then verify them whether or not to meet our definition of hot topic. We will finally get the list of hot topics for the observation point. What's more, according to its energy value curve, we can get a clear picture of each topic's rise and down.

Candidate Topic Discovery. We try to get the candidate topics in the time slot. We set the currently topic set as T, which initialized as $NULL$. Then we use the following method: determine which topic in topic set T is the most similar with the new thread d in current time slot. If their similarity exceeds the predefined $threshold_{sim}$, we merge the thread d into this topic and update the topic vector, otherwise this thread d is considered as a new topic t', and put t' into the topic set T. When all of the threads in current time slot have been dealt, we will get the topic set of the current time slot T. $threshold_{sim}$ needs to be identified in the training data.

Topic Energy Calculation. After candidate topics are discovered in the time slot, each topic should be given an energy valule.

First, we need to measure how much nutrition each post contains.High quality posts usually have more nutrition, which means it can offer more energy. The nutrition of a post is calculated by the Formula 1:

$$getNutrition(p) = z \cdot \frac{\lg(y+2)}{\lg(y+2)} \tag{1}$$

where
$getNutrition(p)$: indicates the nutrition the post p contains
z: denotes the content similarity between the post and the topic
y: denotes the reference number by others posts

Secondly, for all new generated posts in the time slot of the topic, calculate their nutrition and transform the nutritional value into the energy value, then add it to the cumulative energy value of the topic.

At the end of each time slot, we will get a cumulative energy value. It can be divided into two parts. One is the energy value of the last time slot. The other is energy variation in the current time slot. There are a 3-step calculation process:

1. Use $energyFunction^{-1}(e_t^{i-1})$ to convert the topic t 's energy value e_t^{i-1} of last time slot i into nutrition value n_{t-1}.
2. Get the nutrition from all the posts of the topic in the time slot, which noted as n_d (Formula 2), and add it to the legacy value n_{t-1} using α as nutrition transferred factor which is defined in the aging theory (Formula 3). α need be identified from the training data.

$$n_d = \sum_{p \in d}^{p.time \subseteq i} getNutrition(p) \tag{2}$$

$$n_t = n_{t-1} + \alpha \cdot n_d \tag{3}$$

3. Calculate the cumulative energy value at the end of current time slot, using the formula: $energyFunction(n_t)$. The specific form of $energyFunction()$ is defined as Formula 4, analogous to that used in [12]:

$$energyFunction(x) = \begin{cases} \dfrac{1}{1+x}, & x \geq 0, \qquad (4) \\ 0, & x < 0. \qquad (4') \end{cases}$$

Thirdly, perform energy decay on topics. Energy of topics increases with upcoming posts adding to topics, but also decays with the time passing by. The topic energy gets modified by a decay factor that represents the decay in each time slot. If in a certain time slot, the energy value which the topic obtains is less than that it decays, then the topic will show signs of recession. When the energy of a topic is below the predefined threshold, it is supposed to be in a "death" state. So we remove it from the set of survival topics T and add it to the set of removed topics T_{rem}. It will not used in the following steps.

Hot Topic Ranking. We get hot topic candidates by rank the topics according to their energy value. Then, according with the definition of the hot topic, it should have the feature of bursting. In our experiment, if a topic lasts for more than 24 hours, and its standard variation of the energy value is less than 0.05, this topic is considered as no bursting. These topics are most probably composed with very common threads titled of "how about..., how to get to ...", which are frequently appeared all day long. After filtering out the topic whose standard variation is less than 0.05, we get hot topics with highest energy value.

4 Experiment Analysis and Result

In order to evaluate the proposed method, we make experiments on the BBS data.

4.1 Corpus and Parameter Settings

At present, there is no public BBS corpus, so we create a crawler to get the daily posts in the board "NewExpress" on SMTH BBS which is one of most popular BBS in China. System notifications on BBS are ignored by our crawler. All the data were published from March 14,2009 to March 30,2009. There are up to 14,807 threads which totally including 74,506 posts. In average, there are 871 threads including 4,382.7 posts per day. All posts have been already preprocessed.

The corpus was divided into two data sets:

Training Set: Published from March 14 to March 21, which contains 6,903 threads, 34,036 posts, and 10 topics are manually labeled.

Test Set: Published from March 22 to March 30, which contains 7,904 threads, 40,470 posts.

4.2 Hot Topic Detection Experiment on BBS

Hot topics on BBS usually last for a short-term period and are more concentrated. BBS users concern more about hot topics at the point when logging on to BBS. Therefore, it is worth-while to find out accurate hot topics happened recently. To the best of our knowledge, there is no general standard for hot topic evaluation on BBS. Here we adopt "Top-5 Hot Topics" issued by original BBS board as a "*BaseLine*" to measure our results. The experiment is divided into two parts, respectively, hot topics detection on one day and on three days.

Before the experiment, all parameters are settled optimized according to the training data. After validation, it gets to the best result when $threshold_{sim}$ equals 0.2213. α, β are determined according to the method used in [13]: For each topic, a proportion r_1 of the total nutrition corresponds to a proportion s_1 of the total energy. α, β can be determined using two points(r_1,s_1), (r_2, s_2). At last, the final parameter values are the averages of α, β of 10 topics: α=0.118659, β=0.079687. Then the threshold for the standard variation of energy value V is set as 0.05. In addition, each time slot is represented as 3 hours and the observation point is the last minute of each day.

Hot Topic for One Day: Firstly, we perform our experiment to detect hot topics each day during March 22 to March 30. The experimental result is validated by "*Baseline*" of each day. For convenience, the result generated using our method will be noted as "*RUA*"(Result Using Aging theory).

Other symbols used in the result column are predefined as follows:

➠:Hot topic generated by *RUA* is almost equal to the *Baseline* at the same rank.

⇑ : Hot topic discovered by both *Baseline* and *RUA*, but this topic ranks higher in *RUA*.

⇓ : Hot topic discovered both by *Baseline* and *RUA*, but this topic ranks lower in *RUA*.

Table 1. Daily top-5 hot topics on BBS detected on March 26. Comparative results between the *RUA* and *Baseline* are shown at Result column "Re.".Topics are listed in the descending order of hotness.

March 26			
Topic	*RUA*	*Baseline*	Re.
1	①Be a Buddhist monk is not a job(13) ②Where are Buddhist nun from?(30) ③How hard to be vegetarians as Monks(15) ④Why are some Monks married?(31)	①What can you guess is the name of first Chinese Aircraft carrier?	**N**
2	①What can you guess is the name of first Chinese Aircraft carrier? ② Name the first Aircraft carrier ③Original concept of Aircraft carrier ④ China would build Aircraft carrier? ⑤The first Chinese Aircraft carrier should be named as 'Never supremacy'	① ShingTung Yau, a big lie?□	**M**
3	①Haizi's in a low status in poetry community ② 20th anniversary for Haizi ③20 years after Haizi Suicide	① Some middle school in Beijing,very bossy□	**M**
4	①Tianqiao Chen and Yuzhu Shi, who is the best	①Turned down for many times,But see my love today	⇑
5	①Turned down for many times, But see my love today	① ShingTung Yau, no logical?□	⇓

N(ew): Hot topic generated by *RUA*, but no corresponding topic appears in *Baseline*.
M(erge): Hot topic in *RUA* merges two or more topics in *Baseline*.
□ : Hot topic appears in *Baseline*, but not in *RUA*.

The result on March 26 is listed in Table 1.Thread titles are listed to represent certain topics. Energy variation for each topic is also shown in Figure 1. The rest result of other 8 day is summarized in the following analysis.

We find three interesting discoveries through this experiment. First of all, *RUA* has the basic coverage of hot topics which retrieved by *Baseline*. These marked with ➠, ⇑, ⇓, or **M** mean topics are both detected by *Baseline* and *RUA*, only in different ranks or different styles of organization. After manual analysis, among 45 topics retrieved in 9 days, there are 35 topics are labeled with marks above, accounting for 77.8% of all. Moreover, there are 5 in above 35 topics marked with '**M**', that means *RUA* can not only find the topic, but also be able to enhance the coherence of the topic by associated with related threads of the topic. This will show the outline of the topic for users. For instance, in result on March 26 showed in Table 1, the second topic detected by *RUA* not only found the hot topic "What, can you guess, is the name of first Chinese Aircraft carrier?", but also organized threads which related to this topic such as "The first Chinese Aircraft carrier should be named as Never supremacy!"etc.

Secondly, *RUA* is competent for detecting hot topics which can not obtained by *Baseline*. It's known that the discussion of a topic may be scattered in a

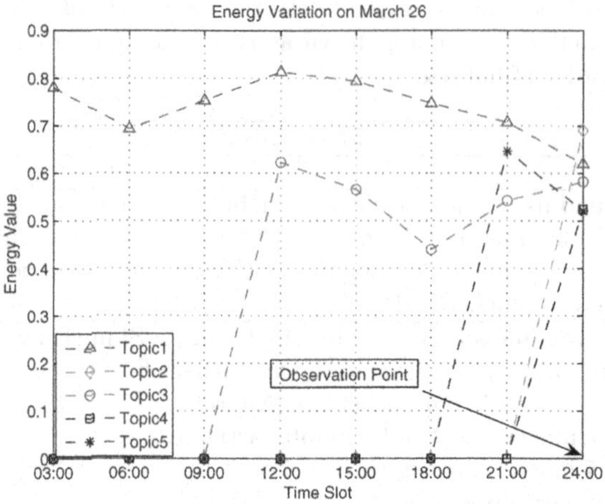

Fig. 1. Energy distribution of daily hot topics on BBS detected by *RUA* on 26 March

number of threads, if none of these threads gather abundant posts, the *Baseline* method can hardly find it. Instead, *RUA* can find these topics(which labeled '**N**' in Table 1) such as topic 1 in Table 1. Though the posts number of each thread is not too much (the posts number is shown after each thread title), accumulation of all these posts reaches up to a large number. From the corresponding energy variation in Figure 1, we know that this topic just starts to thrive in a short period of time. According to the definition of hotness, it is believed as a hot topic. There are 5 in 45 topics marked as '**N**', accounting for 22.2%. Except one topic, 4 other topics are believed as hot topics through manually validating.

Finally, *RUA* is capable of finding more time sensitive topics. Some topics labeled as '□', are detected in the *BaseLine*, but do not appear in *RUA*, such as topic 2 of *BaseLine* in Table 1. After looking into all the threads of this topic, we know that it was discussed a lot before 12:00 on the 26th, but received little concern afterwards. Therefor, it is not suitable to be a hot topic for our observation point at the last minute on the day. *RUA* ranks the energy at the observing point which ensures the timeliness of hot topics.

Hot Topic for Three Days. We also perform an experiment to detect hot topics for three days long during March 22 to March 30. The experimental result on the time period of March 28 to March 30 is shown in Table 2, and corresponding energy variation in Figure 2. The other result are not listed here due to length of the article. Through this experiment, we find that *RUA* can also effectively detect hot topics which last for a relatively long time period. For example, in Figure 2, topic 1 which discussed about holiday policy of Labor Day happened one and half a day ago, and followed with a period receiving little attention. Then it got to be ultimately culminated. Those topics last such long time are hardly detected by *BaseLine* method.

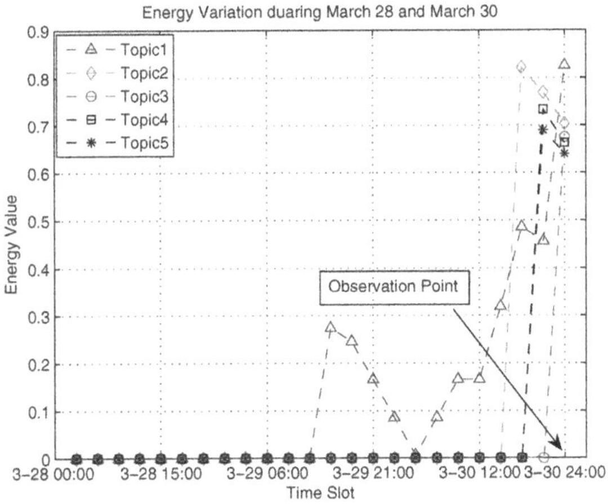

Fig. 2. Energy distribution of hot topics on BBS during 22 March to 30 March

Table 2. Hot topics on BBS of three days long during 28 March to 30 March.Topics listed in descending order of energy value.

March 28 to March 30	
Topic	Experimental Result
1	① NEWS:Guangdong,Cancel the Labor Day Plan!! ②Recover the Golden Week of Labor Day in Guangdong ③Guangdong Government is pig Brain... ④ Professor Cai's Holiday policy report
2	①Is it good to be a teacher at Tsinghua? ②Who can tell me how to get along with students? ③Professor Cai jumped from NaiKai univ.
3	①If civilian workers give up farming? ②Finally understand the reason of farmer's low payment
4	①Let's see CCTV8, LiZhu jump on Xiaoru Fang!
5	Who is a handsome scholar or a pretty girl worldwide?

5 Conclusion and Future Work

Based on aging theory, our method can find the hot topics in any period of time, such as hot topics of the day, hot topics of 3-days and so on. It can improve hot topics retrieval on BBS. Users can know what's going on clearly and quickly.

The main contributions of the paper are as follows. First, we analyze the characteristics of hot topic in BBS. Second, we propose a effective way to validate the post importance. We apply aging theory on BBS posts in order to get the hot topics. After comparison with the method in [8] which employs quite a few parameters to measure the hotness of the topic, we summarize our method: (1) With the help of energy calculation, many spam posts are avoided. (2) Large training data is not needed. It saves human labor. (3) The choice of the observation time point will have an impact on hot topics detected by our method.

Our next step will focus on how to combine BBS, Blog, even Twitter[3] to find hot topics.

References

1. Allan, J., Carbonell, J., Doddington, G., Yamron, J., Yang, Y.: Topic detection and tracking pilot study: Final report. In: Proc. of the DARPA Broadcast News Transcription and Understanding Workshop. OMG Press, Needham (1998)
2. Canhui, W., Min, Z., Liyun, R., Shaoping, M.: Automatic online news topic ranking using media focus and user attention based on aging theory. In: Proceeding of the 17th ACM conference on Information and knowledge management, pp. 1033–1042. ACM Press, New York (2008)
3. Yang, Y., Pierce, T., Carbonell, J.: A study of retrospective and online event detection. In: Proceedings of the 21st annual international ACM SIGIR conference on Research and development in information retrieval, pp. 28–36. ACM Press, New York (1998)
4. Zhu, M., Hu, W., Ou, W.: Topic Detection and Tracking for Threaded Discussion Communities. In: IEEE/WIC/ACM International Conference on Web Intelligence and Intelligent Agent Technology, pp. 77–83. IEEE Press, Washington (2008)
5. Huimin, Y., Wei, C., Guanzhong: Design and implementation of online hot topic discovery model. J. Wuhan University Journal of Natural Sciences 11(1), 21–26 (2006)
6. He, T., Qu, G., Li, S., Tu, X., Zhang, Y., Ren, H.: Semiautomatic Hot Event Detection. In: Xue, J., Osmar, L., Zhanhuai, R., Xi'an, L, eds. (2006)
7. Chen, K.Y., Luesukprasert, L., Chou, S.c.T.: Hot Topic Extraction Based on Timeline Analysis and Multidimensional Sentence Modeling. In: IEEE Transactions on Knowledge and Data Engineering, pp. 1016–1025. IEEE Press, Piscataway (2007)
8. Lan, Y., Yongping, D., Jiayin, G., Xuanjing, H., Lide, W.: BBS based Hot topic retrieval using backpropagation Neural Network. In: Su, K.-Y., Tsujii, J., Lee, J.-H., Kwong, O.Y. (eds.) IJCNLP 2004. LNCS (LNAI), vol. 3248, pp. 139–148. Springer, Heidelberg (2005)
9. Robert, D.N., Lina, Z.: Social Computing and Weighting to Identify Member Roles in Online Communitie. In: Proc. of the 2005 IEEE/WIC/ACM International Conference on Web Intelligence (2005)
10. ZhiLi, W., Chunhung, L.: Topic Detection in Online Discussion Using Nonnegative Matrix Factorization. In: Proc. of the 2007 IEEE/WIC/ACM International Conferences on Web Intelligence and Intelligent Agent Technology Workshops (2007)
11. Tuulos, V., Tirri, H.: Combining Topic Models and Social Networks for Chat Data Mining. In: Proc. of the 2004 Web intelligence International Conference, pp. 206–213. IEEE Press, Washington (2004)
12. Chen, C.C., Chen, Y.T., Sun, Y., Chen, M.C.: Life Cycle Modeling of News Events Using Aging Theory. In: Lavrač, N., Gamberger, D., Todorovski, L., Blockeel, H. (eds.) ECML 2003. LNCS (LNAI), vol. 2837, pp. 47–59. Springer, Heidelberg (2003)

[3] http://www.twitter.com/

A Heuristic Algorithm Based on Iteration for Semantic Telecommunications Service Discovery

Dong Cao, Xiaofeng Li, Xiuquan Qiao, and Luoming Meng

State Key Laboratory of Networking and Switching Technology
Beijing University of Posts and Telecommunications, Beijing, 100876 China
caodfangyc@yahoo.com.cn, {xfli,qiaoxq,lmmeng}@bupt.edu.cn

Abstract. How to discover the suited telecommunications service more quickly is important for improving the user experience. In the field of semantic web service most of discovery algorithms pay attention to the precision of service discovery, while few algorithms concentrate on the efficiency of service discovery. Therefore, this paper presents a heuristic algorithm based on iteration for semantic telecommunications services discovery, which can improve the efficiency of services discovery effectively. We also found the implied relevance during the matching process, which is not concerned by others. At last we have built a prototype and evaluated the performance of algorithm proposed by this paper. The experimental result shows the new algorithm can improve the efficiency of services discovery effectively and ensure the precision of web services discovery.

Keywords: Semantic telecommunications service, Heuristic algorithm, Iteration computation, matching space.

1 Introduction

The next generation telecommunications network will be a convergent network in future. It will provide a seamless communication environment for the user who does not need to consider the network where the service is from, namely, the service logic is separate with the service implementation in the semantic telecommunications network. The converged service will be derived from two or more different networks or service providers. So how to discover the qualified service more quickly among these networks or providers will be important for user experience.

In the field of semantic web service, the service discovery includes the precision and efficiency of discovery process. Most of existing researches focus on the precision of semantic web service discovery while there are few discussions to aim at the efficiency of the discovery. Also, the implied relevance during the matching process is not concerned by others. We found that if the matching degree of the previous subspace is lower then the matching degree of the next subspace is lower too.

Therefore, we have drawn lessons from the heuristic search and introduced the idea to the algorithm based on iteration of semantic web service discovery. The paper presents a heuristic algorithm for semantic web services discovery, which can

W. Liu et al. (Eds.): WISM 2009, LNCS 5854, pp. 139–148, 2009.

improve the efficiency of services discovery effectively on the condition that the algorithm does not reduce the precision.

The article is as follows: Section 2 introduces the related research on semantic web service discovery in brief. Section 3 defines some concepts in the paper. Section 4 explains the algorithm to service discovery in detail. Section 5 verifies the performance of the proposal algorithm. Section 6 is the summary for the full paper.

2 Related Research

In recent years, there are many researches on semantic web service discovery. Paolucci, M. et al. [1], [2] compared functional attributes of the semantic web service and partitioned the matching degree into four layers from high to low: Exact, Plug in, Subsumes and Fail, to provide the foundation to measuring the match degree. Brogi, A. et al. [3] proposed a new technique for the discovery of web services that account for the need of composing several services to satisfy a client query. The proposed algorithm makes use of OWL-S ontologies, and returns the sequence of atomic process invocations that the client must perform in order to achieve the desired result. Wei, D.P. et al. [4] proposed a new method to enhance the semantic description of semantic web service by using the semantic constraints of service I/O concepts in specific context. Ma, Q. et al. [5] presented to define QoS data into service descriptions and adopted the ontology reasoning to change previous syntactic matchmaking into a semantic way. Ma, J.G. et al. [6] used a novel approach to partition a large set of search results into a set of smaller groups by employing a clustering approach. Also, there are many other researches for semantic web service discovery [7], [8], [9], [10], [11], [12].

From the above, we found in the field of semantic web service that: (1) most researches focuses on the precision of discovery but efficiency; (2) there is few research which consider the implied relevance during the matching process, which is if the matching degree of the previous subspace is lower, the matching degree of the next subspace is lower too. There is an implied relation between the previous subspace and next one. That is the reason why this paper proposes the heuristic algorithm based on iteration.

3 Definition of Concept

Semantic web service discovery is a process of searching proximal services published by service providers against the request service.

Definition 1. Matching Space is a set which consists of services published by service providers.

$MS=(AS_1 \quad AS_2 \quad \cdots \quad AS_n)$, where: AS_i is the advertisement service i . Further, AS_i is defined as: $AS_i=(SCS_i \quad FAS_i \quad NFAS_i)^T$, where: SCS_i is the service category attribute of AS_i ; FAS_i is the functional attribute of AS_i ; $NFAS_i$ is the nonfunctional attribute of AS_i , so,

$$MS = \begin{pmatrix} SCS_1 & \cdots & SCS_n \\ FAS_1 & \ddots & FAS_n \\ NFAS_1 & \cdots & NFAS_n \end{pmatrix} = [SCS \quad FAS \quad NFAS] \ . \tag{1}$$

From the analysis, matching space can be divided into three subspaces: SCS, FAS and NFAS. The algorithm proposed by the paper is based on these subspaces.

Definition 2. Semantic Similarity of Ontology (SSO) is used to measure the similarity degree between two ontologies. The paper transforms the inherent logic of domain ontology repository into an ontology tree, and each node corresponds to a concept in the domain ontology repository, which is as shown in Figure1.

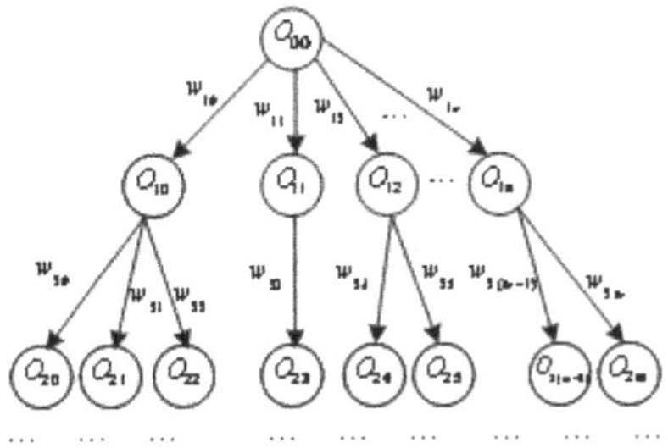

Fig. 1. The tree structure of domain ontology repository

Each child node has different importance to the father node, so the paper improves the measure method [13] to the semantic similarity and assigns weight to edges of tree, to make the measure of semantic similarity more exact.

If the ontology $o_2 \subset o_1$, then the concept o_1 is regarded as the superordinate concept of o_2. The assembly of all superordinate concepts for the ontology o is defined as the superordinate concept set, labeled as OS. $W_R(o_{ij})$ is labeled as the relative weight between the node o_{ij} and root node o_{00}. Therefore, the semantic similarity $sim(o_{ij}, o_{st}) \in [0,1]$ between the concept o_{ij} and o_{st} is as follows.

$$sim(o_{ij}, o_{st}) = \begin{cases} 1 & O_{ij}S \equiv O_{st}S; \\ \dfrac{|o_{st}s \cup o_{00}s| W_R(o_{ij})}{|o_{ij}s \cup o_{00}s| W_R(o_{st})} & \phi \subset O_{ij}S \cap O_{st}S \subset U, |o_{st}s \cup o_{00}s| W_R(o_{ij}) < |o_{ij}s \cup o_{00}s| W_R(o_{st}); \\ \dfrac{|o_{ij}s \cup o_{00}s| W_R(o_{st})}{|o_{st}s \cup o_{00}s| W_R(o_{ij})} & \phi \subset O_{ij}S \cap O_{st}S \subset U, |o_{ij}s \cup o_{00}s| W_R(o_{st}) < |o_{st}s \cup o_{00}s| W_R(o_{ij}); \\ 0 & O_{ij}S \cap O_{st}S = \phi \end{cases} \ . \tag{2}$$

Definition 3. Heuristic Function is an iteration function used to select the qualified service which will be matched in the next matching subspace.

$$f_{AS_i}(t) = \frac{\mathrm{sgn}\left[\prod_{k=0}^{2} match_{(k)}(RS, AS_i) - t\right] + 1}{2} \ . \tag{3}$$

Where: $k = \{0,1,2\}$ respectively corresponds to three subspaces. $match_{(k)}(RS, AS_i)$ is defined as the matching degree between the request service and advertisement service. $t \in [0,1]$ is called the matching threshold.

The above function shows that we consider not only the matching value of current subspace, but also the matching result of the former subspace. If the matching degree of the former subspace is low, the degree of the next one is also low and the probability of service which becomes the final qualified service is little, namely, the former matching result has an influence on the whole matching process.

4 Heuristic Algorithm Based on Iteration for Service Discovery

4.1 Introduction of the Prototype System

We have built the semantic telecommunications service system which consists of five parts: semantic web service registration center, agent, service provider, and service portal as well as ontology repository. The system is shown as Figure 2.

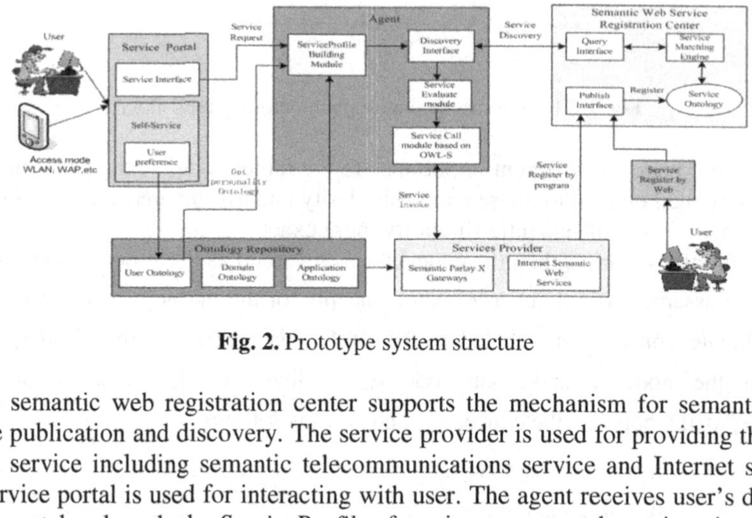

Fig. 2. Prototype system structure

The semantic web registration center supports the mechanism for semantic web service publication and discovery. The service provider is used for providing the converged service including semantic telecommunications service and Internet service. The service portal is used for interacting with user. The agent receives user's demand via the portal and sends the *ServiceProfile* of service request to the registration center. Using the discovery algorithm proposed, the registration center automatically selects matched services, which are returned to the agent in the form of list. According to the *ServiceProfile* of returned services, the agent invokes and performs the service published by service providers to respond the user demand.

The algorithm proposed by this paper runs on the service matching engine. To import the heuristic information into the discovery process, we divided the matching space into three subspaces in matching order: service category subspace, functional attribute subspace and nonfunctional attribute subspace. Three subspaces are basic features of semantic telecommunications service, so we divided the matching space into three ones. Before matching service in the next subspace, candidate services are selected to match in the next subspace according to the heuristic information, which can reduce the unnecessary computation for discovery and improve the efficiency for matching.

4.2 Matching in the Service Category Subspace

The matching in this subspace is a process to find the advertisement service AS_i whose service category attribute is matched to the corresponding attribute of request service. $SCS = (SCS_1 \quad SCS_2 \quad \cdots \quad SCS_n)$ is the aggregation of different service categories. Because there are inherent logical relations among these service categories, these relations can be expressed in the form of tree called service category tree. Each service category SCS_i corresponds to a node in the tree. Thus, the matter on the matching in this subspace is transformed to the search problem in the tree.

Because the service category attribute SC_{RS} and SCS_i are ontology type, the matching degree for service category equals to the semantic similarity between SC_{RS} and SCS_i. We get the matrix M_{oss} for semantic similarity by the Depth-First Search.

$$\begin{aligned} M_{SCsim} &= \left(sim(SC_{RS}, SCS_1) \quad sim(SC_{RS}, SCS_2) \quad \cdots \quad sim(SC_{RS}, SCS_n)\right) \\ \Leftrightarrow M_{SCmatch} &= \left(match_{SCS}(RS, AS_1) \quad match_{SCS}(RS, AS_2) \quad \cdots \quad match_{SCS}(RS, AS_n)\right) \end{aligned} . \tag{4}$$

Importing the heuristic function $f_{AS_i}(t)$, we can get the matching degree matrix $M'_{SCmatch}$ with heuristic information, which is used to select the candidate service matching in the next subspace. The advertisement service AS_i corresponding to non-zero element in matrix $M'_{SCmatch}$ is the candidate service to be matched in the next subspace.

$$M'_{SCmatch} = f_{AS_i}(t) \cdot M_{SCmatch} = \frac{\operatorname{sgn}\left[match_{(0)}(RS, AS_i) - t\right] + 1}{2} \cdot M_{SCmatch} . \tag{5}$$

4.3 Matching in the Functional Attribute Subspace

Matching in the functional attribute subspace is a process to find AS_i whose functional attribute is matched to the corresponding attribute of request service. The advertisement service AS_i is from candidate services in service category subspace.

The functional attribute consists of input $FASI$ and output $FASO$, where, $FASO = (FASO_1 \quad FASO_2 \cdots \quad FASO_n)$ shows the output of advertisement service in functional attribute subspace. $FASI = (FASI_1 \quad FASI_2 \quad \cdots \quad FASI_n)$ shows the input of advertisement service. The functional attribute FA_{RS} of request service is labeled as $FA_{RS} = (FAO_{RS} \quad FAI_{RS})$. So the matching for functional attribute can be divided into input matching and output matching.

During the output matching, outputs $FASO_i$ and FAO_{RS} of service usually have many parameters. The matching degree $match_{FAO}(FASO_i, FAO_{RS})$ between $FASO_i$ and FAO_{RS} is defined as the geometric mean of maximum semantic similarity between each element in FAO_{RS} and corresponding element in $FASO_i$. The matching degree $match_{FAO}(FASO_i, FAO_{RS})$ is labeled by matrix M_{FAsim}. The maximum of row i in the matrix M_{FAsim} is labeled as $sim_{\max}(output_{iRS}, output_j)$, so

$$match_{FAO}(FASO_i, FAO_{RS}) = \left(\prod_{i=1}^{n} sim_{\max}(output_{iRS}, output_j) \right)^{\frac{1}{n}} . \tag{6}$$

Therefore, we get the matching degree between FA_{RS} and FAS_i, also get the matching degree matrix between FA_{RS} and functional attribute is M_{FAS} as follows.

$$match(FAS_i, FA_{RS}) = \left(match_{FAI}(FASI_i, FAI_{RS}) \cdot match_{FAO}(FASO_i, FAO_{RS}) \right)^{\frac{1}{2}} \tag{7}$$

$$M_{FAS} \Leftrightarrow M_{FASmatch}\left(match_{FAS}(AS_1, RS) \quad match_{FAS}(AS_2, RS) \cdots \quad match_{FAS}(AS_n, RS) \right) . \tag{8}$$

Importing the heuristic functional $f_{AS_i}(t)$, we can get the matching degree matrix $M'_{FASmatch}$ with heuristic information. The advertisement service AS_i corresponding to nonzero element in matrix $M'_{FASmatch}$ will be the candidate service to match in the next subspace.

$$M'_{FASmatch} = f_{AS_i}(t) \cdot M_{FASmatch} = \frac{\operatorname{sgn}\left[match_{(0)}(AS_i, RS) \cdot match_{(1)}(AS_i, RS) - t \right] + 1}{2} \cdot M_{FASmatch} . \tag{9}$$

4.4 Matching in the Non-functional Attribute Subspace

Matching in nonfunctional attribute subspace occupies an important position in the process for discovering service. For example, network type, terminal type, charging way and QoS are all belong to the nonfunctional attribute of telecommunications service. These attributes are benefit to help users to select more suitable service.

The advertisement service AS_i in the nonfunctional attribute subspace is from the candidate service of the functional attribute subspace. In the paper, the matching degree between NFA_{RS} and $NFAS_i$ is defined as the geometric mean of semantic similarity between each element in NFA_{RS} and the corresponding element in $NFAS_i$. For example, $NFAS_i = (NT_i \quad TT_i \quad CI_i \quad QoS_i)^T$ respectively indicates network type, terminal type, charging way and QoS, which are expressed by ontology. The matching degree $match_{NFAS_i}$ between NFA_{RS} and $NFAS_i$ is as follows.

$$match(NFA_{RS}, NFAS_i) = \left(sim(NT_{iRS}, NT_i) \cdot sim(TT_{iRS}, TT_i) \cdot sim(CI_{iRS}, CI_i) \cdot sim(QoS_{iRS}, QoS_i) \right)^{\frac{1}{4}} . \tag{10}$$

Importing the heuristic functional $f_{AS_i}(t)$, we can get the matching degree matrix $M'_{NFASmatch}$ with heuristic information

$$M'_{NFASmatch} = f_{AS_i}(t) \cdot M_{NFASmatch} = \frac{\operatorname{sgn}\left[match_{(0)}(RS, AS_i) \cdot match_{(1)}(RS, AS_i) \cdot match_{(2)}(RS, AS_i) - t\right] + 1}{2} \cdot M_{NFASmatch} \quad (11)$$

The advertisement service AS_i corresponding to nonzero element in matrix $M'_{NFASmatch}$, which is regarded as matching results, will be returned to service requester to finish the whole process of service discovery. The heuristic function is important to the service discovery. By adjusting the threshold t of heuristic functional $f_{AS_i}(t)$, we remove unqualified services and reduce the redundant computation for matching. Meanwhile, the algorithm sufficiently has considered the essential factor, such as service category, functional attribute and nonfunctional attribute, and ensures the precision ratio for semantic web service discovery.

5 Estimate of Algorithm Performance

On the runtime environment including Windows operating system, 3.4GHz CPU and 3GB Memory, the semantic web service registration center was running and we have selected 100 semantic web services enrolled in the registration center, which are divided into some types. We have randomly selected 20, 40, 60, 80, 100 services respectively to build the matching space and test the algorithm performance.

Figure 3(a) shows the contrast of service discovery time between the threshold t=0 and t=0.5. When t=0, it means no heuristic information and is equivalent to most of current algorithms which have not heuristic information. Under the same experimental conditions, the efficiency of service discovery can be approximatively represented by the service discovery time.

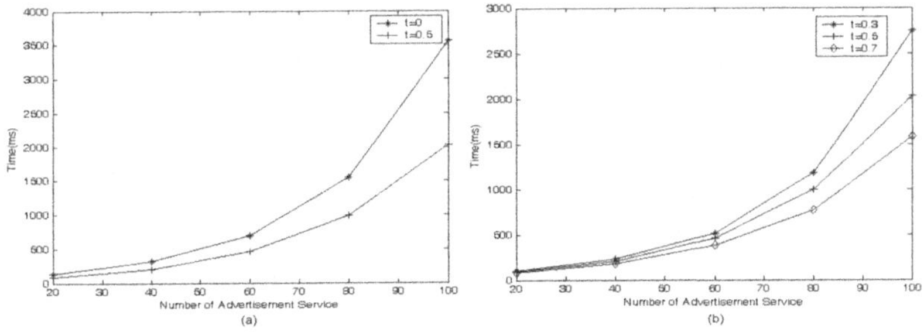

Fig. 3. The contrast of service discovery time

The abscissa of figure 3(a) shows the quantity of advertisement service participating in service matching and the ordinate of figure 3(a) shows the service discovery time. The time is defined as the time difference from starting to process the service request to returning the search result. On the whole, the time for service discovery after importing the heuristic information is shorter than the time without importing the heuristic information. The reason is that by importing the heuristic information, we remove the unqualified service, reduce the redundant computation for service matching and improve the efficiency. Figure 3 (b) shows the service discovery time with different thresholds, t=0.3, t=0.5 and t=0.7, to verity the relationship between the efficiency of service discovery and the threshold t.

From the figure 3 (b), we find the time for service discovery decreases with the increase of threshold t. Because the demand for matching degree is higher when the threshold t increases, the qualified service which be matched in the next subspace is fewer and the efficiency for service matching will be improve.

Table 1 shows the contract for the precision of service discovery between the algorithm proposed by this paper and methods presented by [14] and [15].

Table 1. Comparison for the precision of service discovery

Methods	SAS in [14]	KS in [14]	Ref. [15]	t=0.7 in this paper	t=0.5 in this paper
Precision	86.5%	76%	78%	83.3%	80.7%

To verify the precision of algorithm proposed by this paper, we compare the algorithm with algorithms in [14] and [15]. When the threshold value $\theta = 0.56$ in [14], the precision of one method called SAS (Semantic association Search) is 86.5% and the other called KS (Keyword Search) is 76%. The precision of the method in [15] is 78%. Meanwhile, the precision proposed by this paper is 83.3% and 80.7% respectively when threshold t=0.7 and t=0.5.

The KS method in [14] uses the traditional keyword search, so the KS method presented by [14] is beneath with the one proposed by us. The SAS method in [14] adopted semantic association search, especially focus on QoS-constrained expressions, so [14] has a little higher precision than this paper. The method proposed by [15] does not consider the non-functional attribute, such as QoS, so [15] has lower precision than this paper.

From the above, we found if the heuristic information is introduced to the process of matching, the time of matching reduce effectively. It is described clearly by the figure 3(b). Meanwhile, we contrasted five different cases about the precision of services discovery. Obviously, discovery algorithms supported by semantic is superior to those without semantic, such as the traditional keyword search. Algorithms with semantic which consider more factors or features are better than those which concern less factors. So the algorithm with heuristic information which considers more factors proposed by this paper can improve the efficiency of service discovery and ensure the precision of service discovery.

6 Conclusion

The paper shows the efficiency of service discovery is important for improving the user experience in semantic telecommunications service. But fewer existing algorithms concentrate on the efficiency of services discovery. We also found the former matching degree has an influence on the next one. So the heuristic algorithm based on iteration for service discovery introduces matching space, heuristic information and iterative computation into the discovery to improve the efficiency of service discovery. We divide the matching space into three subspaces in order: service category subspace, functional attribute subspace and nonfunctional attribute subspace. The heuristic information is imported into the matching result of the subspace and candidate services are gotten, which will be matched in the next subspace. At last, we build the prototype and verify the algorithm proposed by the paper and experimental results show the new algorithm can improve the efficiency of services discovery effectively and do not reduce the precision of web services discovery.

Acknowledgements. This work is supported by National Natural Science Foundation of China (No. 60802034, No.60672122), Specialized Research Fund for the Doctoral Program of Higher Education (No.20070013026) and Beijing Nova Program (No.2008B50).

References

1. Paolucci, M., Kawamura, T., Payne, T.R., Sycara, K.: Semantic matching of web services capabilities. In: Horrocks, I., Hendler, J. (eds.) ISWC 2002. LNCS, vol. 2342, pp. 333–347. Springer, Heidelberg (2002)
2. Srinivasan, N., Paolucci, M., Sycara, K.: Semantic Web Service Discovery in the OWL-S IDE. In: 39th Hawaii International Conference on System Sciences, p. 109.2. IEEE Press, New York (2006)
3. Brogi, A., Corfini, S., Popescu, R.: Semantics-based composition-oriented discovery of Web services. ACM Transactions on Internet Technology 8(4), Article No. 19 (2008)
4. Wei, D.P., Wang, T., Wang, J., Chen, Y.D.: Extracting Semantic Constraint from Description Text for Semantic Web Service Discovery. In: Sheth, A.P., Staab, S., Dean, M., Paolucci, M., Maynard, D., Finin, T., Thirunarayan, K. (eds.) ISWC 2008. LNCS, vol. 5318, pp. 146–161. Springer, Heidelberg (2008)
5. Ma, Q., Wang, H., Li, Y., Xie, G.T., Liu, F.: A Semantic QoS-Aware Discovery Framework for Web Services. In: The 2008 IEEE International Conference on Web Services, pp. 129–136. IEEE Press, New York (2008)
6. Ma, J.G., Zhang, Y.C., He, J.: Web Services Discovery Based on Latent Semantic Approach. In: The 2008 IEEE International Conference on Web Services, pp. 740–747. IEEE Press, New York (2008)
7. Skoutas, D., Sacharidis, D., Kantere, V., Sellis, T.: Efficient Semantic Web Service Discovery in Centralized and P2P Environments. In: Sheth, A.P., Staab, S., Dean, M., Paolucci, M., Maynard, D., Finin, T., Thirunarayan, K. (eds.) ISWC 2008. LNCS, vol. 5318, pp. 583–598. Springer, Heidelberg (2008)
8. Kovács, L., Micsik, A., Pallinger, P.: Handling User Preferences and Added Value in Discovery of Semantic Web Services. In: The 2007 IEEE International Conference on Web Services, pp. 225–232. IEEE Press, New York (2007)
9. Lin, L., Arpinar, I.B.: Discovery of Semantic Relations between Web Services. In: The 2006 IEEE International Conference on Web Services, pp. 357–364. IEEE Press, New York (2006)
10. Stollberg, M., Hepp, M., Hoffmann, J.: A Caching Mechanism for Semantic Web Service Discovery. In: ISWC/ASWC, pp. 480–493 (2007)
11. Kourtesis, D., Paraskakis, I.: Combining SAWSDL, OWL-DL and UDDI for Semantically Enhanced Web Service Discovery. In: Bechhofer, S., Hauswirth, M., Hoffmann, J., Koubarakis, M. (eds.) ESWC 2008. LNCS, vol. 5021, pp. 614–628. Springer, Heidelberg (2008)

12. Rao, J.H., Dimitrov, D., Hofmann, P., Sadeh, N.: A Mixed Initiative Approach to Semantic Web Service Discovery and Composition: SAP's Guided Procedures Framework. In: The 2006 IEEE International Conference on Web Services, pp. 401–410. IEEE Press, New York (2006)
13. Xia, H., Li, Z.Z., Chen, Y.P.: Research of Semantic Web Service Matching Based on Concept Lattice. Journal of Beijing University of Posts and Telecommunications 29(Sup.), 185–188 (2006)
14. Li, A.H., Du, X.Y., Tian, X.: Towards Semantic Web Services Discovery with QoS Support Using Specific Ontologies. In: 3rd International Conference on Semantics, Knowledge and Grid, pp. 358–361. IEEE Press, New York (2007)
15. Nayak, R., Lee, B.: Web Service Discovery with additional Semantics and Clustering. In: The IEEE/WIC/ACM International Conference on Web Intelligence, pp. 555–558. IEEE Press, New York (2007)

An Improved Storage and Inference Method for Ontology Based Remote Sensing Interpretation System

Xiaoguang Jia[1], Zhengwei Lin[1], and Ning Huang[2]

[1] Dept. of Computer Science
[2] Dept. of Reliability Engineering
Beihang University, Beijing, China (100191)
xiaoguangjia@gmail.com

Abstract. As the incredibly expanding volumes of remote sensing archives, and the number of objects necessary to identify in remote sensing pictures increasing, the conventional process of remote sensing interpretation is becoming more and more inefficient, and it seems impossible to finish all interpretation tasks in time. This paper applies ontology techniques to this process. It uses ontology to describe the domain knowledge, and brings forward a new hybrid ontology mapping model and an effective inference method without using any ontology inference engine. Finally it constructs an interpretation system, using which the work efficiency can be improved greatly.

1 Introduction

The techniques of remote sensing have developed greatly these years. During the traditional Special Target Remote Sensing Interpretation Process (STRSIP), experts use their experience and knowledge to obtain the final interpretation results. As the incredibly increasing volumes of remote sensing archives, this process is becoming more and more inefficient. So, obtaining and sharing the domain and expert knowledge by ontology techniques is meaningful among all the staffs, since it is promising to alleviate workload and accelerate the present interpretation process.

On the drive of semantic web and knowledge engineering, ontology technique is becoming more and more widely used, because it can formally describe conceptual models of one domain, and it realizes the share of the domain knowledge. This paper uses ontology to describe and share the conceptual models of the domain of remote sensing interpretation, and constructs the Large scale Special target remote sensing interpretation Ontology (LSO). However, as the work of applying ontology in the domain of remote sensing interpretation is in the initial stage, there are still various problems, such as the efficiency of large-scale ontology storage and inference. Conventional ontology mapping model and ontology inference engine are not competent for the storage and inference of LSO. In this paper, a new hybrid ontology mapping method called OOM/ORM is created to generate the storage model of the LSO. Instead of using any ontology inference engine, eight groups of rules are extracted according to the semantics of the OWL Lite language to reason implicit knowledge of the LSO. Finally an experiment is carried out, verifying that the method presented in this paper is a good solution to the efficiency problem of ontology storage and inference.

W. Liu et al. (Eds.): WISM 2009, LNCS 5854, pp. 149–157, 2009.

2 Related Work

Several foreign research institutions have been doing the research on the application of the domain knowledge in dealing with remote sensing information. Since the year of 1980, European Space Agency in Germany has carried out a series of projects[16,17,18], among which the KES and KEO projects use ontology to describe and store the domain knowledge, and obtain a good effect. As the expanding of remote sensing information, Yves Voirin[19], a researcher of Remote Sensing Application Centre in Canada, uses Jess, which is based on CLIPS, as ontology inference engine, and employs the language supported by Jess as the description language of ontology.

However, there are still many problems during the process of applying ontology to the domain of remote sensing interpretation. The most prominent one is the efficiency of large-scale ontology storage and inference. In order to resolve these problems, people conduct many researches and acquire some known research results, such as LAS[2], IS[3], DLDB[4], Sesame[5] and so on. These systems use RDBMS as the base storage mechanism to store and manage the large scale ontology, and use ontology inference engines to reason implicit knowledge. In allusion to OWL Lite[7] ontology, Hstar[6] designs a storage model based on File System. This model summarizes eighteen semantic rules of OWL Lite, and then applies these rules to the ontology successively and independently when the ontology is loaded in order to reason implicit knowledge. Implicit and obviously defined knowledge will be stored in a tree structure together, which improves the speed of accessing ontology at the premise of not reducing the reasoning capability of ontology.

However, these systems all have deficiencies when dealing with LSO: (1) Hstar is applicable to ontology of small scale. When dealing with large-scale ontology, the efficiency of Hstar is not satisfying. (2) There is a bottleneck between access speed and reasoning capability of ontology. These systems improve the performance of one side at the cost of the other one. However, during the application of remote sensing interpretation, both sides are important.

This paper brings forward a new storage and reasoning method of ontology, which increases the speed of accessing ontology as well as ontology reasoning capability is not affected. Finally we construct a system, which realizes methods we have presented.

3 The Description and Storage of Remote Sensing Interpretation Knowledge

3.1 Ontology Description Language

OWL, recommend by w3c, is a description language specification of semantic web ontology. It contains three sublanguages [7]: OWL Lite, OWL DL and OWL FULL. It has strong description capability, and a universal application. However, it is not suitable for ontology reasoning. Paper [8] details OWL language components which influence ontology reasoning ability greatly but rarely used, and creates a sublanguage of OWL Lite called OWL Lite- by limiting the use of these components. Accordingly it ensures the reasoning process is completed in polynomial time as well as the language's description ability is not weakened. This paper uses restricted ontology language to describe LSO.

3.2 Ontology Mapping Method

This paper uses RDBMS as the bottom storage mechanism of ontology. In order to use advantages of RDBMS in data storage, service management and accessing control, the mapping mode between ontology and relational database has to be determined.

In allusion to the characteristic of LSO, and comparing with the five conventional mapping methods[4, 9], this paper presents an improved hybrid ontology mapping method called OOM/ORM. We use this method to create storage structure of LSO. The basic idea of this method is that, ontology is mapped to POJO (Plain Old Java Object) [10, 12] object model at first, and then the object model is mapped to relational model using the technique of ORM (Object Relational Mapping)[10]. This method is called OOM/ORM for short.

OOM mode realizes the mapping from ontology to POJO object model. It contains four sub-modes: OOM accessing mode, OOM element mode, OOM reasoning mode and OOM storage mode.

In OOM accessing mode, ontology classes and object properties are mapped to POJO classes, data-type properties are mapped to elements of POJO classes, and inclusion relations among ontology classes and object properties are mapped to inheritances of POJO classes.

Based on OOM accessing mode, OOM element mode contains the relation between ontology class and object property, and property characteristics, which are symmetric property, transitive property and inverse property.

OOM reasoning mode removes inclusion relations among ontology classes, and consequently ontology reasoning can be conducted efficiently.

OOM storage mode removes all the inclusion relations kept in OOM accessing mode in order to provide more flexibility for managers.

POJO object needs to be mapped to relational database so as to ontology mapping is completed finally. ORM is a technique in order to resolve the mismatch between object and relational database [10]. ORM mode realizes the conversion from POJO object model to relational model, and it has two sub-modes: ORM element mode and ORM storage mode.

In ORM element mode, OOM element mode is mapped to relational database. Ontology metadata is stored in the database in order for the use of ontology query and reasoning.

ORM storage mode is used to create table structure corresponding to object model. Table 1 shows relations of each element in ORM storage mode mapping.

Table 1. This table shows the relations of each element in ORM storage mode mapping

Ontology	POJO	Database
class	class	table
single-value property	property	line
multi-value property	property	table/FK
object property	property	table
inclusion relation	inherence	FK

4 Ontology Reasoning Method

The implicit knowledge in this paper refers to inclusion relations among ontology classes and object properties, and also TBox[11] and ABox[11] knowledge implicitly contained in inverse property, symmetric property and transitive property. TBox describes knowledge in terms of controlled vocabularies, for example, a set of classes and properties. ABox describes knowledge associated with instances of ontology classes[11].

Instead of using any ontology inference engine, eight groups of rules are extracted according to the semantics of OWL Lite and OWL Lite- to reason implicit TBox and ABox knowledge of the LSO.

The zero group of rules declares the base of reasoning. The implicit knowledge reasoning is based on obviously defined knowledge in ontology.

The first group to the fourth group of rules are used to reason TBox knowledge, and the other three groups to reason ABox knowledge.

4.1 Reasoning Rules of TBox Knowledge

The first group declares that inclusion relations of ontology classes have transitive property. The second group declares that inclusion relations of object properties have transitive property. The third group declares that, suppose P and Q are two distinct object properties, if P is inverse of Q, then Q is inverse of P. The fourth group declares that one subclass can inherit properties from its parent.

4.2 Reasoning Rules of ABox Knowledge

The fifth group of rules declares that instances of one subclass are also its parent's. The sixth group declares that, suppose P and Q are two distinct object properties, if two individuals have P property, and Q is parent of P, then these two individuals have Q property. The seventh group has many rules, which are used to reason implicit knowledge contained in inverse property, symmetric property and transitive property separately.

4.3 Ontology Reasoning Process

This section explains how to use TBox and ABox reasoning rules to reason implicit knowledge.

Ontology TBox reasoning is to reason implicit TBox knowledge. The reasoning process is as follows. Firstly we use the zero group of rules to obtain the base of our reasoning, and then the first to the third group are executed successfully. The fourth group of rules is used in the process of data loading rather directly in the reasoning process.

Ontology ABox reasoning is to reason implicit ABox knowledge. Data-type property can be ignored because it doesn't participate in the reasoning process. At this point, ontology can be regarded as a directed graph, in which ontology classes and object properties are nodes, and inclusion relations and inverse relations of object properties are edges. If there are no inverse properties in ontology, the directed graph is with no loop. If there is, we can unite nodes representing object properties having inverse relation, so as to we can get a new directed graph with no loop.

Based on topological ordering, this paper presents ABox knowledge reasoning process. At first we need to transform the ontology to a directed graph, and topological ordering is conducted on the transformed ontology to obtain a topological ordering sequence. Then we take out every element successively from the sequence to reason implicit ABox knowledge using rules of the fifth to seventh groups.

5 Remote Sensing Interpretation System

This paper has completed an interpretation system, which realizes ontology storage and inference methods we present.

Our system uses layered structure. They are Presentation Layer, Logic Layer, Core Layer, Data Access Layer and Data Layer from the top down. Presentation Layer contains many clients which can be used to operate the system. Logic Layer is composed of a group of services published in the form of Serlvet, which can be called by clients in Presentation Layer. We realize the storage and inference method presented in this paper in Core Layer. Data Access Layer contains several tools to access system data. Data Layer, which contains an ontology base and a database, is the data source of our system.

6 Efficiency Analysis

In this section, we first introduce criterion this paper uses to evaluate ontology reasoning capability and access speed. And then the process and results of experiment is presented. Finally we use the criterion to evaluate the efficiency of our system.

6.1 Brief Introduction to LUBM

LUBM[13] is a virtual criterion brought forward by Lehigh University of America to evaluate the capability of large scale OWL ontology Knowledge Base System (KBS). LUBM uses a university ontology named Univ-Bench described by OWL Lite language as the evaluation base.

According to the characteristics of university ontology, LUBM designs 14 base queries to evaluate the system's capability from 5 aspects, which are loading time, space complexity, accuracy rate and full rate, response time and general indicator.

Accuracy rate and full rate of base queries are evaluated by F- value. Its calculation method is shown in Formula (1), where F_q denotes F value of query q, C_q is full rate, S_q is accuracy rate, and β is the weight of S_q relative to C_q.

$$F_q = \frac{(\beta^2+1)*C_q*S_q}{\beta^2*C_q+S_q}. \tag{1}$$

The P- value is defined in LUBM to evaluate capability of the query. Its calculation method is shown in Formula (2), where T_q is the response time of query q, N is the data scale related to query q, and a, b are parameters.

Based on F- value and P- value, a general criterion called CM- value is defined to generally evaluate capability of ontology reasoning and access speed. Its calculation method is shown in Formula (3), where M is the number of base queries, w_q is the weight of query q, and α is the weight of F_q relative to P_q.

$$P_q = \frac{1}{1+e^{a*T_q/N-b}} . \tag{2}$$

$$CM = \sum_{q=1}^{M} w_q \frac{(\alpha^2+1)*P_q*F_q}{\alpha^2*P_q+F_q} . \tag{3}$$

6.2 Process and Results of the Experiment

This paper conducts a comparative experiment between two systems, Hawk1.5[14] and our system. Hawk1.5 system is evolved from DLDB.

The experiment rewrites the 14 base queries using ontology query language offered by Hawk1.5, then loads the university ontology of different scale, and finally executes the 14 base queries.

Similarly to Hawk1.5, we rewrite the 14 base queries using HQL/SQL[10] language in our system, and then construct the mapping from the university ontology to RDBMS using OOM/ORM method. Next we load university ontology of different scale, conduct ontology reasoning, and finally execute the 14 base queries.

Part of the results is shown in Table 3 and Table 4.

Table 3. The querying results of the two systems under ontology scale LUBM (1, 0). We can get the exact results from paper [13].

	LUBM(1,0)		
	Hawk	Our System	Exact Result
#1	4	4	4
#2	0	0	0
#3	6	6	6
#4	34	34	34
#5	719	719	719
#6	7790	7790	7790
#7	67	58	67
#8	7790	7790	7790
#9	208	208	208
#10	4	4	4
#11	0	224	224
#12	0	15	15
#13	1	1	1
#14	5916	5916	5916

6.3 Analysis of Reasoning Capability

Because accuracy rate is as important as full rate, β is assigned with value 1 when calculating F- value. We calculate accuracy rate and full rate of Hawk1.5 and our system depending on data shown in Table 3, and then calculate F- value of each query.

This paper uses a weighted average F- value based on F- value as our general criterion to evaluate ontology reasoning capability. From the point of view of remote sensing interpretation, all the base queries are the same important, so the weighted average F- value can be expressed as the average value of the 14 queries' F- values.

After calculating, the weighted average F- value of Hawk1.5 is 85.71, and our system's is 99.49. Apparently, the reasoning capability of our system is better than Hawk1.5 under the criterion of LUBM.

Table 4. The response time (ms) of each system under different ontology scales. We can see from this table that the response time of our system is much shorter than Hawk's.

	LUBM(1,0)		LUBM(15,0)	
	Hawk	Ours	Hawk	Ours
#1	1051.6	57.8	24771.8	231.2
#2	1135.9	65.6	21962.7	1264.1
#3	1070.4	26.5	15631.2	132.7
#4	229.6	68.9	1770.2	482.9
#5	754.7	60.9	10867.1	167.2
#6	639.1	278.2	20153.4	7746.8
#7	623.4	28.1	2921.8	314
#8	1257.7	544	2428.3	2670.5
#9	1273.6	275.1	22603	22521.9
#10	28.1	0	123.5	165.6
#11	65.6	15.5	359.3	196.8
#12	39.2	1282.8	48.5	848.4
#13	310.7	26.5	5614.2	660.9
#14	440.9	68.9	11431.1	4621.9

6.4 Analysis of Query Speed

When calculating P- value, we assign the number of university ontology to N, 0.001 to parameter a, and 0 to b. According to data in Table 4, we can calculate the P- value of every base query. Similar to the weighted average F- value, we define the weighted average P- value to evaluate query speed of one system. The result is shown in Table 5.

We can see from the table that the weighted average P- value of our system is much larger than that of Hawk1.5. Consequently, the query efficiency of our system is better than Hawk1.5 under the criterion of LUBM.

Table 5. The weighted average P- value of two systems under different ontology scales

	The weighted average P- value	
	Hawk1.5	Ours
LUBM (1,0)	0.7051	0.9060
LUBM (5,0)	0.7229	0.9219
LUBM (10,0)	0.7259	0.9236
LUBM (15,0)	0.7032	0.9087
LUBM (20,0)	0.6724	0.9001
LUBM (25,0)		0.9942
LUBM (30,0)		0.9011

6.5 General Evaluation

It is considered that query speed is as important as reasoning capability in the application of remote sensing interpretation, so when calculating the general criterion CM-value, value 1 is assign to parameter α.

After calculating, the CM- value of Hawk1.5 is 0.67, and our system's is 0.94. So our system is better than Hawk1.5 in terms of general criterion.

From the results and analysis we can see that our system has better ontology reasoning capability and query response speed.

7 Conclusions and Future Work

This paper restricts the usage of some time consuming but unnecessary parts of OWL Lite language when constructing the LSO, which makes the implicit knowledge inference of the LSO to be a polynomial –time complexity. Then a new hybrid ontology mapping method called OOM/ORM is created to complete the mapping from the LSO to RDBMS. Instead of using any ontology inference engine, eight groups of rules are extracted according to the semantics of OWL Lite to reason implicit TBox and ABox knowledge of the LSO, followed by the process of ontology reasoning. Then an interpretation system is constructed based on the storage and inference methods presented above to help experts complete the interpretation tasks. Finally an experiment is conducted on the system, the result of which verifies that our system has advantages in ontology reasoning and query speed.

Our future work is to make a more intensive study of methods we have presented in this paper, improve them in order to resolve other problems existing in the domain of ontology storage and reasoning. Consequently the efficiency of the remote sensing interpretation work can be improved more.

References

1. Xian-Chuan, Y.: Remote Sensing Image Interpretation. Electric Industry Press (2003)
2. Chen, C., Haarslev, V., Wang, J.: LAS: Extending racer by a large abox store. In: Proc. of the Int. Description Logic Workshop (DL 2005), pp. 41–50 (2005)

3. Horrocks, I., Li, L., Turi, D., Bechhofer, S.: The Instance Store: DL reasoning with large numbers of individuals. In: Proc. of the Int. Description Logic Workshop (DL 2004), pp. 31–40 (2004)
4. Pan, Z., Heflin, J.: DLDB: Extending Relational Databases to Support Semantic Web Queries. In: Fensel, D., Sycara, K., Mylopoulos, J. (eds.) ISWC 2003. LNCS, vol. 2870, pp. 109–113. Springer, Heidelberg (2003)
5. Broekstra, J., Kampman, A.: Sesame: A Generic Architecture for Storing and Querying RDF and RDF Schema. In: Horrocks, I., Hendler, J. (eds.) ISWC 2002. LNCS, vol. 2342, pp. 54–68. Springer, Heidelberg (2002)
6. Chen, Y., Ou, J., Jiang, Y., et al.: HStar-a semantic repository for large scale OWL Documents. In: Mizoguchi, R., Shi, Z.-Z., Giunchiglia, F. (eds.) ASWC 2006. LNCS, vol. 4185, pp. 415–428. Springer, Heidelberg (2006)
7. McGuinness, D.L., van Harmelen, F.: OWL Web Ontology Language Overview, http://www.w3.org
8. de Bruijn, J., Polleres, A., Fensel, D.: OWL Lite-: WSML Working Draft, http://www.wsmo.org
9. Wan, T., Hong-Yan, Y.: Design of OWL Ontology Storage Schema in Relational Database. Computer Technology and Development 17(2), 111–114 (2007)
10. Hibernate, http://www.hibernate.org
11. Baader, F., McGuinness, D.L., Nardi, D., et al.: The Description Logic Handbook: Theory, implementation, and applications. Cambridge University Press, Cambridge (2003)
12. Richardson, C.: POJOs in Action. Manning Publications Co. (2005)
13. Guo, Y., Pan, Z., Heflin, J.: LUBM: A Benchmark for OWL Knowledge Base Systems. Journal of Web Semantics 3(2-3), 158–182 (2005)
14. Hawk1.5, http://swat.cse.lehigh.edu/projects/index.html
15. Zheng-Wei, L.: The Research of Ontology Application in Military Target Remote Sensing Image Interpretation System. Beihang University, Beijing (2007)
16. Durbha, S.S., King, R.L.: Semantics-Enabled Framework for Knowledge Discovery From Earth Observation Data Archives. IEEE Transactions on Geoscience and Remote Sensing 43(11), 2563–2572 (2005)
17. Schröder, M., Rehrauer, H., Seidel, K., et al.: Interactive Learning and Probabilistic Retrieval in Remote Sensing Image Archives. IEEE Tran. on Geoscience and Remote Sensing 38(5), 2288–2298 (2003)
18. Colapicchioni, A.: KES-KIMV Final Presentation, http://earth.esa.int
19. Voirin, Y., et al.: A Forest Map Updating Expert System based on the Integration of Low Level Image Analysis and Photointerpretation Techniques. In: Geoscience and Remote Sensing Symposium (IGARSS 2002), vol. 3, pp. 1618–1620 (2002)

SASL: A Semantic Annotation System for Literature*

Pingpeng Yuan, Guoyin Wang, Qin Zhang, and Hai Jin

Service Computing Technology and System Lab
Cluster and Grid Computing Lab
School of Computer Science and Technology
Huazhong University of Science and Technology, Wuhan, 430074, China
hjin@hust.edu.cn

Abstract. Due to ambiguity, search engines for scientific literatures may not return right search results. One efficient solution to the problems is to automatically annotate literatures and attach the semantic information to them. Generally, semantic annotation requires identifying entities before attaching semantic information to them. However, due to abbreviation and other reasons, it is very difficult to identify entities correctly. The paper presents a *Semantic Annotation System for Literature* (SASL), which utilizes Wikipedia as knowledge base to annotate literatures. SASL mainly attaches semantic to terminology, academic institutions, conferences, and journals etc. Many of them are usually abbreviations, which induces ambiguity. Here, SASL uses regular expressions to extract the mapping between full name of entities and their abbreviation. Since full names of several entities may map to a single abbreviation, SASL introduces Hidden Markov Model to implement name disambiguation. Finally, the paper presents the experimental results, which confirm SASL a good performance.

Keywords: Semantic Annotation, Name Disambiguation.

1 Introduction

Currently, many search engines for scientific literatures, such as Google Scholar, DBLP, CiteSeer, the ACM Portal, and the IEEE Digital Library are available. Although these search services allow users to retrieve detail information of publications, due to ambiguity, they may not search the right publications. There are several reasons for ambiguity. One reason is abbreviations. Different from news etc, named entities in scientific literatures which are used to describe entities are generally abbreviations. One abbreviation may have multiple full names. For example, the full name of CAS may be: Central Authentication Service, Channel Associated Signaling, Chemical Abstracts Service, Code Access Security, Computer Arts Society, Cycle Accurate Simulator, and Computer Algebra System. When users want to search Central Authentication Service using keyword CAS, the search engines may return many wrong literatures. Ambiguity also occurs when authors, organizations who use different name variations or when there are multiple entities with the same name.

* This work is funded by National 973 Basic Research Program of China under grant No.2003CB317003.

W. Liu et al. (Eds.): WISM 2009, LNCS 5854, pp. 158–166, 2009.

Although ambiguity induces incorrect search results, in fact the problems can be solved if computer can understand the data as man do. How to make the data computer understandable is the work of semantic annotation. The semantic annotation offered here is a specific metadata generation and usage schema targeted to enable new information access methods and extend existing ones [1, 2]. The metadata provides both class and instance information about the entities referred in the literatures. In other words, semantic annotation is considered as assigning to the entities in the text links to their semantic descriptions [4].

This paper presents SASL, a *Semantic Annotation System for Literature.* SASL utilizes the semantic metadata from Wikipedia as knowledge base to annotate scientific literatures. So it guarantees knowledge base correct and timely. Because Wikipedia does not provide the mapping relationship between full names and theirs abbreviations, SASL uses regular expressions and hidden Markov model to determine the exact mapping relationships. SASL is integrated with SemreX [12], a literature sharing system we developed and demonstrates good performance and scalability while yielding good quality results.

The remainder of this paper is organized as follows. In section 2 the related work is presented. Section 3 describes the architecture of SASL in detail. SASL is consisted of following components: knowledge base, conversion tool, named entity identification, and name disambiguation. Section 3 describes some approaches adopted in those components. The experiments and the evaluation of the performance of SASL are presented in section 4. Finally, conclusions are given in section 5.

2 Related Work

Generally, according to approaches adopted by semantic annotation systems, semantic annotation systems can be classified as manual, semi-automatic, and automatic annotation [1]. Manual annotation requires users to manually annotate a document content using a predefined ontology. An example of the annotation tools is the OntoMat-Annotizer tool [5]. An apparent drawback of manual annotation is that it is prone to errors due to that annotators are unfamiliar with a domain. In addition, manual annotation requires much time and effort. In order to relieve annotators' burdens, semi-automatic annotation systems identify concept instances from text and then relate them to corresponding concepts. However, due to ambiguous terms, human intervention is required to clarify them. An example of this type of annotation is SemTag [6]. Although Reeve and Han claim that there do not exist automated annotation tools [1], some researchers try to develop the tools, for example C-PANKOW [7].

Reeve and Han also classify annotation systems based on the type of annotation method used, which are pattern-based, machine learning, and multi-strategy-based [1]. Pattern-based annotation finds patterns for a defined initial set of entities. The process repeats until no more entities discovered or the user stops the process. This process can also use manual rules to find entities in text. Machine learning-based annotation uses two methods: probability and induction. Probabilistic annotation tools use statistical models to locate entities within text. Induction tools use either linguistic or structural analysis to perform wrapper induction [1]. Multi-strategy annotation tools combine both pattern-based and machine-based methods. However, Reeve and

Han claim that until now no system exists that implements the multi-strategy annotation method [1].

Typical annotation systems include MnM [2], *Knowledge and Information Management* (KIM) platform [3], MUSE [8] etc. MnM is an annotation tool which can both automated and semi-automated annotate web pages. Before annotation, MnM requires training texts which are fed into a wrapper induction system [2]. Then some induced rules that be used to extract information from texts generated.

In KIM, the semantic annotation process relies on KIMO, which is a pre-built lightweight ontology. KIM identifies named entities found during the annotation process and matches them in the knowledge base. KIM introduces alias mechanism to solve the mapping between full name and abbreviation to a certain extent, but it cannot solve the case that multiple different full names have same abbreviation.

MUSE deals with different types of documents separately, and can eliminate ambiguity. However, the named entities which MUSE annotates are organization/corporation, person (stars, leaders, CEO), events, big cities (such as the capital of nation, the capital of province), architect, river, lake, mountain, money etc. The meaning of those entities is less ambiguity than the entities of scientific literatures. Thus, MUSE does not perform well on scientific literature.

3 Semantic Annotation System for Literature

3.1 Architecture of SASL

Although many annotation tools are available, those tools mainly annotate news, history events etc. The named entities annotated by those tools, for example corporation names, international organizations, location (including city, country, scenic spots), time, currency are unambiguous. The named entities from scientific literatures are generally abbreviations which may map onto many full names. For example, the full name of CAS may be: Central Authentication Service, Channel Associated Signaling, Chemical Abstracts Service, Code Access Security, Computer Arts Society, Cycle Accurate Simulator, and Computer Algebra System. Moreover, an author may use different name variations and there are multiple other authors with the same name. All the cases introduce ambiguity and make semantic annotation for literature difficult.

To solve the problems, SASL adopts schema matching to identify named entities. Furthermore, if some named entities are not disambiguated, SASL adopts regular expression combined with hidden Markov model to disambiguate named entities. SASL consists of following components: knowledge base, conversion tool, named entity identification, and name disambiguation. Figure 1 shows architecture of SASL.

There exist different paper formats. The common paper formats include PDF, TXT, DOC and so on. In order to annotate paper, the Conversion Tool of SASL needs convert many paper formats into a single format. Since SASL is a web based system, the target format is HTML. The Conversion Tool of SASL converts paper format into HTML text.

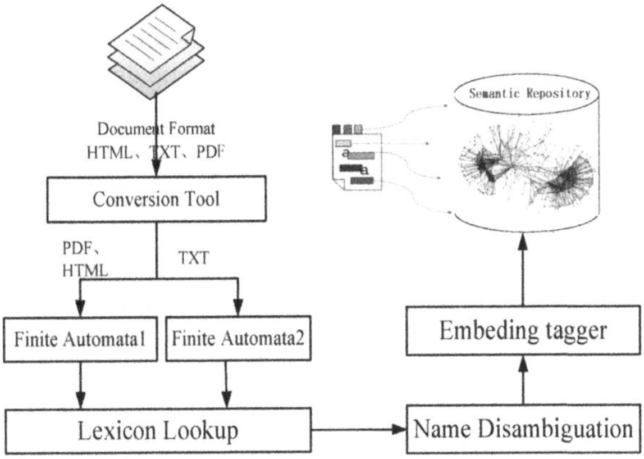

Fig. 1. Architecture of SASL

Finite automata are used to identify named entities from papers. Considering the semantic completeness of content, SASL adopts the left-most matching approach to identify named entities. Some identified named entities are abbreviations. As mentioned above, an abbreviation may map into many full names. In our cases, we combine regular expression and hidden Markov model to determine the right identity of named entities. After the right identity of named entities is known, SASL looks up knowledge base to find the semantic information of name entities and embeds the semantic information into HTML text using tag.

3.2 Knowledge Base for Semantic Annotation

Wikipedia is the largest encyclopedia in the world today and has more than a billion words spread over 10 million articles in 250 languages, including 2.5 million articles in English [9]. Wikipedia is edited by many domain experts. Thus, data from Wikipedia are reliable and are updated continuously. Thus, utilizing Wikipedia as knowledge base to annotate literatures, SASL can guarantee its knowledge base comprehensive, correct, and latest.

Wikipedia organizes the terminologies and categories of domains as semantic-associated graphs. In Wikipedia, there exist following relations among categories and terminologies: a terminology is subordinate to a category; a category is father category of another category or a category is the sub-category of another category. Each article in Wikipedia belongs to at least one category, and hyperlinks between articles capture their semantic relations, specifically, the represented semantic relations are equivalence (synonymy), hierarchical (hyponymy), and associative. Besides, categories are nested in a hierarchical organization. The resulting hierarchy is a directed acyclic graph, where multiple categorization schemes co-exist [10].

SASL mainly extracts those entities of computer science from Wikipedia and stores them into knowledge base of SASL. The entities are mainly computer academic terminology (such as Natural Language Processing, Information Extraction), international academic institutions entities (IEEE, ACM, and W3C et al), international academic conferences entities (such as International Joint Conference on Artificial

Intelligence, European Conference on Artificial Intelligence), and so on. There are more than 108,948 entities in knowledge base. Each entity has five properties as follows: urid, entity name, the definition of entity or semantic description of entity, corresponding link in Wikipedia data, and the category of entity.

3.3 Context Based Mapping Abbreviation to Full Name

For the purpose of simple presentation, most academic terminologies are usually replaced by their abbreviation in literature. However, Wikipedia does not provide the mapping relation between full names and theirs abbreviations. SASL must find the mapping relations among abbreviations and concepts of knowledge base.

Generally, we can observe three cases where full names and theirs abbreviations occur. The first case is that full name always follows with parentheses, and corresponding abbreviations of full names are put into parentheses, such as "Artificial Intelligence (AI)". There also exist the other cases. One case is that full names of entities always follow with abbreviation, and authors use 'or' to indicate the relationship between them. Some authors may not adopt the above two means to express the relation between full names of entities and theirs abbreviation, but state their relationship directly using language. For instance, if some authors want to tell readers the relationship between Artificial Intelligence and AI, they can say "Artificial Intelligence generally abbreviated to AI". Regardless of which cases, we can conclude that majority of the full name and abbreviation have a same feature: the first letter of the abbreviation and first letter of full name are same and are generally uppercase.

3.3.1 Combining Regular Expression and Local Context for Disambiguation

According to the above analysis, in order to find the relationship between full names and abbreviations, SASL defines three regular expressions to extract the mapping between full name and abbreviation as follows:

```
(1) ((\\p{Upper})[\\w-]+(\\s+[\\w-]+){1,7}?)
\\s*\\((\\s*\\2\\p{Upper}+\\s*)\\)
(2) ((\\p{Upper})[\\w-]+(\\s+[\\w-]+){1,7}?) \\s+or\\s+(\\2\\p{Upper}+)
(3) ((\\p{Upper})[\\w-]+(\\s+[\\w-
]+){1,7}?)\\s+((was|is)\\s+)?(commonly\\s+)?abbreviated\\s+(to|as)\\s+(\\2\\p{U
pper}+)
```

We use the above regular expressions to extract the mapping relationships between full name of entity and abbreviation. There are about 5,014 pairs in computer literatures. Among them, 673 abbreviations map more than one full name. One typical case is CAS mentioned above which has 8 different full names. We calculate an abbreviation may map to 2.4 full names in average. Thus, it introduces ambiguity. In the following section, we will introduce how to eliminate those ambiguities.

Moreover, it is not always true to use the above rules to identify relationship between full name and abbreviations. For example, Web Ontology Language is commonly abbreviated to OWL, and abbreviation is not always consisted of upper case. Although there exist those complicated cases mentioned above, it is always true that full name of entity appears along with abbreviation when abbreviation of entity appears for the first time in literature. Moreover, literature always focuses on specific topic, thus, it restricts possible full name of abbreviations. For instance, it is not possible for "Information extraction (IE)" and "Information engineering (IE)" to appear in the same text. Considering those cases, furthermore, we use the identified abbreviations of literature as local context to identify other abbreviations of literature. SASL stores local context local cache. At the same time, the content of local context is also kept into knowledge base as global cache. The content of the local cache is only effective on the current literatures. Generally, most mapping relationships between full names and abbreviations can be found in local cache.

3.3.2 Hidden Markov Model Based Disambiguation

If full name can not be found in local cache, SASL looks up knowledge base namely global cache. SASL combines global cache with hidden Markov model to achieve disambiguation. Generally, there exist certain relations among adjacent entities. For instance, they may belong to the same category, have a common parent category, or co-occur frequently. According to the observation, SASL combines global context with hidden Markov model to achieve disambiguation.

The definition of a HMM [11] is as follows: $\lambda=\{A, B, \pi\}$, where A is a transition array, storing the probability of state j following state i: $A=\{a_{ij}\}$. The states are categories or full names of entities. B is the observation array, storing the probability of observation k (namely full name) being produced from the state j, independent of t: $B=\{b_{jk}\}$. π is the initial probability array of categories and named entities which occur with their full names: $\pi=\{\pi_i\}$. In order to use HMM for name disambiguation, we need estimate the model parameters that best describe that process. We adopt supervised learning to do this task. After the parameters are determined, we can use the model to disambiguate named entities. Then the algorithm for name disambiguation is as follows:

Algorithm 1. Name disambiguation
Input: $A=\{a_{ij}\}$ and $B=\{b_{jk}\}$
Output: the probabilities of the observation: $P(O|\lambda)$
Step 1: Defines the forward variable as the probability of the partial observation sequence $o_1, o_2, \ldots, o_T$, when it terminates at the state i. $\alpha_t(i)=P(o_1\ldots o_t, q_t=s_i|\lambda)$, and is initialized as formula (1):

$$\alpha_1(i)=\pi_i b_i(o_1) \quad 1\le i\le N \tag{1}$$

Step 2: Recursively calculate:

$$\alpha_t(j)=\left[\sum_{i=1}^{N}\alpha_{t-1}(i)a_{ij}\right]b_j(o_t) \quad 2\le t\le T, 1\le j\le N \tag{2}$$

Step 3: Calculate the corresponding probability of entity and category:

$$P(O\mid\lambda)=\sum_{i=1}^{N}\alpha_T(i) \tag{3}$$

Step 4: Calculate the corresponding probability of full name and abbreviation pairs and rank them. If the difference of the probability of top 2 pairs is greater than threshold δ, the full name which has larger probability is chosen as right full name of the corresponding abbreviation, else go to step 1.

4 Experimental Results

We have implemented SASL which can annotate multiple format text, including PDF, Word, and HTML etc. Figure 2 is a screenshot of annotated text using SASL. Although SASL mainly annotates literatures from computer science, it does not mean SASL can only annotate computer literatures. In fact, SASL can annotate texts from any domain due to two reasons: one reason is that SASL adopts Wikipedia as knowledge base. Wikipedia is the largest encyclopedia in the world today. The second reason is that the approaches we develop in SASL is independent and not related with any special domain.

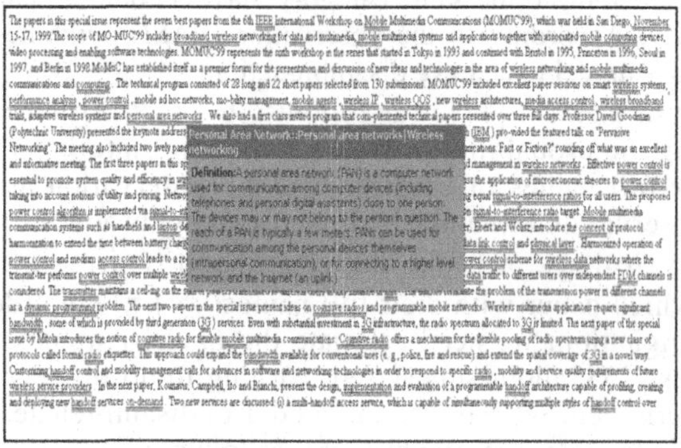

Fig. 2. Screenshots of annotated text by SASL

Since few data sets for semantic annotation are available, in order to evaluate the performance of SASL, we randomly download 26 papers from Web. The length of each paper is in the range of eight to thirty pages, and each paper contains 300~500 entities. There are 7,800 ~ 1,300 named entities in those papers.

Because no training data set for semantic annotation are available, we utilize the text annotated by SASL as training set. For the purpose of improving the performance of SASL, we evaluate the performance under different threshold δ, for example δ=0.1, 0.05 and 0.1/n, respectively, where n is the number of full name which have the same abbreviation. The results are shown in Figure 3.

We can find SASL achieve the best performance when δ=0.1/n. The reason is that the same abbreviation may map to different full names. If SASL uses the same δ, the performance of SASL can not be guaranteed.

When δ is set as 0.1/n, the performance of SASL is compared with KIM, MUSE. The experimental results are shown in Figure 4. The precision and recall of SASL are 0.92 and 0.95, respectively. However, the precision and recall of KIM and MUSE are 0.86 and 0.82, 0.935 and 0.923, respectively. According to the results, SASL achieves better F-Measure than KIM or MUSE does.

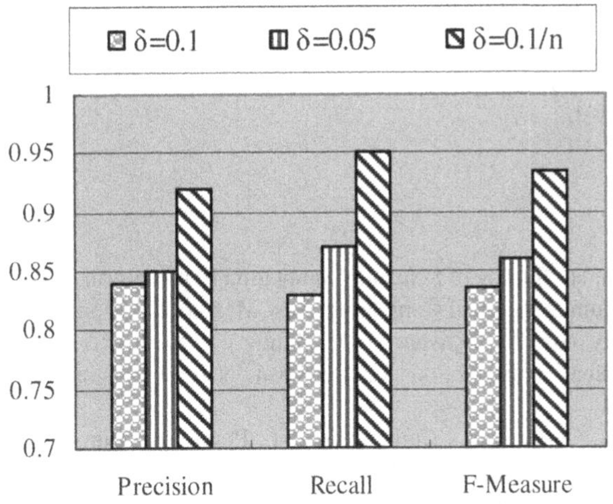

Fig. 3. The performance of SASL when δ=0.1, 0.05 and 0.1/n

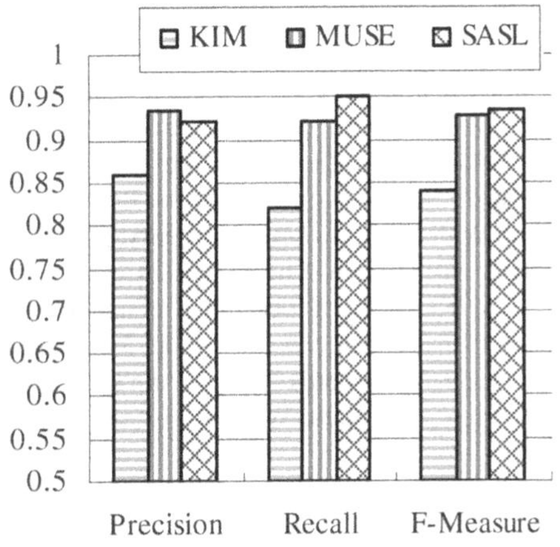

Fig. 4. Performances of SASL, KIM, MUSE

5 Conclusion

It is very important for researchers to retrieve scientific literatures efficiently. However, due to ambiguity of named entities, search engine may return many irrelevant results. In order to let computer understand the meaning of named entities, it is necessary to annotate named entities so as to attach semantic information to them. *Semantic Annotation System for Literature* (SASL) can automatically annotate named entities of literatures. Due to the correctness, popularity and comprehensive of Wikipedia, SASL utilizes Wikipedia as knowledge base. For the purpose of identifying entities correctly, SASL adopts regular expression and HMM. Experimental results show SASL achieves good performance comparing with KIM and MUSE.

Moreover, one efficient way to achieve semantic web is to automatically annotate existing or new web pages or documents and attach the semantic information into them. Since SASL is independent on specific domain, SASL can promote the realization of semantic web.

References

1. Reeve, L., Han, H.: Survey of Semantic Annotation Platforms. In: Proceedings of the 2005 ACM Symposium on Applied Computing, New Mexico, USA, pp. 1634–1638 (2005)
2. Popov, B., Kiryakov, A., Ognyanoff, D., Manov, D., Kirilov, A.: KIM: a Semantic Platform for Information Extraction and Retrieval. Natural Language Engineering 10(3-4), 375–392 (2004)
3. Bontcheva, K., Kiryakov, A., Cunningham, H., Popov, B., Dimitrov, M.: Semantic Web Enabled, Open Source Language Technology. In: Proceedings of EACL 2003 Workshop on Language Technology and the Semantic Web, Budapest, Hungary, pp. 335–343 (2003)
4. Collier, N., Takeuchi, K., Kawazoe, A.: Open Ontology Forge: an Environment for Text Mining in a Semantic Web World. In: Proceedings of International Workshop on Semantic Web Foundations and Application Technologies, Nara, Japan, March 12, pp. 128–140 (2003)
5. OntoMat-Annotizer, `http://annotation.semanticweb.org/ontomat/simple.html`
6. Dill, S., Eiron, N., Gibson, D., Gruhl, D., Guha, R., Jhingran, A., Kanungo, T., Rajagopalan, S., Tomkins, A., Tomlin, J.A., Zien, J.Y.: SemTag and Seeker: Bootstrapping the Semantic Web via Automated Semantic Annotation. In: Proceedings of the 12th International Conference on World Wide Web, Budapest, Hungary, pp. 178–186 (2003)
7. Cimiano, P., Ladwig, G., Staab, S.: Gimme' The Context: Context-driven Automatic Semantic Annotation with C-PANKOW. In: Proceedings of the Fourteenth International World Wide Web Conference, Chiba, Japan, May 10-14, pp. 332–341 (2005)
8. Cimiano, P., Handschuh, S., Staab, S.: Towards the Self-Annotating Web. In: Proceedings of the 13th International Conference on World Wide Web, New York, USA, pp. 462–471 (2004)
9. Lih, A.: The Wikipedia Revolution: How a Bunch of Nobodies Created the World's Greatest Encyclopedia. In: Hyperion, 1st edn., March 17 (2009)
10. Gabrilovich, E., Markovitch, S.: Overcoming the Brittleness Bottleneck Using Wikipedia: Enhancing Text Categorization with Encyclopedic Knowledge. In: Proceedings of the Twenty-First National Conference on Artificial Intelligence, Boston, MA, pp. 1301–1306 (2006)
11. Rabiner, L.R.: A Tutorial on Hidden Markov Models and Selected Applications in Speech Recognition. Proceedings of the IEEE 77(2), 257–286 (1989)
12. SemreX, `http://www.semrex.cn`

Multi-issue Agent Negotiation Based on Fairness

Baohe Zuo[1], Sue Zheng[2], and Hong Wu[2]

[1] School of Software Engineering, South China University of Technology, Guangzhou 510006, China
[2] School of Computer Science & Engineering, South China University of Technology, Guangzhou 510006, China
zuobh@scut.edu.cn, sue_zheng@163.com, wuhong40@tom.com

Abstract. Agent-based e-commerce service has become a hotspot now. How to make the agent negotiation process quickly and high-efficiently is the main research direction of this area. In the multi-issue model, MAUT(Multi-attribute Utility Theory) or its derived theory usually consider little about the fairness of both negotiators. This work presents a general model of agent negotiation which considered the satisfaction of both negotiators via autonomous learning. The model can evaluate offers from the opponent agent based on the satisfaction degree, learn online to get the opponent's knowledge from interactive instances of history and negotiation of this time, make concessions dynamically based on fair object. Through building the optimal negotiation model, the bilateral negotiation achieved a higher efficiency and fairer deal.

Keywords: Agent, satisfaction, fairness, interactive instances.

1 Introduction

E-Commerce is growing quickly from the rapid development of Internet, which can help people doing businesses effectively, saving cost and improving productivity.etc [1]. Nowadays the Web environment becomes more and more complex and product providers are uncertain. All of these reasons need a new technology which reflects intelligently and dynamically the changes of the information, and then it will satisfy people's demanding of the matured E-Commerce.

Agent has the intelligence, autonomy, mobility, interactive and collaborative characteristics. They can divide works, cooperate with others without human intervention and complete most tasks together, which results in agents superseding negotiator to carry out negotiation autonomously [2], [3]. Negotiation is an important means of Internet E-Commerce and the key target of software agent design. One of issues to be settled urgently in the multi-Agent system is how to make the agent enhance the capacity in the interaction of negotiation process.

Multi-Agent negotiation is carried through in an environment that is unknown and irregularly changing. It is not sufficient to deal with the dynamic changes of the surrounding environment by its own initial belief. If there is no knowledge of opponent's information prior to the negotiation, negotiation time will be increasing and success rate will become smaller. Agent auto-negotiation model based on the history of interaction[4],[5] can gain knowledge from opponent's interactive history prior to

W. Liu et al. (Eds.): WISM 2009, LNCS 5854, pp. 167–176, 2009.

the negotiation, form the initial belief in negotiation, estimate opponent's strategy more effectively, shorten the negotiation time and increase the success rate of negotiation, but they hardly consider fairness. On the other hand, fuzzy logic-based Agent auto-negotiation model [6], [7] can carry through bilateral negotiation with less initial faith, but the negotiation time is often too long. This paper combining the advantages of negotiation theories [4], [5], [6], [7] and learning the opponent's offer to further study the opponent's strategy, proposed a optimal multi-issue bilateral negotiation framework based on the fairness.

The rest of this article is organized as follows. Section 2 describes the definition of the agent negotiation model. Section 3 describes the process of agent negotiation. Section 4 shows the effectiveness of the proposed method by experiments. Finally, some conclusions are put forward in section 5.

2 Agent Negotiation Model

2.1 Formal Definition of Negotiation Model

Jennings believes that negotiation has three main elements [8]: *negotiation protocols, negotiation objects, negotiation strategies. Negotiation protocols* give the basic framework of negotiation and the rules that agents in the interaction process comply with together; *negotiation objects* determine when the negotiation to terminate, achieving balanced negotiation objects; *negotiation strategies* are the basic principles used to achieve the objects. Summarizing the contents of these three aspects, this paper give a history-based negotiation model. It is defined as follows:

$Model =< G, X, V, F, W >$

Where

$G : G = \{a_1, a_2, a_3, \cdots, a_n\}$ is agents set that involved in negotiation, and n denotes the number of agent.

$S : S = \{< s_1, z_1 >, < s_2, z_2 >, \cdots, < s_M, z_M >\}$ is a set of propositions that participate in negotiation. For a random binary group $< s_m, z_m > \in X$, $1 \le m \le M$, s_m is the theme of this proposition, which include proposition name, logo and otherwise; $z_m =< p_1, p_2, \cdots, p_K >$ is vector of issues that involve in the negotiation.

$V : V = \{V_{a_1}^{s_m}, V_{a_2}^{s_m}, \cdots, V_{a_n}^{s_m}\}$ is a set of issue vector about proposition s_m that proposed by agents. For $\forall a_i \in G(1 \le i \le n)$ at t^{th} $(1 \le t \le T$, T is the maximal negotiation time) interaction, the issues vector that proposed by agent a_i is $V_{a_i} =< v_{p_1}^{a_i}, v_{p_2}^{a_i}, \cdots, v_{p_K}^{a_i} >$, where $v_{p_k}^{a_i}$ is the value of issue x_k . At the same time, there also need reservation value vector $\overline{V}_{a_i} =< \overline{v}_{x_1}^{a_i}, \overline{v}_{x_2}^{a_i}, \cdots, \overline{v}_{x_k}^{a_i} >$ for each negotiation issue. The value of issue cannot less than the reservation value during negotiation process.

$F: F = \{F_{a_1}^{s_m}, F_{a_2}^{s_m}, \cdots F_{a_n}^{s_m}\}$ is a set of reservation value vectors that forecast about opponent. It indicates the result that an agent predicts about opponent agent's reservation value. For $\forall a_i \in G(1 \le i \le n)$, $F_{a_i}^{s_m} =< f_{p_1}^{a_i}, f_{p_2}^{a_i}, \cdots, f_{p_K}^{a_i} >$ denotes the reservation value vector of all issues in proposition s_m that predicted by agent a_i.

$W: W = \{w_1^{s_m}, w_2^{s_m}, \cdots, w_K^{s_m}\}$ is a set of weight about the proposition s_m that indicated by agent a_i, and $\Sigma_{k=1}^{K} w_k = 1$.

2.2 Utility Evaluation

2.2.1 Utility Evaluation of Multi-issue

Multi-issue negotiation needs to consider the partial order of every issue, that is, the weights of all issues also have impact on the utility function. The literature[4]give an utility function of multi-issue negotiation, defined as follows:

$$U(V,T) = \Sigma_{i=1}^{K} U(v_i, t_i) \tag{1}$$

Where v_i denotes the offer value of issue x_i in proposition s_m, t_i denotes the time negotiation success and $U(v_i, t_i)$ is the utility function of issue x_i, described as follows:

$$U(v_i, t_i) = \begin{cases} (\bar{v}_i - v_i) \times w_i \times (\delta_i)^{t_i} & \text{buyer } Agent \\ (v_i - \bar{v}_i) \times w_i \times (\delta_i)^{t_i} & \text{seller } Agent \end{cases} \tag{2}$$

Where δ denotes time discount rate, when $\delta > 1$ shows that the agent is patient type, the efficiency increases through time, otherwise, it shows that the agent is non-patient type, the efficiency decrease through time.

2.2.2 Utility Evaluation Based on Fairness

The above method bases on negotiation time, the values of both negotiation agents' utility function has a large difference, which only can meet the interests of the buyer or seller and cannot reach a fair balance between the two negotiators. This paper references the literature [7], makes an improvement on utility function, bring forward the satisfaction evaluation function based on satisfaction of both sides:

$$\Phi(V,T) = \Sigma_{i=1}^{K} \Phi(v_i, t_i) \tag{3}$$

Where $\Phi(v_i, t_i)$ is the satisfaction degree (SD) of issue x_i, defined as:

$$\Phi(v_i, t_i) = (threshold + (v_i - \bar{v}_i) \times \frac{(1 - threshold)}{(v_{i0} - \bar{v}_i)}) \times w_i \times (\delta_i)^{t_i} \tag{4}$$

Where *threshold* is the minimal SD, v_{i0} is the initial value of the issue x_i. It is assumed that the SD of each issue is 1 when the issue's value is v_{i0}.

When we use the primary function to evaluate the utility, the better the function value is higher. But the evaluation method proposed in this article is based on fairness of both negotiators, which considers the proximity of both agents' aggregated satisfaction value (ASV). The nearer ASV leads to better negotiation results.

3 Negotiation Process

Agents of the bilateral negotiation follow the principles of alternating-offers protocol. In offer exchange process, negotiators propose and respond offers alternatively, until agreement or negotiation deadline is reached. The action of a negotiator after it takes over an offer include: *accept, reject and propose a count offer, reach deadline and quit* [1].The intelligence of the agent can be embodied by the process of generating an initial offer and count offer, evaluating the received offer whether is acceptable and learning online about negotiation strategy of opponent agent. Fig.1 shows an Agent's internal negotiation process.

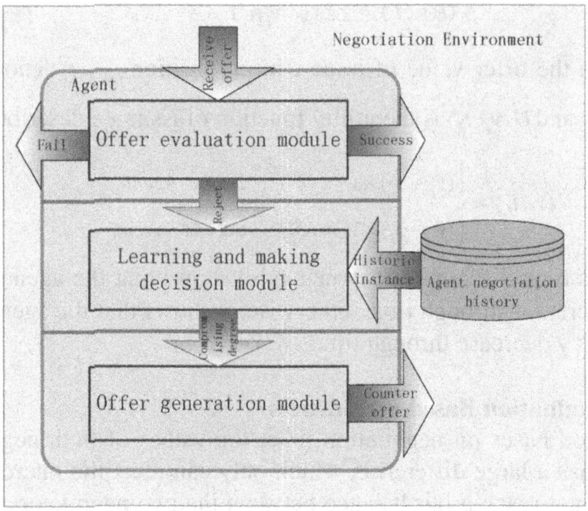

Fig. 1. The process of agent negotiation

When the buyer has selected a suitable seller, they began a negotiation on the issues which they concerned on. In a certain round of negotiation, negotiation agent can prepare a count offer in advance of the next round. The negotiation model can divide into three modules: *the evaluation module, learning and making decision module, offer generate module*.

3.1 Offer Evaluation Module

The model of this paper evaluates a negotiation process with the satisfaction degree of both negotiators. In the evaluation of an offer, it will be more consistency if it is measured by satisfaction degree. On the assumption that the initial value of each issue has the maximal SD (set 1); reservation value has the minimal SD (set $threshold$). Values of all issues change along the direction that SD decreasing.

After agent g receives an offer x from opponent a, it calculates the SD to decide whether to accept the offer, which is described as:

$$\begin{cases} \varphi_g(x) \geq \varphi_a(x) & accept \\ \varphi_g(x) < \varphi_a(x) & reject \end{cases} \tag{5}$$

Where $\varphi_a(x)$ is the aggregated satisfaction value (ASV) of opponent agent a about offer x, and $\varphi_g(x)$ is the ASV of agent g about offer x:

$$\varphi_g(x) = \sum_{i=1}^{K} (threshold + (v_i - \overline{v}_i) \times \frac{(1 - threshold)}{(v_{i0} - \overline{v}_i)}) \times w_i \tag{6}$$

That is, when the satisfaction degree for opponent's offer is greater than opponent's own satisfaction degree, agent accepts the offer, and if the negotiation time does not exceed the deadline, negotiation succeeded. If the negotiation time exceeded the deadline, negotiation failed neither accepts nor rejects. Otherwise, it comes to the learning and making decision module to decide how to make a concession to produce an offer which is acceptable to both negotiators.

3.2 Learning and Making Decision Module

The learning and making decision module begins to deal with the offer after agent rejects the offer. In the proceeding of alternating-offers, the information in offers is limited. It's difficult to accurately predict opponent's strategy, so it is better when the forecast information is closer. The module can be designed as an expert system, which is mainly due to fact that people's preference is abstract and uncertain. If the traditional mathematical methods to learn opponent's strategy are used, it will be very complicated and the time and space complexity will be growing dramatically. The learning of opponent's strategy can predict opponent's linear satisfaction degree function with lower time and space. The process is described as follows:

Step 1: Agent g receives an offer X from Agent a, if it is the first time that g receive offer from a, go to *Step 2*, otherwise go to *Step 3*.

Step 2: Agent g checks the negotiation interaction history, calculates each issue's expected value $h_i = \sum_{j=1}^{L} \frac{v_{i,j}}{L}$ (L is the success time) and expected value of SD $Args = \sum_{j=1}^{L} \varphi_j \times p(\varphi_j)$ ($p(\varphi_j)$ is the historical success probability) from the success negotiation instances. The agent estimates the satisfaction degree of opponent's issue by the received offers, this degree is described as:

$$f_a^1(v_i) = k \times v_i + b \tag{7}$$

Where $b = Args - k \times h_i$, $k = {(\varphi_a(x) - Args)}/{(v_i - h_i)}$. The cross-point of this function and the satisfaction degree function $f_g(v_i)$ of Agent g is regarded as the common SD $\varphi_{ag}^1(v_i)$, and then go to *Step 4*.

Step 3: Proposing that the weights of opponent's each issue w_i are the same. Agent g gets opponent's offers of $(t-1)^{th}$ round and this round, recalculates the opponent's satisfaction degree function. Then it can achieve a new satisfaction degree function by reference the original function, the new function is defined as follows:

$$f_a^t(v_i) = k \times v_i + b \tag{8}$$

The cross-point $\varphi_{ag}^t(v_i)$ can be figured out and then go to *Step 4*.

Step 4: Agent g checks whether cross-point $\varphi_{ag}^t(v_i)$ is greater than the satisfaction degree threshold. When $\varphi_{ag}^t(v_i) < threshold$, considered there is no overlapping zone between the reservation value of the two agents, agent sets *threshold* as the SD, the difference of SD between the two agents about issue i is set as the concession degree. This concession degree can be defined as:

$$\gamma_i = \alpha(\varphi_g^t(v_i) - threshold) \tag{9}$$

Where α is concession parameter that defined beforehand. When $\varphi_{ag}^t(v_i) \ge \varphi_g^t(v_i)$, set γ_i as 0, where $\varphi_g^t(v_i)$ is SD of g itself. When $threshold \le \varphi_{ag}^t(v_i) \le \varphi_g^t(v_i)$, set $\gamma_i = \alpha(\varphi_g^t(v_i) - \varphi_{ag}^t(v_i))$ and go to *Step 5*.

Step 5: Agent sums up the concession degree which is calculated from *Step 4* to γ. Then γ will be sent to the offer generate module as a parameter.

3.3 Offer Generation Module

After the agent works out the concession degree, it can generate a new count offer according to the negotiation concession strategy. This strategy can be defined as:

$$\ell_{v_i} = \{v_i^{t+1} \mid (\varphi(v_i^{t+1}) = \varphi(v_i^t) - \gamma_i) \Lambda (\varphi(v_i^{t+1}) \ge threshold)\} \tag{10}$$

Where γ_i is the product of γ and the weight w_i, $\varphi(v_i^t)$ is SD of issue i at t^{th} round, $\varphi(v_i^{t+1})$ denotes SD of new value. This new value which comes from the new SD, can be defined as:

$$v_i^{t+1} = \overline{v}_i + (v_{i0} - \overline{v}_i) \times {\gamma_i}/{(1 - threshold)} \tag{11}$$

The agent brings off the new ASV of the $(t+1)^{th}$ round from all issues' new value. At the end, it sends the new count offer to the opponent agent.

4 Experiment

An experimental prototype system has been designed to verify the effectiveness of the model. The negotiators in this system are designed as Agent *seller* and Agent *buyer*, and the negotiation proposition is designed as s_m.

The concession degree in the negotiation process is affected by the parameter α, the Fig.2 shows the changes in the number of interactive round when parameter value changes.

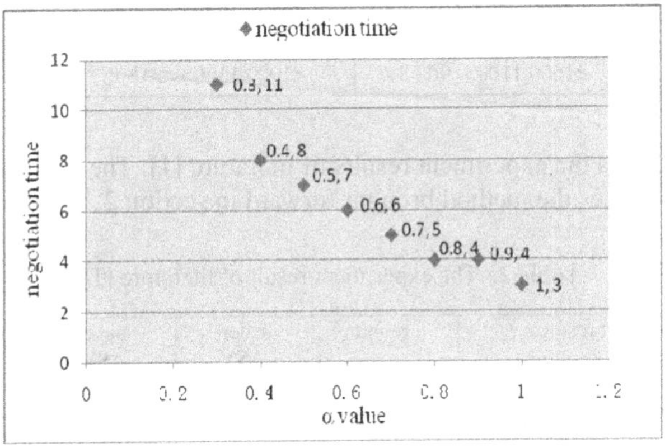

Fig. 2. The changes of negotiation time with α value

Fig.3 shows the changes of the difference between the negotiation agents' ASV when the parameter value changes.

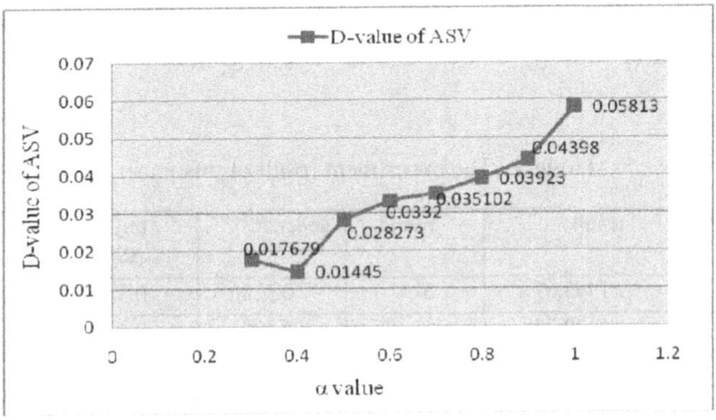

Fig. 3. The changes of D-value with α value

As it can be seen from the figures 2, 3, the greater the value of α, the smaller the number of interactive round, but the greater the difference of two agents' ASV. The value of α is set to0.8 in the following experiments.

For the purpose of convenience, the number of negotiation round t takes the place of negotiation time in experiments. Considering the fairness of negotiation model, this paper cites the literature [1] as a reference and compares the results. Table 1 show the data used in experiments.

Table 1. The data for experiment

series	seller $< V, \bar{V} >$	Buyer $< V, \bar{V} >$	weight
S_1	<1600,1400><90,60>	<1300,1500><50,80>	<0.5,0.5>
S_2	<1650,1350><90,58>	<1250,1530><60,84>	<0.5,0.5>
S_3	<1600,1230><88,55>	<1150,1450><50,75>	<0.5,0.5>
S_4	<1570,1200><88,62>	<1100,1450><56,80>	<0.5,0.5>
S_5	<1600,1100><90,58>	<1000,1360><50,80>	<0.5,0.5>

Table 2 describes the experiment results of literature [1]. The calculation of buyer's and seller's ASV uses the method brought forward in section 2.

Table 2. The experiment result of literature [1]

series	result	round	seller ASV	buyer ASV	D-value
S_1	<1428,69>	5	0.298	0.427	0.129
S_2	<1383,65>	6	0.2479	0.6925	0.4446
S_3	<1227,65>	6	0.2327	0.6145	0.3818
S_4	<1270,67>	7	0.2717	0.5752	0.3035
S_5	<1118,65>	7	0.2776	0.54	0.2624

In this paper, agent uses the information of historical negotiation and interaction instances of this negotiation, estimates opponent's satisfaction function and coordinates the negotiation strategy of next round. The negotiation outcome is summarized in Table 3.

Table 3. The experiment result of this paper

series	result	round	seller ASV	buyer ASV	D-value
S_1	<1455,67>	5	0.3286	0.3377	0.091
S_2	<1439,72>	5	0.4204	0.4304	0.01
S_3	<1324,69>	4	0.3652	0.3969	0.0318
S_4	<1372,70>	6	0.3585	0.3398	0.0187
S_5	<1305,68>	6	0.3113	0.284	0.0273

From the experimental results of Table 2 and Table 3 it can be seen that there is small difference of negotiation rounds between this paper's model and the literature [1] model. But in reducing the difference value of negotiators' satisfaction degree, namely, fairness, this paper's model has a greatly improvement.

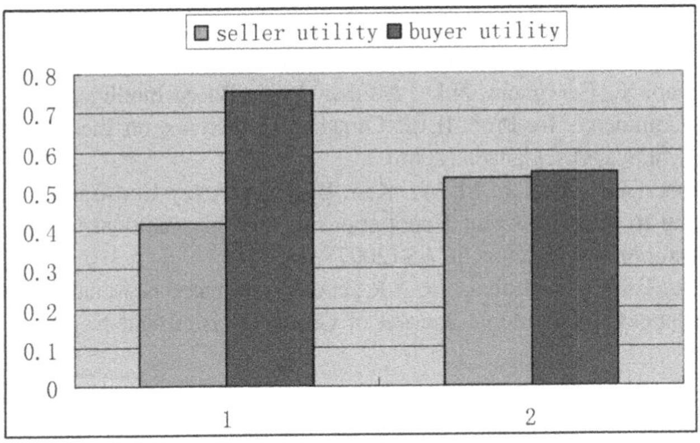

Fig. 4. Comparisons of two method's utility

The Fig.4 shows the utility of these two models. In the model of this paper, the difference of utility value about the two negotiators is narrow clearly.

The above experiment results show that using negotiation information of history and interaction process can lead to estimate the opponent's negotiation strategy more accurately. The process speed and fairness can be improved.

5 Conclusions

This paper presents an optimal bilateral and multi-issue negotiation framework based on the fairness. The agent obtains the historical information from the negotiation environment and analyzes the opponent's success negotiation instances to predict opponent's initial belief and negotiation strategy. Besides, the negotiators in the fact environment always are greed, how to balance the greed and the fairness yield need the further research.

References

1. Al-Axhmaway, W.H., El-Sisi, A.B., Nassar, H.M., Ismail, N.A.: Bilateral Agent Negotiation for E-Commerce Based on Fuzzy Logic. In: ICCES 2007International Conference on Computer Engineering and Systems (2007)
2. Faratin, P., Sierra, C., Jennings, N.R., Buckle, P.: Designing Responsive and Deliberative Automated Negotiators. In: Proc. AAAI Workshop on Negotiation: Settling Conflicts and Identifying Opportunities, Orlando, pp. 12–18 (1999)

3. Lonmuscio, A., Wooldridge, M., Jennings, N.: A classification scheme for negotiation in electronic commerce. International Journal of Group Decision and Negotiation 12(1), 31–56 (2003)
4. Fatima, S.S., Wooldridge, M., Jennings, N.R.: Multi-issue negotiation under time constraints. In: Proceedings of the First International Conference on Autonomous Agents and Multiagent Systems (AAMAS 2002), Bologna, Italy (2002)
5. Fatima, S.S., Wooldridge, M., Jennings, N.R.: An agenda- based framework for multi- issue negotiation. Artificial Intelligence 152(1), 1–45 (2004)
6. Wang, X., Shen, X., Georganas, N.D.: A Fuzzy Logic Based Intelligent Negotiation Agent (FINA) In ECommerce. In: Proc. IEEE Canadian Conference on Electrical and Computer Engineering, Ottawa, ON, Canada (2006)
7. Yu, T.-J., Robert Lai, K., Lin, M.-W., Kao, B.-R.: A Fuzzy Constraint-Directed Autonomous Learning to Support Agent Negotiation. In: 3rd International Conference on Autonomic and Autonomous Systems, ICAS (2007)
8. Jennings, N.R., Faratin, P., Lomuscio, A.R., et al.: Automated negotiation: prospects, methods and challenges. International Journal of Group Decision and Negotiation 10(2), 199–215 (2001)

A Hemodynamic Predict of an Intra-Aorta Pump Application in Vitro Using Numerical Analysis

Bin Gao, Ningning Chen, and Yu Chang

School of Life Science and BioEngineering, Beijing University of Technology, Beijing, 100124, P.R. China
changyu@bjut.edu.cn

Abstract. The Intra-Aorta Pump is a novel LVAD assisting the native heart without percutaneous drive-lines. The Intra-Aorta Pump is emplaced between the radix aortae and the aortic arch to draw-off the blood from the left ventricle to the aorta. To predict the change of pressure drop and blood flow along with the change of pump speed, a nonlinear model has been made based on the structure and speed of the Intra-Aorta Pump. To do this, a nonlinear electric circuit for the Intra-Aorta Pump has been developed. The model includes two speed dependent current sources and flow dependent resistant to simulate the relationship between the pressure drop of the Intra-Aorta Pump and the flow through the pump along with the change of pump speed. The pressure drop and blood flow is derived by solving differential equations with variable coefficients. The parameters of the model are determined by experiment, and the results of the experiment show that these parameters change along with the change of the pump speed distinctness. The accuracy of the model is tested experimentally on a test loop. The comparison of the prediction data derived from the model with the experimental data shows that the error is lest than 15%. The experimental results showed that the model can predict the change of pressure drop and blood flow accurately.

Keywords: Hemodynamic Predict, Intra-Aorta Pump, Numerical Analysis.

1 Introduction

Currently heart failure is a severe and steady growing disease that remains a leading cause of morbidity and mortality in the world [1]. The blood pump was firstly transplanted into human body and achieves success in 1971 by Debakey. Then Some artificial hearts or blood pumps became a new choice to therapy or a bridge to transplant for patients such as Hemopump [2], Micromed Debakey [3, 4], Jarvike2000[5,6], and Incor [7] .

Generally, these blood pumps all need percutaneous drive-lines to transport energy and control signal which usually caused some problems such as infection [13] and the low living quality. In 1996 Guorong Li proposed a novel left ventricular assist device--the dynamic aortic valve (DAV) [8, 9, 14] which did not need the percutaneous drive-lines. And the flow field of DAV was more suits for the native one of the circulating system [10].

W. Liu et al. (Eds.): WISM 2009, LNCS 5854, pp. 177–185, 2009.

An Intra-Aorta pump was produced by Biomedical Center of Beijing University of Technology. The Intra-Aorta pump inherited the advantages of the DAV and its simple construction. It consisted of two parts: dynamic system and the blood pump. There were not lines to conduct them. The blood pump was consisted of bear frame and impeller that was made of permanent-magnet. The driver system utilized the permanent-magnet that was drove by a brushless motor to generate alternating magnetic field. The energy was transported into human body utilizing the couple of the magnetic field. The diameter and length of the Intra-Aorta pump is 20mm and 30mm, respectively. The Intra-Aorta pump was placed between the radix aortae and the aortic arch as shown in Figure. 1.

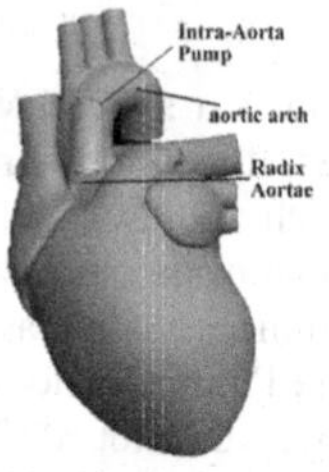

Fig. 1. Anatomical fitting of the Intra-Aorta pump

The pump drew-off the blood from the left ventricle to the aortic arch. Comparing with the DAV, the Intra-Aorta pump did not trauma the valve of the native heart; therefore it was fit for early heart failure to recover. As the dynamic system was placed in vitro, the temperature rise that due to dicing the magnetic field has been eliminated. As the Intra-Aorta Pump was a novel pump, the influence of it to the native heart was unknown, therefore the model of the pump with parameters of the vascular is needed to predict the pressure at the radix aortae and the outlet of the Intra-Aorta Pump and the blood flow through the Intra-Aorta Pump to help engineers improve the control strategy for a special patient.

In this paper a nonlinear model of the Intra-Aorta pump was proposed to predict the pressure drop and blood flow along with the change of the pump speed. And experiments were done to identify the parameters of the model and test the accuracy of the prediction.

2 Model Description and Analysis

The blood flow in a vascular is essentially similar to fluid flow in a pipe and is governed by Poiseuille's law [10]

$$P = f(Q) \tag{1}$$

where P is the pressure drop across the pipe, Q is the fluid flow in the pipe, and f (ȼ) is typically a continuously differentiable nonlinear function. In the absence of knowledge of the exact relationship between P andQ, recent work [10, 15] has suggested the use of a quadratic polynomial, instead of linear, approximation of .That is

$$P = R_0 Q + R_1 Q^2 \tag{2}$$

where R_0 and $R_1$are polynomial coefficients. Because the function is typically unknown, the coefficients can be determined experimentally by fitting expression (2) to measurements of pressure drop and blood flow in the vascular. However, the blood flow in the Intra-Aorta Pump, the equation (1) and (2) do not yield a good model. Because the pressure drop between the outlet and inlet of the pump becomes bigger and bigger along with the blood flow bring down. Hence, there is a constant$C$ that represent the pressure when the blood flow through the Intra-Aorta Pump equal zero.

$$P = C - R_0 Q - R_1 Q^2 \tag{3}$$

Where$C$, R_0 and R_1 are polynomial coefficients that can be determined experimentally by fitting expression. And an electric circuit analog to (3) in which pressure is represented by the voltage and the blood flow by current and which still the model $P = RQ$ can be derived,

$$P_P(\omega) = (R_{PS}(\omega) + k(\omega)Q_{PO})(Q_{PS}(\omega) - Q_{PO}) \tag{4}$$

Where the $Q_{PO}$is the blood flow that is ejaculated from the outlet of the pump in to the aortic arch and the $Q_{PS}$is the blood flow that is pushed out by the impeller. Clearly, the blood pump is modeled as a speed-dependent current source.$P_P(\omega)$ along with the flow-dependent internal resistance denoted by

$$R = R_{PS}(\omega) + k(\omega)Q_{PO} \tag{5}$$

Which also depends on the rotational speed of the blood pump ω. As mentioned earlier that the parameters$R_{PS}(\omega)$, $k(\omega)$ and Q_{PS} in (4) and (5) are typically determined experimentally. The impedance of the other part of the vascular is represented by Z that is defined a constant regardless of the rotation speed and the blood flow through the Intra-Aorta Pump.

Using this theory, the Intra-Aorta Pump system consisting of the radix aortae, Intra-Aorta Pump, and the arteria as well as the effect of the LV, which is substituted by another blood pump, on the system through the cardiovascular hemodynamics can be modeled with the equivalent electric circuit shown in Figure 2.

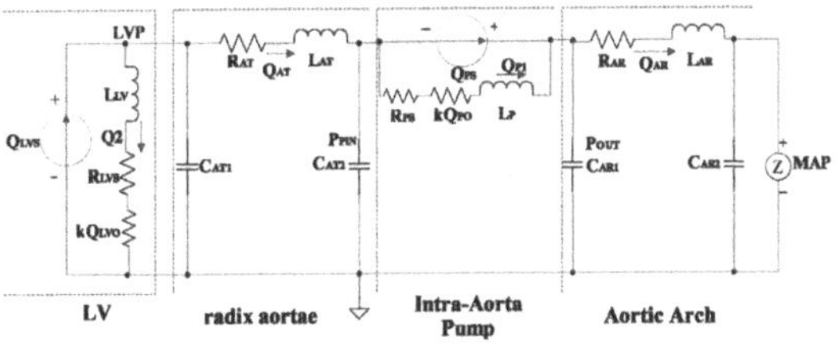

Fig. 2. Electric circuit equivalent of the Intra-Aorta Pump model with vascular and heart hemodynamics

2.1 The Model of the Intra-Aorta Pump

The Intra-Aorta Pump is modeled with the equivalent electric circuit shown in Figure 3. According to (4), the $R_{PS}(\omega)$ is the basic internal resister of the Intra-Aorta Pump that is only determined by the physical construction and the rotation speed of the

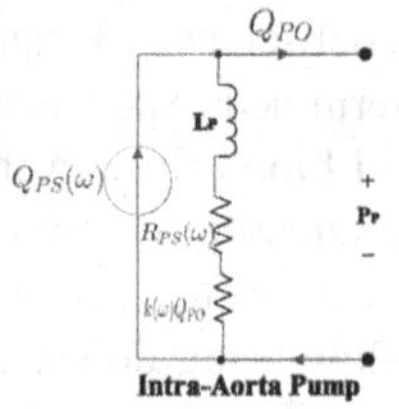

Fig. 3. Electric circuit equivalent of the pump

Intra-Aorta Pump. The $k(\omega)Q_{PO}$ is the variable resistance of the Intra-Aorta Pump that is dependent on the rotation speed and the blood flow through the pump. The L_P is represented the inertance of the pump [11].

To determine the parameters of the electric circuit, we apply basic circuit theory (Kirchhoff's voltage and current law) to derive the following relationship:

$$P_P = R_{PS}(\omega)Q_{PS}(\omega) - (R_{PS}(\omega) - k(\omega)Q_{PS}(\omega))Q_{PO} - k(\omega)Q_{PO}^2 \quad (6)$$

Compared with (3), the parameters of (6) can be determined as:

$$\begin{cases} R_{PS}(\omega)Q_{PS}(\omega) = C \\ R_{PS}(\omega) - k(\omega)Q_{PS}(\omega) = R_0 \\ k(\omega) = R_1 \end{cases} \quad (7)$$

2.2 The Model of the Aorta

According to the model of hemopump proposed by Xiong Li [12], the aorta can be modeled with passive network that consist of resisters, inductances and capacities which is represented the blood flow impedance, inertance and compliance of the arterial respectively. The model of the aorta is shown in Figure 4.

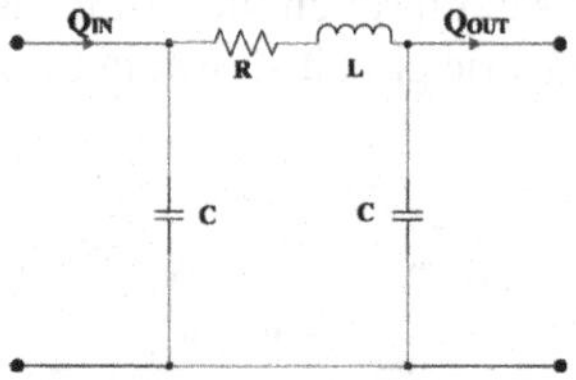

Fig. 4. The electric circuit equivalent of the aorta

2.3 The State-Space Model of the Systemic Circulation with the Intra-Aorta Pump

According to the Figure 2, the state-space function of the systemic circulation is denoted by

$$\dot{X} = AX + BU \quad (8)$$

$$Y = CX + DU \quad (9)$$

Where X is represented the quantity of state: $X = (Q_2, LVP, Q_{AT}, P_{PIN}, P_{P1}, P_{OUT}, Q_{AR}, MAP)^T$; U is represented input quantity: $U = (Q_{LVS}(\omega), Q_{PS}(\omega))^T$; Y is the output of the system: $Y = (LVP, P_{PIN}, P_{OUT}, Q_{AR}, MAP)^T$.

Since our objective is to predict the steady state mean blood flow and pressure in the system. The effect of blood inertia and vascular compliance are neglected, the state-space model is linearization by order$\dot{X} = 0$. Then the functions is simplified as

$$\begin{cases} AX + BU = 0 \\ CX + DU = Y \end{cases} \tag{10}$$

The key point value is derived by solving the system of equations:$Y = (-CA^{-1}B + D)U$. The blood flow and the pressure are contained in the$Y$.

3 Experimentation

In order to test the model, amount of experiments were conducted to derive data for identifying the parameters in the model and test the accuracy of the model in predicting the change of the blood flow and the pressure in several key points of the system along with the change of the pump speed.

The experimental setup consist of a test loop shown in Figure 5, which include a preload chamber; a pump as the left ventricle (LV), aΦ20mm silicone tube as the radix aortae that is 10cm, a Intra-Aorta Pump, aΦ20mm silicon tube as the aortic arch that is 10cm and a clamp to adjust the blood flow manually. A flowmeter is placed after the aortic arch to measure the outflow through the Intra-Aorta Pump. And two pressure transducers are placed on the middle of the radix aortae and the aortic arch respectively to measure the pressure at the inlet and the outlet of the Intra-Aorta Pump. The clamp can be adjusted by manual to achieve the desired set point for flow rate or MAP. The test fluid in the system was a blood analog which consisted of a solution of 35% glycerol and 65% water (by volume) to simulate the viscosity of the blood at 37°C. The height of the fluid in the preload camber was to simulate the LAP, and in this experiment the pressure at the inlet of the LV was considered to approximate to 0 mmHg.

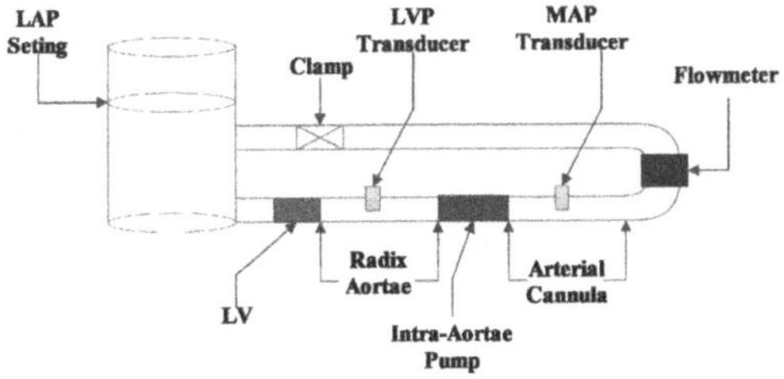

Fig. 5. The schematic of the systemic circulation for the experiment

3.1 Identification of the Systemic Circulation Model Parameters

The experimental facility shown in Figure 4 was used to identify the parameters of the radix aortae and aortic arch. The Intra-Aorta Pump was operated at a constant speed of 5000RPM and the clamp was used to adjust the blood flow from 0 to 5L/min with 1L/min increment base on the readings from the flowmeter. The pressure drop across the vascular at each set point was measured by using pressure transducer and recorded to identify the parameters of the vascular.

The similar test was conducted to derive the relationship between the blood flow and pressure rise of the Intra-Aorta Pump and the LV. In this test the differential pressure transducers were place in the outlet and the inlet of the pumps, respectively. Measurements were obtained at pumps speed of 3000,3600,4200,4800 and 5400 RPM. At each pumps' speed setting the clamp was adjust to derive the flow rate value from 0L/min to the maximum flow at 1L/min increment. The resulting pressure rise and blood flow were recorded to identify the pump model parameters$R_{PS}(\omega)$, $Q_{PS}(\omega)$ and$k(\omega)$.

Table 1. List of intra-aorta pump model parameters identified from least-squares fit of the experiment data

	$C^P(\omega)$	$R_0^P(\omega)$	$R_1^P(\omega)$	$\omega[R/s]$	r^2
1	117.5784	15.5161	-0.7047	90	0.9764
2	80.0360	7.0592	0.0527	80	0.9755
3	73.2118	14.4192	-1.1206	70	0.9773
4	39.8216	2.2608	0.5513	60	0.9741
5	34.0329	5.9447	-0.0899	50	0.9755

Table 2. List of LV model parameters identified from least-squares fit of the experiment data

	$C^{LV}(\omega)$	$R_0^{LV}(\omega)$	$R_1^{LV}(\omega)$	$\omega[R/s]$	r^2
1	236.25	375.41	-169.54	70.18	0.9498
2	464.0948	374.4934	-114.85	90.08	0.9454
3	1205.183	232.1932	-122.70	104.17	0.9443
4	1725.175	-81.5695	-60.983	109.28	0.9473
5	1848.351	-72.4005	-61.848	120.77	0.9466

The speed dependent Intra-Aorta Pumps model parameters $C^P(\omega)$, $R_0^P(\omega)$ and $R_1^P(\omega)$ were determined in least-squares fit procedure to the experiment data. The identified model parameters were listed in Table 1. The speed dependent LV model parameters $C^{LV}(\omega)$, $R_0^{LV}(\omega)$ and $R_1^{LV}(\omega)$ were determined in least-squares fit procedure to the experiment data. The identified model parameters were listed in Table 2. The computing results show that the coefficient of determinations (r^2) in Table 1 and Table 2 are very close to 1 implying a very good fit of the model to the experimental data [10].

3.2 The Prediction of the Pressure and Blood Flow by Applying the Model

The pump preloads were set at 5mmHg, and the LV speed was operated at 4200, 5400, 6000, 6600, 7200 RPM. The Intra-Aorta Pump speed was operated at 3000, 3600, 4200, 4800 and 5400. At each pump speed the clamp was adjusted to derive the desired pressure from 20mmHg to maximum, respectively. The pressure and blood flow in each key point was recorded by pressure transducer and flowmeter, respectively to compare with the predicted pressure and blood flow derived from the model by implementing (10) with the appropriate $\mathsf{A}; \mathsf{B}$ and$\mathsf{C}$, according to the model $\mathsf{D} = 0$.

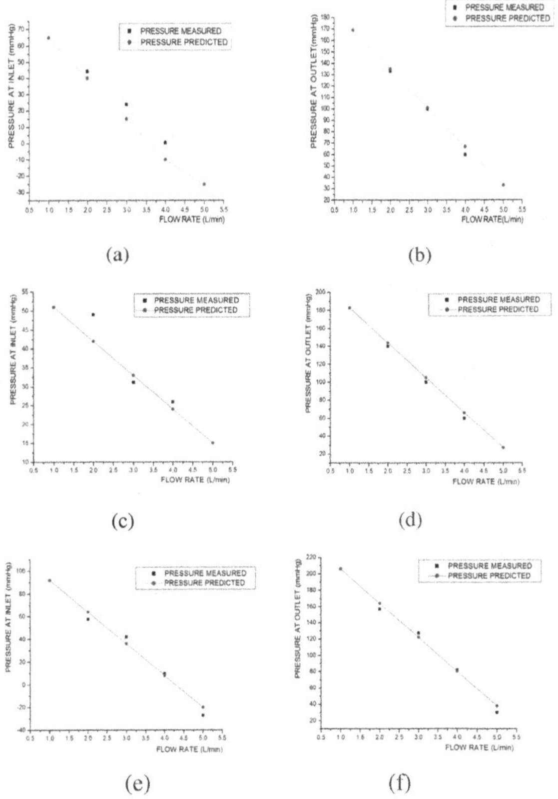

Fig. 6. Pressure prediction versus measurement with various blood flows at different places: (a) the pressure was measured at inlet of the Intra-Aorta Pump; (b) the pressure was measured at outlet of the Intra-Aorta Pump; (c) The pressure was measured at inlet of the Intra-Aorta Pump; (d) The pressure was measured at outlet of the Intra-Aorta Pump;(e) and (f) were the pressure measured at inlet and outlet of the Intra-Aorta Pump, respectively

Figure 5 show plots of the predicted pressure at the inlet and outlet of the Intra-Aorta pump versus the measured one at the pump speed is 5400RPM and the blood flow was adjusted from 2L to 5L with 1L increment. Figure 5(a)-(b) show that the speed of the LV is 6556RPM; the Figure 5(c)-(d) shows that the speed of the LV is 6960RPM; and the Figure 5(e)-(f) was measured when the LV speed was changed at 7547RPM. We note that the predicted pressure (full line) very closes to the measure pressure (dot), that means the model can predict the change of the pressure and blood flow accurately; and although the speed of the LV was changed from 6556 to 6960 and to 7547, when the speed of Intra-Aorta Pump remains constant, the parameters of the model was changed distinctly. That means the speed of the pump was a very important parameter to the model.

4 Conclusion

A nonlinear model was developed to predict the pressure and blood flow of the Intra-Aorta Pump. The predicted data were derived by solving the system of equations which parameters depended on the speed of the pump. The equation is get from the electric circuit analysis techniques and is based on a model for the vascular which includes the relationship between pressure drop and blood flow. As the model is focused on the stationary state of the system, the compliance of the vascular and the inertance of the blood are neglected. Therefore there is no need to modify the model when the situation of the circulating system was changed. The model was tested experimentally on a test loop with the Intra-Aorta Pump and was able to predict the steady state pressure and blood flow in the system. And the parameters of the model changed distinctly along with the speed change of the pump. That meant the model could cover the entire range of the system, it could be used to predict the pressure and blood flow of the system in whole range.

This model can be used to predict the change of pressure and blood flow, before the Intra-Aorta Pump was place into the human body, to modify the control strategy for a specific patient. The model can be integrated in the DSP or MCU control system for the Intra-Aorta Pump control system. And the model also can be used to evaluate the performance of the Intra-Aorta Pump system for improvement.

5 Discussion

The experimental results demonstrated that the model can predict the change of pressure and blood flow in the circulating system accurately. However the result is derived based on the test loop in vitro, therefore the condition of the test loop is different from the native circulating system of human body and the parameters derived from these experiments can not be used for patient directly. The parameters of the model will be measured from the patient in the future for design a special model for him [11, 12, 13]. On one hand the object is focus on the steady state of the pressure and blood flow in the loop, therefore the compliance of the vascular and the inertance of the blood are ignored to simplify computation. When the model is used in clinical experiment, the two of components has been considered, and the parameters in the model are derived by solving systems of differential equations with variable coefficients. As it is difficult to solve the differential equations with variable coefficients, the parameters is determined by checking look-up table established previously according to the physiological parameters of the patients. On the other hand as the Intra-Aorta Pump has two parts that is only connected by magnetic coupling, which means the situation of the blood pump transplanted in human body can not be derived by the controller, such as the rotation speed of the impeller and the controller can only control the Intra-Aorta Pump by way of open loop that assume the speed of the impeller is same as the one of the driver motor [9, 10]. Utilizing the model we can derive the impeller speed by solving the equation such as $R_{PS}(\omega) = R_{computation}$ that only associate with the speed of impeller and make the speed of the impeller as a state variable added to the control strategy.

As the model is established in vitro and it is focused on the steady state pressure and blood flow, the pulsatility generated by native heart will be consider adding into the model in the future for clinical experiment.

Acknowledgment. This work partly sponsored by the Beijing Nova Program (33015998200701), and Scientific Research of Beijing Municipal Commission of Education (JC015999200802).

References

1. Henning, E.: New devices under investigation. In Mechanical Circulatory Support. In: Mechanical Circulatory system, pp. 155–184. Springer, Heidelberg (1997)
2. Debacey, M.E.: Left ventricular bypass pump for cardiac assistance. Clinical experience, Am. J. Cardiol. 27(1), 3–11 (1971)
3. Bulter, K.C., Moise, J.C., Wampler, R.K.: The Hemopump-a new cardiac prothesis device. IEEE Trans. Biomed. Eng. 37(2), 193–196 (1990)
4. Tayama, E., Olsen, D.B., Ohashi, Y., et al.: The Debakey ventricular assist device: Current status in 1997. Artif. Organs. 23(12), 1113–1116 (1999)
5. Noon, G.P., Morley, D., Irwin, S., et al.: Development and clinical application of the MicroMed Debakey VAD. Curr. Opin. Cardiol. 15(3), 166–171 (2000)
6. Westaby, S., Katsumata, T., Houel, R., et al.: Jarvik 2000 heart: Potential for bridge to myocyte recovery. Circulation 98, 1568–1574 (1998)
7. Frazier, O.H., Myers, T.J., Jarvik, R.K., et al.: Research and development of an implantable, axial-flow left ventricular assist device: The Jarvik 2000 heart. Ann. Thorac. Surg. 71(3), 125–132 (2001)
8. Pedro, B., Paulo, B.G., Ida'gene, A., et al.: Mechanical behavior and stability of the internal membrane of the InCor ventricular assist device. Artif. Organs 25(11), 912–921 (2001)
9. Li, G.R., Ma, W., Zhu, X.: Development of a new left ventricular assist device: The dynamic aortic valve. Asaio J. 47(3), 257–260 (2001)
10. Li, G.R., Zhao, H., Zhu, X.D., et al.: Preliminary in vivo study of an intra-aortic impeller pump driven by extracorporeal whirling magnet. Artif. Organs 26(10), 890–893 (2002)
11. Yu, Y.-C., Simaan, M.A., Mushi, S.E.: Performance Prediction of a Percutaneous Ventricular Assist System Using Nonlinear Circuit Analysis Techniques. IEEE Transactions on Biomedical Engineering 55(2), 419–429 (2008)
12. Qiu, A.-Q., Bai, J.: Multiple modeling in the study of interaction of hemodynamics and gas exchange. Computers in Biology and Medicine 31, 59–72 (2001)
13. Li, X., Bai, J., He, P.: Simulation study of the Hemopump as a cardiac assist device. Medical & Biological Engineering & Computing 40, 344–353 (2002)
14. DeBakey, M.E., Thomas, J., Fossum, T.W.: Complications common to ventricular assist device support are rare with 90 days of DeBakey VAD. ASAIO Journal 47(3), 288–292 (2001)
15. Li, G.-R., Zhu, X.-D.: Preliminary study of a new concept for left ventricular assistance dynamic aortic valve. Journal of Biomedical Engineering 16(1), 116–119 (1999)
16. Michael, S., Martin, M., Hans-H, S.: The influence of a nonlinear resistance element upon in vitro aortic pressure tracings and aortic valve motions. ASAIO Journal 50(5), 498–502 (2004)

The Sampling Synchronization of the Chaotic System with Different Structures Based on T-S Model

Aiping Li[1], Dongsheng Yang[1], Aimin Wu[2], and Enjia Wang[3]

[1] School of Information Science and Engineering
Northeastern University, Shenyang, China
{liaiping,yangdongsheng}@mail.neu.edu.cn
[2] Gudianpuyang Thermal & Power Co. Ltd, Puyang, China
gdprwam@163.com
[3] School of Information Science and Engineering
Northeastern University, Shenyang, China

Abstract. This paper deals with the problem of the generalized synchronization via T-S fuzzy models for synchronization with non-identical chaotic system via lyapunov functional method and matrix inequality techniques. A new kind of fuzzy sampling controllers and the stability criterion of the synchronization error system were proposed. The simulation results show the effectiveness of the method.

1 Introduction

Synchronization technology based on the automatic control theory and computer technology for the complex system control have achieved rapid developments. Based on self-similar structure of the synchronization problem, studying the state synchronization problem of two chaotic systems with different structures and different initial states has wider applications than the general sense of chaos synchronization[1]. Based on the study of the T-S fuzzy model data controller, the sampling period of the study that designing the fuzzy sampling data system observer [2, 3]in the use of hopping system theory and the multirate fuzzy robust controller[4] in the use of cycle lyapunov function may be constant or integer times of the corresponding. Therefore, it is restricted in practice. This paper accesses to better synchronize error curves for fuzzy sampling controllers designed in the synchronization problem of two fuzzy chaotic systems with different structures.

2 Problem of Synchronous Sampling Control

Assuming the matrix has a match dimension, the unit matrix and zero matrix can be said I and 0 respectively. The sign is on behalf of the symmetric block of the symmetric matrix, the standard symbols>(<) is used to say a positive (negative definite)matrix. The Inequality $X > Y$ says that the matrix $X - Y$ is a positive matrix. For the T-S fuzzy model chaotic system, the differences in structure perform the differences in the last variables and the parameters behind.

W. Liu et al. (Eds.): WISM 2009, LNCS 5854, pp. 186–194, 2009.

When we use the singleton fuzzifier and weighted average defuzzifier, the output of the above fuzzy driving system is inferred as follows:

$$\dot{x}_d(t) = \sum_{i=1}^{l} \mu_i(z(t)) A_i x_d(t) \tag{1}$$

where

$$\mu_i(z(t)) = \frac{m(z(t))}{\sum_{i=1}^{l} m(z(t))}, \mu_i(z(t)) \geq 0,$$

$$\sum_{i=1}^{l} \mu_i(z(t)) = 1, m_i(z(t)) = \prod_{j=1}^{l} M_{ij},$$

$$\sum_{i=1}^{l} m(z(t)) \geq 0, m(z(t)) \geq 0.$$

The response system can be represented as follows:

$$\dot{x}_r(t) = \sum_{i=1}^{l} h_i(\hat{z}(t)) \hat{A}_i x_r(t) + u(t), \tag{2}$$

where

$$h(\hat{z}(t)) = \frac{n_i(\hat{z}(t))}{\sum_{i=1}^{l} n(\hat{z}(t))}, h(\hat{z}(t)) \geq 0$$

$$\sum_{i=1}^{l} h(\hat{z}(t)) = 1, n_i(\hat{z}(t)) = \prod_{i=1}^{\infty} F_{ij},$$

$$\sum_{i=1}^{l} n_i(\hat{z}(t)) \geq 0, n_i(\hat{z}(t)) \geq 0.$$

Assuming driving system and response system have the same rules, however, their structures have a difference that is $A_i \neq \hat{A}_i$.We can design response system controller $u(t)$ and make the closed-loop error system consisting of the driving system and response system achieve stability in the asymptotic.

From (1) and (2), the driving system and response system simultaneous errors can be defined as:

$$\dot{e}(t) = \sum_{i=1}^{l} \mu_i(z(t)) A_i x_d(t) - \sum_{i=1}^{l} h_i(\hat{z}(t)) \hat{A}_i x_r(t) - u(t) \tag{3}$$

3 Synchronous Sampling Control of the Chaotic System with Different Structures

For (3), controller designed enables global asymptotic stability. When using the sampling control, assuming its former variables on line can be measured in a row. So we can place the zero-order holder in each sub-controller, select the controller which can be represented as:

$$u(t) = u_d(t) + u_r(t). \tag{4}$$

The driving controller $u_d(t)$ global model can be described as:

$$u_d(t) = -\sum_{i=1}^{l} \mu_i(z(t)) K_i e(t_k) \tag{5}$$

The response controller $u_r(t)$ global model can be described as:

$$u_r(t) = -\sum_{i=1}^{l} h_i(\hat{z}(t)) H_i e(t_k) \tag{6}$$

The $e(t_k)$ is the state vector of the sampling moment t_k, with the function of the zero-order holder, for the fuzzy systems with sub-right in a row of time-varying delay $\tau(t) = t - t_k, (t \in [t_k, t_{k+1}])$.

From (7), (9) and (11),

$$u(t) = -\sum_{i=1}^{l} \mu_i(z(t)) K e(t - \tau(t)) - \sum_{i=1}^{l} h(\hat{z}(t)) H e(t - \tau(t)) \tag{7}$$

Then the error system from (3) by adding the delay can be described as:

$$\begin{aligned} \dot{e}(t) = &\sum_{i=1}^{l} \mu_i(z(t)) \left\{ A_i e(t) + K_i e(t - \tau(t)) \right\} \\ &+ \sum_{i=1}^{l} h_i(\hat{z}(t)) \left\{ \hat{A}_i e(t) + H_i e(t - \tau(t)) \right\} + \varpi(t) \end{aligned} \tag{8}$$

Assuming $t_{k+1} - t_k \le h, \forall k \ge 0$, and h is on behalf of the largest sampling interval. Obviously $\tau(t) < t_{k+1} - t_k$, then $\tau(t) < h$. Error system model (8) is essentially fuzzy model with uncertain delay without boundary. The following equivalent form can be described as:

$$\begin{aligned} E\dot{\bar{e}}(t) = \begin{bmatrix} \dot{e}(t) \\ 0 \\ 0 \end{bmatrix} = &\sum_{i=1}^{l} \mu_i(z(t)) \left\{ \begin{bmatrix} 0 & I & 0 \\ A_i + K_i & -I & \frac{1}{2} I \\ 0 & 0 & I \end{bmatrix} \bar{e}(t) - \begin{bmatrix} 0 \\ K_i \\ 0 \end{bmatrix} \int_{-\tau(t)} y_1(s) ds \right\} \\ &+ \sum_{i=1}^{l} h_i(\hat{z}(t)) \left\{ \begin{bmatrix} 0 & I & 0 \\ \hat{A}_i + H_i & -I & \frac{1}{2} I \\ 0 & 0 & I \end{bmatrix} \bar{e}(t) - \begin{bmatrix} 0 \\ H_i \\ 0 \end{bmatrix} \int_{-\tau(t)} y_2(s) ds \right\}, \quad t > h, \end{aligned} \tag{9}$$

where $E = diag\ [I, 0, 0]$, $\bar{e}(t) = [e^T(t), y^T{}_\Delta(t), \omega^T(t)]^T$.

The initial condition is $e(t)=\phi(t)$, $(t\in[-h,0])$, but $\phi(t)$ is a continuous function. When it multiples the former condition $\sum_{i=1}^{l}\mu(z(t))$, we have $y_\Delta(t)=y_1(t)$. When it multiples the former condition $\sum_{i=1}^{l}h_i(\hat{z}(t))$, we have $y_\Delta(t)=y_2(t)$.

From the lemma [5]: For any $a\in R^n, b\in R^{2n}, N\in R^{2n\times n}$, $T\in R^{n\times n}, Y\in R^{n\times 2n}$ and $Z\in R^{2n\times 2n}$, we have:

$$-2b^TNa\le\begin{bmatrix}a\\b\end{bmatrix}^T\begin{bmatrix}T & Y-N^T\\ * & Z\end{bmatrix}\begin{bmatrix}a\\b\end{bmatrix},\begin{bmatrix}T & Y\\ * & Z\end{bmatrix}\ge 0.$$

Theorem 1. If the matrixes ($n\times n$ $P_1>0, P_2, P_3, P_4, P_5, P_6, Z_1, Z_2, Z_3, Z_4, Z_5, Z_6$) and the matrix $T>0$, also matrixes ($P_7>0, P_8, P_9, P_{10}, P_{11}, P_{12}, Z_7, Z_8, Z_9, Z_{10}, Z_{11}, Z_{12}$) and the matrix $\hat{T}>0$ meet the following matrix inequalities. so for all the samplings which meet $t_{k+1}-t_k\le h$, error system (9) is able to be controlled calm by Gain matrix K_i, H_i.

We have

$$\psi_1<0,\qquad \begin{bmatrix}T & [0 \quad K_i^T \quad 0]P\\ * & Z\end{bmatrix}\ge 0$$

$$\psi_2<0,\qquad \begin{bmatrix}\hat{T} & [0 \quad H_i^T \quad 0]\hat{P}\\ * & \hat{Z}\end{bmatrix}\ge 0$$

where

$$\psi_1=P^T\begin{bmatrix}0 & I & 0\\ A_i+K_i & -I & \frac{1}{2}I\\ 0 & 0 & I\end{bmatrix}+\begin{bmatrix}0 & I & 0\\ A_i+K_i & -I & \frac{1}{2}I\\ 0 & 0 & I\end{bmatrix}^T P$$
$$+hZ+\begin{bmatrix}0 & 0 & 0\\ 0 & hT & 0\\ 0 & 0 & 0\end{bmatrix},$$

$$\psi_2=\hat{P}^T\begin{bmatrix}0 & I & 0\\ \hat{A}_i+H_i & -I & \frac{1}{2}I\\ 0 & 0 & I\end{bmatrix}+\begin{bmatrix}0 & I & 0\\ \hat{A}_i+H_i & -I & \frac{1}{2}I\\ 0 & 0 & I\end{bmatrix}^T \hat{P}$$
$$+h\hat{Z}+\begin{bmatrix}0 & 0 & 0\\ 0 & h\hat{T} & 0\\ 0 & 0 & 0\end{bmatrix},$$

$$i=1,\cdots,l \qquad P=\begin{bmatrix} P_1 & 0 & 0 \\ P_2 & P_4 & 0 \\ P_3 & P_5 & P_6 \end{bmatrix}, \qquad Z=\begin{bmatrix} Z_1 & Z_2 & Z_4 \\ 0 & Z_3 & Z_5 \\ 0 & 0 & Z_6 \end{bmatrix},$$

$$\hat{P}=\begin{bmatrix} P_7 & 0 & 0 \\ P_8 & P_{10} & 0 \\ P_9 & P_{11} & P_{12} \end{bmatrix}, \qquad \hat{Z}=\begin{bmatrix} Z_7 & Z_8 & Z_{10} \\ 0 & Z_9 & Z_{11} \\ 0 & 0 & Z_{12} \end{bmatrix}.$$

It proves that if the Lyapunov-Krasovskii function can be represented as:

$$V(t)=\overline{e}^T(t)EP_\Delta\overline{e}(t)+\int_{t-h}^{t}\int_{s}^{t} y_1^T(v)Ty_1(v)dvds + \int_{t-h}^{t}\int_{s}^{t} y_2^T(v)\hat{T}y_2(v)dvds$$

When it multiples the former condition $\sum_{i=1}^{l}\mu_i(z(t))$, we have $P_\Delta = P$.When it multiples the former condition $\sum_{i=1}^{l}h(\hat{z}(t))$, we have $P_\Delta=\hat{P}$ .So we can get:

$$\begin{aligned} \dot{V}(t) &= 2\overline{e}^T(t)P_\Delta^{\ T}\begin{bmatrix} \dot{e}(t) \\ 0 \\ 0 \end{bmatrix} + hy_1^{\ T}(t)Ty_1(t) \\ &- \int_{t-h}^{t} y_1^{\ T}(s)Ty_1(s)ds + hy_2^{\ T}(t)\hat{T}y_2(t) \\ &- \int_{t-h}^{t} y_2^{\ T}(s)\hat{T}y_2(s)ds \end{aligned} \tag{10}$$

$$=\sum_{i=1}^{l}\mu_i(z(t))\left\{\begin{array}{l} 2\overline{e}^T(t)P^T\begin{bmatrix} 0 & I & 0 \\ A_i+K_i & -I & \frac{1}{2}I \\ 0 & 0 & I \end{bmatrix}\overline{e}(t) \\ -2\int_{t-\tau(t)}^{t}\overline{e}^T(t)P^T\begin{bmatrix} 0 \\ K_i \\ 0 \end{bmatrix}y_1(s)ds + hy_1^{\ T}(s)Ty_1(s) - \int_{t-h}^{t}y_1^{\ T}(s)Ty_1(s)ds \end{array}\right\}$$

$$+\sum_{i=1}^{l}h_i(\hat{z}(t))\left\{\begin{array}{l} 2\overline{e}^T(t)\hat{P}^T\begin{bmatrix} 0 & I & 0 \\ \hat{A}_i+H_i & -I & \frac{1}{2}I \\ 0 & 0 & I \end{bmatrix}\overline{e}(t) \\ -2\int_{t-\tau(t)}^{t}\overline{e}^T(t)\hat{P}^T\begin{bmatrix} 0 \\ H_i \\ 0 \end{bmatrix}y_2(s)ds + hy_2^{\ T}(t)\hat{T}y_2(t) - \int_{t-h}^{t}y_2^{\ T}(s)\hat{T}y_2(s)ds \end{array}\right\}$$

If we set $Y_i=[0 \quad K_i^T \quad 0]P, \hat{Y}_i=[0 \quad H_i^T \quad 0]\hat{P}$, according to the application of the lemma [5], $\dot{V}(t)$ can be described as:

$$\dot{V}(t) \le \sum_{i=1}^{l} \mu_i(z(t))\overline{e}^T(t)\left\{ P^T \begin{bmatrix} 0 & I & 0 \\ A_i + K_i & -I & \frac{1}{2}I \\ 0 & 0 & I \end{bmatrix} + \begin{bmatrix} 0 & I & 0 \\ A_i + K_i & -I & \frac{1}{2}I \\ 0 & 0 & I \end{bmatrix}^T P + hZ + \begin{bmatrix} 0 & 0 & 0 \\ 0 & hT & 0 \\ 0 & 0 & 0 \end{bmatrix} \right\} \overline{e}(t)$$

$$+ \sum_{i=1}^{l} h_i(\hat{z}(t))\overline{e}^T(t)\left\{ \hat{P}^T \begin{bmatrix} 0 & I & 0 \\ \hat{A}_i + H_i & -I & \frac{1}{2}I \\ 0 & 0 & I \end{bmatrix} + \begin{bmatrix} 0 & I & 0 \\ \hat{A}_i + H_i & -I & \frac{1}{2}I \\ 0 & 0 & I \end{bmatrix}^T \hat{P} + h\hat{Z} + \begin{bmatrix} 0 & 0 & 0 \\ 0 & h\hat{T} & 0 \\ 0 & 0 & 0 \end{bmatrix} \right\} \overline{e}(t)$$

From the assumption that the above-mentioned formula, we have

$$\psi_1 = P^T \begin{bmatrix} 0 & I & 0 \\ A_i + K_i & -I & \frac{1}{2}I \\ 0 & 0 & I \end{bmatrix} + \begin{bmatrix} 0 & I & 0 \\ A_i + K_i & -I & \frac{1}{2}I \\ 0 & 0 & I \end{bmatrix}^T P + hZ + \begin{bmatrix} 0 & 0 & 0 \\ 0 & hT & 0 \\ 0 & 0 & 0 \end{bmatrix},$$

$$\psi_2 = \hat{P}^T \begin{bmatrix} 0 & I & 0 \\ \hat{A}_i + H_i & -I & \frac{1}{2}I \\ 0 & 0 & I \end{bmatrix} + \begin{bmatrix} 0 & I & 0 \\ \hat{A}_i + H_i & -I & \frac{1}{2}I \\ 0 & 0 & I \end{bmatrix}^T \hat{P} + h\hat{Z} + \begin{bmatrix} 0 & 0 & 0 \\ 0 & h\hat{T} & 0 \\ 0 & 0 & 0 \end{bmatrix}.$$

Obviously, for any non-zero $\overline{e}(t)$, if $\psi_1 < 0$, we have $\psi_2 < 0$, $\dot{V}(t) < 0$ .Due to the existence of certain non-linear, condition (10) is not LMI, but it can be proved that we can get LMI by linearization. When $A_i = \hat{A}_i$, the structure of the driving system is the same to the structure of the response system, theorem 1 applies to synchronization of chaotic systems with the same structure.

4 Simulation Study of the Chaotic Sampled Control

As figure 1(a), the system is chaotic. Assuming $x \in [-d, d]$, and $d > 0$ then put the *Lorenz* system as the driving system. From *Lorenz* system phase diagram$|x_1| \le 15$, if we set d=15, we have:

$$A_1 = \begin{bmatrix} -10 & 10 & 0 \\ 28 & -1 & -20 \\ 0 & 20 & -8/3 \end{bmatrix}, A_2 = \begin{bmatrix} -10 & 10 & 0 \\ 28 & -1 & 20 \\ 0 & -20 & -8/3 \end{bmatrix}.$$

To consider figure 1(b), the system is *Rössler* system. Assuming $\hat{x} \in [-d, d]$, and $d > 0$, then put the *Rössler* system as the response system. From the system phase diagram$|\hat{x}_1| \le 10$.we have

$$\hat{A}_1 = \begin{bmatrix} 0 & -1 & -1 \\ 1 & 0.34 & 0 \\ 0.4 & 0 & -10 \end{bmatrix}, \hat{A}_2 = \begin{bmatrix} 0 & -1 & -1 \\ 1 & 0.34 & 0 \\ 0.4 & 0 & 10 \end{bmatrix}.$$

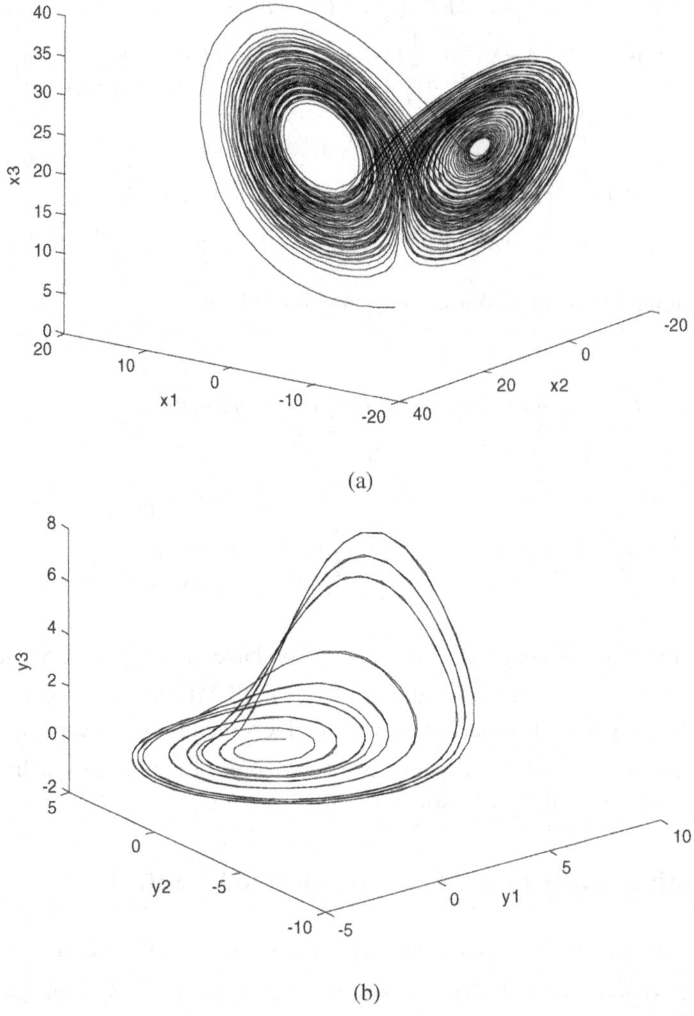

Fig. 1. The phase-plane trajectory of fuzzy Lorenz system (a) and the phase-plane trajectory of fuzzy Rosser system (b)

In the following of simulation, we set the initial condition of the driving system $x_d(0) = [0.2, 0.6, 0.1]$, the initial condition of response system $[0.8; -0.1; 0.23]$, and we can set $\gamma = \hat{\gamma} = 0.002, h = 0.5$. So we can solve the LMI condition(17), get the process parameters in order to gain control parameters. We also can make fuzzy driving system and response system simultaneously. the error in figure 2 (a) (b) (c) below:

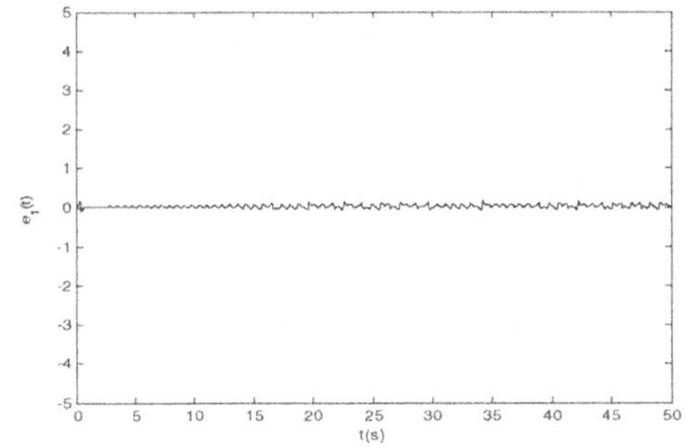

(a) $e_1(t)$

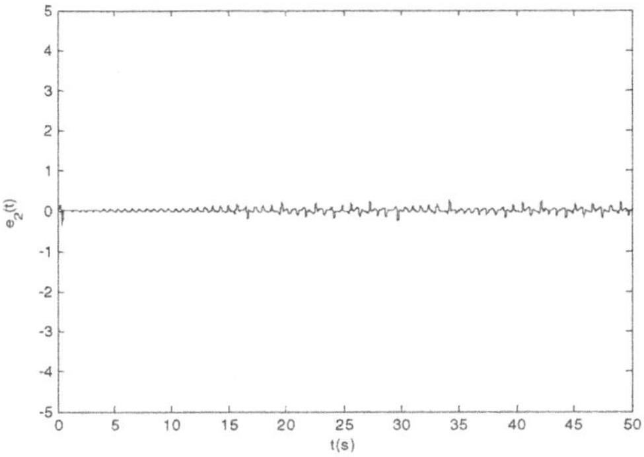

(b) $e_2(t)$

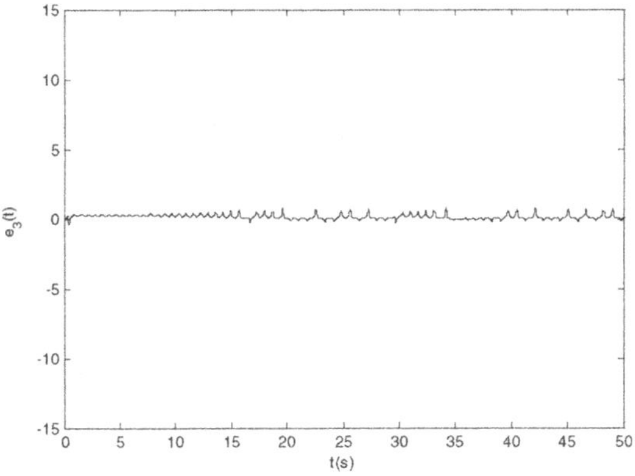

(c) $e_3(t)$

Fig. 2. The synchronization error component

5 Conclusion

This paper uses the way of inputting delay and designs sampling controller of the T-S fuzzy model, gains the sufficient condition of the chaotic synchronization expressed by LMI, realizes the sampling synchronization of chaotic system with different structures. The simulation results show that the controller has a strong ability to control for synchronization error system with two different structures, and can adapt to the different sampling intervals, can applied to non-uniform uncertain sampling situation, has a strong Anti-decipher ability. And thus has a wider application.

Acknowledgment

This work was supported by the National Natural Science Foundation of China (20080416, 60534010, 60572070, 60504006), the Program for Changjiang Scholars and Innovative Research Team in University(IRT0421).

References

[1] Yang, D., Zhang, H., Li, A., Meng, Z.: The chaotic system with different structures based on the fuzzy model synchronism. Physics 56(6), 3121–3126 (2007)

[2] Kim, D.W., Park, J.B., Lee, H.J., Joo, Y.H.: Discretisation of continuous-time T-S fuzzy system: global approach. IEE Proceedings Control Theory Applications 153(2), 237–246 (2006)

[3] Lo, J.C., Hsieh, L.T.: Observer-based control for sampled-data fuzzy systems. In: The 2005 IEEE International Conference on Fuzzy Systems, pp. 383–388 (2005)

[4] Fridman, E., Seuret, A., Richard, J.-P.: Robust sampled-data stabilization of linear systems: An input delay approach. Automatica 40, 1441–1446 (2004)

[5] Lo, J.C., Su, C.H.: Current observer for sampled-data fuzzy systems. In: 2005 American Control Conference, Portland, OR, USA, June 8-10, pp. 4210–4214 (2005)

Web Information Systems for Monitoring and Control of Indoor Air Quality at Subway Stations

Gi Heung Choi[1], Gi Sang Choi[2], and Joo Hyoung Jang[2]

[1] Dept. of Mechanical Systems Engg., Hansung University
389 Samsun-dong 3-ga, Sungbuk-gu, Seoul, 136-792, Korea
gihchoi@hansung.ac.kr
[2] Dept. of Elec. & Computer Engg. Univ. of Seoul
90 Jeonnong-dong, Dongdaemoon-gu, Seoul, Korea
simpson@uos.ac.kr

Abstract. In crowded subway stations indoor air quality (IAQ) is a key factor for ensuring the safety, health and comfort of passengers. In this study, a framework for web-based information system in VDN environment for monitoring and control of IAQ in subway stations is suggested. Since physical variables that describing IAQ need to be closely monitored and controlled in multiple locations in subway stations, concept of distributed monitoring and control network using wireless media needs to be implemented. Connecting remote wireless sensor network and device (LonWorks) networks to the IP network based on the concept of VDN can provide a powerful, integrated, distributed monitoring and control performance, making a web-based information system possible.

Keywords: Web-based Information System; In-door Air Quality (IAQ); Safety; LonWorks; Zigbee.

1 Introduction

In crowded subway stations in-door air quality (IAQ) is a key factor for ensuring the safety, health and comfort of passengers. Standard definition of IAQ that is specific to such an environment is not available. However, in terms of ensuring the safety, health and comfort of passengers and occupants in a more generic building environment, IAQ can be defined as the physical, chemical and biological properties that indoor air in the subway stations must have in order not to cause or aggravate illness of passengers or occupants, and to secure high level of safety, health and comfort to them in the performance of the designated activities for which the station has been intended and designed [1, 2, 3].

In subway stations intensive and customized information has to be provided to operators and also to passengers. In this study, a framework for monitoring and control of IAQ in subway stations is considered. Specifically, a basic framework for wireless sensor network using LonWorks/IP Gateway/Web server is suggested to better perform monitoring and control of IAQ. In view of safety, health and comfort

W. Liu et al. (Eds.): WISM 2009, LNCS 5854, pp. 195–204, 2009.

management in VDN environment, a method to guarantee interoperability between devices is also suggested.

2 Web-Based Monitoring of Indoor Air Quality (IAQ)

2.1 Need of Sensor Measurement at Multiple Locations

The important factors affecting IAQ in subway stations are identified as [4] particulate matters, CO, NO_2 , ozone, VOCs (volatile organic compounds), radon, electromagnetic fields, chemical agent, and environmental tobacco smoke (ETS) and biological agents. In the previous study [5, 6] the average values of particulate (PM_{10}) were measured at multiple locations in several subway stations. Particulate in indoor air appears, as shown in Table 1, to be denser than in outdoor air. Tunnel and platform are the weakest locations for the health of passenger. The results suggest that improving IAQ of subway station needs multiple measurements and monitoring in multiple locations. In order to make subway station more comfortable, ventilation methods must also be upgraded in tunnels and platforms.

Table 1. Measurement of average particulate density (PM_{10}, $\mu g/m^3$) [5]

	Measurement			
	OA	PL	PF	TN
A Line	84.08	126.04	139.12	156.38
B Line	110.68	138.88	152.47	170.98
C Line	80.81	133.61	133.12	163.56
Average	91.88	132.88	141.57	163.64

OA, outdoor air; PL, passenger lounge; PF, platform; TN, tunnel.

In monitoring and controlling IAQ as well as comfort in subway stations following issues must be taken into account in the system design [4]: concentrations, exposures, health effect, comfort of passengers and occupant, costs, effectiveness of monitoring and control system, and energy use. These issues imply that large amount of data need to be collected and processed effectively in the field. Real-time feedback on the conditions of pollutant sources, indoor/outdoor air quality and activities in the stations as well as device, equipment and facilities is also required for efficient management of IAQ.

Problems with the traditional data collection systems are, however, personnel costs, geographical distances, lack of sensor and data integration, poor connectivity, and complexity and cost of custom data acquisition system [4]. Remote control is not allowed in many cases and only one way data collection is normally performed.

2.2 VDN-Based Web Information System

Recently, internet access is increasingly available and affordable, and along with the "Internet" is the backbone of modern data networks. By integrating device network with IP network, the Internet can be used for distributed control of a remote system. By connecting device network into IP network, multiple sites/locations can be simply

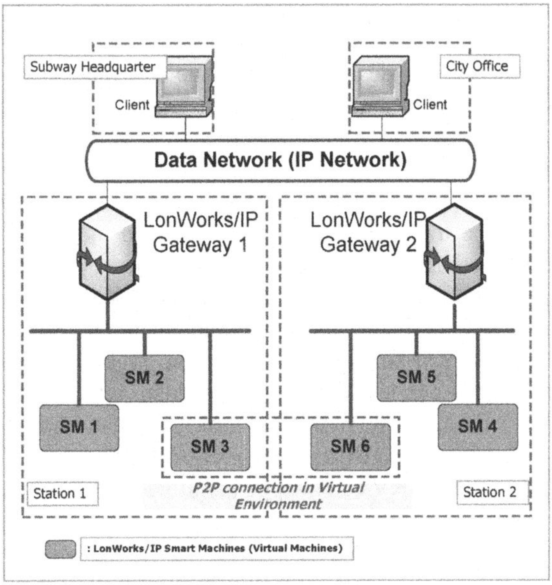

Fig. 1. VDN based web information system realized in distributed server-client environment using LonWorks/IP Gateway/Web server

integrated into a seamless "Virtual Device Network" (VDN) [7, 8]. The VDN includes one or more remote sites connected with one or more monitoring/control applications located on the Internet. The general architecture of a VDN is shown in Fig. 1. One example is virtual machine (VM) or smart machine (SM), which utilizes the virtual device network (VDN). In this configuration, monitoring and control IAQ, safety and health management and predictive maintenance, for example, can be performed both in the subway stations and in remote site through internet.

Considering the fact that physical properties, pollutants such as particulate, VOCs and biological agents are closely monitored and controlled in multiple locations in the subway station, concept of distributed monitoring and control network such as LonWorks network technology needs to be implemented. LonWorks over IP gateway then utilizes both an Ethernet capability and LonWorks compatibility. The Ethernet connection can support user to access IP network, and the LonTalk (LonWorks protocol) adapter can support user to access LonWorks network from any workstation with a TCP/IP connection.

The other key concept to this architecture is the peer-to-peer communication. Fire accident in 2003 in a subway station in the city of Daegu in Korea which killed nearly 200 passengers indicates that such configuration is particularly effective in ensuring the safety of passengers. Communication and transfer of monitoring and control data between two adjacent stations, ventilation control for example, needs to be assured to secure proper safety management and to save lives of passengers in case of fire, or any disaster.

2.3 Framework for Wireless Sensor Network

The suggested framework was designed to deal with various communication media such as twisted pair (TP), power line (PL) and Zigbee wireless media. Fig. 2 shows the concept of LonWorks/IP smart machine using various communication media for monitoring and control of IAQ. Wireless media such as Zigbee was suggested for ease of installation. Simple tests, however, showed that electromagnetic interference due to power line of the train especially when the train moves along the platform severely degrades the real-time transmission performance of the wireless media. In that case, twisted pair communication media must be used. Power line media was considered for the case where twisted pair communication channel can hardly be installed due to complexity of structure of the station. Zigbee to Lonworks converter (ZL converter) transforms wireless communication data into network variable (NV) data, then transmitted through LonWorks network. Any data must be transmitted to other devices in the LonWorks network or other client through LonWorks/IP server in the form of NV. The values of NV can then be monitored in real time.

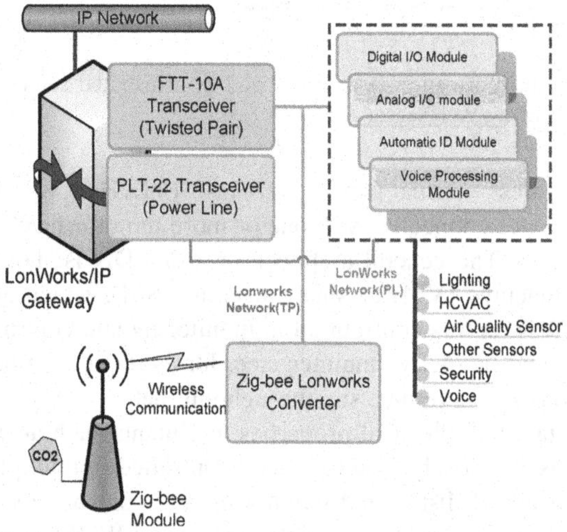

Fig. 2. Concept of LonWorks/IP smart machine using various communication media for monitoring and control of IAQ

2.4 Network Variable Binding in LonWorks Network

Different nodes in LonWorks network can communicate with each other by means of network variables. A network variable can be propagated on the network and received by other nodes. Two types of network variables, i.e., input variables and output variables are used. These variables can be bound to each other, allowing output variables to be propagated to the input variables. Fig. 3 shows how an input network variable is bound to an output network variable in other node across the internet.

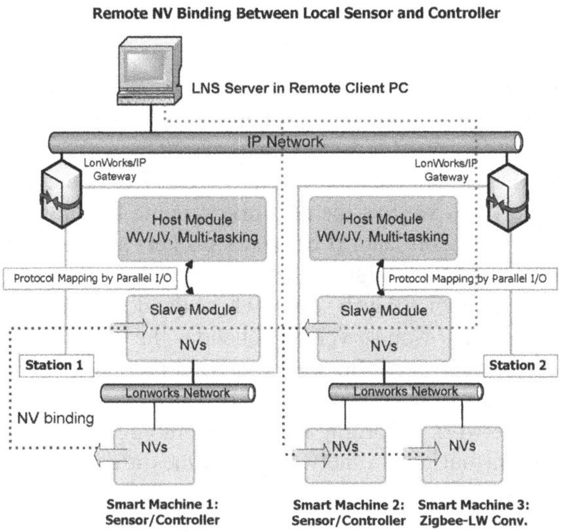

Fig. 3. Binding of the network variables between web server and LonWorks devices

Fig. 4 shows the web page for monitoring the temperature, the humidity, the level of CO, CO2 and particulate in the subway station.

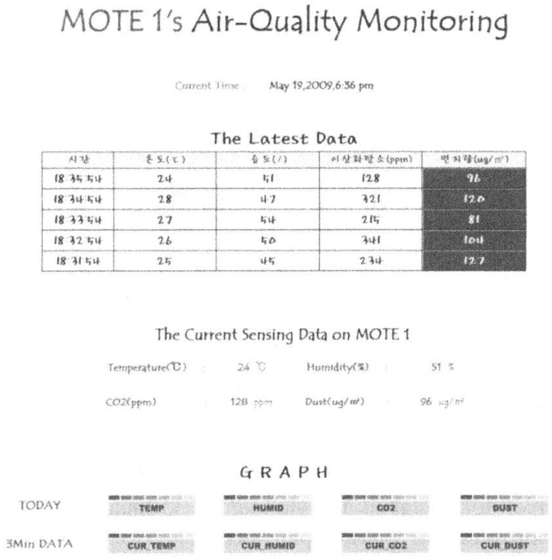

Fig. 4. LonWorks/IP web server supplied web page for monitoring physical properties at a single sensor node in the subway station

3 Security Issues in the IP Network

LonWorks/IP data transmission is performed by the tunneling technique which is the core part of EIA/CEA-852 standard established by CISCO and Echelon. Fig. 5 depicts a security check logic described in EIA/ LonWorks/IP tunneling standard. A tunneling technique encodes the communication packet of LonWorks including address information device networks as well as the network variable binding information into the data packet of an IP protocol. Data (LonWorks packet) and the 128 bit secrete word are then encoded using MD 5 algorithm to construct an IP packet. This packet is transmitted to the receiving L/IP web server where the decoding process is performed. The receiving LonWorks/IP web server first encodes the transmitted data with its own 128 bits secrete word to check if the packet is from the authorized web server. Once the security check is passed, it transmits the LonWorks data to the destination device on the lower device network according to the NV binding information it received. An L/IP web server on virtual device networks becomes a connection pass, which can transmit and receive the data/information of device networks safely in this method.

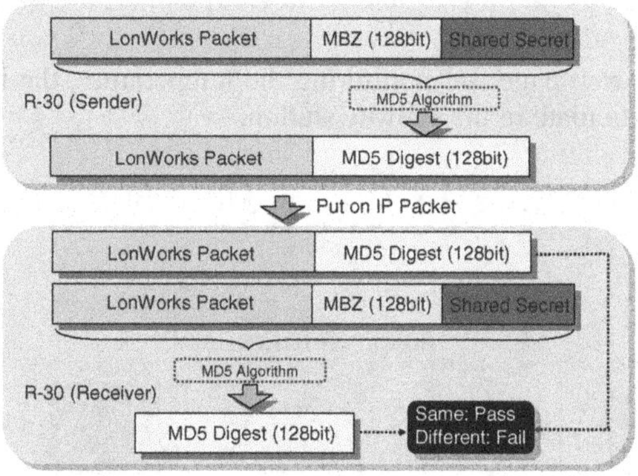

Fig. 5. MD 5 authentication as specified by EIA/CEA 852 LonWorks/IP tunneling standard

4 Experiments

Preliminary test results in a metro subway station of Line 5 showed that data loss occurred as the distance between two adjacent sensor nodes becomes longer [1]. The number of data received for a Zigbee module located 35 meters away from the sensor node on the same floor in Fig. 6, for example, was around 70% of all data transmitted. It is, however, noticed that as the node moves two to three meters away from the stairway to the lower floor data reception was impossible.

In case of inter-floors communication, the communication distance becomes much shorter than planar case and is estimated to be less than 30 meters. The maximum transmission power of Zigbee node used was 1 mW which requires the ad-hoc use of such nodes to guarantee the usefulness of wireless data communication in metro

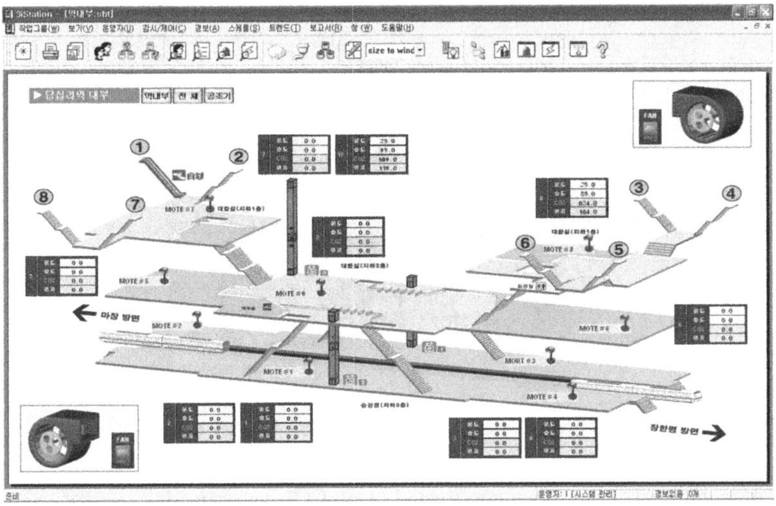

Fig. 6. A sample web information page for a metro subway station in Seoul. This web page shows multiple sensor readings at different locations in the station.

station environments. Another possible way to guarantee the whole reception of sensor reading is to use wired network for inter-stairs communication and wireless communication for each floor.

A simple networking test in a laboratory was performed to assess the transmission characteristics of wireless communication using Zigbee modules under busy and hostile environment. The sensor node was set to generate the sensing data every 10 sec. Each sensor node was composed of sensor module linked to a Zigbee wireless communication module so that the node transmits the sensor readings to the receiver node which is Zigbee to LonWorks converter (ZL converter) in Fig. 2. Fig. 7 shows the composition of each sensor node. Two sets of network configuration were used to test the networking performance.

Set 1: In this networking as shown in Fig. 8, the sensor data were gathered in local sensor node which is connected to local Zigbee node, transmitted to ZL converter through wireless media and only LonWorks network was used between ZL converter and the destination node as the reference.

Set 2: In this networking as shown in Fig. 8, the wireless device network was composed of LonWorks/IP web server, I/O devices, Zigbee to LonWorks protocol converter. The sensor data were gathered in local sensor node which is connected to local Zigbee node, and transmitted to ZL converter through wireless media and then to LonWorks/IP server. LonWorks/IP packet then transmitted to other LonWorks/IP server which in turn passes the packet data to destination I/O device bound to it. The network interface in this case is called "the virtual network interface (VNI) in VNN environment. Destination device, in general, actuates the HVAC facilities to control IAQ. Any client on the IP network can access to the local network and monitor the sensor data in remote site.

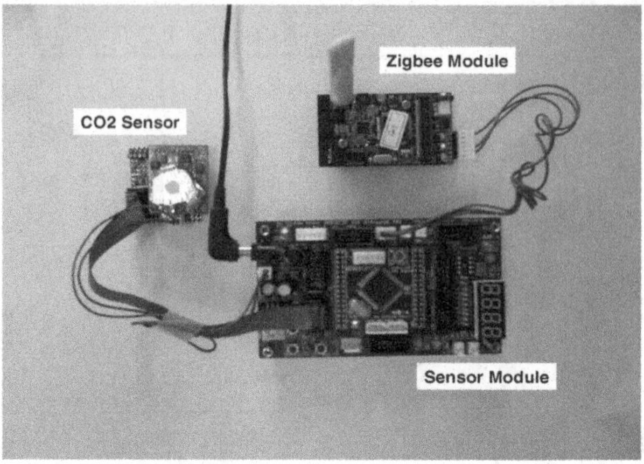

Fig. 7. A example of sensor node constructed using CO2 sensor module, sensor module and Zigbee wireless communication module

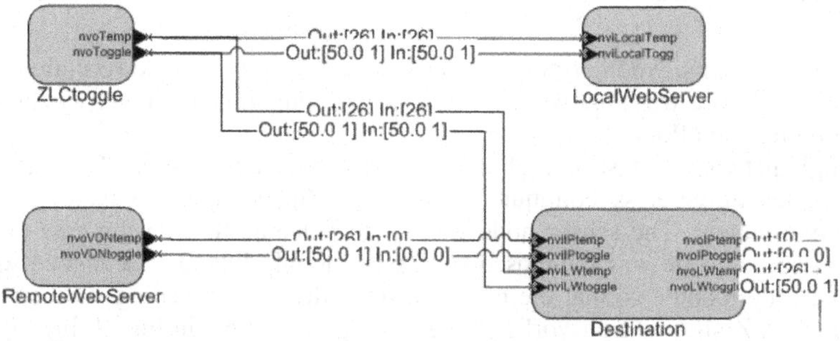

Fig. 8. Network configuration of LonWorks/IP virtual device network in network Set1 and Set2

Next, 1Hz pulse input was used to test the time delay in each network set. The output signals in destination device in four network sets were measured in oscilloscope. A thousands pulse input signals were generated and their delay were measured. In Fig. 9, the transmission delay in wireless media appears to be within a narrow range of 150~160ms. Fig. 10 shows the histogram of the delay of received signals in Set1 and Set2. For Set1, delay of 190~240ms were measured, where as 230~530ms were for Set2. Additional delay of up to 380ms in Set2 was due to delay in IP network and data processing in LonWorks/IP server. It is interesting to note, however, that the measured delay is widely and evenly spread over the 240 to 430 ms range in Set 2 where the sensor data of discrete nature were transmitted.

Previous study [3] indicates the similar results in unidirectional transmission time (UTT) test where the delay was measured in RNI configuration. In this case, the type of transmitted data was fixed and the delay time is limited around 300ms. Transmission delay is due both to the network channel and to the calculation time for protocol conversion in the web server. It takes time for the LonWorks/IP web server to convert the LonWorks network variable to the IP packet and back to NV in destination

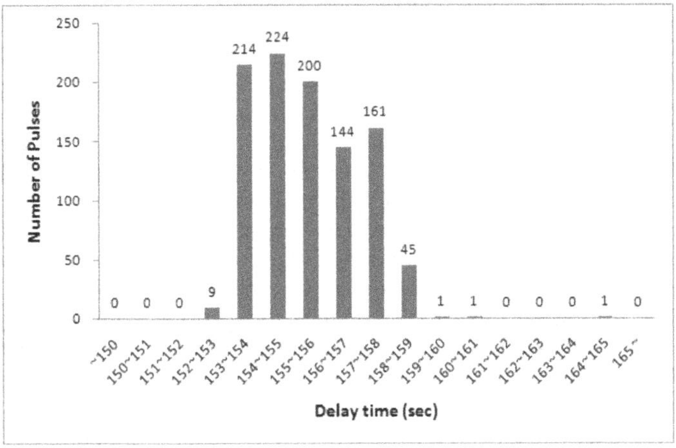

Fig. 9. Network delay in transmission of data through wireless media

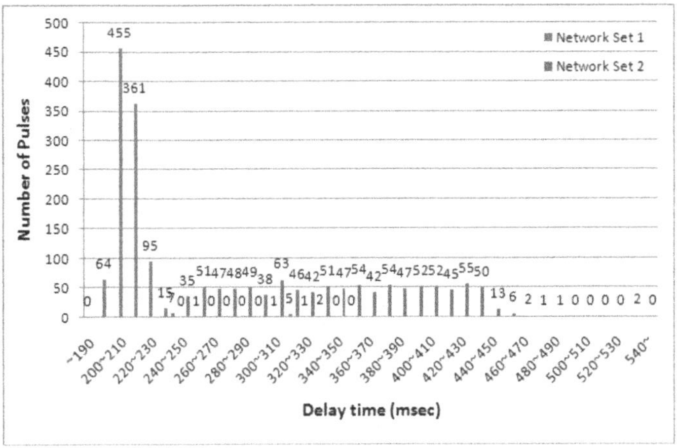

Fig. 10. Network delay in transmission of data in network Set 1 and Set 2

LonWorks/IP server. Delay in transmitting the IP packet is minimal and does not surpass that in local LonWorks network in relatively short distance within a city and data conversion appears to consume more time. Attention., therefore, needs to be paid to efficient and fast conversion of transmission data in both ZL converter and LonWorks/IP server. Compared to 200 seconds of sensing interval used for this type of systems in practice, delay in this range, however, does not appear to be a major setback.

5 Conclusions

A basic framework for web-base information system in VDN environment that can be applied to IAQ management in subway station was suggested. Since the physical variables need to be closely monitored and controlled in multiple locations in the subway stations, the concept of distributed monitoring and control network need to be implemented. The suggested framework was designed to deal with various communication media used in twisted pair (TP) local network, wide area IP network and Zigbee wireless network. Simple preliminary test indicates that delay in transmitting the data is not a severe setback in reality. Attention, however, needs to be paid to efficient and fast conversion of transmission data in both ZL converter and LonWorks/IP server. Peer-peer binding of network variable enables not only the monitoring of sensor data in remote client systems but also the control between peer devices. Any safety activity between adjacent stations in case of emergency such as fire is, therefore, possible in a timely manner. This raises the efficiency and the productivity of operation, and eventually, results in innovation in web information service to public using subway stations.

Acknowledgment. This research was supported by the City of Seoul under Grant No. GS070154 (Implementation of Metropolitan Railway Safety System Based on Broadband Convergence Network) in the year 2008.

References

1. Choi, G.H., Choi, G.S.: A Framework for Web-based Monitoring and Control of Indoor Air Quality (IAQ) in Metro Subway Station. In: Proceedings of International Symposium on Safety Science and Technology, Beijing, China, pp. 2514–2518 (2008)
2. Choi, G.H., Choi, G.S.: Design of Service System Framework for Web-based Monitoring and Control of Indoor Air Quality (IAQ) in Subway Stations. In: Proceedings of NISS, Beijing, China (2009)
3. Choi, G.H., Choi, G.S., Jang, J.H.: A Framework for Wireless Sensor Network in Web-based Monitoring and Control of Indoor Air Quality (IAQ) in Subway Stations. In: Proceedings of International Conference on Computer Science and Information Technology, Beijing, China (2009)
4. Bronsema, B., et al.: Performance Criteria of Buildings for Health and Comfort, Report of ISIAQ-CIB, Task Group TG42 (2004)
5. Lee, H.W., Park, J.-k., Jang, N.-S., Lee, H.-R., Kim, H.-M.: Analysis of Ambient Air Quality Level in Subway Area in Busan Metropolitan City. J. Environmental Science 2, 207–215 (2003)
6. Chang, J.-W., Cho, J.-J., Choi, W.-G., Park, D.-S., Jung, W.-S., Kim, T.-O.: A Study of PM-10 and Heavy Metal Characteristics in the air at the each site of subway station. J. Environmental Science 4, 125–129 (2005)
7. Song, K.W., Choi, G.S., Choi, G.H., Kim, J.S.: LonWorks-based Virtual Device Network (VDN) for Predictive Maintenance. International Journal of Applied Electromagnetics and Mechanics 18(1-3), 67–80 (2003)
8. Choi, G.H.: Lonworks Based Virtual Device Network (VDN) for Predictive Maintenance. In: Proc. KOSOS Fall Conference, Samchuk, pp. 324–327 (2008) (in Korean)

A Layered Overlay Multicast Algorithm with PSO for Routing Web Streams

Yuhui Zhao[1], Junwei Wang[1], Yuyan An[2], and Fan Xia[3]

[1] Northeast University at Qinhuangdao, Qinhuangdao, 066004, China
[2] Hebei Vocational College of Foreign Language
[3] Hebei Vocational and Technical College of Building Materials
yuhuizhao@mail.neuq.edu.cn, wjw0331@163.com,
yuyanan12345@126.com

Abstract. For satisfying the request of multi-constrained QoS (Quality of Service) routing of large-scale multi-domain web streams, a new architecture named Layered Overlay Multicast Network (LOMN) is adopted to improve the service capabilities, which used a layered overlay multicast algorithm with PSO (Particle Swarm Optimization). The proposed algorithm can solve the problem's NP-completeness of the routes' QoS and meeting with the fuzziness of the network status description, and find the QoS-satisfied routes using an effective mathematical model and the PSO. Simulated implementations and performance evaluations have been done over some actual and virtual network topologies, showing that the method can obtain a near optimal solution for building the core multicasting tree.

Keywords: Web Streams, QoS, Overlay Multicast, Particle Swarm Optimization, Layered Overlay Multicast Network (LOMN).

1 Introduction

There are many web-based streaming media applications, such as video conferencing, video on demand, real-time control, on-line shopping, gaming, stock exchange, and so forth, which request lower delay, higher QoS, and more transmission efficiency. So a more effective multicast architecture and algorithm are expanded to support the stream media based on web applications. This kind of Multicast is a QoS routing problem of sending a message from a single source to multiple destinations, which has two kind of solution, one is based on IP, and the other on application-layer.

IP multicast makes it fast, resource efficient to support very large multicast groups. It is still far from being widely deployed because there are the protocol complexity, network and end-system heterogeneity and limited multicast IP address[1]. Application-layer multicast(ALM), also called overlay multicast, ALM's multicast-related features are implemented at end hosts, data packets are transmitted between end hosts via unicast, and are replicated at end hosts, so can be easily deployed. But most solutions are higher delay, and lower Experience of the Quality (QoE).

For solving the above problems, we suggest a multicast architecture called Layered Overlay Multicast Network (LOMN), which is different the existing ALM and adopts

W. Liu et al. (Eds.): WISM 2009, LNCS 5854, pp. 205–213, 2009.

the PSO-based layered overlay multicast algorithm(PLOM) to build the trees. LOMN is composed of Service brokers (SvBs). In this paper, we use a PSO method to build the core multicast tree (CMT) and route the web streams to all destinations. Because the CMT with multi-constraint can be reduced to the Minimal Steiner Tree Problem (MSTP) which has been shown to be NP-Complete, it is necessary for CMT to adopt a heuristic routing method to form the core overlay multicasting tree and balance overlay traffic load on SvBs and overlay links.

The rest of the paper is organized as follows. The related works are discussed in Section 2. In Section 3, we introduce Layered Overlay Multicast Network. Then, the PLOM Algorithm for Building the Core Multicast Tree under Delay Constraint for LOMN is described in Section 4. And a series of simulations and results are discussed in Section 5. Finally, we draw the conclusions in Section 6.

2 Relative Works

Recently the efforts on overlay networks have been very active. In the proposals, Some is for specific application such as Host multicast [2], content distribution networks [3], peer-to-peer file sharing [4]; and the others aiming at developing generic overlay service networks for a variety of applications. For example, Yoid [5] is a generic overlay architecture, which is designed to support a variety of overlay applications that are as diverse as net news, streaming broadcasts, and bulk email distribution. Another similar effort is the Planet-lab [6] experiment whose goal is to build a global testbed for developing and accessing new network services. X-Bone [7] operates at the IP layer and based on IP tunnel technique. OverQoS [8] is an architecture proposed to provide Internet QoS using overlay networks.

A delay-constrained unicast routing algorithm is proposed in [9], which computes the route by constructing delay vector and cost vector respectively. In [10], those links not satisfying the user bandwidth requirements are pruned at first, and then SPF gets the route with the shortest delay. In [11], SPF is used along with fuzzy tower, and the routes with the user QoS requirements satisfied are found. In [12], based on the probability theory, multiple parameters, including delay and cost, etc., are transformed into a comprehensive one, and thus the QoS routing problem is simplified significantly. In [13], genetic algorithms are used respectively, and the specific optimal or sub-optimal routes are found.

3 Layered Overlay Multicast Network

LOMN is hierarchical and consists of service brokers (SvB) that are strategically deployed by the LOMN provider who dimensions its overlay network according to end user requests and sells its multicast services to group coordinators via a service contract. Most of the Internet domains can have one or more core SvBs that depends on the needs of live media communication.

SvBs are specialized nodes that can be placed in the Internet to provide generic overlay service and support to overlay multicast applications. These SvBs are interconnected over the transfer layer to form the LOMN. SvBs can be placed either at the edge of a domain or in the core and subscribe high bandwidth connections to the Internet backbone. Usually they are designed to different levels according to the relationship of the ASs. This addresses knowledge of the SvBs in the neighboring domains can be incorporated during deployment or by exchange of messages. The SvBs are also responsible for encapsulation and decapsulation of the outgoing and incoming packets of the overlay network, respectively.

Fig.1. is the Layered Overlay Multicast Network architecture. There are two kind of SvB, the core service broker(corSvB) and the access service broker(acSvB). All the brokers form a hierarchical mesh, the core brokers form the core mesh and the access brokers form the access mesh. This can be formulated as follows.

If the physical network topology is modeled as an undirected graph $G = (V,E)$, where V and E denotes the sets of network nodes and physical links respectively, the overlay dimensioning problem can be formulated as follows: given a set of groups $\{M\}$ with group member distribution and bandwidth requirement, and a physical network topology $G = (V,E)$, find a virtual topology G' on top of G so that G' can accommodate all the groups while keep the cost of G' minimum under the bandwidth waste threshold. Here, we assume the multicast group set $\{M_i\}$ is obtained from the service contract or from long-term measurement in the steady state of the network and group dynamics. The G' is the LOMN network.

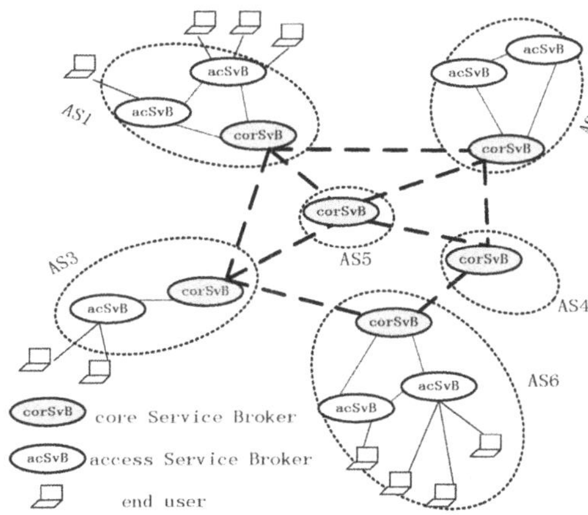

Fig. 1. The Layered Overlay Multicast Network architecture

4 Mathematical Models the PLOM

According the LOMN graph, Using D_x, denotes a ASx, the set $\{ D_x, x \in N \}$, $G = (V, E)$, V, E are the set of the SvBs and the edges of the overlay edges. V_i denotes the node SvB$_i$, E_{ij} denotes the overlay edge between the node SvB$_i$ and the node SvB$_j$, $|V|$ denotes the numbers of the SvBs, $|E|$denotes the numbers of the connections from the source to the destination SvB.

For each node $v_i \in V(i=1, 2, 3, ..., |V|)$, consider the following parameters: delay, delay jitter and error rate. Just for simplicity, the parameters of the node are merged into those of its downstream edge. For each edge $e_{ij} \in E(i, j=1, 2, 3, ...,|V|)$, consider the following parameters: available bandwidth bw_{ij}, delay del_{ij}, delay jitter jt_{ij} and error rate ls_{ij}.

Assume that the source node is $v_s \in V$ and the destination nodes are $\{v_{di}\} \in V(i=1, 2, 3, ..., |V|)$, find the specific routes p_{sd} between v_s and $\{v_{di}\}$, satisfying the following constraints for every path:

A1) The available bottleneck bandwidth of p_{sdi} is not smaller than the bandwidth requirement bw_req_i of v_{di}.

A2) The total delay of p_{sdi} is not bigger than the delay requirement del_req_i of v_{di}.

A3) The total delay jitter of p_{sdi} is not bigger than the delay jitter requirement jt_req_i of v_{di}.

A4) The total error rate of p_{sd} is not bigger than the error rate requirement ls_req_i of v_{di}.

The objective of the proposed algorithm is try to maximize $Q(p_{sd})$, that is, the QoS satisfaction degree to the found p_{sd}.

The corresponding mathematical model is described as following:

$$\text{Maximize} \quad \{Q(p_{sd})\} \tag{1}$$

$$\text{s.t.} \quad \min_{e_{ij} \in p_{sd}} \{bw_{ij}\} \geq bw\_req_i \tag{2}$$

$$\sum_{e_{ij} \in p_{sd}} del_{ij} \leq del\_req_i \tag{3}$$

$$\sum_{e_{ij} \in p_{sd}} jt_{ij} \leq jt\_req_i \tag{4}$$

$$1 - \prod_{e_{ij} \in p_{sd}} (1 - ls_{ij}) \leq ls\_req_i \tag{5}$$

5 PLOM Algorithm Design

PSO is based on the concept of colony and fitness. In PSO, the position of the particle represents the feasible solution to the specific problem, and the suitability of the particle to the problem is evaluated by its fitness. Each particle tries to fly to the global best position of the particle swarm according to its own and other's local best positions, so that the optimal solution to the specific problem can be reached.

Assume that there are *n* particles in the D-dimensional searching space. The position of the *i*th particle is represented as a vector $X_i = (x_{i1}, x_{i2}, \ldots, x_{iD})$ and its local best position is denoted by $P_i = (p_{i1}, p_{i2}, \ldots, p_{iD})$. The global best position of the particle swarm is represented as $P_g = (p_{g1}, p_{g2}, \ldots, p_{gD})$. The velocity of *i*th particle is represented as $V_i = (v_{i1}, v_{i2}, \ldots, v_{iD})$. The following formulas are used to update the particle velocities and position:

$$V_{id}(t+1) = w * V_{id}(t) + c_1 * rand() * [P_{id}(t) - X_{id}(t)] + c_2 * rand() * [P_{gd}(t) - X_{id}(t)] \tag{6}$$

$$X_{id}(t+1) = X_{id}(t) + V_{id}(t) \tag{7}$$

Here, $1 \le i \le n$, $1 \le d \le D$, *rand*() is a [0, 1] random number generator, c_1 and c_2 are called cognitive and social parameters respectively, and the parameter w is called inertia weight.

5.1 Path Evaluation

According on the fuzzy theory, the membership degree functions for the path available bottleneck bandwidth, delay, delay jitter and error rate are constructed to meet with the fuzziness of the network status.

The path available bottleneck bandwidth suitability membership degree function $g_1(bw, bw\_req)$ is defined as follows:

$$g_1(bw,bw\_req) = \begin{cases} 0 & bw < bw\_req \\ \varepsilon & bw = bw\_req \\ \left(\dfrac{bw-bw\_req}{b-bw\_req}\right)^k & bw\_req < bw < b \\ 1 & bw \ge b \end{cases} \tag{8}$$

The path delay suitability membership degree function $g_2(Jp, del, del\_req)$ is defined as follows:

$$g_2(del,del\_req) = \begin{cases} 0 & del > del\_req \\ \varepsilon & del = del\_req \\ 1 - e^{-(\frac{del\_req-del}{\sigma_1})^2} & del < del\_req \end{cases} \tag{9}$$

The path delay jitter suitability membership degree function $g_3(Jp, jt, jt\_req)$ is defined as follows:

$$g_3(Jp,jt,jt\_req) = \begin{cases} 0 & jt > jt\_req \\ \varepsilon & jt = jt\_req \\ 1 - e^{-(\frac{jt\_req-jt}{\sigma_2})^2} & jt < jt\_req \end{cases} \tag{10}$$

The path error rate suitability membership degree function $g_4(Jp, ls, ls\_req)$ is defined as follows:

$$g_4(Jp, ls, ls\_req) = \begin{cases} 0 & ls > ls\_req \\ \varepsilon & ls = ls\_req \\ 1 - e^{-(\frac{ls\_req - ls}{\sigma_3})^2} & ls < ls\_req \end{cases} \tag{11}$$

In the above formula (8)-(11), ε is a positive pure decimal fraction and is much smaller than 1; bw, del, jt and ls are the available bottleneck bandwidth, delay, delay jitter and error rate of the path; k, b, σ_1, σ_2 and σ_3 all are positive constant coefficients, $k > 1$.

Thus, an evaluation matrix $R=[g_1, g_2, g_3, g_4]^T$ for each candidate path can be obtained. A weight matrix $W=[w_1, w_2, w_3, w_4](0<w_1, w_2, w_3, w_4<1)$ is given, and w_1, w_2, w_3, w_4 represent the relative importance of bandwidth, delay, delay jitter and error rate on the user QoS requirement respectively, $w_1+w_2+w_3+w_4=1$. Then, the QoS satisfaction degree of the user to the candidate path is derived as follows:

$$Q = W \times R \tag{12}$$

The bigger the value of Q is, the higher the suitability of the candidate path to the QoS requirements will be.

5.2 Algorithm Descriptions

For the description of PSO-based Layered Overlay Multicast (PLOM) steps, there need some definitions.

Definition 1. Assume *S* is a nonempty queue. A bijection from *S* to *S* is called a displacement of *S*.

Definition 2. For a queue S with *n* elements, the set of the *n!* displacements of *S* is denoted as S_n.

Definition 3. Assume $\Psi_1 \in S_n$ and $\Psi_2 \in S_n$, $\oplus$ is a binary operation on S_n, and $\Psi_1 \oplus \Psi_2$ represents doing Ψ_1 and Ψ_2 on S sequentially.

Definition 4. $\otimes$ is the reverse operation of $\oplus$.

Consider the candidate path p_{sd} as a queue, and the order of its nodes represents their sequence along p_{sd}. Doing the displacement operation on the p_{sd} can generate a new candidate path. By comparing the user QoS satisfaction degrees to these paths, the optimal route from v_s to v_d can be found.

According to the formula (6) and (7), the update of the particle velocity and position is computed as follows:

$$\begin{aligned} V_{id}(t+1) = w * V_{id}(t) \oplus c_1 * rand() * [P_{id}(t) \\ \otimes X_{id}(t)] \oplus c_2 * rand() * [P_{gd}(t) \otimes X_{id}(t)] \end{aligned} \tag{13}$$

$$X_{id}(t+1) = X_{id}(t) \oplus V_{id}(t) \tag{14}$$

The PLOM algorithm steps are described as follows:

The position and the velocity of the *i*th particle and the *i*th user QoS satisfaction degree of the candidate path in the *n*th iteration are denoted by the X_{ni}, V_{ni} and $Q(X_{ni})$ respectively. Its local best position and the global best position of the particle swarm are denoted as P_i and P_g respectively.

Step 0. Initialization: the particle swarm size is set to be *I*, the iteration times is set to be *N*, the threshold times of which the P_g is not updated is set to be *F*, *n*=1, *i*=1, *f*=1.

Step1. Generate *N* particles randomly, and compute the $Q(X_{n1}), Q(X_{n2}), \ldots, Q(X_{nI})$ according to the formula (12).

Step 2. Let $P_i = X_{ni}$, $P_g = \{P_i \mid Q(P_i) = \max\{Q(X_{n1}), Q(X_{n2}), \ldots, Q(X_{nI})\}\}$.

Step 3. Let *n*=*n*+1, and compute V_{ni} and X_{ni} according to the formula (13) and (14).

Step 4. Determine whether the corresponding path of the X_{ni} can satisfy the constraints (1)-(5). If so, go to Step 7, otherwise go to Step 5.

Step 5. Compute $Q(X_{ni})$. If $Q(X_{ni}) < Q(P_i)$, go to Step 7, otherwise let $P_i = X_{ni}$ and go to Step 6.

Step 6. If $Q(X_{ni}) \geq Q(P_g)$, let $P_g = X_{ni}$ and *f*=1.

Step 7. Let $i = i + 1$. If $i > I$, go to Step 8, otherwise go to Step 3.

Step 8. Let $f = f + 1$. If $n > N$ or $f > F$, output the corresponding route of the P_g and the algorithm ends, otherwise go to Step 3.

6 Performance Evaluations

Simulations have been done on NS (Network Simulator) 2 [14] platform. The SPF-based unicast routing scheme, GA (Genetic Algorithm)-based QoS routing scheme [15] and the proposed scheme have been simulated over some actual and virtual network topologies and performance evaluations have been done. Three example topologies used in the simulation have been shown in Fig.2 a, Fig.2 b and Fig.2 c. For the proposed scheme, the particle swarm size is set to be 25, the initial velocity of the particle is set to be 0, and the iteration times is set to be 25. For the GA-based one, the population size is set to be 25, the crossover probability is set to be 0.87, the mutation probability is set to be 0.1, and the roulette selection policy is adopted.

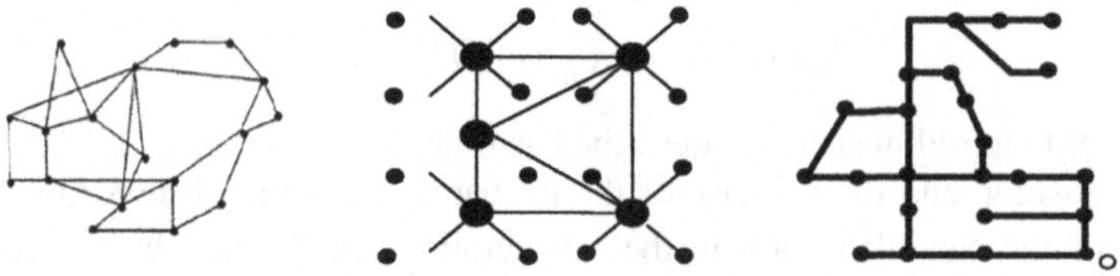

a Topology 1 b Topology 2 c Topology 3

Fig. 2. The Simulated Topology

Table 1. Request Succeeded Rate

Algorithm \ Topology	Topology 1	Topology 2	Topology 3
SPF-based unicast routing scheme	90.97%	90.35%	87.31%
GA-based QoS unicast routing scheme	98.13%	98.69%	95.93%
PLOM	98.76%	99.67%	97.80%

Table 2. User QoS Satisfaction Degree

Algorithm \ Topology	Topology 1	Topology 2	Topology 3
SPF-based unicast routing scheme	80.41%	79.68%	76.62%
GA-based QoS unicast routing scheme	81.38%	82.54%	83.96%
PLOM	82.04%	83.61%	85.30%

Comparisons of the request succeeded rate and the user QoS satisfaction degree among the SPF-based routing scheme, GA-based QoS routing scheme and the proposed scheme are shown in Table 1 and Table 2 respectively. It has shown that the PLOM algorithm is efficient and usually superior to the other ones.

7 Conclusions

In this paper, a PSO-based overlay multicast algorithm is proposed for the LOMN, considering the QoS requirements of the user (such as bandwidth, delay, delay jitter and error rate) and the fuzziness of the network status. With the suitability membership degree functions of the path available bottleneck parameters are defined, the routes from the source to the destination nodes with the maximum user QoS satisfaction degree can be found. Simulations have shown that the proposed algorithm is feasible and effective.

Acknowledgements

This work is based on the project 200801451101 which Supported by Doctoral Program Foundation of Institutions of Higher Education of China.

References

1. Diot, C., Levine, B., Lyles, J., Kassem, H., Balensiefen, D.: Deployment issues for the IP multicast service and architecture. IEEE Network (January 2000)
2. Zhang, B., Jamin, S., Zhang, L.: Host multicast: A framework for delivering multicast to end users. Presented at the INFOCOM 2002, New York (June 2002)
3. Krishnamurthy, B., Wills, C., Zhang, Y.: On the use and performance of content distribution networks. Presented at the ACM SIGCOMM Internet Measurement Workshop, San Francisco, CA (November 2001)
4. Stoica, I., Morris, R., Karger, D., Frans Kaashoek, M.: Chord: A scalable peer-to-peer lookup service for internet applications. In: Proc. ACM SIGCOMM, August 2001, pp. 149–160 (2001)
5. Francis, P.: Yoid: Extending the Internet Multicast Architecture, `http://www.aciri.org/yoid/docs/index.htm`
6. Duan, Z., Zhang, Z., Hou, Y.T.: Bandwidth provisioning for service overlay networks. Presented at the SPIE ITCOM Scalability and Traffic Control in IP Networks (II) 2002, Boston, MA (July 2002)
7. XBone, `http://www.isi.edu/xbone`
8. Subramanian, L., Stoica, I., Balakrishnan, H., Katz, R.H.: Over QoS: Offering Internet QoS using overlays. Presented at the HotNet-IWorkshop, Princeton, NJ (October 2002)
9. Douglas, S., Hussein, F.: A Distributed Algorithm for Delay-Constrained Unicast Routing. IEEE/ACM Transactions on Networking 8(2), 239–250 (2000)
10. Wang, Z., Crowcroft, J.: QoS Routing for Supporting Resource Reservation. IEEE Journal on Selected Areas in Communications 14(7), 1228–1234 (1996)
11. Wang, X.W., Yuan, C.Q., Huang, M.: A fuzzy-tower-based QoS unicast routing algorithm. In: Yang, L.T., Guo, M., Gao, G.R., Jha, N.K. (eds.) EUC 2004. LNCS, vol. 3207, pp. 923–930. Springer, Heidelberg (2004)
12. Kim, M., Bang, Y.-C., Choo, H.: New parameter for balancing two independent measures in routing path. In: Laganá, A., Gavrilova, M.L., Kumar, V., Mun, Y., Tan, C.J.K., Gervasi, O. (eds.) ICCSA 2004. LNCS, vol. 3046, pp. 56–65. Springer, Heidelberg (2004)
13. Barolli, L., Koyama, A.: A genetic algorithm based routing method using two QoS parameters. IEEE Computer Society 8(1), 7–11 (2002)
14. Network-Simulator Task Force, NS Tutorial (2000), `http://www.isi.edu/nsnam/ns/tutorial/nsindex.html`
15. Xuan, G.N., Cheng, R.W.: Genetic algorithms and engineering design, pp. 72–105. Science Press, Beijing (2000)

Solving University Course Timetabling Problems by a Novel Genetic Algorithm Based on Flow

Zhenhua Yue[1], Shanqiang Li[2], and Long Xiao[3]

[1] Dept. of Material Physics, Harbin University of Science and Technology, Harbin, China
[2] Dept. of Applied Mathematics, Harbin University of Science and Technology, Harbin, China
[3] Dept. of Electronic Information Science and Technology, Harbin University of Science and Technology, Harbin, China
liufengqiuhit@126.com

Abstract. Since the University Course Timetabling Problem (UCTP) is a typical sort of combinatorial issues, many conventional methods turn out to be unavailable when confronted with this complex problem where lots of constraints need to be satisfied especially with the class-flow between floors added. Considering the supreme density of students between classes, this paper proposes a novel algorithm integrating Simulated Annealing (SA) into the Genetic Algorithm (GA) for solving the UCTP with respect to the class-flow where SA is incorporated into the competition and selection strategy of GA and concerning the class-flow caused by the assigned timetable, a modified fitness function is presented that determines the survival of generations. Moreover, via the exchange of lecturing classrooms the timetable with minimum class-flow is eventually derived with the values of defined fitness function. Finally, in terms of the definitions above, a simulation of virtual situation is implemented and the experimental results indicate that the proposed model of classroom arrangement in the paper maintains a high efficiency.

Keywords: University course timetabling, Genetic algorithm, Simulated Annealing algorithm, Flow.

1 Introduction

University course scheduling is a NP-hard problem, which is complicated to solve through conventional methods, such as backtracking, 0-1 programming or general evolutionary algorithms, and the total amount of its computation increases exponentially with the problem size. Additionally, being a typical kind of combinatorial optimization problem, University Course Timetabling Problems (UCTPs) was previously studied by several universities and scientific research agencies and some practical results were derived [1]. However, most problems they analyzed are far from generalization in addition to that most of the papers [1~7] related to this problem have not taken the class-flow into account caused by their obtained timetable, where class-flow problems refer to the excessive flow between floors during the class break resulting from the unreasonable course arrangement to which the unfeasibility of the conventional methods [7] listed above accelerates the research and

W. Liu et al. (Eds.): WISM 2009, LNCS 5854, pp. 214–223, 2009.

exploration of novel algorithms. 0-1 programming method [7] is used previously and the obstacle encountered is hard to tackle resulting from that numerous codes need to be modified each time we implement the optimization of class arrangement. Consequently, other advanced methods are considered as the development of intelligent algorithms comprising genetic algorithm (GA), tabu search, simulated annealing (SA), ant colony algorithm and dynamic evolutionary method. These referred algorithms have largely improved the search ability of optimal results in each specific problem and hence in order to avoid the flaws in each method the hybrid of intelligent algorithms gradually becomes a prevailing trend resulting from its inter-complementation that maintains a high practicality and can generate more promising results. Thus, in our paper the SA algorithm are incorporated into the selection and competition strategies of the considered GA method, thus initially expanding the search space of the algorithm and shrinking the search domains gradually in the medium time interval that eliminates the involvement of local optimal space plus the acceleration of convergence to the optimal results finally.

In the paper each defined chromosome stands for a certain timetable, the events of which call for homogeneous classrooms, yet the assigned classroom may not live up to their requirements. The position of designated chromosome, whose contained information represents the teaching locations, denotes the arranged classes, and via the exchange of teaching locations, namely chromosome mutation, the optimal timetable is eventually derived with the values of fitness function with respect to class-flow.

The paper is structured as follows. After the introduction section, algorithm construction for solving university course timetabling problems based on flow is proposed in Section II. In Section III, the proposed approach is applied to the virtual situation to illustrate the ideas and results above. Section IV concludes this paper.

2 Algorithm Construction

During the process of algorithm construction, the encoded method, chromosome initialization, information base and the evaluation function plus the block diagram are established.

2.1 Encoded Method

Chromosome Representation. Generally speaking, the classrooms in a teaching building can be classified into two categories: common and multimedia, which can both embrace 1, 2, 3 or 4 classes. Simultaneously, the required classroom of courses is also restricted to the classification above and, however, the volume of an assigned classroom can be larger than that required, namely it is also feasible to provide a multimedia classroom with four-class volume to the course that requires the one with three-classroom volume. Thus, it is more convenient to consider in terms of courses resulting from that the classroom type required by a certain course is unique and inversely the provided classroom type is various.

Subsequently, we use one chromosome to store all the course information in classrooms of a certain type and consequently 8 chromosomes are proposed to present the classroom types including respective course arrangement information: 1. one-class common classroom; 2. two-class common classroom; 3. three-class common classroom; 4. four-class common classroom; 5. one-class multi-media classroom; 6. two-class multimedia classroom; 7. three-class multi-media classroom; 8. four-class multi-media classroom. On the basis of the class-flow we considered, another 8 chromosomes are used to represent the information in the second class.

Finally, a complete generation comprises 16 chromosomes whose different positions denote distinctive courses with genes of classroom information included. ("C" and "M" respectively denote the common and the multimedia classroom, plus numbers like 1, 2, 3 and 4 stand for the maximum number of accommodated classrooms in addition to that A(or a) and B(or b) represent the first and the second class respectively).

Genes. There are two sorts of information in the genes: one is the type of classroom assigned to courses; the other serves as the specific classroom corresponding to the classroom type.

Gene Base. In terms of the information of genes, the locations of relative classrooms can be searched, namely all locations of classrooms are stored in the gene base. For instance, the length of chromosome C1a is 6 provided there are 6 courses that require one-class common classrooms in the first class. According to figure 1, the first position of chromosome C1a states the information of class k, indicating that class k is proceeded in the first classroom of type C2 (table 1) and it requires a common classroom with the volume of more than one class. We can easily know that the assigned classroom to class k is located in the third floor.

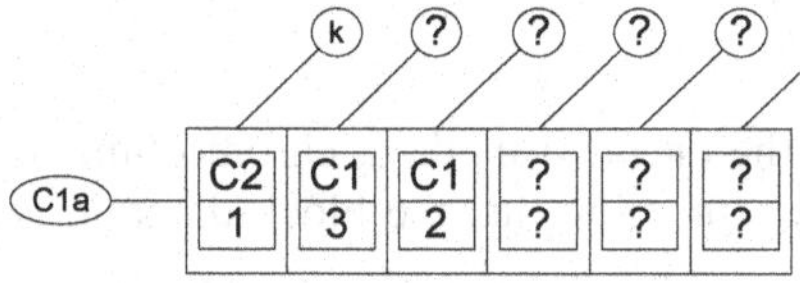

Fig. 1. Chromosome C1a

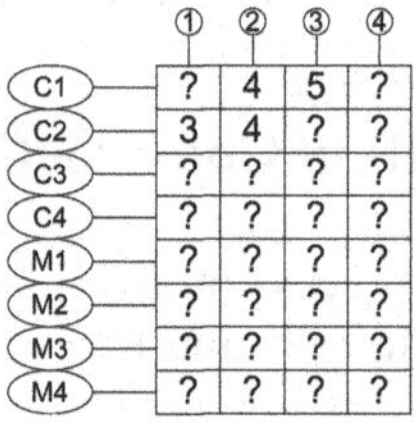

Fig. 2. Gene base

2.2 Chromosome Initialization

Chromosome initialization means to impose initial values on the chromosomes, namely randomly generate the classroom arrangement plans following the principles below.

Constraint 1: No class can be assigned to more than one course at any time slot;

Constraint 2: No more than one course is assigned at the same time in each classroom;

Constraint 3: The assigned classroom is big enough that satisfies all the features required by the course.

Moreover, the random method is used because of the initialization of diverse generations whose initial values should be set different which form a population, through the evolution of which the optimal results can be eventually obtained.

Step 1: the initialization of chromosome C4a. Uniformly choose a classroom from the C4 -type ones to place at the first position of chromosome C4a corresponding to the existent class there, and then from the left classrooms, choose one to place at the second position of chromosome C4a…

Step 2: the initialization of chromosome C3a. Uniformly choose a classroom from the C3 –type and the left C4 -type ones to place at the first position of chromosome C4a corresponding to the existent class there , and then from the left C3 –type and the left C4 –type classrooms, choose one to place at the second position of chromosome C4a…

Step 3: the initialization of chromosome C2a and C1a is implemented similarly.

2.3 Information Base

All the data bases in the optimization process are stored in the information base comprising the gene base (introduced), the course base of classes and their location base where the first two ones are constant while the location base needs to be updated. Thus, the obtained information can be looked up in the information base.

The Course Base of Classes. The course base of classes serves as a data base where the course information of certain classes can be found out through their codes which demonstrates the classroom type and the specific position on the corresponding chromosome.

Codes of classes	The first class		The second class	
?	?	?	?	?
16	C2	1	M4	5
26	Null	NaN	M2	8
?	?	?	?	?

Fig. 3. Some parts of the course base of classes

Figure 3 indicates that the class encoded 16 attends the course located at the first position of chromosome C2a at the first class and afterwards it attends the event at the fifth position of that chromosome. Similarly, the class with 26 encoded has no class in the first class and in the second it attends the lecture lying at the eighth position of chromosome M2b.

The Location Base of Classes. The location base of classes is also a kind of data base where the changing process of classes during the class break can be known by their assigned codes.

Codes of classes	The first class	The second class
?	?	?
16	4	9
26	5	NaN
?	?	?

Fig. 4. Some parts of the location base of classes

Figure 4 indicates that the class encoded 16 is located on the fourth floor previously and in the next class it transfers to the ninth floor while the one encoded 26 attends a class on the fifth floor and then it has no class subsequently.

2.4 The Statistics of the Class-Flow and Its Evaluation Function

The statistics of the class-flow during the class break is to derive the class-flow situation at each floor in terms of classes' transferring process and their respective number of students in the location base of classes. The main set parameters are given as follows:

- The entrance class-flow $f_{in,k}$ where k denotes the located floor;
- The exit class-flow $f_{out,k}$ and the surrounding class-flow including $f_{up,k}$ and $f_{down,k}$ that stand for the class-flow up and down the floor respectively.

From the referred parameters above, the evaluation function of class-flow, which is not simply the summation of the considered four parameters, can be defined as follows:

$$f_k = \left(\sum_j \alpha_{j,k}\left(1-\alpha_{j,k}\right)\right)\left(\sum_j f_{j,k}\right)^2 \tag{1}$$

where

$$\alpha_{j,k} = \frac{f_{j,k}}{f_{in,k} + f_{out,k} + f_{up,k} + f_{down,k}} \quad (j = in, out, up, down) \tag{2}$$

The square in equation (1) is formed by the least square method to avoid the difference among the defined four kinds of class-flow, which lessens the variance of each f_k so as to make each sort of class-flow become comparable and immune from local excessive class-flow. From (2), we can know that $\alpha_{in,k}$, $\alpha_{out,k}$, $\alpha_{up,k}$ and $\alpha_{down,k}$ all belongs to the interval $[0,1]$, which satisfies the formula below.

$$\alpha_{in,k} + \alpha_{out,k} + \alpha_{up,k} + \alpha_{down,k} = 1 \tag{3}$$

where it is known that when

$$\alpha_{in,k} = \alpha_{out,k} = \alpha_{up,k} = \alpha_{down,k} = 0.25$$

α_k approaches its maximum value.

Furthermore, $\alpha_k \neq 0$ when $\alpha_{in,k} = 0$ and $\alpha_{out,k} = \alpha_{up,k} = \alpha_{down,k} \neq 0$, which eliminates the error generated when there is no class-flow in a certain direction. In terms of the large value of f_k, the final class-flow evaluation function is presented for the sake of computational convenience, which is given by

$$F = \lg \sum_k f_k \tag{4}$$

2.5 The Competition and Selection Strategy

Natural selection plays a significant role in the process of creatures' evolution whose primary strategy can be summarized as "survival of the fittest", which makes the average fitness of the population increase with generations. The competition and selection strategy makes it possible that the desirable attributes (with high fitness) of the child will be passed on to future generations beneficial for the outcome of optimal ones.

The Internal Competition and Its Selection Strategy. The internal competition: a parent P generates two generations $F_{1,1}$ and $F_{1,2}$ through the interchange and shift mutation, and then there exists an internal competition among P, $F_{1,1}$ and $F_{1,2}$, which can be evaluated by their respective value of class-flow function where only one of them can survive. The selection strategy of the internal competition is regulated as follows:

- In the same family, if the minimum value of class-flow function belongs to the generation, then the generation survives while the rest is abandoned.
- In the same family, if the minimum value of class-flow function belongs to the parent, then the selection result depends on the modified roulette wheel method.

The survival probability of each generation is defined with the incorporation of the SA algorithm into the selection strategy of the roulette wheel method. Let the current temperature is T , the parent's value of class-flow function is F_P, the two generations' values of class-flow function are $F_{F_{1,1}}$ and $F_{F_{1,2}}$ respectively, and then the influence of $F_{F_{1,1}}$ and $F_{F_{1,2}}$ on F_P can be described by

$$\Delta F = F_{F_1} - F_P \tag{5}$$

Their weights can be expressed by

$$W_{F_1} = e^{-\frac{\Delta F}{T}} \tag{6}$$

$$W_P = \frac{1}{2}\left(2 - \left(W_{F_{1,1}} + W_{F_{1,2}}\right)\right) \tag{7}$$

Based on equation [5], [6] and [7], we can know that if F_P, $F_{F_{1,1}}$ and $F_{F_{1,2}}$ are constant the weight of generations is larger than that of the parent when T is high while the parent outweighs the generations provided T is low, which is the core spirit of the SA algorithm. Namely, the search space can be expanded when the temperature is high while the search domain decreases when the temperature gradually falls, and subsequently the low temperature makes it approach convergence to the optimal results. Finally, the survival one is determined by the generation of random values in $0 \sim W_P + W_{F_{1,1}} + W_{F_{1,2}}$ following uniform distribution.

- P survives when the generated random value belongs to $0 \sim W_P$;
- $F_{1,1}$ survives when the generated random value belongs to $W_P \sim W_P + W_{F_{1,1}}$;

- $F_{1,2}$ survives when the generated random value belongs to $W_P + W_{F_{1,1}} \sim W_P + W_{F_{1,1}} + W_{F_{1,2}}$.

The External Competition and Its Selection Strategy. The external competition: each parent P in numerous families (denoted by n) generates two generations $F_{1,1}$ and $F_{1,2}$ through the interchange and shift mutation, and then there exists an internal competition among P , $F_{1,1}$ and $F_{1,2}$ in addition to the external competition with other families where only n of them can survive. The selection strategy of the competition is regulated as follows:

- The first step: Find out the generation with the minimum value of class-flow evaluation function and it is entitled to survive.
- The second step: Calculate the subtraction of the obtained optimal value from that of each generation denoted by $\Delta F_i = F_i - F_{C\min}$
- The third step: Calculate the weight of each generation with $W_i = \exp(-\Delta F_i / T)$
- The fourth step: Randomly choose the survival generation through the roulette wheel method from the undetermined generations.
- The fifth step: Repeat the fourth step until the number of chosen generations reaches n .

2.6 The Block Diagram of the Proposed Algorithm

The proposed intelligent method is a hybrid of the core spirit in GA and SA mainly composed of three blocks that are initialization, the internal and the external competition respectively.

Initialization

- Determine the population size and the initial temperature;
- Construct the gene base;
- Create initial population;
- Chromosome initialization;
- Construct the course base of classes;
- Construct the location base of classes;
- Calculate the class-flow evaluation function and the statistics of class-flow.

The internal competition

- Generate generations through the interchange and the shift mutation;
- Update the location base of classes according to the mutation;
- The statistics of the class-flow and its evaluation function according to the mutation;
- Select the survival generation through the strategy of the internal competition.

The external competition

- Generate generations through the interchange and the shift mutation;
- Update the location base of classes according to the mutation;
- The statistics of the class-flow and its evaluation function according to the mutation;
- Select the survival generation through the strategy of the external competition.

Resulting from the blocks illustrating the algorithm process, users can construct freely according to their wish. Presently, we propose a fine block diagram (figure 5), whose idea is to make the population relatively stable through numerous internal competitions at a certain temperature and then by the external competition select excellent and potential generations, and proceed the circulation through the decrease of temperature.

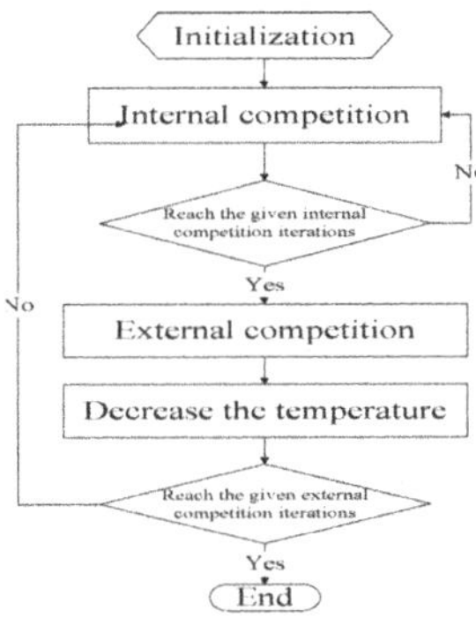

Fig. 5. The block diagram of proposed algorithm

3 Simulations and Results

In order to testify the practicality of the proposed algorithm in the optimization of classroom arrangement, a simulation is implemented to the virtual situation which presents an optimal classroom arrangement with respect to the class-flow. A teaching building in a certain university comprises 7 floors, the first three of which are common classrooms and the rest is multimedia classroom with a total number of 92 classrooms and 184 classes. Then the simulation starts based on the situation that the classrooms are used completely.

Let the population size is 100 and the initial temperature is 1. The decrease of temperature is classified into 3 sections: (1) the temperature starts from 1, decreasing by 0.11 every time during which 35 internal competitions are proceeded plus one external competition; (2) the temperature starts from 0.01, decreasing by 0.0001 each time during which 26 internal competitions are proceeded in addition to one external competition; (3) the temperature maintains 1.00e-4 and the simulation starts. Adopting the methods above, the results shown in Figure 6 are obtained, which denote the optimal and average values of class-flow evaluation function respectively.

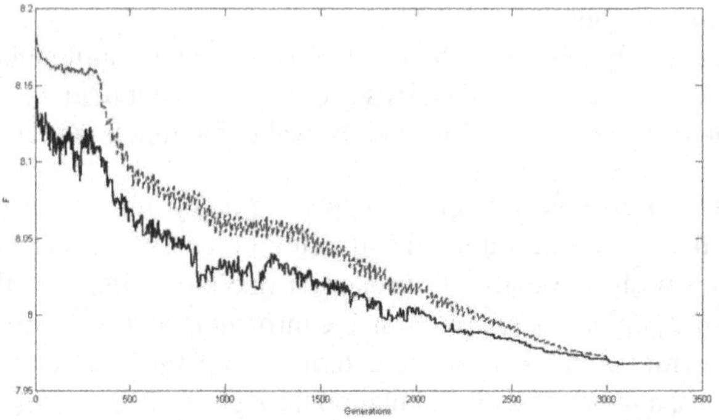

Fig. 6. The optimal (the curve below) and average values of class-flow evaluation function

Table 1. The number of transferred classes in each floor

Transferred Floors	**0**	**1**	**2**	**3**	**4**	**5**	**6**	**7**
Classes bef. opti.	19	48	46	45	31	23	16	6
Classes aft. opti.	60	44	40	25	35	16	9	5

Table 2. The statistic data of the class-flow

Floor	$f_{in,k}$	$f_{in,k}$	$f_{up,k}$	$f_{down,k}$
1	2075	2122	0	0
2	1019	1134	1566	1698
3	890	789	1555	1654
4	869	858	1132	1215
5	769	724	715	691
6	378	369	377	373
7	373	377	0	0

Table 3. The statistic data of the class-flow in in each floor before optimization each floor after optimization

Floor	$f_{in,k}$	$f_{out,k}$	$f_{up,k}$	$f_{down,k}$
1	1457	1416	0	0
2	579	692	1236	1085
3	615	604	1390	1327
4	836	857	947	1001
5	576	569	378	453
6	256	187	191	226
7	226	220	0	0

In the classroom arrangement without optimization (opti.), many classes need to transfer while after optimization with the proposed algorithm fewer ones should move. Simultaneously, the number of transferred classes changes from 16 to 60 after optimization, namely from 8.12 to 25.64%.

Moreover, the statistic data of the class-flow in each floor are presented in table 2 and 3.

Through the comparison of table 2 and 3, we can easily know that the class-flow on each floor decreases sharply, for instance, the class-flow on the first floor decreases by 31.55%. Overall, the value of F changes from 8.2229 to 7.9675, when transformed to its original form its value decreases by 44.46%, showing the high efficiency of the proposed algorithm in lessening the class-flow on each floor.

4 Conclusions

Based on the intelligent algorithms, the proposed model of classroom arrangement in the paper maintains a high efficiency especially in the large scale problems. Moreover, the convergence rate when searching for the optimal results can be easily controlled by the temperature, suitable for the classroom arrangement of different situations. In addition, the proposed algorithm can adapt to problems of distinctive scales with the modification of the number of internal and external competitions.

Our development covers the consideration of class-flow caused by the classroom scheduling and the incorporation of the SA algorithm into the competition and selection strategy of GA and additionally the proposition of the internal and external competition is another highlight. Thus, the presented method can be regarded as a novel algorithm that can be well used in the course timetabling of universities and colleges with respect to the class-flow.

References

1. Ghaemi, S., Vakili, M.T., Aghagolzadeh, A.: Using genetic algorithm Optimizer to Solve University Timetable Scheduling Problem. In: Proc. of the 2007 IEEE (2007)
2. Jat, S.N., Yang, S.: A Memetic Algorthm for the University Course Timetabling Problem. In: Proc. of the 2008 IEEE International Conference on Tools with Artificial Intelligence (2008)
3. Abramson, D., Abela, J.: A Parallel Genetic Algorithm Solving the School Timetabling Problem. Appeared in 15 Australian Computer Science Conference, Hobart, February 1992, pp. 1–11 (1992)
4. Abdullah, S., Turabieh, H.: Generating University Course Timetable Using Genetic Algorithm and Local Search. In: Proc. of the 2008 IEEE International Conference on Convergence and Hybrid Information Technology (2008)
5. Weare, R., Burke, E., Elliman, D.: A Hybrid Genetic Algorithm for Highly Constrained Timetabling Problems. Computer Science Technical Report No. NOTTCS-TR-1995-8
6. Rahoual, M., Saad, R.: Solving Timetabling Problem by Hybridizing Genetic Algorithm and Tabu Search. In: PATAT 2006, pp. 467–472 (2006) ISBN 80-210-3726-1
7. Blöchliger, I.: Modeling Staff Scheduling Problems. European Journal of Operational Research 158, 533–542 (2004)

Indexing Temporal XML Using FIX

Tiankun Zheng, Xinjun Wang, and Yingchun Zhou

School of Computer Science and Technology
Shandong University
Ji'nan, China
ztk5912@163.com, wxj@sdu.edu.cn, ychzhou@yahoo.cn

Abstract. XML has become an important criterion for description and exchange of information. It is of practical significance to introduce the temporal information on this basis, because time has penetrated into all walks of life as an important property information .Such kind of database can track document history and recover information to state of any time before, and is called Temporal XML database. We advise a new feature vector on the basis of FIX which is a feature-based XML index, and build an index on temporal XML database using B+ tree, donated TFIX. We also put forward a new query algorithm upon it for temporal query. Our experiments proved that this index has better performance over other kinds of XML indexes. The index can satisfy all TXPath queries with depth up to K(>0).

Keywords: temporal XML; TFIX index; feature-based.

1 Introduction

Temporal XML database preserves the historical records of modification, so that you can recover it to the state of any time before. Figure 1 shows a temporal XML data fragment, which record basic information about the previous case manager. Paper [1] proposed a temporal XML query language TXPath, for example, It may have following form of queries for data in Figure 1, company/manager[@from>05 and @to<06]/name.

In the researches of indexing non-temporal XML field, many types of indexes have been proposed. And these can be classified into two main categories, node-record-style index [2-3] and structural summary-style index [4-5]. In the field of temporal XML index, most of them create a new physical version each time an update occurs, leading to large overheads especially when processing temporal queries that span multiple versions. While some others draw attention to non-temporal XML database index, which commonly extends them with algorithm of processing temporal attribute (e.g. TempIndex[6]). And indexes proposed in papers [7,8,9,11] mainly using idea of structural summary-style to simplify the path information.

W. Liu et al. (Eds.): WISM 2009, LNCS 5854, pp. 224–231, 2009.

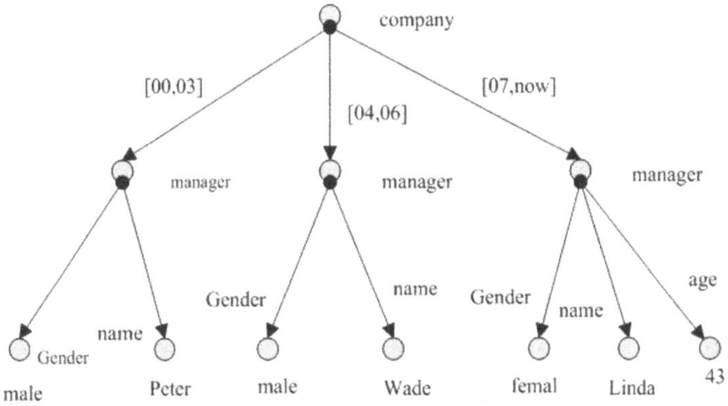

Fig. 1. Temporal XML document

In this paper, we advise a new feature vector to characterize the temporal XML documents on the basis of FIX which is a kind non-temporal index based on feature vector [10], and build an index on temporal XML database using B + tree, donated TFIX. The feature vector proposed here also involves the valid time of the root node of the document besides the ones contained by FIX. We also put forward a new query algorithm to process temporal query. The TFIX index makes good use of the containment nature of matrix eigenvalues to obtain Intermediate candidates, a subset of document chips, reducing the unnecessary searching and improving query performance. In the process of index establishment, XML documents are enumerated into small fragments of depth up to K, a given positive number, so it can satisfy all queries within the depth of K. And our experiment shows that while processing query, the index filters out many impossible document chips to gain intermediate candidates that may contain final results. It avoids unnecessary search effectively, and has better performance.

The rest of the paper is organized as follows: In section 2 we introduce some background knowledge which is theoretical support for us. In section 3 we introduced the structure of TFIX index, and the process of indexing algorithm. In section 4 we gives query algorithm in the temporal XML database using TFIX index, and finally in Section 5 experimental results will be showed.

2 Theoretical Basis

We will present theoretical basis for TFIX index briefly in this section, some specific proof to the conclusions referred to can be found in the corresponding literature.

2.1 Bisimulation Graph

We will introduce bisimulation concept for obtaining structure summary to simplify the presentation of the XML tree. For every XML tree, there is a unique minimum bisimulation graph that preserves all structural constraints in the tree. The bisimulation graph defined in this paper is based on the bisimilarity notion [12] as following.

Definition 1: Given an XML tree $T(V_t,E_t)$ and a labeled graph $G(V_g,E_g)$, an XML tree node $u \in v_t$ is bisimilar to a vertex $u \in v_g$ ($u \cong v$) if and only if all the following conditions hold:

- u and v have the same label name.
- If there is an edge (u,u_0) in E_t, then there is an edge (v,v_0) in E_g such that $u_0 \cong v_0$.
- If there is an edge (v,v_0) in E_t, then there is an edge (u,u_0) in E_g such that $u_0 \cong v_0$.

Then we define bisimulation graph as following,

Definition 2: Graph G is a bisimulation graph of T if and only if G is the minimum graph leading to the fact that every vertex in G is bisimilar to a vertex in T.

It is obvious that the bisimulation graph of a tree is a directed acyclic graph (DAG for short). Otherwise, if the bisimulation contains a cycle, the tree must also contain a cycle based on the definition which is contradictory to the definition of tree. An XML document and a TXPath query both can be viewed as a tree, so that corresponding bisimulation graphs exist. Figure 2 shows the bisimulation graph for Figure 1 regardless of the temporal information first.

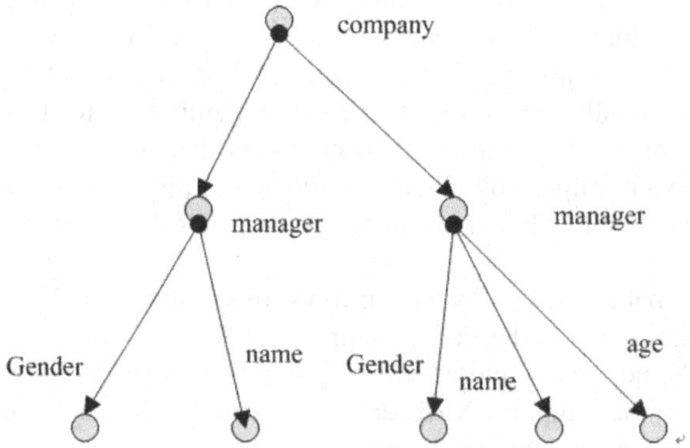

Fig. 2. Bisimulation graph of figure 1

2.2 Containment Nature for Eigenvalues of Matrices

We know that a DAG can be used to describe an anti-symmetric matrix. The non-zero-value matrix elements reflect weights of different edges, and the symbol of value reflects the directive nature of the edge. For example, the element $M(i,j)(>0)$ of an anti-symmetric matrix M tells us that the weight of the edge between vertex j and j is $M(i,j)$ and its direction is from vertex i to j. While every XML document or TXPath query can be viewed as a tree, corresponding anti-symmetric matrix for them can be obtained easily.

There are n eigenvalues for every n-order matrix allowing arbitrary number of them to be equal to each other. We also consider them the eigenvalues of the graph reflected by the matrix. Throughout the rest of this paper, we denote the maximum and minimum eigenvalues as λ_{max} and λ_{min}, and denote $\lambda(G)$ of graph G as the eigenvalue of the matrix representation of G.

An important theorem is put forward in paper [13] about the relationship between the eigenvalues of graph and its induced subgraph as is shown bellow.

Theorem 1: Let G and H be two DAGs, and M_G and M_H be the anti-symmetric matrix representations of G and H respectively. If H is an induced subgraph of G, then the following containment property holds, $\lambda_{min}(M_G) \leq \lambda_{min}(M_H) \leq \lambda_{max}(M_H) \leq \lambda_{max}(M_G)$

The main idea of the proof is to convert the anti-symmetric matrix to a symmetric matrix with the help of Hermitian matrix which is symmetric matrix in the complex domain, and use the same proof idea for symmetric matrix.

2.3 Containment Nature of Valid Time

In a valid temporal XML document, there is a basic precondition between the valid time of each node and its sub-nodes that every sub-node's valid time is contained by its parent-node. Here we forward a conclusion base on that to be used in our index.

Theorem 2: If An XML document chip contains final results of a TXPath query, the following relationship established, the valid times involved in the TXPath query as $T_i, i = 1..n$, and T represent the valid time of the root node in the document chip, then $T \supset T_i, i = 1..n$.

It is intuitive to prove this Theorem. If there exist any valid time T_i in the TXPath query is not contained by that of root-node T, then T_i would not be contained by any valid time in the document chip due to the precondition. That is, there will not be any result available in that document chip.

3 TFIX Index

3.1 Construction Algorithm

Firstly, we enumerate all document chips from temporal XML documents with depth up to positive number K, a given number. Then we convert these document chips to corresponding bisimulation graph without considering the temporal information here, and calculate λ_{max} and λ_{min} of the anti-symmetric matrix representing its bisimulation graph, and form our new feature vector $(r, \lambda_{max}, \lambda_{min}, [begin, end])$ for each document chip with label of root node r, the maximum and minimum eigenvalues, and the valid time of root node. At last, we insert the features into the B+ tree as keys. The detailed algorithm steps shows bellow.

Algorithm 1: Convert XML document to bisimulation graph

Input: XML document

Output: bisimulation graph

1. Define an empty stack S, the stack element representing document nodes contains two sub-structures as follows: label name N of the node and the signature SIG marking information about all child nodes. If both sub-structures of two nodes are equal to each other respectively, the corresponding nodes of them in the bisimulation graph are the same.
2. Define an empty graph G for preservation of the generated bisimulation graph
3. Produce a stack element for each node in the document through traversal of all the descendants of that node to get its signature, and push it into S.
4. If N and SIG of the top-element in S already exist in G, then turn to step 5, if not, then turn to Step 6.
5. The top-element of S exists in G, and then releases it. if S is not empty then turn to step 4, otherwise algorithm ends, and return G
6. The corresponding node in bisimulation graph for the top-element of S doesn't exist in G yet, then add it into G, and generate edges from it to its descendants according to the SIG. If S is not empty then turn to step 4, otherwise algorithm ends, and return G.

Algorithm 2: Construct the index

Input: temporal XML document

Output: TFIX index

1. For current node(the root-node of document at the beginning), if there exist a sub document chip with depth K(> 0) considering the current node to be the root node, extract them as a document chip, donate D. Otherwise, generate D including all the descendents of the node. Then Algorithm 1 is called to get the bisimulation graph, and turn to step 2. Algorithm ends when finishing traversal, or redo Step 1.
2. Convert the bisimulation graph to an anti-symmetric matrix M, and compute $\lambda_{\max}$ and $\lambda_{\min}$, then turn to step 3.
3. Record the valid time of the root node, donate $[\text{begin}, \text{end}]$.If there is no obvious valid time of the root node, take the valid time of the nearest ancestor node in the original document then turn to step 4.
4. Generate the feature vector $(r, \lambda_{\max}, \lambda_{\min}, [begin, end])$, whose *r* represent the label of the root node in D, and insert it into the B+ tree as a key.

3.2 Completeness of Index

We show the process of constructing the index in the previous subsection. At first, we need to restrict the number K to an acceptable value before enumerating sub document chips in the XML tree. However, the index loses some expressive power with this construction that it can only answer TXPath query with depth up to k. In fact, the tradeoff between expressive power and efficiency is common [12] and won't invalidate the benefit of building the index.

The number of the sub document chips enumerated is exactly the number of the nodes in the XML tree, which Guarantee that any TXPath query with depth up to k can be answered by our index. In another words, the following Theorem holds.

Theorem 3: If the TFIX index is built with depth limit k, a TXPath query with depth up to K is not contained in the temporal XML document, if it is not contained in the index.

3.3 Query Processing

When processing query using TFIX index, the major step is divided into two: The first step is to filter out many impossible nodes in B+ tree, and get candidate sub document chips with the help of the feature vector. Second step is refinement. That is to focus on the candidates to search the final results. For TXPath queries with "//" in the middle, we have pretreatment on the situation. We decomposed them into several sub-queries. Take the query "// manager [// secretary [name] [email]] / salary [@ from> 01, @ to <03] " for example, it is decomposed into two sub-queries "// manager /salary [@ from> 01, @ to <03] " and "// secretary [name] [email] ". Query processing algorithm is given below:

Algorithm 3: Query processing
Input: TXPath query
Output: Final results

1. For current TXPath query, valid time requirements is extracted from it and saved into an array, donate *TARRAY* .The *TARRAY* is empty if there is no such requirement, which means the query is non-temporal.
2. If "//" is contained in middle of the TXPath query, it is decomposed into sub-queries, and the first one is selected whose root node is the same as the original query for computing λ_{max} and λ_{min} .Otherwise, the whole query is use for computing.
3. Organize the feature vector $(r, \lambda_{max}, \lambda_{min}, TArray)$, and search candidates in the B+ tree. Any candidate must have equal r to the TXPath's, satisfy the theorem 2, and each valid time in $TArray$ is contained by the TXPath's.
4. Traverse the candidate document chips to obtain final results.

The cost of query processing algorithm consists of three parts: CPU cost of converting a TXPath query into its bisimulation graph, converting the graph into a matrix, and computing the eigenvalues. These costs are negligible if K is a reasonable value. The main I/O cost includes searching the B+ tree and reading the document chips while our index is clustered index. The cost of searching the B+ tree has been well studied for a long time.

4 Experiment

In our simulation experiment we chose a relatively typical temporal XML index TempIndex for comparison, and the non-indexed query method (DOM) is also included. Our experiment is carried out on a PC with Pentium IV 3.00GHz CPU, 1GB memory, and 120G IDE disk running Windows XP system. The experiment data is a temporal XML document including multiple employees with most of the staff attributes inconsistent. The data is structure-rich, fairly deep and flat, therefore, the structures are less repetitive. The size of the document is 35M with 133240 nodes in it. The index is implemented as clustered index in order to show better performance. The size of TFIX index built is 0.9M and the value of K is set to be 5. The result graph figure 3 showed bellow reflects the comparison among the performance of TFIX, TempIndex, and DOM method in processing the TXPath query "// manager [secretary [name] [email]] / salary [@ from> 01, @ to <03] ".

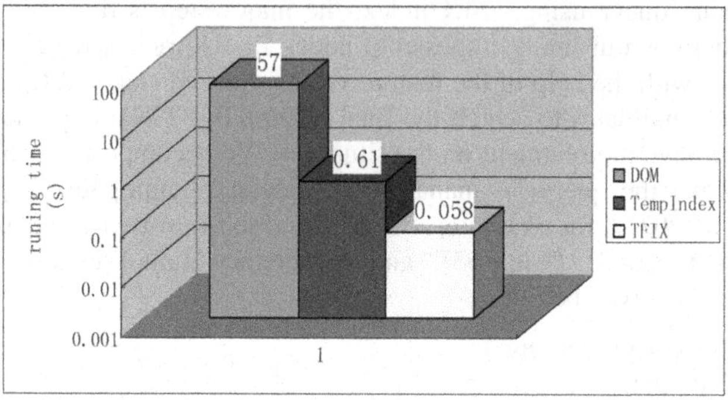

Fig. 3. Performance comparison in processing TXPath query

From figure 3 we can see that TFIX index is much more efficient than the other two methods in processing the TXPath query. DOM is a state of non-index query. It needs to traverse the entire document for each node. And when the number of nodes increases, its performance is often unacceptable, because of its much unnecessary traversal searching and disk access. And for TempIndex, it cannot achieve the desired effect due to the structure-rich data we choose. TempIndex also need to traverse many nodes when processing such kind of TXPath that takes some predicates. But for the TFIX index, it just traverse the candidates generated using the feature vector. And the value of K is reasonable which guarantee the final search is efficient. The size of bisimulation graph is a little less than its corresponding document chip owing to the structure-rich document. So the nodes in B+ tree reference to reasonable size of document chips. So TFIX index performance better.

5 Conclusion

In this paper, we advise a new feature vector to characterize the temporal XML documents, and build an index on temporal XML database using B + tree, donated TFIX. We involve the valid time of the root node in the feature vector and put forward a new query algorithm to process temporal query. Our simulation experiments show that the index has better performance. The value of K will impact on the performance directly. This paper hasn't yet focused on its influence through experiments, which may be the future work of our research.

Acknowledgment

Supported by the National Natural Science Foundation of China under Grant No.60673130;No.90818001, the Natural Science Foundation of Shandong Province of China under Grant No.Y2006G29; No.Y2007G24; No.Y2007G38, Key Technology R&D Program of Shandong Province under Grant No.2007GG10001009; No.2008GG30001005, Science and Technology Development Program of Shandong Province under Grant No.2006GG2201052; No.2007GG1QX01036, the Special Fund for Postdoctoral Innovative Project of Shandong Province under Grant No. 200703084.

References

1. Vaisman, A.A., Mendelzon, A.O., Molinari, E., Tome, P.: Temporal XML: Model, Language and Implementation
2. SKratky, M., Pokorny, J., Snasel, V.I.: Indexing XML data with UB-trees. In: Manolopoulos, Y., Návrat, P. (eds.) ADBIS 2002. LNCS, vol. 2435, pp. 155–164. Springer, Heidelberg (2002)
3. Rao, P., Moon, B.: PRIX: Indexing and querying XML using prüfer sequence. In: Titsworth, F. (ed.) Proc. of the 20th Int'l Conf. on Database Engineering (ICDE), pp. 288–300. IEEE Computer Society, Boston (2004)
4. Chung, C., Min, J., Shim, K.: APEX: An adaptive path index for XML data. In: Franklin, M.J., Moon, B., Ailamaki, A. (eds.) Proc. of the 2002 ACM SIGMOD Int'l Conf. on Management of Data (SIGMOD), pp. 121–132. ACM Press, Madison (2002)
5. Kaushik, R., Bohannon, P., Naughton, J.F., Korth, H.: Covering indexes for branching path queries. In: Franklin, M.J., Moon, B., Ailamaki, A. (eds.) Proc. of the 2002 ACM SIGMOD Int'l Conf. on Management of Data (SIGMOD), pp. 133–144. ACM Press, Madison (2002)
6. Mendelzon, A.O., Rizzolo, F., Vaisman, A.: Indexing Temporal XML Documents. In: Proceedings of the 30th VLDB Conference, Toronto,Canada (2004)
7. Kaushik, R., Bohannon, P., Naughton, J.F., Korth, H.F.: Covering indexes for branching path queries. In: ACM SIGMOD. Wisconsom, Madison, pp. 133–144 (2002)
8. Cooper, B., Sample, N., Franklin, M.J., Hjaltason, G.R., Shadmon, M.: A fast index for semistructured data. In: VLDB, Rome, Italy, pp. 341–350 (2001)
9. Li, Q., Moon, B.: Indexing and querying XML data for regular path expressions. In: VLDB, Rome, Italy, pp. 361–370 (2001)
10. Zhang, N., Özsu, M.T., llyas, I.F., Aboulnaga, A.: FIX: Feature-based Indexing Technique for XML Documents. In: Proceedings of the 32nd international conference on VLDB (2006)
11. Chen, Q., Lim, A., Ong, K.W.: D(k)-index: An Adaptive Structural Summary for Graph-Structured Data. In: SIGMOD 2003, pp. 134–144 (2003)
12. Henzinger, M.R., Henzinger, T.A., Kopke, P.W.: Computing Simulation on Finite and Infinite Graphs. In: FOCS 1995, pp. 453–462 (1995)
13. Bollobás, B.: Modern Graph Theory. Springer, Heidelberg (1998)

Semantic Structural Similarity Measure for Clustering XML Documents

Ling Song[1,2], Jun Ma[1] , Jingsheng Lei[3], Dongmei Zhang[1,2], and Zhen Wang[2]

[1] School of Computer Science &Technology, Shandong University, 250101, China
[2] School of Computer Science &Technology, Shandong Jianzhu University, 250101, China
[3] College of Computer, Nanjing University of Posts and Telecommunications, Nanjing, 210003, China
song_ling@sina.com

Abstract. Clustering XML documents semantically has become a major challenge in XML data managements. The key research issue is to find the similarity functions of XML documents. However, previous work gave more importance to the topology structure than to the semantic information. In this paper, the computation of similarity between two XML documents is based on both structural and semantic information. Then a minimal spanning tree clustering method is used to cluster XML documents. The experiment results show that the new method performs better than baseline similarity measure in terms of purity and rand index.

Keywords: XML document similarity, Clustering, Minimal spanning tree.

1 Introduction

With the widespread diffusion of XML documents, much research effort is devoted to support the storage and retrieval of large collections of XML documents. The computation of similarity between XML documents plays an important role in the management of the XML data. XML clustering is a process of grouping the similar XML documents together. Clustering techniques have been around for several years for grouping the flat data such as text. However, clustering of XML data significantly differs from the clustering of the flat data, since XML embedded the semantic and structural aspects to document contents, resulting in the semi-structured and hierarchical data. An XML document is an ordered, labeled tree. The nodes of the tree are elements.

Several clustering methods have been suggested. One of them is based on the graph-based divisive clustering algorithm, which developed a framework for representing a set of multi-dimensional data as a minimal spanning tree (MST) [1-3]. Through this MST representation, a multi-dimensional clustering problem can be converted to a tree partitioning problem. By the elimination of any edge from the MST, subtrees are got which correspond to clusters. Dalamagas etc. modeled the XML documents as rooted ordered labeled trees and apply MST clustering algorithms using edit distances [4]. There is considerable previous work on similarity functions for XML documents. Because XML documents can be labeled as tag trees of element

W. Liu et al. (Eds.): WISM 2009, LNCS 5854, pp. 232–241, 2009.

nodes, majority of these methods relied on the notion of tree edit-distance to find common structures [5]. Assuming that XML documents use a similar set of paths will likely have a similar structure, some work introduced a measure of structural similarity that is based on the comparison of sets of paths [6-7]. Assuming that documents use a similar set of element names will likely be similar, [8] and [9] measured the structural similarity between XML documents by considering their common element names. Some work have taken into account not only element names, but also contextual constraints [10-11].

However, previous work gave more importance to the trees or paths' topology than to the semantic information contained in the element names. In fact, elements could not be regarded separately without considering the semantic information among elements and the structural information that shows the relationships between elements. *PathSim* [12] calculated similarity between two paths combining their corresponding lists of element names and the position that the elements located in paths. But elements comparing only took into account some string similarity, which ignored the semantic similarity based on some kind of ontology. In this paper, we propose a novel approach to measure similarity between two XML documents, which takes element matching, path matching and document matching into account. The computational process of element matching focuses on both string similarity and WordNet based semantic similarity between elements. The computational process of path matching focuses on contextual constraints of elements. Then a MST clustering method is used to cluster XML documents.

2 XMLSIM: An Approach to Coumpte Similarity between XML Documents

The same information may be represented in XML documents by different users, leading to partial or total overlap of information. XML label tree can be represented by a set of paths. Each path expression can be viewed as a sequence of elements following the edges from the root node to a leaf node. We give the formally descriptions as following.

Definition 1. Given a XML document, it can be labeled as a tag tree composed of a set of paths $\{p_1, p_2, \ldots, p_m\}$, where m is the number of unique paths in the tree.

Definition 2. Given a path $p = e_1, e_2, \cdots e_d$, where e_i is the ith element in the path, specially, e_1 is the root element , e_d is a leaf element. For every $e_i e_{i+1}$ in the path, there is a direct parent/child relationship.

In order to determine the similarity between two XML documents quantitatively, two sets of paths corresponding to two XML documents are compared, which aims at finding the maximal similar paths between two sets of paths. So the path comparing problem should take into account two key factors: similarity between every pair of elements in two paths and contextual description similarity between two paths. In this section, instead of exact element equality, we relax it with element linguistic similar. Instead of exact elements order equality of paths, we relax it with contextually similar, which handles cases that elements are in the relative order but not necessarily contiguous.

2.1 ELEMSim: An Approach to Compute Similarity between Two Elements

Element name similarity measures how close each element with an element in the other path are. This is assuming that two similar paths use a set of similar pairs of elements. This section addresses the issue that the elements may have name differences, and they may represent non-identical but similar content.

Nayak and Iryadi combined the semantic and syntactic relationships to calculate the linguistic similarity between two elements [7]. Levenshtein distance [13] and Longest common subsequence (LCS) [18] are often used comparing string similarity. In this section, we calculated LCS of two elements to find how string similar is [18]. $lcs(s,t)$ can be used to translated to a string similarity value normalized in the interval [0, 1].

$$StringSim\ (s,t) = \frac{lcs(s,t)}{\min(|s|,|t|)} \tag{1}$$

Then we removal delimiters to elements and parse the compound element name into a set of tokens using the use-defined special dictionary. e.g., "J_name"→{Journal name}. Next we calculate the semantic similarity of two tokens. Since we have developed a fuzzy semantic similarity measure based on ontology [15-17], in this paper we use these measure to calculate the semantic similarity between a pair of words based on WordNet[14].

If two compound elements e_1 and e_2 are processed into two sets of tokens $s(e_1)$ and $s(e_2)$, let the semantic similarity of $s(e_1)$ and $s(e_2)$ be the average of the best similarity of each token with a token in the other set:

$$\text{Setsim}_{fuzzy}(s(e_1), s(e_2)) = \frac{\sum_{n_1 \in N_1}\left[\max_{n_2 \in N_2} \text{WordSim}_{fuzzy}(s(e_1), s(e_2))\right] + \sum_{n_2 \in N_2}\left[\max_{n_1 \in N_1} \text{WordSim}_{fuzzy}(s(e_1), s(e_2))\right]}{\max(|s(e_1)|, |s(e_2)|)} \tag{2}$$

Where $|s(e_1)|$ and $|s(e_2)|$ are the number of tokens in the set $s(e_1)$ and $s(e_2)$.

Linguistic similarity between two elements is defined as following: Let e_1, $e_2 \in$ a set of elements, $s(e_1)$ and $s(e_2)$ are their token sets. The linguistic similarity function is defined:

$$Esim(e_1, e_2) = \max(SetSim_{semantics}(s(e_1), s(e_2)), \text{StringSim}(e_1, e_2)) \tag{3}$$

Definition 3. If $Esim(e_1, e_2) \geq \alpha$,then we say e_1 and e_2 is a similar element pair.

2.2 NPathSim: An Approach to Compute Similarity between Two Paths

Now we consider the path context of XML elements, since the path "issue/articles/author/address" is different from "journal/article information/writer/affiliation", but they are similar in both structural and semantic information. Let us introduce an approach *NPathSim* to capture the degree of similarity of two paths that coming from two XML documents.

Given a sequence $X=\{x_1,x_2,\ldots,x_m\}$, call the sequence $Z=\{x_1,x_2,\ldots x_k\}$ is a subsequence of X if there exists a strictly increasing sequence $<i_1,i_2,\ldots,i_k>$ of indices of X such that for all $j=1,2,\ldots k$, we have $x_{i_j} = z_j$.

Definition 4(similar sequence). Let $Z=\{z_1,z_2,\ldots,z_k\}$and $W=\{w_1,w_2,\ldots,w_k\}$ be a pair of similar sequences, if they have k pairs of similar characters, where k characters come from a sequence, k characters come from another sequence. That is to say, for each pair of similar characters $(z_l, w_l)(1 \leq l \leq k)$, there exist $sim(z_l, w_l) \geq \alpha$.

Definition 5(Maximal Similar Subsequence). Given two sequences X and Y, $X=\{x_1,x_2,...x_m\}$, $Y=\{y_1,y_2,...y_n\}$, $Z=\{z_1,z_2,...z_k\}$ and $W=\{w_1,w_2,...w_k\}$ are two subsequences of X and Y, The Maximal-Similar-Subsequence (MSS) problem is maximizing similarity value of Z and W.

Let a sequence represents a path, a character in a sequence represents an element, the linguistic similarity of any pair of elements that come from two paths can be calculated, and then the solution of MSS problem can be used to solve our path similarity problem. Furthermore, According to the description of subsequence, we know that ancestor/descendant relation in XML paths satisfy partial order. For arbitrary elements in sub-path still satisfies partial order. For instance, "*department/name/first name*" is a sub-path of the path "*department/gradstudent/name/first name*", which still satisfies a partial order.

In addition, we assign a proper weight to each element when two corresponding elements are compared in order to evaluate how much the contexts of two paths are similar. In order to reflect elements located in different levels contribute different impacts to similarity, an ancestor node should gain more weight than a descendant node [19].

Definition 6. A weighted path is a path where each element is assigned non-negative integer weight. Given an element x in path p, the depth of the path is d, the height that x located is l, then x's weight is set to $r^{d-l}(0<r\leq 1)$.

Given a sequence, $X=\{x_1,x_2,...x_m\}$, we define the ith prefix of X, for $i=0,1,...,m$, as $X_i=<x_1,x_2,...,x_i>$. X_0 is the empty sequence. For two paths P_a and P_b, they are represented by sequences $P_a=\{e_{a1}/e_{a2}/,..../e_{am}\}$ and $P_b=\{e_{b1}/e_{b2}/,..../e_{bn}\}$, Let $s[i,j]$ to be similarity value of the sub-path P_{ai} and P_{bj}. Then $PathSim_{linguistic}(P_a, P_b)$ is defined as $s[m,n]$.

$$s[i,j]=\begin{cases} 0 & \text{if } i=0 \text{ or } j=0 \\ Max(s[i-1,j],\, s[i,j-1],\, s[i-1,j-1]+Esim(e_{ai}, e_{bj})\times\gamma^{m-i}\times\gamma^{n-j}) & \text{if } i,j>0 \text{ and } sim(x_i, y_j)\geq\alpha \\ Max(s[i-1,j],\, s[i,j-1]) & \text{if } i,j>0 \text{ and } sim(x_i, y_j)<\alpha \end{cases} \quad (4)$$

Where $Esim(x_i, y_j)$ comes from formula (3), which shows the similarity degree of two elements x_i and y_j.

Finally, to obtain a score in [0,1], we normalize the maximal-semantic-Similarity of two paths by the maximal length of two paths as following:

$$NPathSim(P_a, P_b) = \frac{PathSim_{linguistic}(P_a, P_b)}{\min(|P_a|, |P_b|)} \quad (5)$$

2.3 XMLSim: An Approach to Compute Similarity between Two XML Documents

To measure the structural similarity between two XML documents, the similarity of each pair of tags (or element names) of two XML documents and tags' structural relative order are measured and aggregated to form a similarity measure between them.

NPathSim is used to measure structural and contented similarity between two paths. The XMLSim is used to measure structural and contented similarity between two XML documents. Instead of counting the number of common paths or subsumed paths, we aims at finding the maximally similar paths that are common and similar for two given XML documents, using path matching techniques in section 2.2. Let A and B be two XML documents, let p^A and p^B be their two sets of paths in definition 1. If paths match more between two XML documents, they become more similar. So similarities between A and B based on path matching is the average of the best similarity of each path with a path in the other path set:

$$\mathrm{XMLSim}(A,B)=\frac{\sum_{p_1\in p^A}\left[\max_{p_2\in p^B}\mathrm{NPathSim}(p_1,p_2)\right]+\sum_{p_2\in p^B}\left[\max_{p_1\in p^A}\mathrm{NPathSim}(p_1,p_2)\right]}{\left|P^A\right|+\left|P^B\right|} \tag{6}$$

Where $|p^A|$ and $|p^B|$ are the number of paths in the path set p^A and p^B, respectively.

3 XML Documents Clustering

The previous work demonstrated that though the inter-data relationship is greatly simplified in the MST representation, no essential information is lost for the purpose of clustering[1,3]. Two key advantages in representing a set of multi-dimensional data as an MST are [2]: (1) the simple structure of a tree facilitates efficient implementations of clustering algorithms; (2) as an MST based clustering does not depend on detailed geometric shape of a cluster, it can overcome many of the problems faced by classical clustering algorithms. Because MST is based on distance, similarity from formula (6) should be translated into distance.

$$D(A,B)=1-XMLSim(A,B) \tag{7}$$

The process of MST clustering method is as following. Let $X=\{\boldsymbol{x}_1, \boldsymbol{x}_2,\ldots, \boldsymbol{x}_N\}$ be a set of the XML documents which we want to distribute in different clusters. Let $D_{i,j}=D(\boldsymbol{x}_i, \boldsymbol{x}_j)$ be the distance defined between any $\boldsymbol{x}_i$ and $\boldsymbol{x}_j$, we form a fully connected graph G with n vertices and $n\times(n-1)/2$ weighted edges. The weight of an edge corresponds to the distance between the vertices that this edge connects. We use Prim's algorithm for constructing a MST, removed the edges that the value of distance is larger than β, which breaks up the tree into several connected subtrees. The points of each connected subtree are the members of a cluster. This simple algorithm works very well as long as the inter-cluster (subtree) edge-similarities are clearly smaller than the intra-cluster edge-similarities. The algorithms were implemented in C++. *When* β *decreases from 1 to 0, the classification will become more and more refined.*

4 Experimental Evaluation and Discussion

We carried out the experiments with two steps. First, we tune the parameters on the training data set D_0 and discover the optimal parameters. Second, the identified optimal parameters are used to cluster XML documents in test data set D_1.

4.1 Determining the Optimal Parameter Value of α and γ

In order to discover the optimal parameter value of α and γ in formula (4), we performed path retrieval. The idea behind the experiments is to investigate whether the proposed path similarity measures can achieves higher quality in path retrieval. The data used in experiments mainly come from XML Data Repository [22], sigomod[23] and xmark[24]. These domains contain a number of documents that have structural and semantic differences. Parser reads XML documents and converts them into DOM objects that can be accessed with JavaScript. 303 paths were extracted from DOM trees, which are used as our training data set D_0. The path query set was obtained with the help of 6 undergraduates. The query set was generated as follows: The students were presented a few examples of some information that were included in XML documents but were unstructured and devoid of any kind of mark-up. The students were instructed to generate query paths from the example data, freely choosing the structure and the element names. 94 queries were generated.

We use MRR (Mean Reciprocal Rank) and P@*N*(the mean of the precision of the first *N* documents retrieved) to evaluate the performance. The MRR is the average of the reciprocal ranks of results for a sample of queries Q:

$$MRR = \frac{1}{|Q|}\sum_{i=1}^{Q}\frac{1}{rank_i} \quad (8)$$

Where $1/rank_i$ if $rank_i > 0$; and $1/rank_i$ is 0 if no answer is found, and Q is total number of queries. Table 1 is the performance of retrieval with different threshold α under same γ. As shown in table 1, the *NPathSim* raised the performance of the queries when α dropped from 1 to 0.6, we obtained the highest retrieval performance with the MRR of 0.944 and P@1 of 97.2% when α is set to be 0.6.This is because with the decreasing of α, the linguistic similarity between such a pair of elements may be accumulated, which may detect difference of similarity between adjacent rank values (e.g. rank=1 and rank=2) in rank list. But the *NPathSim* reduced the performance of the queries when α dropped from 0.6 to 0.4, this is because it is still may be accumulated when α is small enough when such a pair of elements are not linguistically similar. So the threshold α needs a tradeoff. From table 1, we can conclude that 0.6 is optimal parameter to α.

Table 1. Performance of retrieval with different threshold α under same γ

NPathSim	MRR	P@1	P@2	P@5	P@10
α=0.4 γ=1	0.910	90.6%	89.2%	43.7%	29.9%
α=0.5 γ=1	0.923	93.3%	91.8%	45.7%	33.4%
α=0.6 γ=1	0.944	97.2%	97.2%	46.9%	32.6%
α=0.7 γ=1	0.929	96.7%	96.7%	53.3%	35.7%
α=0.8 γ=1	0.922	95.2%	96.6%	55.8%	41.4%
α=0.9 γ=1	0.903	95.2%	96.2%	60.6%	42.1%
α=1 γ=1	0.879	92.5%	94.7%	61.0%	42.1%

Table 2 is performance of retrieval with different threshold γ under same α. As shown in table 2, the performance achieved better with $\gamma<1$ than with $\gamma=1$, which means γ plays an important role in path comparing. It indicates that with γ decreasing, the performance keeps unchanging. However, if γ is too small, the score of similarity is small even if two paths are perfect matching, which can not reflect human's judgments. So for threshold setting of γ, it also needs a tradeoff. We select 0.9 as optimal parameter to γ.

Table 2. Performance of retrieval with different threshold γ under same α

NPathSim	MRR	P@1	P@2	P@5	P@10
α= 0.6 γ=1	0.944	97.2%	97.2%	46.9%	32.6%
α= 0.6 γ=0.9	0.952	98.1%	98.3%	57.3%	43.6%
α=0.6 γ=0.85	0.952	98.1%	98.3%	57.3%	43.6%
α=0.6 γ=0.6	0.952	98.1%	98.3%	57.3%	43.6%

4.2 Experiments on Clustering XML Documents

We performed XML documents clustering in the test data set D_1, which mainly came from NIAGARA Experimental Data [20], XML Data Repository [22] and IBM XML generator [25]. These domains contain a number of documents that have structural and semantic differences. There are 105 XML documents in test data set D_1. The documents are from various domains such as bibliography, club, company profiles, department, personal information, movie and actors. The number of tags varies from 9 to 28 in these sources. The average depth varies from 2 to 6. Parser reads XML documents and converts them into DOM objects that can be accessed with JavaScript. Then the XML documents are separated into a bag of paths. Let α= 0.6, γ=0.9, similarity between two documents is calculated by *XMLSim*. And we cluster these documents in D_1 with MST clustering algorithm.

The performance of clustering is evaluated using the standard criteria namely purity and precise [21].Purity is a simple and transparent evaluation measure. To compute purity, each cluster is assigned to the class which is most frequent in the cluster, and then the accuracy of this assignment is measured by counting the number of correctly assigned documents and dividing by *N*. Formally:

$$purity(\Omega, C) = \frac{1}{N}\sum_k \max_j |\omega_k \cap c_j| \tag{9}$$

Where $\Omega = \{\omega_1, \omega_2, \cdots \omega_k\}$ is the set of clusters and $C = \{c_1, c_2, \cdots c_j\}$ is the set of classes. We interpret ω_k as the set of documents in ω_k and c_j as the set of documents in c_j.

We also use The Rand Index (RI) measure to evaluate how well the clustering matches a set of gold standard classes. The destination of clustering is assigning two documents to the same cluster if and only if they are similar. A true positive (TP) decision assigns two similar documents to the same cluster, a true negative (TN) decision assigns two dissimilar documents to different clusters. There are two types of errors we can commit. A false positive (FP) decision assigns two dissimilar documents to the same cluster. A false negative (FN) decision assigns two similar documents to different clusters. The Rand index (RI) measures the percentage of decisions that are correct [21].

$$\mathrm{RI} = \frac{TP + TN}{TP + FP + FN + TN} \tag{10}$$

As a base line for clustering experiments we have taken XSim as similarity function to calculate similarity between XML documents, which take PathSim as similarity function to calculate the similarity between paths [12]. Compared with the purity of XSim clustering, the purity of XMLSim clustering is as figure 1. Figure 2 is the RI value of clustering with XSim and XMLSim on D1. Figure 3 shows a comparison of the time taken to cluster the XML documents with XMLSim and XSim.

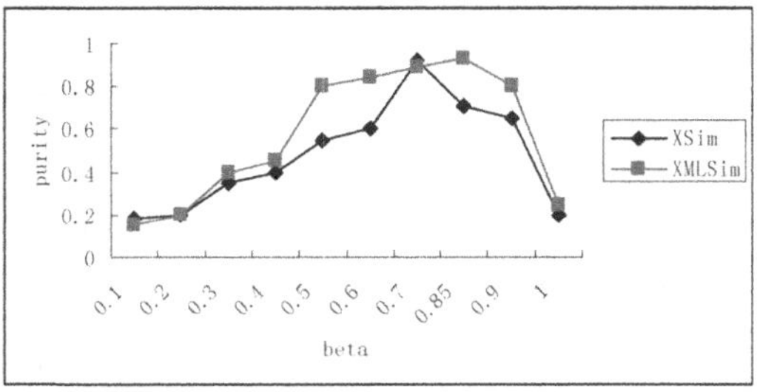

Fig. 1. The purity performance of XSim and XMLSim on XML documents

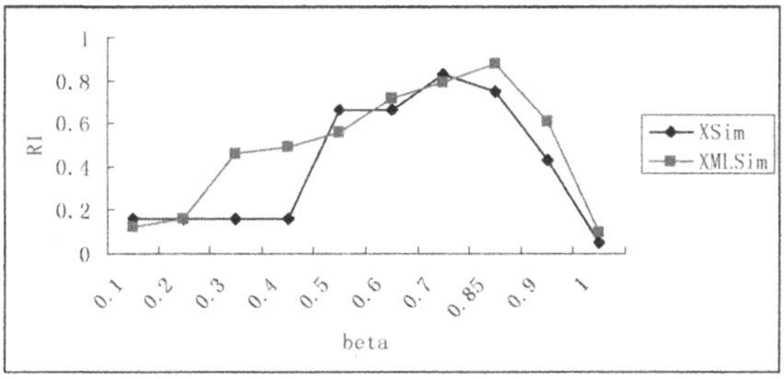

Fig. 2. The RI performance of XSim and XMLSim on XML documents

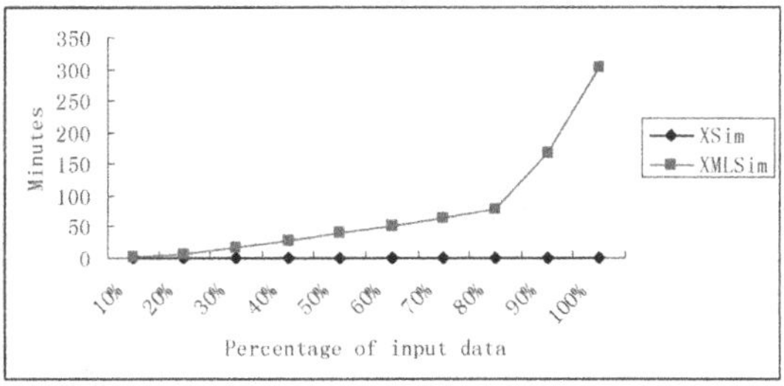

Fig. 3. Running time of XSim and XMLSim on XML documents

From the figure 1-3, we reach the following observations:

1. *XMLSim* performs better than *XSim* in *purity*. The *purity* goes up when β changes from 0 to 0.85, we obtain the highest *purity* 0.93 when β is set to be 0.85.
2. *XMLSim* performs better than *XSim* in *RI*. The *RI* reaches the maximum value 0.88 at β=0.85.
3. The computation time increasing extremely large as the number of XML documents increase with our XMLSim measure.

5 Conclusion and Future Work

This paper addresses a reasonable and quantitative approach to measure the similarity between any pair of XML documents. Then a MST clustering method is used to cluster XML documents. The experiment results show that the new similarity method performs better than baseline similarity measure in terms of purity and rand index when they are used in clustering.

In this paper we only consider the parent/child (ancestor/descendant) relationships between the elements when separating the documents into a bag of paths, ignoring the sibling relationships. Furthermore the generation of similarity matrix may be computationally expensive when dealing with large data due to the need for the pair wise similarity matching among diverse documents. In future, the algorithm will be improved by considering the more structural information and lowering the computationally complex.

References

1. Xu, Y., Olman, V., Xu, D.: Minimum Spanning Trees for Gene Expression Data Clustering
2. Gower, J.C., Ross, G.J.S.: Minimum spanning trees and single linkage cluster analysis. Applied Statistics 18, 54–64 (1969)
3. Page, R.L.: A Minimal Spanning Tree Clustering Method. Communications of the ACM 17(6), 321–323 (1974)
4. Dalamagas, T., Cheng, T., Winkel, K.-J., Sellis, T.: A Methodology for Clustering XML Documents by Structure. Information Systems 31(3), 187–228 (2006)
5. Zhang, K., Statman, R., Shasha, D.: On the editing distance between unordered labeled trees. Inform. Process. Lett. 42(3), 133–139 (1992)
6. Joshi, S., Agrawal, N., Krishnapuram, R., Negi, S.: A Bag of Paths Model for Measuring Structural Similarity in Web Documents. In: SIGKDD 2003, Washington, DC, USA, August 24-27 (2003)
7. Nayak, R., Iryadi, W.: XML schema clustering with semantic and hierarchical similarity measures. Knowledge-Based Systems 20, 336–349 (2007)
8. Nierman, A., Jagadish, H.V.: Evaluating structural similarity in XML documents. In: Fifth International Workshop on the Web and Databases (2002)
9. Nayak, R.: Investigating Semantic Measures in XML Clustering. In: Proceedings of the 2006 IEEE/WIC/ACM International Conference on Web Intelligence (WI 2006 Main Conference Proceedings), WI 2006 (2006)

10. Bertinoa, E., Guerrinib, G., Mesiti, M.: A matching algorithm for measuring the structural similarity between an XML document and a DTD and its applications. Information Systems 29(1), 23–46 (2004)
11. Lee, M.L., Yang, L.H., Hsu, W., Yang, X.: XClust: Clustering XML Schemas for Effective Integration. In: Proceedings of 7th Conf. on Information and Knowledge Management, pp. 292–199. ACM, New York (2002)
12. Vinson, A.R., Heuser, C.A., da Silva, A.S., de Moura, E.S.: An Approach to XML Path Matching. In: Workshop on Web Information And Data Management. Proceedings of the 9th annual ACM international workshop on Web information and data management. SESSION: XML and semi-structured data, pp. 17–24 (2007) ISBN:978-1-59593-829-9
13. Левенштейн, В.И.: Двоичные коды с исправлением выпадений, вставок и замещений символов. Доклады Академий Наук СССР 163(4), 845–848 (1965); Appeared in English as: Levenshtein, V.I.: Binary codes capable of correcting deletions, insertions, and reversals. Soviet Physics Doklady 10, 707–710 (1966)
14. `http://wordnet.princeton.edu/`
15. Ling, S., Jun, M., Lian, L., Zhumin, C.: Fuzzy similarity from conceptual relations. In: Proceedings of 2006 IEEE Asia-Pacific Conference on Services Computing, APSCC 2006, pp. 3–10 (2006)
16. Ling, S., Jun, M., Dongmei, Z., Li, L., Zhumin, C.: Fuzzy semantic similarity methods and their application to information retrieval. Journal of Computational Information Systems 3(3), 917–924 (2007)
17. Song, L., Ma, J., Liu, H., Lian, L., Zhang, D.: Fuzzy Semantic Similarity between Ontological Concepts. In: Advances and Innovations in systems, computing sciences and software engineering, pp. 275–280. Springer, Heidelberg, ISBN:978-1-4020-6263-6
18. Cormen, T.H., Leiserson, C.E., Rivest, R.L.: Introduction to Algorithms, 2nd edn. The Massachusetts Institute of Technology Press, ISBN : 0-262-03293-7.2001
19. Yang, C.C., Liu, N.: Measuring Similarity of Semi-structured Documents with Context Weights. In: Annual ACM Conference on Research and Development in Information Retrieval, Proceedings of the 29th annual international ACM SIGIR conference on Research and development in information retrieval, Seattle, Washington, USA, pp. 719–720 (2006) ISBN:1-59593-369-7
20. `http://www.cs.wisc.edu/niagara/data/`
21. Manning, C.D., Raghavan, P., Schütze, H.: An Introduction to Information Retrieval. Cambridge University Press, Cambridge (2006)
22. `http://www.cs.washington.edu/research/xmldatasets`
23. `http://www.acm.org/sigs/sigmod/record/xml`
24. `http://monetdb.cwi.nl/xml`
25. `http://www.alphaworks.ibm.com/tech/xmlgenerator`

A Bloom Filter Based Approach for Evaluating Structural Similarity of XML Documents*

Dunlu Peng, Huan Hou, and Jing Lu

School of Optical-Electrical and Computer Engineering,
University of Shanghai for Science and Technology, Shanghai, 200093, China
dlpeng@usst.edu.cn, huoh@usst.edu.cn, jinglu76@gmail.com

Abstract. Evaluating Similar structure of XML is a key issue for building the core algorithms for XML document clustering, XML classification and the extraction of schema or DTD from a corpus of XML documents. This evaluation is based on the structural similarity between XML documents. This work employs Bloom filter to represent an XML document with two structures: one is Tag-based Bloom filter (TBF) which describes an XML document with the tags of elements, and the other is Path-based Bloom filter (PBF) which describes hierarchical structure of the XML document. Based on this two structures, an approach is developed to evaluate the similarity of XML documents. A group of experiments was conducted to investigate the performance of the proposed approach.

1 Introduction

In recent years, XML has become a de facto standard for data exchange and representation. Classifying and clustering structural similar documents is an efficient way for storage and retrieval of XML documents. Similar structure evaluation is a key to build the core algorithms for the two issues, which are based on the structural similarity between XML documents. Other applications such as the integration of XML data (cleaning XML data)[1],the web site structural analysis [2] and the extraction of schema or DTD from a set of XML documents [3], are relevant to this issue.

It is well known that *Tree Edit Distance* (TED)[2,3,5,6,7] is a commonly used metric for measuring the structural similarity between XML documents. However, the computational cost of TED-based approaches is extremely expensive. Flesca et al. [5] develop an approach by quantifying the structures of two documents and interpreting the results as time series. These two time series are then analyzed and compared using a *Discrete Fourier Transformation* (DFT) based algorithm for comparing two documents. Although this approach is more efficient than the previous two, in the worst case it is still nonlinear.

This paper proposes an XML structure evaluating approach that we are aware of to employ Bloom filters for XML structural similarity evaluating. A Bloom

* This work was partially supported by the Innovation Program of Shanghai Municipal Education Commission under Grant No. 08YZ98 and the Nominated Excellent Youth-Teacher Program of Shanghai Higher Education under Grant No. SLG08012.

W. Liu et al. (Eds.): WISM 2009, LNCS 5854, pp. 242–251, 2009.

filter is an efficient hash-based data structure designed to support membership queries with allowable false positive errors [4]. However, Bloom filters are unable to directly describe the hierarchies of XML documents. To overcome this shortage, inspired by [8], in this work, we represent an XML document as two kinds of Bloom filters: **T**ag-based **B**loom **F**ilters (TBFs) and **P**ath-base **B**loom **F**ilters (PBFs). TBFs are able to describe the tag names of the XML document, and PBFs are used to represent the hierarchical structure of the XML documents. By employing these two structures, we develop an approach for estimating the similarity between XML documents. We also conducted some experiments to verify the performance of our approach.

The rest of the paper is organized as follows. in Section 2, we provide the background. Section 3 gives an introduction of Bloom filter and the definition of TBF and PBF. Section 4 discusses the approach for evaluating structural similarity between XML documents. The experiment results shown in Section 5 illustrate the performance of our approach. Finally, we conclude this work in Section 6.

2 Fundamental and Problem Description

In the section, we first introduce the fundamental of XML documents and the statement of our problem, and then present the related work.

2.1 XML Data Model

In this work, we focus on the evaluation of structural similarity between XML documents, therefore, in our XML data model, we ignore the components that can give few contributions to the structure of XML documents. These components include the values of elements, the comments, processing instructions, namespaces and hyper-links to other documents. Conventionally, an XML document is modeled as a tree structure, here, we name it as XML tree and formally define it as follows.

Definition 1. (XML tree) *An* **XML tree** *is a directed, node-labeled tree* $T = \{V, E, root\}$*, where* V *is the finite set of nested tags of nodes,* E *is the set of paths on* V *where each pair* $(u, v) \in E$*,* $u, v \in V$*.* (u, v) *represents the ancestor-descendent relationship between* u *and* v*. Node* u *is an ancestor of* v *and* v *is one of the descendent nodes of* u*.* $root \in V$ *is the only one node without parent and represents the root of* T*. Every other node of the tree has exactly one parent and it can be reached through a path of edges from the root to it. The nodes having the common parent* u *are siblings.* $|T|$ *is the number of nodes in* T*, or the size of* T*. A node of a tree is corresponding to an element or to an attribute.*

The XML document shown at left side of Fig. 1 is the structure of an international conference database. Its tree model is presented at the right side of Fig. 1, in which the tree *root* is `conference` and the tag set, V={ `conference`, `cname`, `location`, `country`, `state`, `city`, `date`, `sponsor`, `sname`, `amount` }, is formed by

```
<conference>
    <name></name>
    <location>
        <country></country>
        <state></state>
        <city></city>
    </location>
    <date></date>
    <sponsor>
        <name></name>
        <amount></amount>
    </sponsor>
</conference>
```

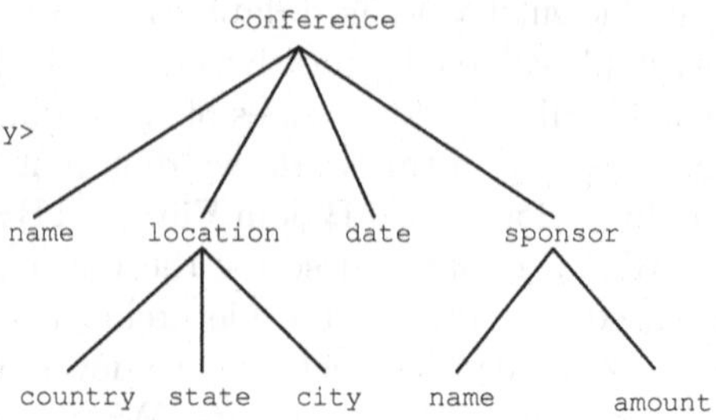

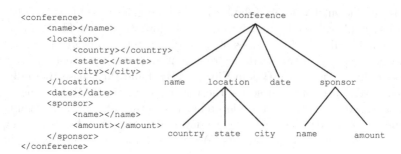

Fig. 1. An example of XML model

the element names of the tree. For every pair of parent-child nodes, there is an edge between them, such as `conference` and `cname`.

Definition 2. ($l - length$ **labeled path**) *An* $l - length$ **labeled path** p *of an XML tree is a sequence of one or more slash-separated labels, denoted as* p=/v_1/v_2/...v_l*, where* v_i *is a node name of the XML tree, "/" represents parent-child relation axes,* $l = |p|$ *is the number of nodes in the path. Specially, the length of the longest path in the tree,* l_{max}*, is the level of the tree, and we represent it as* $l_{max} = level(T)$.

For example, in the example XML tree , `/conference`, `/conference/cname` and `/conference/sponsor/sname` represent a $1 - length$, $2 - length$, $3 - length$ labeled path respectively, see Fig. 1.

2.2 Problem Description

We say two XML documents have similar structure if they have corresponding element names and the way these elements are organized in. Obviously, similarity evaluation is the key to determine whether two documents are structurally similar. To accomplish this evaluation, two sub-tasks should be completed: one is to detect the similarity of the their element names, the other is to evaluate the similarity of elements combination in the two documents. If the XML documents are presented as XML trees, we can use the tree structure similarity representing the documents structure similarity. Hence, we give the following description of our problem.

Problem statement: Given two XML documents D_1 and D_2, whose corresponding XML trees are represented as $T_1(V_1, E_1, root_1)$ and $T_2(V_2, E_2, root_2)$, the similarity between D_1 and D_2(denoted as $Sim(D_1, D_2)$) is corresponding to that of T_1 and T_2 (denoted as $Sim(T_1, T_2)$) which can be divided into two parts: $Sim(V_1, V_2)$ measures the similarity between their element names and $Sim(E_1, E_2)$ measures the similarity between their structures. To evaluating $Sim(D_1, D_2)$, we need to propose an approach for calculating $Sim(V_1, V_2)$ and $Sim(E_1, E_2)$ effectively and efficiently.

3 Using Bloom Filters to Representing XML Tree

3.1 An Overview of Bloom Filter

A Bloom filter [4] is a data structure that was proposed to detect membership of elements. It represents a set $X = \{x_1, x_2, ..., x_n\}$ of n elements as a bit-vector of length m and initially all bits are set to 0. k - independent hash functions $h_1()$, , $h_k()$ with range $\{1, ..., m\}$ are used. For each $x \in X$, the bits $h_i(x)$ $(1 \leq i \leq k)$ are set to 1. If one of the k hash values addresses a bit that is already set to 1, that bit is not changed. Fig. 2 intuitively illustrates the structure of Bloom filter.

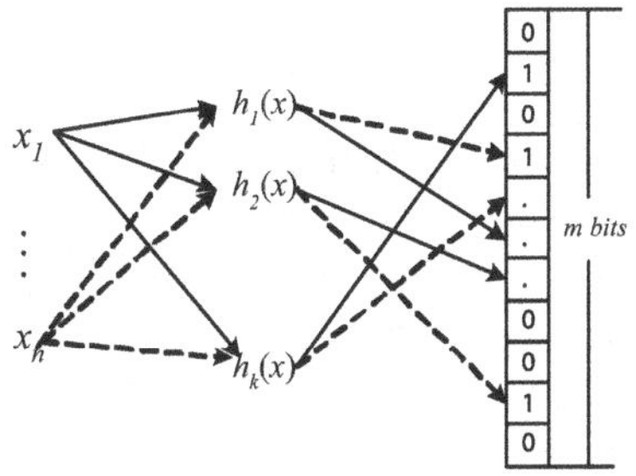

Fig. 2. Example of Bloom filter

Given another set $Y = \{y_1, y_2, , y_l\}$, the Bloom filter algorithm would loop on every element in set Y to check whether an element $y \in Y$ is in set X. We check all bits $h_i(y)$ in the vector are set to 1. If at least one of the bits is 0, then y is declared not to be a member of X. If all the bits are found to be 1, then y is said to belong to set X with a certain probability. If all the bits are found to be 1 and y is actually not a member of X, then y is said to be a false positive. The false positive errors increase as the size of set X, increases, and decrease as the number of bits of the vector, increases.

Bloom filters can be directly used to detect the membership in flat data. However, they can not straightforward represent the hierarchies which are the data structures of XML documents. To address this problem, in following section we first propose two novel hierarchical structures: *Tag-based Bloom filters* (TBFs) and *Path-based Bloom filters* (PBFs), to represent the tags and the hierarchical structure of a XML document respectively. We then discuss how to employ TBFs and PBFs to evaluate the structural similarity between XML documents.

3.2 Sketching XML Document Using Bloom Filters

In Section 2, we discussed that an XML tree $T(V, E, root)$ can be described with two parts: V-the nested tags in the XML document and E-the paths which reflect the ancestor-descendent relationship between the elements in V. As we know, Bloom filters cannot be able to directly describe the hierarchical structure of XML documents, and hence we propose two novel structure TBFs and PBFs based on Bloom filter to sketch the hierarchical structure of XML documents.

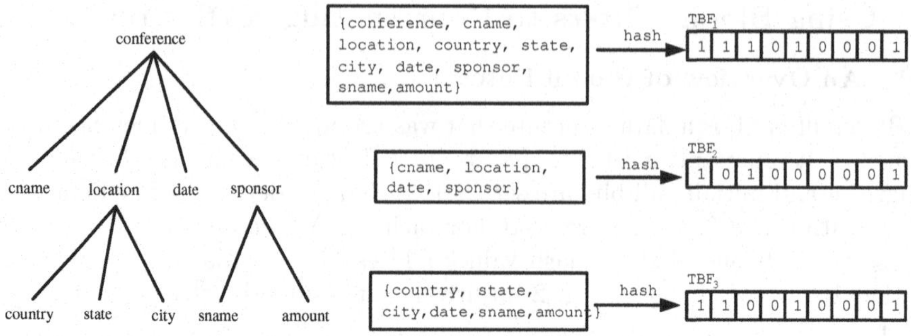

Fig. 3. Tag-based Bloom filter Derived From the Example XML Tree

Definition 3. (Tag-based Bloom filter)(TBF) *Let $T(V, E, root)$ be an XML tree with l levels , and its root level is 1. For the set of elements at level $i(1 \leq i \leq l)$ of T, we hash all its member into a Bloom filter, this Bloom filter is called a* **Tag-based Bloom filter (TBF)** *of T at level i, and denoted as TBF_i. Obviously, all the TBFs, from TBF_1 to TBF_l, give the XML tree a horizontal description.*

Intuitively, for any XML tree with l levels, we need l TBFs, $\{TBF_1, TBF_2, ..., TBF_l\}$, to represent its tags at all levels. For at level 1, there is only one element (the root of the tree), therefore, to make our approach more efficient, we hash all the tags of the tree into TBF_1. Fig. 3 presents the TBFs derived from the example XML tree.

Definition 4. (Path-based Bloom filter)(PBF) *Let $T(V, E, root)$ be an XML tree with l levels. The labeled paths of the tree are divided into l groups according to their length. That is, all $1-length$ labeled paths together form group g_1, $2-length$ labeled paths together form group g_2,..., and till all $l-length$ labeled paths together form group g_l. For any group $g_i(1 \leq i \leq l)$, all its members are hashed into the same Bloom filter, this Bloom filter is named as* **Path-based Bloom filter (PBF)**. *Therefore, we need l PBFs to represent all the possible paths in T. All the PBFs, from TBF_1 to TBF_l, together give T a vertical description, that is, the description of relationship among the elements.*

According to Section 2.1, we know the example XML tree has 3 levels. Therefore, we need 3 PBFs to summarize the structure of the tree. Fig. 4 depicts the three PBFs of the example XML tree.

4 Evaluating Document Similarities

With the discussion in our problem description, we know in order to evaluate the similarity of two XML documents, we need to calculate the similarity between their element names and that of the way their tags are organized in. Therefore,

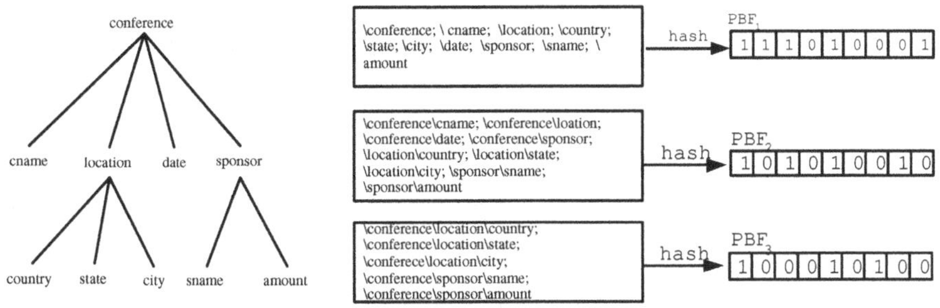

Fig. 4. Path-based Bloom filter Derived From the Example XML Tree

we compute similarity between their tags at first and then if they have similar tag sets, the similarity between their possible paths summarizing their structures will be checked.

4.1 Measuring the Similarity between XML Documents through Tags

For calculating the similarity more effectively, the tags at the same level of the corresponding tree of an XML document form a *sub*-set of the XML. Therefore, for an l level XML tree, we have l *sub*-sets of tags.

Given any two XML documents D_1, D_2 and their level l_1 and l_2, their XML trees are denoted as T_1, T_2. The tag *sub*-set of $T_i (i = 1 \ or \ 2)$ at level j is represented as $S_{ji} (1 \le j \le l_i)$ and the corresponding TBF is represented as $TBF_{ji} (1 \le j \le l_i)$. We use Jaccard similarity to calculate the ratio of tags common to both T_1 and T_2 among all tags in the same level. Jaccard similarity between T_1 and T_2 at level l $(1 \le l \le min(l_1, l_2))$, $Sim_{t_l}(T_1, T_2)$, is defined as

$$Sim_{t_l}(T_1, T_2) = \frac{S_{l1} \cap S_{l2}}{S_{l1} \cup S_{l2}} \tag{1}$$

For example, let S_{l1} ={ `cname, location, date` } and S_{l2} ={`cname, location, date, chair`}, then, $Sim_{t_l} = \frac{3}{4} = 0.75$. If Sim_{t_l} is not less than the threshold δ_t, then we say T_1 and T_2 have similar tags at level l. We define the $t_k - similar$ XML tree as follow.

Definition 5. ($t_k - similar$ **XML trees**) *Given two XML trees, T_1 and T_2, and their levels l_1 and l_2, if T_1 and T_2 have similar tags at level j, $0 \le j \le k \le min(l_1, l_2)$, but do not have similar tags at level $k+1$, then, T_1 and T_2 are called as $t_k - similar$* **XML trees**.

Definition 5 implies that if T_1 and T_2 are $t_k - similar$ XML trees, they must be t_{k-1} similar. This means to evaluate the value of k, we only need to compute the similarity of the tags at the corresponding levels between T_1 and T_2 till arriving the level at which the tag similarity is less than the threshold. To solve this problem, we employ a Bloom filter based approach to improve the performance of the computation. That is, we use TBF defined in Section 3.2 to compute the tag-based similarity at different level between two XML trees. The similarity of between two TBFs at level l is calculated as follows:

$$Sim(TBF_{l1}, TBF_{l2}) = \frac{Number\ of\ 1\ in\ (TBF_{l1}\ AND\ TBF_{l2})}{Number\ of\ 1\ in\ (TBF_{l1}\ OR\ TBF_{l2})} \tag{2}$$

With the overview of Bloom filter (see Section 3.1), we know that there may be some false positives in the result, therefore, $Sim(TBF_{l1}, TBF_{l2}) \geq Sim_{t_l}(T_1, T_2)$ exists. This denotes that we need to keep the false positive in an acceptable bound and use some approaches to filter out the error candidates.

4.2 Measuring the Similarities between XML Documents through Paths

As we aforementioned, after the measurement of the tag similarity, we need to check the structure similarity for two t_k-similar XML trees. The structure here is the possible paths formed with the tags in an XML tree. If any two XML trees have similar possible paths, they also have similar structure. Thus, we calculate the similarity between the possible paths of the two XML trees.

Let T_1 and T_2 be two XML trees, whose levels are l_1 and l_2. $P_1 = \{P_{11}, P_{21}, ..., P_{l_1 1}\}$ and $P_2 = \{P_{12}, P_{22}, ..., P_{l_2 2}\}$ represent the sets of possible paths with different lengths of T_1 and T_2. $P_{ij} \in P_j$, $j \in \{1, 2\}$ denotes the $i - length$ paths of XML tree T_j. If the Jaccard similarity between P_{i1} and P_{i2} is less than the threshold δ_p, then, T_1 and T_2 is said to have $i - length$ similar paths. Corresponding to the $t_k - similar$ XML trees, we give the following definition for $p_k - similar$ XML trees.

Definition 6. *Suppose that T_1 and T_2 are two XML trees. If T_1 and T_2 have $i - length$ similar paths with $1 \leq i \leq k \leq min(l_1, l_2)$, and do not have $(k+1) - length$ similar paths, then, T_1 and T_2 are $p_k - similar$* **XML trees.**

Definition 6 ensures that two $p_k - similar$ XML trees must be $p_{k-1} - similar$ but not $p_{k+1} - similar$ XML trees. Obviously, when comparing two given XML trees T_1 and T_2, we need to compute the similarity between their paths from length 1 to length $k+1$ where $k+1$ is the length of shortest paths whose similarity between T_1 and T_2 is below the threshold δ_p.

Similar to the evaluation of $t_k - similar$ XML trees, in order to improve the efficiency of computing the $i - length$ path similarity between T_1 and T_2 we employ the PBFs defined in Section 3.2. The similarity between two $i - length$ path-based Bloom filter PBF_{i1} and PBF_{i2}, denoted as $Sim(PBF_{i1}, PBF_{i2})$, is measured with the ratio of the number of 1's in the result of bit-wise AND between PBF_{i1} and PBF_{i2} to the number of 1's in the result of bit-wise OR between PBF_{i1} and PBF_{i2}.

5 Experiments

In order to investigate the effectiveness of our techniques, a group of experiments over synthetic and real data has been implemented. We begin with describing the setting and metrics of our evaluation, and then provide the results of our experiments.

5.1 Experimental Setup and Evaluation Metrics

Experimental Setup. In our experiments, we use IBM XML document Generator, which is integrated with Visual Age for Java, to generate a set of 1000 synthetic XML documents conforming to the following DTDs[1]: OrdinaryIssuePage.dtd (DTD_1), ProceedingsPage.dtd (DTD_2), IndexTermsPage.dtd (DTD_3) and SigmodRecord.dtd (DTD_4). The experiments were implemented in Java and performed on a PC with Intel Core 2 Duo 1.8 GHz processor (2GB RAM) running Windows XP professional operating System.

Evaluation Metrics. One of the purposes of similarity evaluation between XML documents is to make the classification of the documents more effectively. Therefore, we make use of structural classification to validate our structural similarity approach. *Precision (Prec)* and *Recall (Rec)* are exploited to evaluate the performance of our approach. A classification approach with high *Precision/Recall* indicates with a good similarity method. In our context, these two metrics are defined as follows:

C_i be the class of XML documents corresponding to DTD_i, d_i be the number of correctly classified documents in C_i, fp_i be the number of documents in C_i but not conformed to DTD_i and fn_i be the number of documents not in C_i but conform to DTD_i. Then, we define

$$Prec = \frac{\sum_{i=1}^{n} d_i}{\sum_{i=1}^{n} d_i + \sum_{i=1}^{n} fp_i}, \quad Rec = \frac{\sum_{i=1}^{n} d_i}{\sum_{i=1}^{n} d_i + \sum_{i=1}^{n} fn_i} \tag{3}$$

where n is the number of classes and $n = 4$ in our experiments.

5.2 Experimental Results

In our experiments, MD5 message-digest algorithm[2] was employed to generate independent hash functions for TBFs and PBFs. We evaluated our approach from two aspects: one is to measure *Prec* and *Rec* being affected by the threshold δ_{t_n} (or δ_{p_n}),where n is the number of classes, and the other is to investigate k of t_k(or p_k) affecting on the metrics.

We classified XML documents by varying the threshold δ_{t_n} (or δ_{p_n}) in the interval $[0, 1]$. At the beginning, we let $\delta_{t_n} = 0.00$ (or $\delta_{p_n} = 0.00$) and assigned all the documents into all the classes, and then we set δ_{t_n} (or δ_{p_n}) to 0.20, 0.40, 0.60, 0.80 and finally 1.00. Fig.5 and Fig. 6 show the results of our experiments.

[1] Available at http://www.dia.uniroma3.it/Araneus/Sigmod/Record/DTD/

[2] Available at http://www.ietf.org/rfc/rfc1321.txt

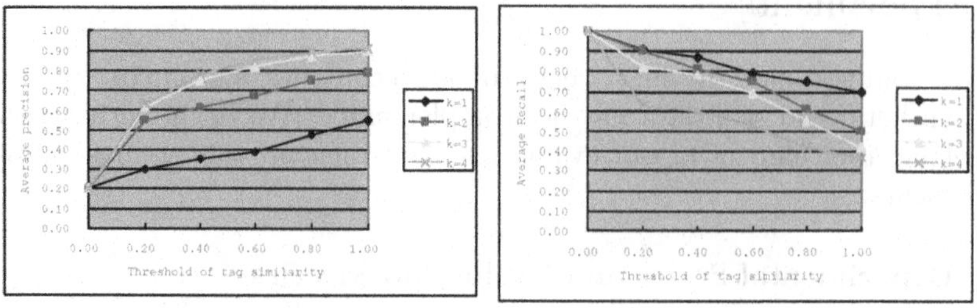

(a) *Precision* (b) *Recall*

Fig. 5. Precision and recall for different k and different tag similarity δ_{t_n}

Fig.5 illustrates the average metrics affected by the similarity threshold δ_{t_n} when k is set to 1, 2, 3 and 4. For a given k, such as $k = 2$, when δ_{t_n} is equal to 0.00, all XML documents in the corpus are assigned into each of the classes corresponding to the DTDs at hand. That is characterized by minimum precision (about 0.20 in our experiments) and maximum recall(1.00 in our experiments). *Prec* increases as δ_{t_n} increases and on the contrary *Rec* decreases while δ_{t_n} increases. This is because during the increment of δ_{t_n}, the false assigned documents are pruned from the classes by degree, eventually the proportion of documents conforming to the corresponding DTDs encompassed in a certain class becomes more. This improves the precision of classification, while the recall of classification is reduced. Fig.6 indicates δ_{p_n} having similar effect on the classification.

Let us study how the value of k affects the experimental results. From Fig.5 and Fig.6, we can see that for a given threshold, the precision of classification is improved as the increase of k and the recall is in contrast. For example, when δ_{t_n} is set to 0.40 (see Fig.5), the average precision of $k = 4$ is about 0.80, while that of $k = 3$, $k = 2$, $k = 1$ is about 0.73, 0.60 and 0.35, respectively; meanwhile, the average recall of $k = 4$ is about 0.60 and that of $k = 3, 2, 1$ is about 0.70, 0.74 and 0.79. This is due to more documents being filtered as k increases and in the filtered documents some are false negative. From the description of Section 3.1, we know this is not the feature of Bloom filter. By investigating the our

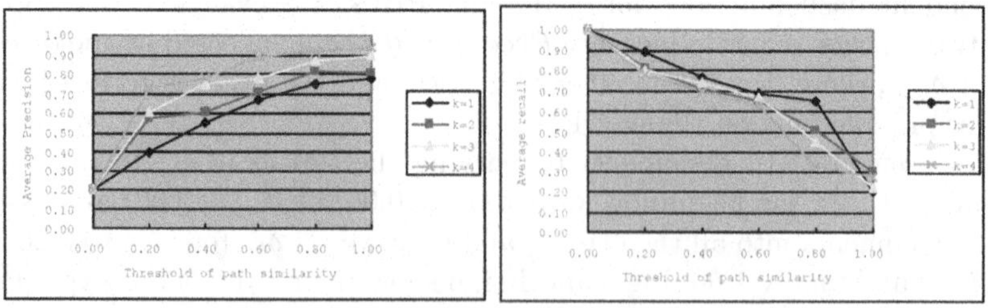

(a) *Precision* (b) *Recall*

Fig. 6. Precision and recall for different k and different path similarity δ_{p_n}

data sets, we found that the maximal length of path for testing documents is distributed from 1 to 7. When k is set to a certain value, the XML documents whose maximal length of path is below k will be filtered. This leads to drop the recall of classication.

6 Conclusion

This work provides a methodology for evaluating the structural similarity between two XML documents. This approach employs Bloom filters for XML structural similarity evaluation. However, Bloom filters are unable to directly describe the hierarchies of XML documents. To overcome this shortage, we represent an XML document as two kinds of Bloom filters, Tag-based Bloom filters (TBFs) and Path-base Bloom filters (PBFs). TBFs are able to describe the tag names of the XML document, and PBFs are used to represent the hierarchical structure of the XML documents. Experimental results indicate that our approach provides effectively representing XML documents and efficiently calculating the similarity between XML documents.

In our future work, we will compare our approach with some existing struc- tural similarity evaluation approaches, such as TED-based approaches, to demonstrate the ability of our approach for actual applications. Also, we will study how our approach scales using larger corpus of XML documents.

References

1. Lee, M.L., Yang, L.H., Hsu, W., Yang, X.: XClus: Clustering XML Schemas for Effective Integration. In: Proceedings of 11th Int. Conf. on Information and Knowledge Management, CIKM 2002 (2002)
2. de Castro Reis, D., Golgher, P.B., da Silva, A.S., Laender, A.H.F.: Automatic Web News Extraction Using Tree Edit Distance. In: Proceedings of 13th Int. WWW Conference, pp. 502–511 (2004)
3. Nierman, A., Jagadish, H.V.: Evaluating Structural Similarity in XML Documents. In: Proceedings of 5th WebDB Conferences, pp. 61–66 (2002)
4. Bloom, B.H.: Space/Time Trade-offs in Hash Coding with Allowable Errors. Communications of the ACM 13(7), 422–426 (1970)
5. Flesca, S., Manco, G., Masciari, E., Pontieri, L., Pugliese, A.: Detecting Structural Similarities between XML Documents. In: Proceedings of 5th WebDB Conference, Andrew, pp. 55–60 (2002)
6. Dalamagas, T., Cheng, T., Winkel, K., Sellis, T.: Clustering XML Documents Using Structural Summaries. In: The Proceedings of EDBT Workshops 2004, pp. 547–556 (2004)
7. Milano, D., Scannapieco, M., Catarci, T.: Structure Aware XML Object Identification. In: Proceedings of the First Int. VLDB Workshop on Clean Databases (2006)
8. Georgia, K., Evagge, P.: Filters for XML-based Service Discovery in Pervasive Computing. The Computer Journal 47(4), 461–474 (2004)

Similarity Based Semantic Web Service Match

Hui Peng[1], Wenjia Niu[2], and Ronghuai Huang[1]

[1] R&D Center for Knowledge Engineering, Beijing Normal University, 100875, Beijing, China
[2] Key Laboratory of Intelligent Information Processing, Institute of Computing Technology, Chinese Academy of Sciences, 100190, Beijing, China
penghui1999@sohu.com, niuwenjia@ics.ict.ac.cn, huangrh@bnu.edu.cn

Abstract. Semantic web service discovery aims at returning the most matching advertised services to the service requester by comparing the semantic of the request service with an advertised service. The semantic of a web service are described in terms of inputs, outputs, preconditions and results in Ontology Web Language for Service (OWL-S) which formalized by W3C. In this paper we proposed an algorithm to calculate the semantic similarity of two services by weighted averaging their inputs and outputs similarities. Case study and applications show the effectiveness of our algorithm in service match.

Keywords: semantic web service, service match, ontology concept, similarity.

1 Introduction

With the proliferation of web service, it is becoming increasingly difficult to find a web service that will satisfy our requirements[1]. While UDDI[9] becomes an appealing registry for web service, its discovery mechanism which based on key word matching does not make use of semantic information of a service yields coarse results.

The semantic web initiative addresses the problem of web data lack of semantic by creating a set of XML based language and ontology. After combining semantic web with web service, researchers have built service ontology to describe capability information about a web service and make use of the description and reason function of the ontology to discovery web service automatically. Ontology Web Language for Service (OWL-S)[8] is a prevalent language which formalized by W3C for building service ontology. OWL-S Profile provides methods to describe service capabilities. The function description in the OWL-S Profile describes the capabilities of the service in terms of inputs, outputs, preconditions and effects (IOPEs). An input is what is required by a service in order to produce a desired output. An output is the result which a service produces. A precondition represents conditions in the world that should be true for the successful execution of the service. And an effect of a service is its influence to its environment.

OWL-S Profile is proposed for both service providers and service requesters. It provides a way to describe the services offered by the providers, and the services

W. Liu et al. (Eds.): WISM 2009, LNCS 5854, pp. 252–260, 2009.

needed by the requesters. A matchmaker can match the IOPEs of a request service with the IOPEs of an advertised service to carry out semantic match between the two services. Because domain ontology concepts are often used to express the semantic of IOPEs of a service, the similarity of IOPEs of two services is concerned with the ontology concepts similarities. Then the problem is to define reasonable similarity of ontology concepts which concerned with service IOPEs.

2 The Similarity Based Algorithm

In OWL-S, the semantic of web service is expressed as IOPEs and the semantic of IOPEs can be expressed as ontology concepts which the service located in. Consider the situation of both request service and advertised service utilize the same domain ontology, the match degree between a request output/input and an advertised output/input is drawn from the subsume relationship of the two ontology concepts which stands for the request output/input and the advertised outputs/inputs each. The subsume relation is the most important binary relation between concepts in an ontology. The subsume relation is on the contrast with subClassof relation which showed in Fig. 1, that is to say, if concept A subsumes concept B, then concept B is the subclass of concept A. If concept A is the same as concept B then concept A subsumes concept B and concept B subsumes concept A.

2.1 Similarity of Ontology Concepts

In this part we define the semantic distance and the similarity of two concepts as following.

Definition 1: The semantic distance of two nodes in an ontology can be defined as:

1. If node vp is the same node as node vq, which marked as $vp = vq$, then $distance(vp, vq) = 0$.
2. If there is no subsume relations between node vp and node vq, which marked as $vp \not\subset vq$ and $vq \not\subset vp$, then $distance(vp, vq) = \infty$.
3. If the total layer of an ontology tree is m, the hierarchy where node vp located is i(the hierarchy of the root of the tree is 0), which marked as vp_i, the distance between node vp_i and its direct father is:

$$distance(vp_i) = \frac{1}{m} + \frac{m^2 - 1}{m^2} * (\frac{m+1}{m})^{-i}, \quad (i \geq 1, m \geq 2)$$

$$distance(vp_i) = 0, \ (i=0)$$

It is easy to prove that $1/m \leq distance(vp_i) \leq 1$, and $distance(vp_i) > distance(vp_{i+1})$ in the same tree when $i \neq 0$.

According to Definition 1, in Fig. 1, m=5, the value of i of each node in the tree is marked in the Fig. 1. The $distance(vp_i)$ is:

$distance$(Thing) =0,

$distance$(Book) =1/5+(24/25)*(6/5)$^{-1}$=1,

$distance$(Computer Book)

= $distance$(Language Book)

= $distance$(Economy Book)

=1/5+ (24/25)*(6/5)$^{-2}$=13/15≈0.87

……

with the same method, the distances between nodes whose relation is direct father and son are shown in Fig. 1.

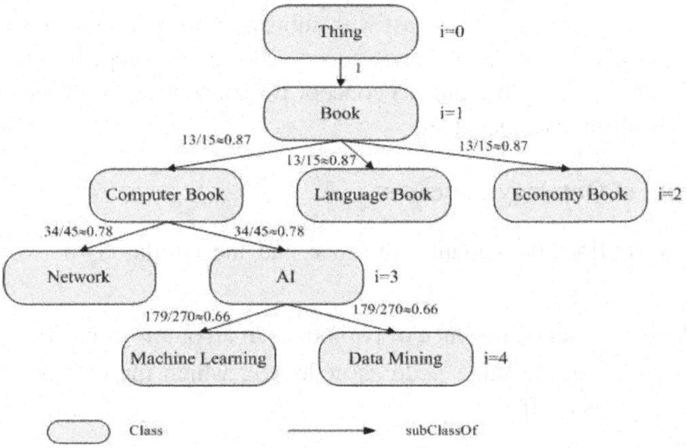

Fig. 1. Distance between Neighbor Nodes

So the distance between two nodes with subsumes relations is

$$distance(vp_i, vq_j) = distance(vp_{i+1}) + distance(vp_{i+2}) + \ldots + distance(vq_j)$$

$$< 1+1+\ldots+1 \leq m+1$$

According to Definition 1, for Fig. 1, there exists

$distance$(Computer Book, Computer Book)=0, $distance$(Computer Book, Machine Learning) =0.78+0.66=1.44,

$distance$(Machine Learning, Computer Book) =0.78+0.66=1.44,

$distance$(Computer Book, Language Book)=∞.

……

Definition 2: The semantic similarity of two nodes in an ontology is defined as

$$sim \prec vp, vq \succ = \begin{cases} 1 & if\ vq = vp \\ \frac{1}{2} + \frac{1}{2} * (1 - distance(vp, vq)/(m-1)) & if\ vp \subset vq \text{ and } vq \neq vp \\ \frac{1}{2} * (1 - distance(vp, vq)/(m-1)) & if\ vq \subset vp \text{ and } vq \neq vp \\ 0 & if\ vq \not\subset vp \text{ and } vp \not\subset vq \end{cases}$$

According to Definition 2, for Fig. 1, there exists

sim(Computer Book, Computer Book)=1,

sim(Computer Book, Machine Learning) =1/2*(1-1.44/4)=0.32,

sim(Machine Learning, Computer Book) =1/2+1/2*(1-1.44/4) =0.82,

sim(Computer Book, Language Book)=0

From Definition 2, we can conclude that:

1) When $vp = vq$, then $sim < vp, vq >= 1$.

2) When node vq subsumes vp, then $sim < vp, vq > \in (1/2, 1)$.

3) When node vp subsumes vq, then $sim < vp, vq > \in (0, 1/2)$.

4) When there is no subsumes relations between vp and vq, then $sim < vp, vq >= 0$.

2.2 Grade Match Algorithm for Semantic Web Services

The grade match algorithm[2] proposed by Paolucci and co-authors is one of the most important algorithms in semantic web services match. It introduces the flexible match mechanism in service discovery and brings the match result into four grades instead of true or false which produced by keyword search. Although the algorithm is very successful in semantic service match, a main disadvantage of the algorithm is that its "grade" match can not reflect the difference of match degree in the same grade. The similarity based algorithm introduced in part 2 of this paper exactly quantify the grade match algorithm and reflect the semantic relations of IOPEs between two service.

In the grade match algorithm, outR stands for an output of a request service and outA stands for an output of an advertised service, the main part of the algorithm can be described as[2]

degreeOfMatch(*outR*,*outA*):
if *outA*=*outR* then return exact
if *outR* is subclassOf *outA* then return exact
if *outA* subsumes *outR* and *outR* is not a direct subclassOf *outA* then return plugIn
if *outR* subsumes *outA* then return subsumes
else return fail

From the above algorithm, we can conclude that the specialties of the grade algorithm are:

1. For two outputs match, if outR and outA is the same concept or outR is the direct subclass of outA, then the degree of match is "exact", which means the advertised output can satisfy the request output completely.

2. For two outputs match, if there is no subsumes relation between outR and outA, then the degree of match is "fail", that means outA does not match outR. Take the domain ontology in Fig.1 as an example, if a advertise service which its output is an instance of "Language Book", it can not satisfy a request which demands an instance of "Computer Book", although "Language Book" and "Computer Book" have the same father node "Book".

3. If outR subsumes outA, then the degree of match is "subsume", in this situation, outA may partly satisfy the request. For example, a advertise service which its output is an instance of "Computer Book" may partly satisfy a request which demands an instance of "Book".

4. If outA subsumes outR and outR is not a direct subclass of outA, then the degree of match is "plugIn". In this situation, outA can satisfy the request more than the situation that outR sunbsumes outA. For example, we can assume a service which outputs are instances of "Book" can satisfy a request which outputs are instances of "AI" book completely because instances of "Book" include all instance of "AI" books. So the level of match is exact, plugIn, subsume and fail from higher to lower.

For two inputs match, the rational match degree generated from the subsume relation between inR and inA should be on the contrast to the outR and outA, That means if inR stands for an input of a request service and inA stands for an input of an advertised service, the degree of match between inR and inA should be:

---

degreeOfMatch(*inA*,*inR*):
if *inR* = *inA* then return exact
if *inA* is a subclassOf *inR* then return exact
if *inR* subsumes *inA* and *inA* is not a direct
subclassOf *inR* then return plugIn
if *inA* subsumes *inR* then return subsumes
else return fail

An advertised service matches a request service when all the outputs of the request are matched by the outputs of the advertised, and all the inputs of the advertised are matched by the inputs of the request. This criteria guarantees that the matched service satisfies the needs of the requester, and that the requester provides to the matched service all the inputs that it needs to operate correctly[2]. Thus the algorithm performs flexible matches to judge the match degree between a request service and an advertise service. In reference[1], the reasonability about the algorithm of grade is introduced in more detail.

2.3 The Relationship of Similarity and Grade

The grade match algorithm has become a well-known algorithm in semantic service matching as its match is based on function of services and its advantage compared with keyword match is recited in work [3]. But the prominent problem of the algorithm is that it can not reflect the difference of match in the same grade. For example in Fig. 1, if the output of a request service is "Machine Learning", no matter the output of an advertised service is "AI" or "Computer Book" or "Book" or "Thing", each result of match is "plugIn", but the degree of match among the four pairs are apparently different, the best should be "AI", the worst should be "Thing". This situation lowers the match quality of the algorithm apparently. It also cuts down the application of this algorithm in many projects of service discovery and composition because when there is more than one service match a request service in the same grade, we do not know how to select the best one for service selection and composition. With the similarity defined in 2.1 we can quantify the grade match algorithm and change the grade match into value match. This value reflects not only the difference of match in different grade but the difference of match in the same grade. With this value, it is convenient to rank the services and select the most satisfying one for applications.

Definition 3: the similarity between two outputs and inputs can be defined as: *valueOfMatch*(outR,outA)=*sim*<outR,outA>,
valueOfMatch(inA,inR)=*sim*<inA,inR>.

From Definition 3, we can conclude that:

1) When outR=outA, then *sim*<outR,outA>=1, that is consistent with the "exact" degree of match in grade algorithm.

2) When outA subsumes outR, then *sim*<outR,outA>$\in$(1/2,1), that is the "plugIn" degree of match in grade algorithm.

3) When node outR subsumes outA, then *sim*<outR,outA>$\in$ (0,1/2), that is the "subsume" degree of match in grade algorithm.

4) When there is no subsumes relations between outR and outA, then *sim*<outR,outA>=0, that is the "fail" degree of match in grade algorithm.

3 Service Match Based on Similarity

In 2.3, we propose the formula which calculates the similarity of two outputs or inputs which one is from a request service and another is from an advertised service. Generally, there are more than one input and output of every service. So we should take some rules to get the similarity of two services from every match value which comes from all pairs of output and input match. In this paper, the following two rules can be adopted to fulfill the match of two services.

Rule 1: for each output of a request service, matching it with every output of an advertised service, then select the maximum value from all results as the match result of this output of the request service. So does each input of an advertised service. That is to say

valueOfMatch(outR)=max(valueOfMatch(outR,outAj)) *j=1,2,...n* *n* is the number of output of an advertised service.

valueOfMatch(inA)=max(valueOfMatch(inRj,inA)) *j=1,2,...m* *m* is the number of input of a request service.

Rule 2: Issue a weighting factor which the value is in [0, 1] to each similarity, the value of each weighting factor means the importance of its similarity. For example, if there are

2 input similarity and 1 output similarity and they are equal important, then the weighting factor should be 1/3 for each similarity. of The result of weighted sum of all similarities can be the match degree of the request service and the advertised service.

According to Rule 1 and Rule 2, the similarity of two services is also a real number which its value is in [0, 1] and the bigger value indicates the higher match degree.

4 The Applications

SWSMathmaker[7] is an ongoing research work, which aims to provide an open reusable infrastructure for essential service composition mechanisms. It is developed to test our service match maker which is composed of OWL-S service register, AI planning method[7], DDL(Dynamic Description Logic) reasoner[10][11] which are developing in our lab. Fig. 2 shows the basic structure of SWSMathmaker.

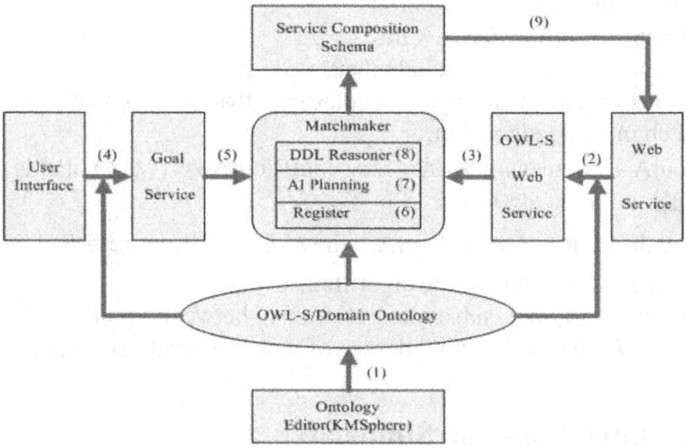

Fig. 2. Structure of SWSMathmaker

The workflow of SWSMathmaker is:

1) Domain Ontology is edited with our knowledge management tool—KMSphere[14].

2) With the help of the tool-- WSDL2OWL-SwithPE[12] developed in our lab, a semantic web service description file(OWL-S file) is generated from WSDL. In the OWL-S file, domain ontology and OWL-S upper ontology are utilized to express the capability of a service. The inputs and outputs of a service are described in ontology concepts, and the precondition and effect are described in SWRL[15][13]. Fig.3 shows the interface of this tool.

3) OWL-S file is published in the register of service matchmaker.

4) The requester's demands are collected from user interface.

5) An OWL-S file is generated from requester demands. Domain ontology and OWL-S upper ontology are also utilized to express the capability of a request service.

6) Advertised services are published in the service register and the service request inquiry the registered service.

7) If there is no single service match the goal service, AI planner decompose the goal according to some planning algorithm such as HTN[16].

8) DDL reasoner produce composition schema according the precondition and effect and select the most satisfying service by inputs and outputs matches.

9) The matched service is invoked.

In the above steps, the step 7) and 8) apply the match algorithm introduced in this paper and is responsible for match inputs and outputs of two services. To improve the efficiency of service discovery, we calculate the similarity of every pair concepts in the ontology and save these similarities in a table in advance. In the process of service match, the similarities can be changed into match value of degrees according to part 2.3 and 3.

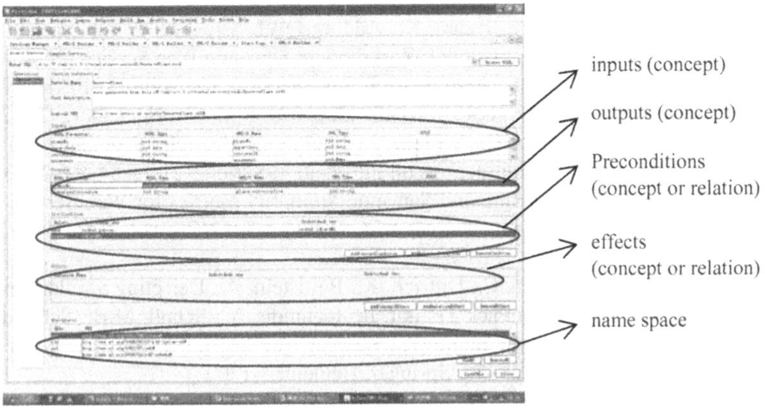

Fig. 3. Interface of WSDL2OWL-SwithPE

5 Related Work

Although there are plenty of work has been done on the similarity of ontology concepts based on the tree distance or graph distance. Not all of them are suitable for services match. And very few of them are suitable to quantify the grade algorithm to match services according to their function. For example, in work [3][5][6], the valuable methods to calculate the similarity of concepts in ontology are proposed, but the similarity defined in these works are not suitable to quantify the grade algorithm because in these works. For any two nodes vp and vq, $sim < vp, vq >$ is always equals to $sim < vq, vp >$ in these work, while according to grade, when $vp \subset vq$ or $vp \subset vq$, the similarity between vp and vq should be $sim < vp, vq > \neq sim < vq, vp >$. In work [4], a method is proposed to calculate similarities between concepts for service match, this similarity is $sim < vp, vq > \neq sim < vq, vp >$ when $vp \subset vq$ or $vp \subset vq$, but its aim is not to quantify the grade algorithm, so it does not cut its match value into four range:0,(0,1/2),(1/2,1),1 and it does not match the different range to different grade in the grade algorithm. Compared our work in [17], the algorithm in this paper take the depth of tree nodes into account when calculate the distance of tree nodes, because the depth of tree nodes will influence the semantic distance between nodes, for example, distance(thing, book) should be longer than distance(book, computer book) in book ontology.

Acknowledgements

This work is supported by the project of designing m-learning terminals for IPv6-basd networking environment([2008]48 from Technological Bureau of Ministry of Education, China.) and the National Science Foundation of China (No. 90604017 and No. 60775035).

References

1. Srinivasan, N., Paolucci, M., Sycara, K.: Adding OWL-S to UDDI, implementation and throughput. In: The First International Workshop on Semantic Web Service (2004)
2. Paolucci, M., et al.: Semantic Matching of Web Services Capabilities. In: Horrocks, I., Hendler, J. (eds.) ISWC 2002. LNCS, vol. 2342, p. 333. Springer, Heidelberg (2002)
3. Ganesan, P., Garcia-Molina, H., Widom, J.: Exploiting hierarchical domain structure to compute similarity. ACM Transactions on Information Systems 21(1), 64–93 (2003)
4. Gunay, A., Yolum, P.: Structure and Semantic Similarity Metrics for Web Service Matchmaking. In: Psaila, G., Wagner, R. (eds.) EC-Web 2007. LNCS, vol. 4655, pp. 129–138. Springer, Heidelberg (2007)
5. Ziegler, P., Kiefer, C., Sturm, C., Dittrich, K., Bernstein, A.: Detecting Similarities in Ontologies with the SOQA-SimPack Toolkit. In: Ioannidis, Y., Scholl, M.H., Schmidt, J.W., Matthes, F., Hatzopoulos, M., Böhm, K., Kemper, A., Grust, T., Böhm, C. (eds.) EDBT 2006. LNCS, vol. 3896, pp. 59–76. Springer, Heidelberg (2006)
6. Bernstein, A., Kiefer, C.: iRDQL Imprecise RDQL Queries Using Similarity Joins. In: K-CAP 2005 Workshop on: Ontology Management: Searching, Selection, Ranking, and Segmentation (2005)
7. Qiu, L., Lin, F., et al.: Semantic Web Services Composition Using AI Planning of Description Logics. In: Proceedings of the 2006 Asia-Pacific Conference on Services Computing (2006)
8. The OWL Services Coalition. OWL-S: Semantic Markup for Web Services (2004), `http://www.w3.org/Submission/OWL-S/`
9. Universal Description, Discovery and Integration (UDDI): `http://www.uddi.org/specification.html`
10. Shi, Z., Dong, M., Jiang, Y., Zhang, H.: A Logic Foundation for the Semantic Web. Science in China, Series F 48(2), 161–178 (2005)
11. Chang, L., Shi, Z., Qiu, L., Lin, F.: Dynamic Description Logic: Embracing Actions into Description Logic. In: Proc. of the 20th International Workshop on Description Logics (DL 2007), Italy (2007)
12. `http://www.intsci.ac.cn/users/pengh/swsbroker/provider.zip`
13. `http://owlseditor.semwebcentral.org/related.shtml`
14. `http://www.intsci.ac.cn/KMSphere/`
15. SWRL: `http://www.w3.org/Submission/SWRL/`
16. Sirin, E., et al.: HTN planning for Web Service composition using SHOP2. Web Semantics: Science, Services and Agents on the World Wide Web 1(4), 377–396 (2004)
17. Peng, H., et al.: Improving Grade Match to Value Match for Semantic Web Service Discovery. In: ICNC 2008 (2008)

Clustering-Based Semantic Web Service Matchmaking with Automated Knowledge Acquisition

Peng Liu[1], Jingyu Zhang[2], and Xueli Yu[2]

[1] Research Center for Grid Technology, Institute of Command Automation, PLA University of Science and Technology, Nanjing Jiangsu, China, 210007
[2] School of Computer and Software, Taiyuan University of Technology, Taiyuan, Shanxi, China, 030024
{milgird,zhangjingyu0428,xueliyu}@163.com

Abstract. The proliferation of Web services demands for a discovery and matchmaking mechanism to find Web services that satisfy the requests more accurately. We propose a clustering-based semantic matching method with Automated Knowledge Acquisition to improve the performance of semantic match based on OWL-S. The clustering method enables service matchmaker to significantly reduce the overhead and locate the suitable services quickly. The Automated Knowledge Acquisition mechanism is used for extending ontology to solve the problem that the user's query word doesn't exist in the ontology model. We present performance analysis of the experiments and obtain the great improvement of precision and recall due to the method.

Keywords: clustering, Semantic Web, matchmaking, OWL.

1 Introduction

The increasing usage of Web services on the Internet has led to much interest in the study of service discovery. Current Web services discovery mechanism is based on XML-based standards including WSDL (Web Service Description Language) and UDDI (Universal Description, Discovery and Integration). However, the current XML-based specifications provide only syntactical descriptions of functionality and lacks semantics. The lack of semantic description requires user intervention in the decision making process. OWL-S [1] is service ontology to support greater automation in service selection and invocation between heterogeneous services. Although UDDI has many features that make it an appealing registry for Web services, its discovery mechanism has two crucial limitations. First limitation is its search mechanism based on keyword and classification. The second shortcoming of UDDI is the usage of XML to describe its data model. OWL-S provides capability-based search mechanism and semantic Web service description. OWL-S Service Profile [1] describes the capabilities of Web services, hence crucial in the Web service discovery. Capability-based search will overcome the limitations of UDDI and would yield better search results. Robotics Institute of Carnegie Mellon University propose OWL-S/UDDI matchmaker in [2, 3, 4]. Our study is based on the work.

W. Liu et al. (Eds.): WISM 2009, LNCS 5854, pp. 261–270, 2009.

Since Web service discovery depends on the service repository and classifications, one direct method to improve service discovery is preprocessing service information in UDDI. Researchers have worked on using service clustering to improve the service discovery in [5, 6 7]. We adopt the Single Linkage algorithm [6] to cluster mass Web services and group similar services. In addition, we employ an Automated Knowledge Acquisition (AKA) method to support memorized optimization and ontology extending, which makes ontology module self-learned according to contextual information from queries by WordNet [8].

Our system employs matchmaking algorithms based on clustering with automated knowledge acquisition (AKAC) to improve the service discovery, and our system extends OWL-S/UDDI matchmaker architecture. In this paper, we also present performance analysis in precision and recall due to the clustering and self-learning capability of the system.

2 Extended OWL-S/UDDI Matchmaker Architecture

Fig.1 shows the architecture of our system. Based on the semantic matchmaker mechanism proposed in [2, 3, 4] and the self-learning mechanism in [11], we augment the UDDI registry with OWL-S matchmaker component and Automated Knowledge Acquisition Module.

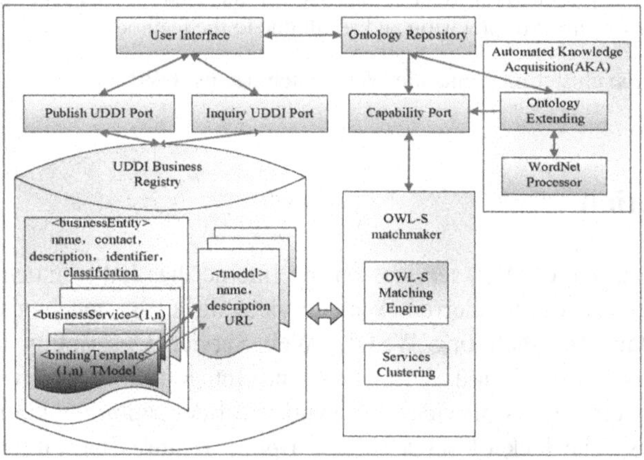

Fig. 1. Extended OWL-S/UDDI Matchmaker Architecture

OWL-S matchmaker is for processing OWL-S profile and extends the UDDI registry by adding a capability port. Automated Knowledge Acquisition (AKA) module is for ontology self-learning by ontology extending. We describe it in detail.

2.1 OWL-S Matchmaker Module

In order to combine OWL-S and UDDI, the OWL-S/UDDI registry requires the embedding of OWL-S profile information inside UDDI advertisements. We adopt the

OWL-S/UDDI mapping mechanism described in [9].The OWL-S Profiles can be converted into UDDI advertisements and published using the same API. We need to augment the UDDI registry with an OWL-S Matchmaking component, for processing OWL-S profile information. On receiving an advertisement through the publish port the UDDI component, in the OWL-S/UDDI matchmaker, processes it like any other UDDI advertisement. If the advertisement contains OWL-S profile information, it forwards the advertisement to the matchmaking component.

Mass services in UDDI have similar functions, although these services have different names and parameters. These services have similar concepts and semantic information. Pre-clustered in advance, services can be matched effectively. The clustering algorithm groups the collections of services listed in UDDI according to semantic similarity. Services in one cluster have great similarity after the services are clustered. Service Clustering makes the needed services located quickly and the search space narrowed, which improves the service discovery.

A client can use the UDDI's inquiry port or capability port to search services in the registry. Extended UDDI API is to access the capability-based search functionality in the OWL-S/UDDI matchmaker. Using the capability port, we can search for services based on the capability descriptions, i.e. inputs, outputs, pre-conditions and effects (IOPEs) of services. The queries received through the capability port are processed by the matchmaker component; hence the queries are semantically matched based on the OWL-S Profile information.

2.2 Automated Knowledge Acquisition Module

Based on the self-learning mechanism proposed in [10], we add the Automated Knowledge Acquisition (AKA) module in our system. Ontology Extending is employed to extend Ontology file such as OWL, and Dictionary Processor is used to obtain better contextual information relevant o the client query. Ontology Extending is built using Jena Toolkit for ontology operation. The module is built using the WordNet2.0 Dictionary. We use JWNL 1.3 API to access the WordNet Dictionary. Initially, if through the capability port, the query words are matched with the Subject, Object and Predicate of each ontology statement or OWL Class et al in OWL concepts. We use Dictionary Processor to obtain synonyms of query words for extending ontology. Considering different senses of a particular word, we can ensure that the selected ontology domain has the closest relevance to the client query words.

For synonym matching, four different search outcomes are possible.

- Neither the query word nor the synonym words are present in any of the ontology models.
- Some of the synonyms are presented, but not the query word.
- The query word is present, but not its synonyms.
- Both the query word and its synonyms are present in the ontology model.

We use above outcomes to extend the ontology models for enriching the ontology model. We have designed the AKA module that can stores the knowledge and information of a previously made query to later queries for predicting more accurate results. If both the query word and its synonyms are not found, the ontology model can't be extended. If both the query word and its synonyms both exist in the ontology model, the ontology model keeps intact. The ontology file is extended when a synonym of the query word exists in the ontology model but not the query word. In this case, we infer that the query word is quite possible to have contextual relevance to the ontology model.

If we have a query "car" and we don't have the keyword "automobile" in current ontology model. Suppose from Dictionary Processor we can infer "automobile" is a synonym of car and "automobile" already exists in the current ontology model. We infer "car" is semantically similar to "car" and should include it in the ontology model. Then we restructure the ontology model and include the keyword "car" in the ontology model by Jena. We update the model and reload it. However, if a keyword in the user's query string is already present in the ontology model, every synonym of it doesn't have to be incorporated into the ontology model. We get "cable car" as a synonym for car from Dictionary Processor. Since "cable car" is not present in the ontology file, we can't extend the ontology because we infer "cable car" is irrespective of the present context of "car".

3 Semantic Service Matchmaking in UDDI

The clustering-based semantic service matchmaking with AKA method is based on the algorithm presented in [2]. The services in UDDI can be grouped into different clusters according to the semantic similarity and the best offers of service can be retrieved using the cluster matching. We also employs an AKA method for optimization to match user queries with services published in UDDI. We describe the method in detail.

3.1 Clustering

We assume that the Web services have been published as advertisements. Suppose OWL-S Profile is mapped to UDDI data structure and have been registered.

For ws1 and ws2, we discuss the similarity calculation of functional attributes (mainly Description, Input and Output parameters). Suppose the descriptions, input parameters, and output parameters of the two services are denoted as Description1/Input1/Output1 and Description2/Input2/Output2 respectively. The method gives different weights to each attributes. The calculation can be through the following steps.

Extracting Words. Terms are extracted from scanning the Service Profile. We extract the notional words useful for Web service discovery and exclude the stop words such as a, the, there, etc and the OWL terms such as description, input, output.

Calculating Semantic Similarity. The Jaccard coefficient is used to calculate the similarity using the terms between two Web services [11]. Let X and Y be the two different Web services. The dissimilarity of these service descriptions is defined as:

$$Jaccard_{ws1-ws2} = \frac{T_{xy}}{T_x + T_y + T_{xy}} \tag{1}$$

$$SimDes_{ws1-ws2} = Jaccard_{ws1-ws2} \tag{2}$$

Where T_{xy} is the number of common terms used for describing X and Y, and T_x and T_y are the number of terms used in X only and Y only respectively. $SimDes_{ws1\text{-}ws2}$ is the similarity coefficient for Description. Similarity coefficients for Input, Output between a pair of services are calculated based on Jaccard coefficient. Concepts of Input and Output are semantically annotated, so we have to consider their semantic distances in the ontology model according to Equ3. $SimInput_{ws1\text{-}ws2}$ is influenced by the two factors, which is shown in Equ4. $SimOutput_{ws1\text{-}ws2}$ is calculated in the same way as $SimInput_{ws1\text{-}ws2}$.

$$Sim_{semantics}(C_1,C_2) = \frac{\alpha\times(l_1+l_2)}{(Dis(C_1,C_2)+\alpha)\times\max(|l_1-l_2|,1)} \tag{3}$$

$$SimInput_{ws1-ws2} = Sim_{semantics}(C_1,C_2)\times Jaccard_{ws1-ws2} \tag{4}$$

here l_1 and l_2 are the hierarchy of C_1 and C_2 in the ontology model respectively. α is a adjustable parameter that its value is greater than 0. Dis (C_1, C2) is the semantic distance in the ontology model.

Each of these components forms their own similarity matrix. The accumulative similarity coefficient between each pair of Web services is calculated based on the following formula where $SimDes_{ws1\text{-}ws2}$, $SimInput_{ws1\text{-}ws2}$and $SimOutput_{ws1\text{-}ws2}$ are similarity coefficients. The whole process is repeated for each pair of services.

$$\begin{aligned} Sim_{ws1-ws2} &= w_1\times SimDes_{ws1-ws2} + w_2\times SimInput_{ws1-ws2} \\ &+ w_3\times SimOutput_{ws1-ws2} \end{aligned} \tag{5}$$

Empirically, w_1, w_2, w_3, are weight parameters assigning significance to each corresponding coefficient. These weight values are preset as $w_1 = 0.5$, $w_2 = 0.3$, $w_3 = 0.2$. This formula gives the terms of the service description a much higher weight than others. The overall similarity coefficient is 1.0.

Having obtained the similarity measure between the services, a similarity matrix is constructed for being used as input for clustering. A similarity matrix $Dm\times m$ is an $m\times m$ matrix containing all the pair wise similarities between the Web services.

Clustering Web service. We use Single Linkage algorithm [6], one kind of the hierarchical agglomerative clustering algorithm, for clustering Web services. In single linkage, the distance between two clusters is computed as the distance between the two closest elements in the two clusters. We omit it because of length of the article.

The clusters (terms representing Web services), the level of each cluster and the similarity coefficients are stored in the UDDI registry database. There are prepared for OWL-S matchmaking engine to match the best offers of services using the cluster matching results.

3.2 Automated Knowledge Acquisition

From Figure1, the Ontology Repository stores the vocabularies for a wide variety of domains. We use OWL as the ontology definition language. The ontology definitions are modeled in the Ontology Matcher using Jena [11]. Jena provides a simple OWL API for processing vocabularies.

The query words are first searched in the ontology models. These models consist of statements where each statement is made up of Subject, Predicate and Object or Classes. If the query words are not included in the model, the ontology model is possible to be extended.

Within ACK module, the Dictionary Processor is used and its features are employed to obtain better contextual information relevant to the client query. Initially, the query words are matched using direct keyword matching with the Subject, Object and Predicate of each ontology statement. We use the Dictionary Processor to employ synonym, hypernym, and hyponym matching techniques. By taking into consideration different senses of a particular word, we can ensure that the selected ontology file has the closest relevance to the client query string. The dictionary matching method uses synonym, hypernym, and hyponym of a particular key word from WordNet2.0 by JWNL1.3API. The synonyms of the query words are fetched and matched with the statements or classed in the ontology model. If the synonym match fails to provide a positive result then hypernyms and hyponyms of the query words are retrieved and matched with the model. On a positive match, the synonym, hypernyms or hyponyms can be added into the ontology model using Jena. The extended ontology is ready for capability searching.

3.3 Semantic Service Matchmaking in UDDI

The matching algorithm is based on the OWL's subsumption mechanism and relies on the results of clustering. When a request is submitted, the algorithm finds an appropriate service by first matching the outputs of the request against the outputs of the published and clustered advertisements, and then, if any advertisement is matched after the output phase, the inputs of the request are matched against the inputs of the advertisements matched during the input phase.

The main control loop of the matching algorithm is shown below. Requests are matched against advertisements in clusters stored by the registry. Advertisements have been grouped into different clusters. If so, it is easier and quicker to find the most suitable advertisements in clusters than in the whole advertisements repository. When a match between the request and any of the advertisement in some cluster is found, it is recorded and scored to find the matches with the highest degree.

Main control loop

```
Match(request) {
   recordMatch=empty list
   Forall cluster(k) in clusters do {
      For adv in cluster(k) {
         if match(request,adv){
             recordMatch.append(request,adv);
         }
         return L(k);
      }
   }
}
```

4 Experimental Evaluation

We conducted experiments on a computer with a Intel(R) Core(TM)2 CPU @1.80GHz and 1.00GB of RAM running Microsoft Windows XP.We employ JUDDI as a UDDI register server. We employ OWL-S/UDDI matchmaker in JUDDI and use OWL-S/UDDI client to access OWL-S/UDDI matchmaker. We use JSP Pages as User Query interface, and use JWNL 1.3 API to access WordNet2.0 for extending ontology. In addition, we store ontology in a persistent database (We use MySQL as the database). Once the ontology is loaded, the ontologies are accessible without needing to reload them from the source.

Experiment 1. Keyword Search
The user searches for services using the UDDI's common inquiry port without the semantic information or the capability description provided by the OWL-S profile information.

Experiment 2. Clustering-Based Keyword Search
Clustering Web services in UDDI makes similar services in a cluster and dissimilar services in different clusters. The method is realized in Services Clustering in Figure1. When the OWL-S matchmaker matches the services, **it** only needs to locate the most relevant cluster, and then continues to search for the suitable services only in one cluster.

Experiment 3. Clustering-based Semantic Matchmaking
Based on clustering services, the user searches for services in UDDI using the capability port based on ontology and OWL-S profile. We use OWL-S API and Jena for manipulating OWL-S and OWL models.

Experiment 4. Clustering-based Semantic Matchmaking with AKAC
Suppose the services in UDDI have been clustered using the **Services** Clustering model. The user searches for services using the capability. Most importantly, when the query word doesn't appear in the current ontology model, the model can be extended using the AKA module. JWNL1.3 API can obtain synonyms, hypernyms, hyponyms et al of the user's query word from WordNet. We consider synonyms as the *EquivalentClasses* of the query word, hypernyms as *SuperClasses* of the query word and hyponyms as *SubClasses* in Ontology model. We use Jena to manipulate it and store the update in the model from the persistent ontology model in database.

Fig. 2 shows the average time for querying services in UDDI in the methods presented in the above experiments. The number of services forms an arithmetic progression whose common difference is 4. We can see that the clustering method can optimize the keyword search. Because the number of services in UDDI is limited, the querying time using keyword search doesn't seem to increase a lot. The curve of clustering-based semantic search shows the overall trend is that the consuming time declines with the ontology model loaded. The consuming time of clustering-based semantic search with AKA sometimes increases, because the user's query word doesn't appear in the loaded ontology model needs to be extended with the AKA.

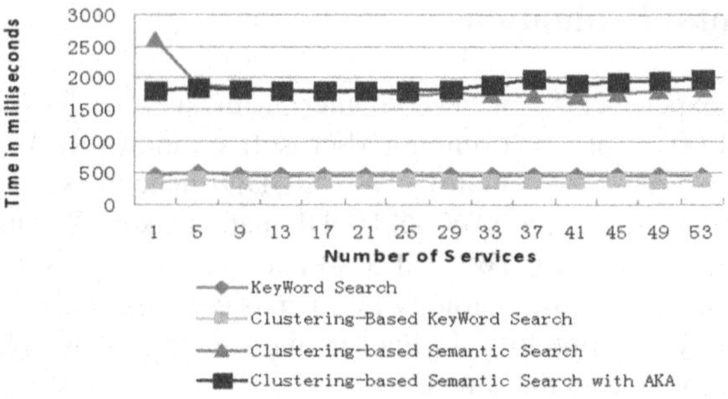

Fig. 2. Querying time for services using different methods

From Fig. 3 it can be seen that the Ontology Load module and Ontology Update consumes most of the processing time using the clustering-based semantic search with AKA. The WordNet Process consumes time to get synonyms et al from WordNet2.0, selects the suitable words appearing the ontology from them in accord with the context information and inserts the new concept into the suitable place in the model. The WordNet Process is vital in the method, but doesn't consume the most time.

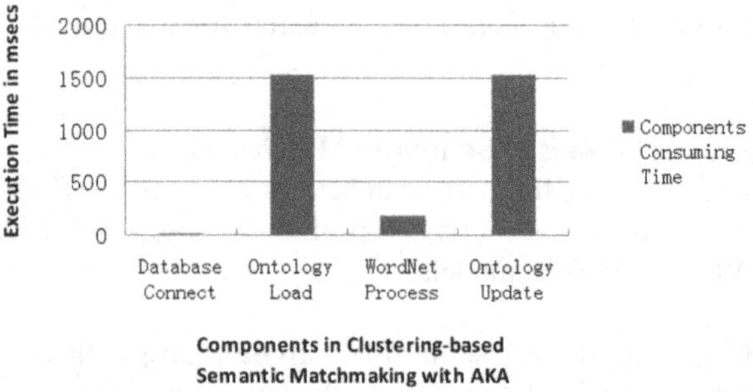

Fig. 3. Execution time taken by components in Experiment 4

We use the precision and recall measurements to study the accuracy of our methods. We define *precision* as ration of the number of relevant services corresponding to a user query and the total number of services returned by the system. We define *recall* as ratio of the number of relevant services retrieved and the total number of the services related to the user's query in UDDI. Table1 shows the average precision and recall of service discovery with the four different search methods in UDDI.

From Table 1, we can compare the performance among the four experiments. Both precision and recall of the semantic search are obviously higher than the keyword search, which indicates that the semantic matchmaker module is vital to the high performance of the system. From Figure3 we can see that clustering method can improve the time efficiency of Web service discovery, whereas Table1 denotes that clustering also can improve the accuracy of service discovery. The AKA module plays an important role when the user's query word doesn't appear in the current

Table 1. Performance of Different Methods

Method	Precision	Recall
Experiment 1	62.83%	56.85%
Experiment 2	72.45%	64.85%
Experiment 3	81.78%	74.51%
Experiment 4	89.63%	80.51%

ontology model and its synonyms, hypernyms, hyponyms et al appear in the ontology model. Tabel1 shows Experimtnt4 improves the greater precision and recall with the AKA module than Experiment3. Clustering and AKA method both help improve the performance of the Web service discovery.

In addition, with the addition of the services related to the user query but not including the user query keyword, both precision and recall increase. With the addition of the services unrelated to the user query, both precision and recall remain unchanged.

The architecture and the algorithm make the requests semantically match the services more accurate and the search results satisfy the user query. The clustering-based semantic matchmaking with AKA method improves the efficiency on the whole.

5 Conclusion and Future work

We have discussed the importance of OWL-S profile in Web service discovery and the disadvantages of the UDDI's discovery mechanism. We try to use OWL-S in combination with UDDI and propose a clustering-based semantic matchmaking method with AKA. The semantic matching algorithm is based on OWL-S service profile. We group similar services with Single-Linkage algorithm so that it is in proper order to match the services against the user's query. When the user's query searches for services through the capability port, we extend the ontology using the AKA module to solve the problem that the user's query word doesn't exist in the ontology model. We implemented it based on the OWL-S/UDDI matchmaker and analyzed the experiment results. Experiment results demonstrate the above approach can improve the search performance. The algorithm proposed in this paper provides a valuable ground for an efficient and scalable implementation of the semantic search in UDDI. We are currently working on how to reduce the time that the semantic search consumes due to the ontology loading, extending and updating.

Acknowledgments. The authors gratefully acknowledge the financial support provided by the China National 863 Hi-Tech R&D Plan (No. 2008AA01A309).

References

1. W3C Member: OWL-S: Semantic Markup for Web Services (2008), `http://www.w3.org/Submission/OWL-S`
2. Srinivasan, N., Paolucci, M., Sycara, K.: An Efficient Algorithm for OWL-S based Semantic Search in UDDI. In: Cardoso, J., Sheth, A.P. (eds.) SWSWPC 2004. LNCS, vol. 3387, pp. 96–110. Springer, Heidelberg (2005)

3. Srinivasan, N., Paolucci, M., Sycara, K.: Semantic Web Service Discovery in the OWL-S IDE. In: To appear in the proceedings of the 39th Hawaii International Conference on System Sciences, Hawaii (2005)
4. Paolucci, M., Kawamura, T., Payne, T.R., Sycara, K.: Importing the Semantic Web in UDDI. In: Proceedings of Web Services, E-business and Semantic Web Workshop (2002)
5. Sudha, R., Thamarai, S.S.: Semantic Grid Service Discovery Approach using Clustering of Service Ontologies. In: Proceedings of IEEE TENCON 2006, November 14-17, pp. 1–4 (2006)
6. Sudha, R., Yousub, H., Huimin, Z.: A Clustering Based Approach for Facilitating Semantic Web Service Discovery. In: Proceedings of the 15th Annual Workshop on Information Technologies & Systems (2006)
7. Nayak, R., Lee, B.: Web Service Discovery with additional Semantics and Clustering. In: Proceedings of the IEEE/WIC/ACM international Conference on Web intelligence, Washington, DC, November 2-5, pp. 555–558 (2007)
8. Miller, G.A.: WordNet: A Lexical Database for the English Language in Comm. ACM (1998)
9. Paolucci, M., et al: Importing the Semantic Web in UDDI. In: Proceedings of Web Services, E-business and Semantic Web Workshop (2002)
10. Gupta, C., Bhowmik, R., Head, M.R., et al.: Improving Performance of Web Services Query Matchmaking with Automated Knowledge Acquisition. In: Proceedings of the IEEE/WIC/ACM international Conference on Web intelligence, Washington, DC, November 2-5 (2007)
11. Nayak, R.: Facilitating and Improving the Use of Web Services with Data Mining. In: Taniar, D. (ed.) Research and Trends in Data Mining Technologies and Application, ch. 12, pp. 309–327 (2007)

An Efficient Approach to Web Service Selection

Xiaoqin Fan[1,2,3], Changjun Jiang[1,2], and Xianwen Fang[1,2]

[1] Key Lab of Embedded System & Service Computing Ministry of Education, Tongji university, Shanghai, 201804, China
[2] Electronics and Information Engineering School, Tongji University, Shanghai, 201804, China
[3] Computer and Information Technology School, Shanxi University, Taiyuan, 030006, China
{fxq0917,fangxianwen,jiangcj}@hotmail.com

Abstract. Under the service-oriented circumstances, developers can rapidly create applications through Web service composition. Owing to existing service selection methods having close relationship with the available service number, the efficiency inevitably decreases with the increase of available services. In the paper, a discrete Particle Swarm Optimization (DPSO) oriented to efficient service selection is designed according to Particle Swarm Optimization (PSO) idea. Furthermore, aiming at the practice service composition requirements, an algorithm used to resolve the dynamic service selection with multi-objective and quality of service (QoS) global optimal is presented based on DPSO and the theory of intelligent optimization of multi-objective PSO. Theoretical analysis and experimental results show that DPSO and multi-objective DPSO (MODPSO) are both feasible and efficient in the process of service selection.

Keywords: Web service, Service selection, Particle swarm optimization, Discrete particle swarm optimization.

1 Introduction

At present, the Service composition has been a popular research, and it is defined as the integration of a variety of existing services according to certain process logic to satisfy the users' requirements more effectively. The composition approach can be classified into manual composition, semi-automatic composition, and automatic composition [1]. It is unrealistic to manually analyze and compose the existing services appearing with tremendous growth recently.

For automatic service composition, state diagram or information diagram should be firstly generated based on matching algorithm[2,3,4], in which not only operation rules in grammar, but also the relationship of component service in semantic and the requirement for the quality of composite service need to be taken into account. Semantic relationships considered in the process of service composition mainly include the consistency between the signatures of function or attribute oriented to text, and the compatibility of the attribute type oriented to data, and the satisfiability between attribute value and constraint condition oriented to numeral and so on. In [2] aiming at

W. Liu et al. (Eds.): WISM 2009, LNCS 5854, pp. 271–280, 2009.

the behavior mismatch problem appearing in service composition, synchronization vector product is defined based on synchronization vector and it can effectively resolve the inconsistency between the signatures of function or attribute. The satisfiability between attribute value and constraint condition is considered [3,4] , in which the constraint relationships are classified into independent global constraint and dependent global constraint, and the former is used to describe the strict equal relationship, while the latter describes the partial relationship such as $\subseteq, \leq$ and so on. In automatic service composition backtracking algorithm is used to determine the composition plan [5, 6] after the state diagram or information diagram is generated. To this kind of service composition, on the one hand, it is more complex. On the other hand, the whole composition will fail once one candidate service is invoked unsuccessfully. Therefore, most of researches focus on semi-automatic service composition.

In semi-automatic service composition, developers should firstly construct the process model, and then the instance services are selected automatically for every abstract task. The service selection method based on QoS can meet the user' global restriction well, therefore, most of the existing service selection methods are all QoS-aware[7,8,9], at the same time there are other methods based on trust and reputation management[10], and PageRank analysis[11] and so on. Zeng et al.[7] investigates the service selection using the approaches of global optimization and local optimization respectively after giving the QoS models of component service and composite service, and the service selection problem with global optimization is formalized as an integer linear programming problem about single objective. The service selection problem with maximizing an application-specific utility function under the end-to-end QoS constraints is transformed to a multi-index multi-choice 0-1 knapsack problem based on combinatorial model, and a multi-constraint optimal path problem based on graph model separately [8]. Danilo Ardagna et al. [9] formalized the service selection with QoS constraints as a mixed integer linear programming. In order to detect and deal with false ratings by dishonest providers and users, a trust and reputation management method is introduced by Vu et al. [10]. Considering the problem of uninformed behavior evolution of component service, Lijun Mei et al. [11] proposed an adaptive service selection approach based on PageRank (PR) analysis. The PR value need to be recomputed after every execution so that the rank of candidate services can be updated dynamically, and in this way the adaptive ability of service selection can be improved. A genetic algorithm is presented for QoS-aware service selection based on a special relation matrix in [12].

The service composition approaches mentioned above, including oriented to semi-automatic and automatic service composition, based on exhaustive [7, 8, 9] and approximating [12] optimization methods, mostly depend on the available service number. While available services will appear rapidly with the development of Web service technology. Therefore, it will inevitably lead to the decrease of service composition efficiency. Moreover, most of the service selection approaches based on QoS often transform multiple objectives to one objective function by means of linear weighted method. So the result is the single optimal solution satisfying the constraint conditions, but in most cases users cares more about a set of acceptable non-inferior solutions in which they can make choice according to practical demands. Therefore, it is important to investigate multi-objective dynamic service composition with the restriction of global optimization.

The remainder of the paper is organized as follows: Section 2 presents an approach to dynamic Web service selection, called DPSO, on the basis of combining Web service selection and PSO idea. Section 3 presents MODPSO which considers the practical requirements. Section 4 conducts a variety of experiments to evaluate the validity and feasibility of algorithms. Section 5 ends the paper with some conclusions.

2 DPSO Oriented to Dynamic Web Service Selection

2.1 Problem Description

For the process model containing n abstract tasks $t_1, t_2, \cdots\cdots t_n$, the service selection is to locate n instance services satisfying the constraint conditions, and the n instance services constitute the solution of the problem, which is denoted as N-tuple $(ws_1, ws_2, \cdots\cdots ws_n)$ where $ws_i (1 \le i \le n)$ is the selected instance service of task t_i. Every N-tuple can be corresponding to one position of a particle in the n dimension space. Now we are sure that the change of particle position is because the values of one or more elements in N-tuple have been changed, that is to say, there are at least one task whose instance service has been changed, and the change of particle position origins in the change of particle velocity. Therefore, the particle velocity in the d-th dimension is defined as the substitution operation of available service of the d-th task, and the meaning of substitution operation is defined as definition 1.

For a certain process model and candidate service classes, there are a lot of composite services. For a certain composite service, there are several QoS indexes which are either positive or negative, and there is big difference among different QoS index values. In order to evaluate all composite services the fitness function in [6] is adopted.

2.2 Basic Conception

Definition 1. (Substitution Operation) For the task $t_i (1 \le i \le n)$ in process model, ws_{ij} is the selected service in the t-th generation, and if the selected service in the $(t+1)$-th generation is ws_{ik}, $ws_{ij} \to ws_{ik}$ is defined as the substitution from the t-th to the $(t+1)$-th generation, in which ws_{ij} is the substituted service, and ws_{ik} is the substituting service. It is denoted as $substitute(t_i^{t+1}) = ws_{ik}$. Specially, there is no operation if ws_{ij} is the same as ws_{ik}, called as NOP operation. For example, for the particle P_a and P_b, if the values in the d-th dimension are ws_j and ws_k respectively, $P_{ad} - P_{bd}$ means $ws_k \to ws_j$.

Definition 2. (Substitution Sequence) The tuple constituting of one or more substitution is defined as substitution sequence, and the order of substitution is meaningful. For example, the particle velocity in every generation is a substitution sequence.

Definition 3. (Alpha (α) Probability Substitution) The α probability Substitution of task t_i means that t_i selects the substituting service with the probability α.

For example, for the task t_i and its substitution $ws_{ij} \rightarrow ws_{ik}$, the α probability Substitution of task t_i can be computed as follows: t_i does not select the substituting service when α is equal to 0, that is to say, there is substitute(t_i)=ws_{ij},and t_i selects substituting service when α is equal to 1, that is to say, there is substitute(t_i)=ws_{ik}, and if there is $0<\alpha<1$ and $r \leq \alpha$ in which r is a randomly generated number obeying uniform distribution, there is substitute(t_i)=ws_{ik},else there is substitute(t_i)=ws_{ij}.

Definition 4. (Max operator)For the task t and its probability substitutions $\alpha\ \text{substitute}_1(t)$ and $\beta\ \text{substitute}_2(t)$, the max operator between $\alpha\ \text{substitute}_1(t)$ and $\beta\ \text{substitute}_2(t)$ is defined as follows:

$$\alpha\ \text{substitute}_1(t) \oplus \beta\ \text{substitute}_2(t) = \{ \begin{matrix} \alpha\ \text{substitute}_1(t), \alpha \geq \beta \\ \beta\ \text{substitute}_2(t), \beta > \alpha \end{matrix} ,$$

Here, $\oplus$ is the max operator.

2.3 DPSO Theory

Particle Swarm Optimization (PSO), as a kind of evolutionary algorithm, preserves and utilities not only the position information but also the velocity information in the process of evolution, while others only use the position information. So based on the "conscious" variation [13] the convergence speed of PSO has been raised remarkably. In the past several years, the PSO has been applied successfully in many continuity optimization problems because it not only has simple concept, easy realization, and less parameter, but also can effectively resolve complex optimization problems. While the research of discrete problem based on PSO is less up to date and generally only the corresponding algorithms are designed oriented to specific application problems. For example, M. Fatih et al. [14] designed a discrete particle swarm optimization algorithm aiming at the TSP problem, and Ayed Salman et al. [15] proposed a task allocation algorithm of parallel and distributed system.

In DPSO every individual in the population is considered as a particle without bulk and quality in the d dimension searching space, and it flies with a certain velocity that can be adjusted according to the flying experiences of individual and population. The i-th particle is denoted as $X_i = (x_{i1}, x_{i2}, \cdots\cdots, x_{id})$, and $V_i = (v_{i1}, v_{i2}, \cdots\cdots, v_{id})$ is the present flying velocity of particle i, and $P_i = (p_{i1}, p_{i2}, \cdots\cdots, p_{id})$ is the best place that particle i has experienced, that is to say, P_i is the place with optimal fitness value, and called as local best position. P_g is the best position that all particles have experienced and called as global best position. The evolution equations of DPSO oriented to Web service selection are as follows:

$$v_{id}^{t+1} = \omega v_{id}^{t} \oplus c_1 r_{1d}(P_{id} - x_{id}) \oplus c_2 r_{2d}(P_{gd} - x_{id}) . \tag{1}$$

$$x_{id}^{t+1} = substitute(t_{id}^{t+1}) \tag{2}$$

Here, subscript "i"represents the particle i, and subscript "d"represents the d-th dimension of the particle i, and t means the t-th generation. ω is the inertia weight which can make particle keep movement inertia and can expand the searching space so as to have ability to explore new position. c_1 and c_2 are accelerating constants, and c_1,also named "cognitive coefficient", reflects the effect of local best position on particle flying velocity, and c_2 also named "social learning coefficient" reflects the effect of global best position on particle flying velocity. $r_1 \sim U(0,1)$ and $r_2 \sim U(0,1)$ are two random variables independent each other.

2.4 DPSO Algorithm

The classical PSO is applicable to the continuous problem, while Web service selection is a kind of discrete problem. A new algorithm combining the specific meaning of service selection is presented based on the above evolution equations.

Input: The t-th generation population (including the velocity, position, and local best position of every individual in the population), Fitness function f, Constraint conditions.

Output: The $(t+1)$-th generation population.

1: For each particle i in the population, inertia weight ω is computed based on equation (3) according to constraint conditions;

2: For each dimension d of particle i, the velocity of $(t+1)$-th generation is computed based on equation (1): $v_{id}^{t+1} = \omega v_{id}^{t} \oplus c_1 r_{1d}(P_{id} - x_{id}) \oplus c_2 r_{2d}(P_{gd} - x_{id})$;

3: For each dimension d of particle i, the position of $(t+1)$th generation is computed based on equation (2): $x_{id}^{t+1} = substitute(t_{id}^{t+1})$;

4: The P_{id} of the particle i is updated according to x_{id}^{t+1} and fitness function f;

5: The P_{gd} of the $(t+1)$-th generation population is updated according to x_{id}^{t+1} of all particles in the population;

6: If the Hopeless/Hopeful Principle establishes, optimize the population again.

Here, ω is defined as follows:

$$\omega = \begin{cases} \dfrac{\sum_i |value_i - a_i|}{\sum_i |value_{i-\max} - a_i|} + 0.8 & \dfrac{\sum_i |value_i - a_i|}{\sum_i |value_{i-\max} - a_i|} \le 0.6 \\ 1.4 & others \end{cases} . \tag{3}$$

and $value_i$ is the true index value corresponding to the i-th constraint condition under the present service selection condition, a_i is the constraint value of the i-th constraint condition, which is given by user or process designer, $value_{i-\max}$ is the max value corresponding to the i-th index among all service selection situations.

Assume the process model includes n tasks, and population size is m, and iteration times are k, then the computational complexity of DPSO is $O(k \times m^2 \times n)$. It shows that the computation cost is related to population size, iteration times and task number, but it has no relevance to available services number, which can also be verified by the experiment in the following section, thus better robustness of the algorithm is guaranteed.

3 MODPSO Algorithm Oriented to Multi-objective Web Service Selection

For the process model involving n abstract tasks and candidate service set S, suppose the services are selected under m constraint conditions $g(X)$ and k objective functions $F(X)$. The MODPSO algorithm aiming at the above problem can be summarized as follows:

Input: Process model with n tasks, Candidate services with QoS, $F(X)$, $g(X)$

Output: Instance Pareto optimal composite services meeting the restriction requirements

1 : Initialize the particle population

Initialize the t-th generation of population involving P particles. For every particle X_i in the population whose dimension is N, and velocity and position are $V_i^t = (v_{i1}^t, v_{i2}^t, \cdots\cdots, v_{iN}^t)$ and $X_i^t = (x_{i1}^t, x_{i2}^t, \cdots\cdots, x_{iN}^t)$ respectively, and the local best position is P_i^t.

2: Obtain the sub-population

The i-th $(i = 1, 2 \cdots k)$ sub-population is composed of the first P/k optimal particles. Every particle in the population is separately evaluated According to $f_i(X)(i = 1, 2 \cdots k)$ by means of individual evaluation method of single objective optimization function, then the population is divided into k sub-population, and the size of all sub-populations are P/k. The optimal particle of each sub-population can be determined by corresponding objective function.

3: Extract the sharing information

The global best position of every sub-population is obtained on the basis of information sharing. In detail, it is determined by the position of optimal particles of other sub-populations.

4: Execute the "fly" operation
For j=1: k
Perform "fly" operation for every particle in the j-th sub-population according to algorithm 1.
End for

5: The $(t+1)$-th generation of population involving P particles is generated by combining the above sub-population.

6: Turn to 2 if terminal conditions are not satisfied.

7: Output Instance composition service if constraint requirements are all met, else suggest that no solution exist.

4 Performance Simulation

In order to evaluate for service composition approach, a great deal of experiments have been performed on a wide set of randomly generated process instances, because of the page constraint, here only part of experiment results are presented. Execution cost and time are two objective criteria, and the reliability of service is taken as constraint condition. The candidate services corresponding to every abstract task are generated according to uniform distribution with the given mathematical expectation and variance.

4.1 Validity Analysis

The CPU cost of locating the Pareto optimal composite services satisfying constraints is showed in the Fig.1.

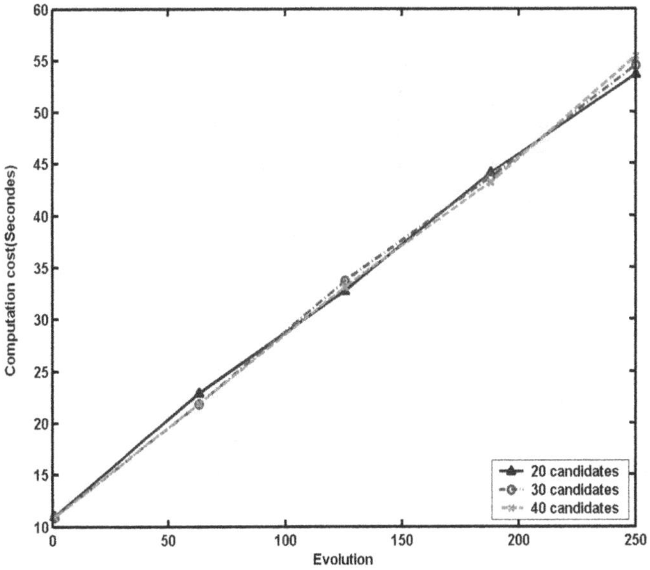

Fig. 1. Computation cost

We can see, for the same iteration times, there is nearly no difference for the CPU cost with the increase of candidate services, and computation cost increases linearly about the different iteration times, which is consistent with the computational complexity analysis in the above. Therefore, it manifests the algorithm has better robustness and stronger applicability. For example, for the process model including 20 tasks, and each with 40 candidate services, the computation cost is 54 seconds through 250 iterations, and such composition scale can meet most of the general composition requirements. Therefore, the validity of the algorithm is obvious.

4.2 Feasibility Analysis

In order to evaluate the feasibility of discovering the Pareto optimal composite services satisfying constraint conditions based on MODPSO, the exhaustive method is used, by which all Pareto optimal solutions and corresponding computation cost can be gotten. The ratio of solution number and computation cost between MODPSO and exhaustive method are showed in the Fig. 2.

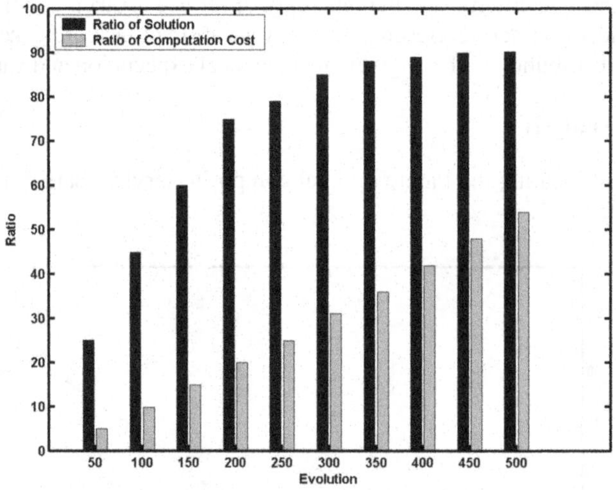

Fig. 2. Ratio of solutions and computation cost

Both the amount of Pareto optimal solutions and computation cost approximately linearly increase, but the ratio of solutions is nearly two times of computation cost for a given iteration times. Specially, it is worth noting that in this experiment the number of tasks and candidates are both little, in fact, with the increasing of task number or candidate services, the performance based on exhaustive method will decrease rapidly. Therefore, when there are a lot of candidate services, based on MODPSO method composite services can be gotten with both low computation cost and high quality.

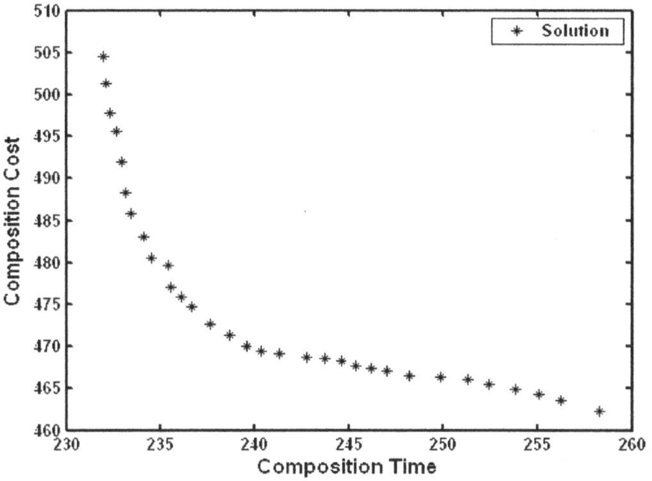

Fig. 3. Pareto optimal solutions

Fig. 3 records a set of feasible solutions produced under the condition of 40 candidate services and 200 iteration times based on MODPSO. These solutions may not optimal from single objective criteria perspective, but they are non-inferior solutions about composition time and cost, and the distribution of solutions is reasonable. Users can select appropriate solutions according to their preferences. So the MODPSO is practical in resolving the service composition problem. Specially, it is significant for the reliable Web service composition in the dynamic Internet environment. For example, the Pareto optimal solutions that are not selected firstly can be used as candidate composition process when the former service composition fails.

5 Conclusions

In the paper, two algorithms, DPSO and MODPSO, are designed for resolving the dynamic Web service selection. Theoretical analysis and experimental results show that they are both feasible and efficient. Compared with the existing methods, one of the obvious advantages is that they are independent of the available service number, so the robustness of algorithms is guaranteed. The other is that, based on MODPSO, a set of Pareto optimal solutions with constraints can be gotten by means of optimizing various objective functions simultaneously. So it can satisfy the users' requirements more effectively.

In the future, we plan to do more experiments, and evaluate our algorithms using emergent properties. We shall study the reliable service composition method based on PSO idea and relative verification technology under the consideration of randomness.

Acknowledgements

We would like to thank the support of the National Natural Science Foundation of China under Grant No.90718012, No.90818023 and No.60803034, the National High-Tech Research and Development Plan of China under Grant No. 2007AA01Z136.

References

1. Majithia, S., Walker, D.W., Gray, W.A.: A framework for automated service composition in service-oriented architectures. In: Bussler, C.J., Davies, J., Fensel, D., Studer, R. (eds.) ESWS 2004. LNCS, vol. 3053, pp. 269–283. Springer, Heidelberg (2004)
2. Carios, C., Pascal, P.: Model-Based Adaptation of Behavioral Mismatching Components. IEEE Transaction on software engineering 34(4), 546–563 (2008)
3. Nalaka, G., et al.: Matching Independent Global Constraints for Composite Web Services. In: WWW 2008, pp. 765–774 (2008)
4. Nalaka, G., et al.: Matching Strictly Dependent Global Constraints for Composite Web Services. In: ECWS 2007, pp. 139–148 (2007)
5. Oh, S.C., Lee, D., Kumara, S.R.T.: Web service Planner (WsPr): An Effective and scalable Web service Web Composition Algorithm. International Journal of Web Services Research 4(1), 1–23 (2007)
6. Adina, S., Jorg, H.: Towards Scalable Web Service Composition With Partial Matches. In: ICWS 2008 (2008)
7. Zeng, L.Z., Boualem, B.: QoS-Aware Middleware for Web services Composition. IEEE Transaction on Software engineering 30(5), 311–326 (2004)
8. Yu, T., Zhang, Y., Lin, K.J.: Efficient Algorithms for Web Services Selection with End-to End QoS Constraints. ACM Transactions on the Web 1(1), Article No. 6 (2007)
9. Danilo, A., Barbara, P.: Adaptive Service Composition in Flexible Processes. IEEE Transaction on Software Engineering 33(6), 369–384 (2007)
10. Vu, L.-H., Hauswirth, M., Aberer, K.: QoS-based service selection and ranking with trust and reputation management. In: Meersman, R., Tari, Z. (eds.) OTM 2005. LNCS, vol. 3760, pp. 466–483. Springer, Heidelberg (2005)
11. Mei, L., Chan, W.K., Tse, T.H.: An Adaptive Service Selection Approach to Service Composition. In: Proceeding of the IEEE International Conference on Web Service, pp. 135–145 (2008)
12. Zhang, C.W., Su, S., Chen, J.L.: Genetic Algorithm on Web Services Selection Supporting QoS. Chinese Journal of computer 29(7), 1029–1037 (2006)
13. Zeng, J.C., Jie, J., Cui, Z.H.: Particle Swarm Optimization [M]. Science Press (2004) (in Chinese)
14. Fatih Tasgetiren, M., Suganthan, P.N., Pan, Q.K.: A Discrete Particle Swarm Optimization Algorithm for the Generalized Salesman Problem. In: Proceedings of the 9th annual conference on Genetic and evolutionary computation (GECCO 2007), pp. 158–167 (2007)
15. Ayed, S., Imtiaz, A., Sabah, A.M.: Particle Swarm Optimization for Task Assignment Problem. Microprocessors and Microsystems, 363–371 (2002)

Constraint Web Service Composition Based on Discrete Particle Swarm Optimization

Xianwen Fang[1,2,*], Xiaoqin Fan[2], and Zhixiang Yin[1]

[1] School of Science, Anhui University of Science and Technology, Huainan 232001, China
[2] Electronics and Information Engineering School, Tongji University, Shanghai, 201804, China
{fangxianwen,fxq0917,zxyin}@hotmail.com

Abstract. Web service composition provides an open, standards-based approach for connecting web services together to create higher-level business processes. The Standards are designed to reduce the complexity required to compose web services, hence reducing time and costs, and increase overall efficiency in businesses. This paper present independent global constrains web service composition optimization methods based on Discrete Particle Swarm Optimization (DPSO) and associate Petri net (APN). Combining with the properties of APN, an efficient DPSO algorithm is presented which is used to search a legal firing sequence in the APN model. Using legal firing sequences of the Petri net makes the service composition locating space based on DPSO shrink greatly. Finally, for comparing our methods with the approximating methods, the simulation experiment is given out. Theoretical analysis and experimental results indicate that this method owns both lower computation cost and higher success ratio of service composition.

Keywords: Discrete Particle Swarm Optimization; Associate Petri net; Web Service Composition; Global Constraint.

1 Introduction

In today's Web application, Web services are created and updated on the fly. It's already beyond the human ability to analysis them and generates the composition plan manually. It is defined as the integration of a variety of existing services by certain process logic to satisfy the users' requirements more effectively [1]. The composition approach can be classified into manual composition, semi-auto composition, and automatic composition. In particular, if no single Web service can satisfy the functionality required by the user, there should be a possibility to combine existing services together in order to fulfill the request. This trend has triggered a considerable number of research efforts on the composition of Web services both in academia and in industry [2].

* Corresponding author. Tel.: +86-013162346379; fax: +86-021-69589864.

W. Liu et al. (Eds.): WISM 2009, LNCS 5854, pp. 281–288, 2009.

They are of two types: local and global constraints. The former restricts the values of a particular attribute of a single service, whereas the latter simultaneously restricts the values of two or more attributes of multiple constituent services. Global constraints can be classified based on the complexity of solving them (i.e. determining the values that should be assigned to their attributes) as either strictly dependent or independent. A (global) constraint is strictly dependent if the values that should be assigned to all the remaining restricted attributes can be uniquely determined once a value is assigned to one. Services that conform to strictly dependent global constraints can be easily located in polynomial time [3].

Any global constraint that is not strictly dependent, is independent. Their location is known to be NP-hard [4]. As a consequence most of the existing matching techniques (for locating composite services) do not consider independent global constraints [5]. Nonetheless, there are some that consider them [3,5] and they use integer programming solutions focusing on local optimizations [4] and AI planners [6] to efficiently locate conforming composite services. However, all the techniques of the latter type are syntactic-based approaches. In literature [5] proposes a semantic-based matching technique that locates services conforming to independent global constraints. The proposed technique that incorporates a greedy algorithm performs better than the existing techniques. Experimental results also show that the proposed approaches achieve a higher recall than syntactic-based approaches. But the proposed greedy algorithm approaches are low efficiencies, and the approximating method can not represent some problem accurately.

Discrete Particle Swarm Optimization (DPSO), as a kind of evolutionary algorithm aiming to discrete problem, preserves and utilities not only the position information but also the velocity information in the process of evolution, while others only use the position information. So based on the "conscious" variation [7] the convergence rate of PSO has been raised remarkably. In past several years the PSO has been applied successfully in many continuity optimization problems because it not only has simple concept, easy realization, and less parameter, but also can effectively resolve complex optimization problems. This paper present independent global constrains web service composition optimization methods based on DPSO and APN.

2 Basic Conception

The concept of associate Petri net (APN) is derived from fuzzy Petri net (FPN). An APN is a directed graph, which contains three types of nodes: places, squares, and transitions where circles represent places, squares represent thresholds of association degree, and bars represent transitions. Each place may contain a token associated with a truth-value between zero and one. Each transition is associated with a trust value between zero and one. Directed arcs represent the relationships between places [8]. The structures, reasoning, and operations of APN are generalized from basic Petri net. Here, we only introduce several conceptions correlating with the paper close, other Petri Nets terms in the literature [10].

Definition 1[8]**.** A 13-tuple $APN=(P,T,S,D,\Lambda,\Gamma,I,O,C,\alpha,\beta,W,Th)$ is called associate Petri net, where

$P=\{P_1,P_2,\cdots,P_n\}$ is a finite set of places,

$T=\{T_1,T_2,\cdots,T_i\}$ is a finite set of transitions,

$S=\{S_1,S_2,\cdots,S_m\}$ is a finite set of supports,

$D=\{D_1,D_2,\cdots,D_n\}$ is a finite set of propositions,

$\Lambda=\{\tau_1,\tau_2,\cdots,\tau_m\}$ is a finite set of thresholds of the supports,

$\Gamma=\{\gamma_1,\gamma_2,\cdots,\gamma_i\}$ is a finite set of thresholds of the confidences,

$P\cap T\cap D=\Phi;|P|=|D|$

$I:T\rightarrow P^{\infty}$ is an input function, a mapping from transitions to bags of places,

$O:T\rightarrow P^{\infty}$ is an output function, a mapping from transitions to bags of places,

$C:T\rightarrow[0,1]$ is the confidence degree of relationship between zero and one,

$\alpha:P\rightarrow[0,1]$ is an associated function ,a mapping from places to real values between zero and one,

$\beta:P\rightarrow D$ is an associated function ,a bijective mapping from places to propositions,

$W:S\rightarrow[0,1]$ is an associated function that assigns a real value between zero to one to each support,

$Th:S\rightarrow\Lambda$ is an associated function that defines a mapping from support to thresholds.

Definition 2[8]**.** In an APN, a transition t_i is enabled (represented as $M[t>)$ if for all $p_j\in I(t_i)$, $\alpha(p_j)\geq\lambda$, $s_m\geq\tau_m$, and $c_i\geq\gamma_i$,where λ,τ_m,γ_i are threshold values and $\lambda,\tau_m,\gamma_i\in[0,1]$. A transition t_i fires by removing the tokens from its input places to pass through all support squares and then depositing one token into each of its output places.

The token value in a place $p_i,p_j\in P$, is denoted by $\alpha(p_j)$, where $\alpha(p_j)\in[0,1]$, If $\alpha(p_j)=y_i$ and $\beta(p_j)=d_i$, then it indicates that the proposition d_j is y_i.

If the antecedent portion or consequence portion of an associate Petri reasoning (APR) contains "and" or "or" connectors, then it is called a composite APR. The composite ARP can be distinguished into the following five basic reasoning rule types. Using this simple mechanism, all the APRs can be mathematically and graphically illustrated. By carefully connecting related place and assigning reasonable trust values to transitions, we can come up with an APN that can make a decision based on the expertise during its construction [8].

3 Constraint Composition Services Modeling Based on APN

The realization of semi-automatic service composition demands that firstly operators make the universal service composition model in accordance to application demands according to given business background. The model that is usually described by graph includes abstract tasks and dependent relation among tasks. In the model, all task nodes only contain the description of function demands, but do not specify a service instance. Then, select an instance service that meets semantic constraints and the users' requirements from a serial of service class which has the same function attributes for all abstract tasks in the model, and finally the composite of service instances can meet the users' requirements.

The composite service acquisition consists of three phases: (1) candidate services acquisition, (2) constraint attributes identification and (3) optimal composite service acquisition. The first phase locates services with different functions description according to abstract task requirements in a composite service template. A candidate service that satisfies the requirement description of an abstract task is a service with a particular function. By locating candidate services, it ensures that the constituent services of a composite service are of appropriate types. The second phase identifies constraint attributes of services conforming to a given independent global constraint and certain model is used to represent the constraint relations. The proposed methods adopt the support nodes of APN to descript the constraint, and it is benefit to analyze multi-constraints problem. In the final phase, according to the QoS requirement, the optimal candidate services that satisfy the function and constraint requirement are obtained to form the composite services.

The proposed approach models independent global constraints for composite Web services by the APN. The modeling methods about the independent global constraints for composite Web services based on the APN are presented in literature [10].

4 Constraint Services Composition Optimization Based on DPSO

For the process model containing n abstract tasks $ta_1, ta_2, \cdots ta_n$, the service selection is to locate n instance services satisfying the constraint conditions, and the n instance services constitute the solution of the problem, which is denoted as N-tuple $(ws_1, ws_2, \cdots\cdots ws_n)$ where $ws_i (1 \le i \le n)$ is the selected instance service of task ta_i. Every N-tuple can be corresponding to one position of a particle in the n dimension space. Now we are sure that the change of particle position is because the values of one or more elements in N-tuple have been changed, that is to say, there exists at least one task whose instance service has been changed, and the change of particle position origins in the change of particle velocity. Therefore, the particle velocity in the d-th dimension is defined as the substitution operation of available service of the d-th task, and the meaning of substitution operation is defined as definition 3.

Definition 3. (Substitution Operation) For the task $ta_i (1 \le i \le n)$ in process model, ws_{ij} is the selected service in the t-th generation, and if the selected service in the (t+1)-th generation is ws_{ik}, $ws_{ij} \rightarrow ws_{ik}$ is defined as the substitution from the t-th to the (t+1)-th generation, in which ws_{ij} is the substituted service, and ws_{ik} is the substituting service. It is denoted as $substitute(t^{i+1}) = ws_{ik}$.

Specially, there is no operation if ws_{ij} is the same as ws_{ik}, called as NOP operation.

For example, for the particle P_a and P_b, if the value is ws_j and ws_k respectively, $P_{ad} - P_{bd}$ means $ws_j \rightarrow ws_k$.

So, the evolution equation of DPSO oriented on Web service selection is as follows:

$$v_{id}^{t+1} = \omega v_{id}^{t} \oplus c_1 r_{1d}(P_{id} - x_{id}) \oplus c_2 r_{2d}(P_{gd} - x_{id}) \tag{1}$$

$$x_{id}^{t+1} = substitute(t_{id}^{t+1}) \tag{2}$$

Here, subscript "i" represents the particle i, and subscript "d" represents the d-th dimension of the particle i, and t means the t-th generation.

ω is the inertia weight which can make particle keep movement inertia and can expand the searching space so as to have ability to explore new position. c_1 and c_2 are accelerating constant, and c_1 ,also named "cognitive coefficient", reflects the effect of local best position on particle flying velocity, and c_2 also named "social learning coefficient" reflects the effect of global best position on particle flying velocity, $r_1, r_2 \in (0,1)$ are two random variables independent each other.

The optimization methods of independent global constraint for composite service based DPSO and APN are as follows:

(1)Initialize the particle population

According to the requirement of composite process, we select fixed number (such as K) firing sequences from APN model. If it exists a firing sequences $\sigma = p_1 t_1 p_2 t_2 p_3 t_3 \cdots p_{n-1} t_{n-1} p_n t_n$, which makes $M_0[\sigma > M_f$, then $\sigma_p = p_1 p_2 \cdots p_n$ (p is corresponding to the matching component services)can be regarded as a particle, here, M_0 is the first marking, M_f is the end marking, $p_1, p_2, p_3, \cdots, p_{n-1}, p_n \in P$, $t_1, t_2, t_3, \cdots, t_{n-1}, t_n \in T$.

(2) For the initial population, computing the optimal best position P_{id} of every particle and the global best position P_{gd} of the all particle.

(3) Computing the fitness value of every particle.

(4) Computing ω, $c_1 r_1$, $c_2 r_2$, and sorting them from small to big.

(5) Executing the substitution operation, for i=1, i.e. for the freedom service, select the random place (component services) from the APN as the substitution, for i=2 to n, select the brother place node of the place p as the substitution, then for j=i+1, select the legal firing path, in order to form a legal firing sequences [9].

(6) According to $v_{id}^{t+1} = \omega v_{id}^{t} \oplus c_1 r_{1d}(P_{id} - x_{id}) \oplus c_2 r_{2d}(P_{gd} - x_{id})$, to compute the v.

(7) According to $x_{id}^{t+1} = substitute(t_{id}^{t+1})$, to compute the x.

(8) For every particle, compute the fitness, and update the optimal best position P_{id} of every particle and the global best position P_{gd} of the all particle.

(9) Determining if the process traps in the local optimal solution, if the process traps in the local optimal solution, then executing the mutation operator [9], update the particles, goto step (3).

5 Experiment Simulation

The methods of independent global constraint service composition based on DPSO and APN is proposed, which not only uses APN superiority in the description multi-attribute multi-constraint problems, but also takes fully Petri net's properties when DPSO locating in the APN model into account. In the theoretical, the method is of great benefit to analyze independent global constraint composite service question, which can avoid high complexity using Petri net methods solely and large randomness using DPSO only. Next, we use experiment simulation so as to analyze feasibility and validity of the methods. Experiment environment: CPU is Intel dual 1.60GHz, Memory is 1.00GB, and operation system is Windows XP, Matlab6.5.

For a given service requirement, using approximate methods presented in the literature [5] and the methods based on DPSO and APN (DPSO&APN) respectively, the experiment effect (in Fig.1) shows the execution time of DPSO&APN methods is less than the approximate methods, and when the more of the service number in the service library, the better is the effect. The reason is that we uses fully APN's properties when DPSO locating in the APN model, which makes locating space lower, and some component services which don't satisfy constraint relations or have minor associate relation need not to match each other. So it can save execution time, and improve the time performance of service composition.

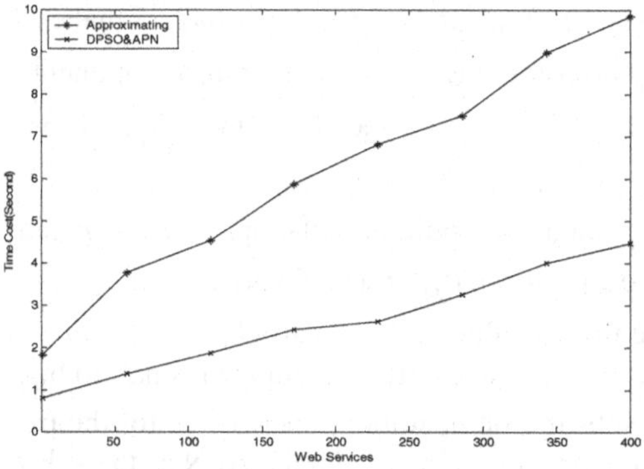

Fig. 1. The execution time of three methods

In Fig.2, we locate feasible candidate services from service library (N=500), and compare the relation graph between the ratio of solution and iteration number. The experiment effect shows the ratio of solution of DPSO&APN methods is higher than the approximate methods. With the iteration number adding, the ratio of solution of DPSO&APN and approximating methods increase quickly, but it is flat when iteration number reaches to some extent. In the DPSO&APN methods, the process of locating the feasible solution is the same as locating the firing sequences in APN model. Because of taking the Petri net legal firing sequences algorithm into account, the methods can guarantee the located solution is feasible. What's more, owing to the selection, crossover and mutation operation of DPSO, the methods can reduce the loss and miss of feasible solution.

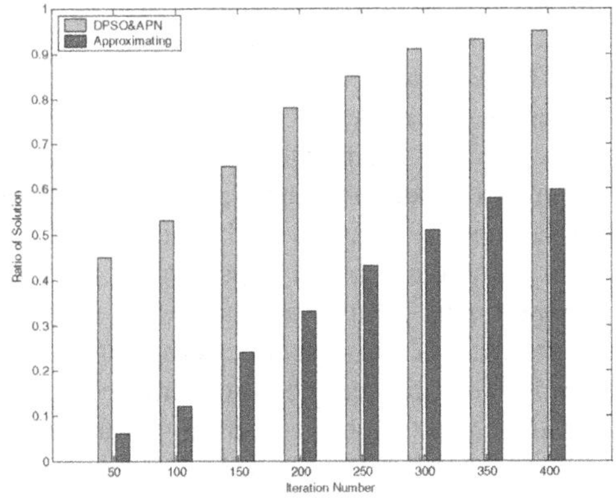

Fig. 2. The relation graph between the ratio of solution and iteration number

In Fig.3, we locate feasible candidate services from service library, and compare the relation graph between the ratio of solution and web services. The experiment effect shows the ratio of solution of DPSO&APN methods is higher than the approximate methods. With the web services number adding, the ratio of solution of DPSO&APN and approximating methods decrease quickly.

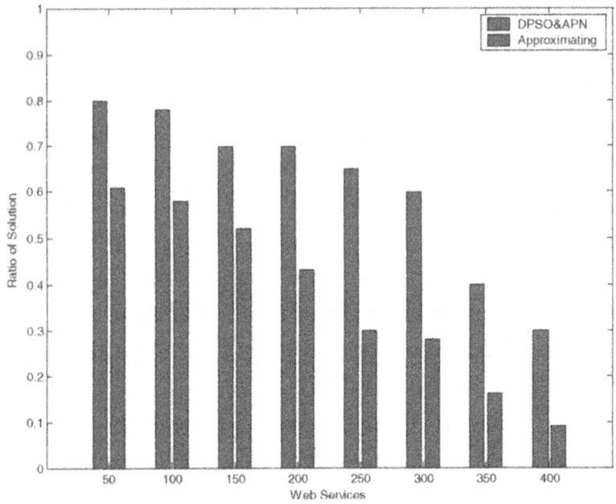

Fig. 3. The relation graph between the ratio of solution and web service

6 Conclusions

For independent global constraint service composition problem, the main research methods are syntactic matching and semantic matching presently. There are index methods and approximate methods to deal with global constraint, which play a good effect in resolving constraint problems, but it is deficient to deal with multi-attribute and multi-constraint problems.

The service composition methods based on DPSO and APN are presented in the paper. Firstly, we model the independent global constraint component services by APN, in APN model, multi-constraint relations are represented using support sets, and the associate relation between component services is obtained by the data mining on the basis of the history calling relation. Secondly, using DPSO to locate legal firing sequences in the APN model, the legal firing sequences are feasible solutions, and then according to the computing rules of APN, an optimal solution is obtained. Finally, we simulate the execution time and the ratio of solution based approximate methods and DPSO&APN methods respectively. Experiment simulation effect shows DPSO&APN has higher execution efficiency and the success rate of service composition.

The study aims at semi-automatic service composition in the paper. In the future, we plan to resolve the independent global constraint for automatic service composition, and propose improved DPSO&APN methods to improve execution efficiency.

Acknowledgments. We would like to thank the support of the National Natural Science Foundation of China under Grant No. 90718012, No.60873144 and No.90818023, the National High-Tech Research and Development Plan of China under Grant No. 2007AA01Z136.

References

1. Rao, J., Küngas, P., Matskin, M.: Logic-based Web services composition: from service description to process model. In: Proceedings of the 2004 International Conference on Web Services, San Diego, USA (2004)
2. Wu, D., Sirin, E., Hendler, J., Nau, D., Parsia, B.: Automatic Web services composition using SHOP2. In: Workshop on Planning for Web Services, Trenton, Italy (2003)
3. Gooneratne, N., Tari, Z., Harland, J.: Matching Strictly Dependent Global Constraints for Composite Web Services. In: Proceedings of the European Conference on Web Services, pp. 139–148 (2007)
4. Ardagna, D., Pernici, B.: Adaptive Service Composition in Flexible Processes. IEEE Transaction on Software Engineering 33(6), 369–384 (2007)
5. Gooneratne, N., Tari, Z.: Matching Independent Global Constraints for Composite Web Services. In: WWW 2008, pp. 765–774 (2008)
6. Sirin, E., Parsia, B.: Planning for Semantic Web Services. In: Proceedings of International Semantic Web Conference, Workshop on Semantic Web Services. IEEE Press, Los Alamitos (2004)
7. Banks, A., Vincent, J., Anyakoha, C.: A review of particle swarm optimization. Part I:background and development. Natural Computing: an international journal 6(4), 467–484 (2007)
8. Shih, D.-H., Chiang, H.-S., Lin, B.: A Generalized Associative Petri Net for Reasoning. Transactions Knowledge and Data Engineering 19(9), 1241–1251 (2007)
9. Jiang, C.: A Polynomial-time Algorithm for the Legal Firing sequences Problem of a Type of Synchronous Composition Petri net. Science in China (series E) 22(1), 116–124 (2002)
10. Fang, X., Jiang, C., Fan, X.: Independent Global Constraints-Aware Web Service Composition Optimization. Information Technology Journal 8(2), 181–187 (2009)

A QoS-Aware Service Selection Approach on P2P Network for Dynamic Cross-Organizational Workflow Development

Jielong Zhou and Wanchun Dou

State Key Laboratory for Novel Software Technology, Nanjing University
Department of Computer Science and Technology, Nanjing University
210093, Nanjing, China
dillanzhou@126.com, douwc@nju.edu.cn

Abstract. A workflow system is often composed of a number of subtasks in its pattern. In services computing environment, a dynamic cross-organizational workflow can be implemented by assigning services to its subtasks. It is often enabled by a service discovery process on Internet. Traditional service discovery approaches are centralized, and suffer from many problems such as one-point failure and weak scalability. Thus, decentralized P2P technique is a promising approach for service publishing and discovery. It is quite probable that there are more than one candidates which have exactly the same function after a service discovery process. It is often a challenging effort to select a qualified service from a group of candidates, especially on P2P networks. In view of this challenge, a QoS-aware service selection approach on unstructured P2P networks is presented in this paper. It aims at discovering and selecting services on two-layered unstructured P2P networks according to QoS parameters of services and preference of service requesters. This approach is applied to a case study based on a simplified P2P network with some virtual services.

Keywords: QoS, service Selection, P2P, workflow development.

1 Introduction

A workflow system expresses flows of subtasks. A subtask is an abstract definition of a piece of work and needs an implement. In a dynamic cross-organizational workflow, web services could be assigned to subtasks as their implements. When searching a service to implement a subtask, it is quite probable that there are more than one candidates which have exactly the same function. Hereby, a group of QoS criteria and the preference of users should be used to choose a best service [7].

However, if a service is assigned to a subtask of a cross-organizational workflow system, it may have more than one end-user.

Generally, before assigning a service to a subtask, some candidate services should be discovered firstly. Traditional service discovery approaches are centralized, base on Universal Description, Discovery and Integration (UDDI). However,

W. Liu et al. (Eds.): WISM 2009, LNCS 5854, pp. 289–298, 2009.

the centralized service registries may suffer from many problems, such as one–point failure, weak scalability, etc. Thus, the UDDI approach is not suitable for the growing amount of services. On the other hand, the Peer-to-Peer (P2P) paradigm emerged as a successful model that is robust and achieves scalability in distributed systems. As a result, adopting existing methods from the P2P paradigm to service publishing and discovery becomes a promising approach. During the last couple of years, a number of attempts on using P2P techniques to facilitate service discovery have been made. P2P is a relatively recent, highly distributed computing paradigm that enables sharing of resources and services through direct communication between peers. The P2P networks can be classified into unstructured networks and structured networks. Unstructured networks are organized by peers which are connected with each other without a global overlay plan. Gnutella [1] and KaZaA [2] are well-known examples of unstructured P2P networks. Structured P2P networks, such as Chord [3], use highly structured overlays and exploit a Distributed Hash Table (DHT) to distribute data and route queries over the network. Structured P2P networks have sophisticated topological structure while unstructured P2P networks are much briefer. Thanks to Distributed Hash Table, structured P2P networks work much more efficient on exact queries [4]. However, the structures need sophisticated mechanism to balance the load and sustain the topology. Furthermore, the vague queries can not be executed on structured P2P networks thanks to the randomicity of DHT. Meanwhile, although not good at exact search, the unstructured P2P networks are easier to sustain and work better on vague or range query. A framework of service discovery on unstructured P2P network was presented in [5]. Mastroianni et al. adopted the super-peer model as the infrastructure of resource discovery, and proved that the method was efficient with experiment [6]. The two-layered unstructured P2P network presented in this paper combines the advantages of these works.

QoS-Aware service selection is also a hot topic during the last few years. There are a lot of papers on how to select the best service from functional similar candidates according to their QoS parameters. In these papers, selection is always made on a service registry or a middleware. However, to the best of our knowledge, very little attention has been paid to service selection on P2P network. If a P2P network is taken as the infrastructure of service publishing and discovery, the service registry or service selection middleware will no longer exist. So the traditional centralized service selection approach can not be used in this situation. In this paper, we propose a novel approach to discover and select services for dynamic cross-organizational workflow subtasks on P2P network. And this approach takes the QoS criteria and the preference of service requesters into account.

The remainder of this paper is organized as follows. Section 2 introduces how to describe the requirements and preference of users on QoS parameters and how to generate the searching and selection request query according to the requirements and preference. Section 3 presents the approach to discover and select services on P2P networks. And then, an application example of our method will be given in section 4. Finally, section 5 summarises the major contribution of this paper and outlines authors' future work.

2 Description of Service Request

A service which is assigned to a cross-organizational workflow often has more than one end user, so the service request query should consider the requirement and preference of every user.

A service request query must contain two segments: the description on functional capabilities and the constraints on Quality of Service (QoS) criteria. In this paper, service discovery and selection will be done at the same time, so the requests also have to contain the preference of service requesters to make selection.

Describing and matching functional capabilities involves semantic techniques. It is beyond the scope of this paper and we will not discuss it.

In this section, we will introduce how to describe the requirements and preference of service requesters on QoS creteria and how to generate the searching and selection request query according to the requirements and preference.

- **Constraints on QoS parameters**

Besides the description of functional capabilities, the constraints on QoS parameters constitute another necessary segment of a service request query. When end users request for a service, they often give some constraints on QoS parameters such as price or duration. E.g., if a customer wants to buy a book, he may desire that the price of the book is not higher than 20$, and the shipment can be accomplished in 3 days. So the constraint on price of purchasing is [0, 20], and the constraint on duration of shipment is [0, 3]. A service can become a candidate for selection only if it satisfies the constraints on QoS parameters. In this paper, a service may have multiple users, so the constraints on QoS parameters should combine the requirements of all users.

- **Preference and weights of QoS parameters**

If there are more than one services which satisfy the functional requirement and the constraints on QoS criteria, a selection has to be made to choose the best one from these candidates. It is almost impossible that every QoS parameter of a service is better than another's. Thus, in order to make selection, overall QoS scores should be computed for all services to compare their quality. Here we use a simple additive weighting technique to compute the overall scores for services [8]. This technique is composed of two phases.

The first phase is scaling phase. In this phase, values for every QoS parameter of every service are computed by following scaling formulae.

$$V_{i,j} = \begin{cases} \dfrac{Q_j^{\max} - Q_{i,j}}{Q_j^{\max} - Q_j^{\min}} & if \quad Q_j^{\max} \neq Q_j^{\min} \\ 0 & if \quad Q_j^{\max} = Q_j^{\min} \end{cases} \tag{1}$$

$$V_{i,j} = \begin{cases} \dfrac{Q_{i,j} - Q_j^{\max}}{Q_j^{\max} - Q_j^{\min}} & if \quad Q_j^{\max} \neq Q_j^{\min} \\ 0 & if \quad Q_j^{\max} = Q_j^{\min} \end{cases} \tag{2}$$

If the QoS criterion is negative, such as price or duration, values are scaled according to Formula 1. If the QoS criterion is positive, such as success rate or reputation, scores are computed according to Formula 2. Q_j^{max} and Q_j^{min} mean the maximum and minimum value of the jth QoS parameter, and V_i means the vector of scaled values of a service.

The second phase is weighting phase. In this phase, the scaled values of QoS parameters of a service will be integrated into one score according to the following formula.

$$Score(S_i) = \sum (V_{i,j} * W_j) \tag{3}$$

W_j means the weight of the jth QoS parameter of service. According to the final scores computed by this formula, selections can be made among candidates which have similar functional capabilities.

However, it is generally very hard for the service requesters to give accurate weights of QoS criteria. Thus, a method should be found by which weights can be calculated from user's preference indirectly. Here we divide the importance of quality into 5 levels: very important, important, normal, unimportant, very unimportant. Each level has a corresponding value from 5 to1. Service requesters only have to describe the levels of importance of QoS criteria, than the weights could be computed according to the following formula.

$$w_i = \frac{I_{i,1+} I_{i,2}}{\sum_j I_{j,1+} I_{j,2}} \tag{4}$$

$I_{i,k}$ means the corresponding value of the importance level of the ith QoS criterion given by the kth user. For example, if the price is very important, duration is important and success rate is very unimportant for a service user, and another user considers that price and duration is important, while success rate is very important. Then the weights of these QoS criteria are 9/23, 8/23 and 6/23 respectively.

- **Format of service request queries**

In this paper, a service request query is a XML document, composed of four sects: functional capability, constraints on QoS parameters, weights of QoS criteria and the information of the temporary best candidate.

As shown in Figure 1, the main label "ServiceRequest" has four sub-labels. "FunctionalCapability" segment describes the function requirement of service requesters. "QoSConstraints" records the constraints on QoS parameters. For each QoS criterion, the query records its maximum and minimum limits. By the way, the examples in this paper only consider three QoS criteria, other criteria can be processed in similar way. "QoSWeights" records the weight of each QoS criterion computed by formula 4. The document also includes a segment labelled "Best Candidate". This segment records the information of the temporary best service among services which have been discovered so far. The information covers the name, URL and QoS parameters of the service. This segment is empty at the beginning of searching process and is updated by every node the request query pass through. When the searching process is finished, the "BestCandidate" segment records exactly the description of the best service. The process of updating will be described more detailedly in next section.

```
<?xml version = "1.0" encoding = "UTF-8"?>
<ServiceRequest>
  <FunctionalCapability>
       <!-- The description of functional capability. -->
       <abstract>online payment.</abstract>
  </FunctionalCapability>
 <QoSConstraints>
       <!-- The constraints on QoS criteria. -->
       <Price Unit = "American dollar">
            <Minimum>15</Minumum>
            <Maximum>25</Maximum>
       </Price>
       <Duration Unit = "Day">
            <Minimum>0</Minimum>
            <Maximum>3</Maximum>
       </Duration>
    <SuccessRate Unit = "Percent">
            <Minimum>95</Minimum>
       <Maximum>100</Maximum>
       </SuccessRate>
  </QoSConstraints>
 <QoSWeights>
       <!-- The weights of different QoS criteria. -->
    <PriceWeight>0.4</PriceWeight>
    <DurationWeight>0.35</DurationWeight>
    <SuccessRateWeight>0.25</SuccessRateWeight>
  </QosWeights>
 <BestCandidate>
       <!-- The information of the best candidate. -->
    <Name>ICDC online bank</Name>
    <QoSParameters>
       <Price Unit = "American dollar">22</Price>
      <Duration Unit = "Day">3</Duration>
      <SuccessRate Unit = "Percent">99</SuccessRate>
       </QoSParameters>
    <URL>www.ICDC.com</URL>
  </BestCandidate>
</ServiceRequest>
```

Fig. 1. An example of service request query

3 A QoS-Aware Service Discovery and Selection Approach on P2P Network

In this section, we will present how to find services on the unstructured P2P networks and how to select the best service from the candidates during the searching process according to the requirements and preference of users.

3.1 Two-Layered Unstructured P2P Network

Our P2P structure is similar to KaZaA or Gnutella v0.6. In this structure, a two-level hierarchy of peers is constructed. Servers are always online and have high bandwidth to provide services, thus they are organized into an unstructured overlay network. In this level, every server connects many other servers randomly, and each server is called a super-peer. On the other hand, clients which are operated by end users have low bandwidth and are not always online. So the clients only connect one super-peer each. Each client is called a leaf. A super-peer and leaves which connect this super-peer constitute a star topological network locally. Figure 2 depicts the structure of a simple two-layered unstructured P2P network.

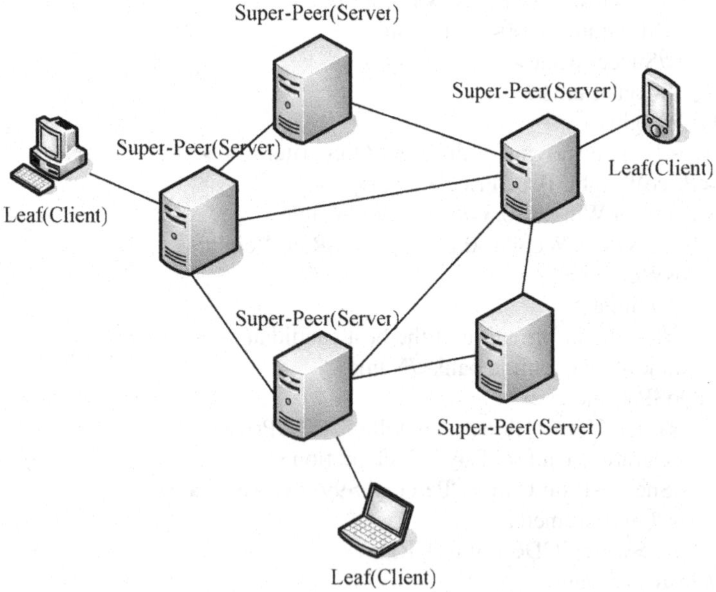

Fig. 2. Two-layered unstructured P2P network

To increase the success rate and amount of results of service discovery, services are published into the network. When a service provider needs to publish a service, the server will send the description document of the service into the super-peer network by random walks or multiple random walks. Every super-peer in the network maintains a list of service descriptions which are published by other super-peers or of its own services.

3.2 A QoS-Aware Service Discovery and Selection Approach on P2P Network

In the remainder of this section, we will introduce how to discover services on the network and how to select the best one from function-similar candidates. The working processes of clients and servers are absolutely different and will be discussed separately.

1) The working process of clients

The working process which should be carried out on the clients to make service discovery and selection takes following steps:

Step 1. Get the description of the functional capabilities of the needed service from service requester.

Step 2. Get the constraints on QoS criteria.

Step 3. Get the importance levels of different QoS criteria.

Step 4. Compute the weights of QoS criteria according to Formula 4 in section 2.

Step 5. Generate the service request query with the information which is gathered in Step 1, 2 and 4.

Step 6. Send the service request query to the super-peer of this leaf (client).

Step 7. Receive the response. The service whose description was in the response is the best service according to the requirement and preference of the requester.

Generally speaking, the process of service discovery and selection which is carried out by clients is constituted of three parts: generates the query, sends the query, and then receives the result. Actually, it is easy to find that the client does nothing on the discovery and selection directly, but only generates and sends the request query into the P2P network. Service discovery and selection are finished by the super-peers in the high-level overlay network.

2) The working process of servers

When a super-peer (server) receives a service request query from another super-peer or a client, following steps should be executed to search local services and select the best one from them.

Step 1. Get the description of functional capabilities from the query.

Step 2. Search the local list of service descriptions for candidates according to the description gotten in Step 1. The set of services which satisfy the requirements on functional capabilities is marked as S1.

Step 3. Get the QoS Constraints from the query.

Step 4. Filter set S1 according to the QoS Constraints. The set of services meet the QoS constraints is marked as S2.

Step 5. Merge S2 and the temporary candidate attached in the query (if exist) , get the set S3.

Step 6. Get the weights of QoS criteria from the query.

Step 7. Using the weights gotten in step 5, compute the integrated scores of services in set S3 according to Formula 1,2,3 in section 2.

Step 8. Select the best scored service as the candidate. Update the query with the information of the candidate.

Step 9. Add 1 to the TTL of the query. Add the number of services in set S1 to the counter of the query.

Step 10. If the TTL or counter of services is less than the threshold, select a connected neighbor randomly and transmit the query to the neighbor.

Step 11. Else, respond to the client with the candidate in the query. This query comes to the end.

As shown above, a super-peer first searches all local services for candidates which satisfy the requirements of users. Then select the best service from these local candidates and the service attached in the query. When a query leaves a super-peer to the next node, the candidate attached to the query is always the best service on all the nodes the query has passed through. When the searching process comes to an end, the selection has also been made.

4 A Running Example

In this section, we will illustrate our approach with an example. We assume that a dynamic cross-organizational workflow contains a subtask of payment. According to the requirements of the payer and payee, the constraint on price is [15, 25], the constraint on duration is [0, 3] and on success rate is [95,100].

Also, the weights of QoS parameters should be calculated according to the preference of the payer and the payee. In this case, the price is very important, duration is important and success rate is very unimportant for the payer. While the payee considers that price and duration is important, while success rate is very important. The weights of these QoS criteria are 9/23, 8/23 and 6/23, according to formula 4 in section 2.

Table 1. QoS parameters of services

	Price	Duration	Success Rate
S	20	3	95
S	18	4	93
S	25	2	98
S	22	3	98
S	24	3	99
S	20	2	90

In this case, there are 6 services on the P2P network satisfy the functional requirement. The QoS parameters of these services (S1-S6) are listed in Table 1.

Figure 3 depicts the structure of the P2P network and the service distribution situation on the network. 6 services are provided by 6 different servers. Each service has been published into the network by random walks algorithm with 2 hops. Services in the brackets are published from other servers. The client connects with the server of S4.

In this case, the client takes random walk algorithm with 3 hops to search for service. And the route of the query is marked in Figure 3 with arrows. First of all, the client generates the service request query according to the functional capability and QoS constraints and weights. Then the query is sent to the super-peer connected with the client, which is also the server of S4.

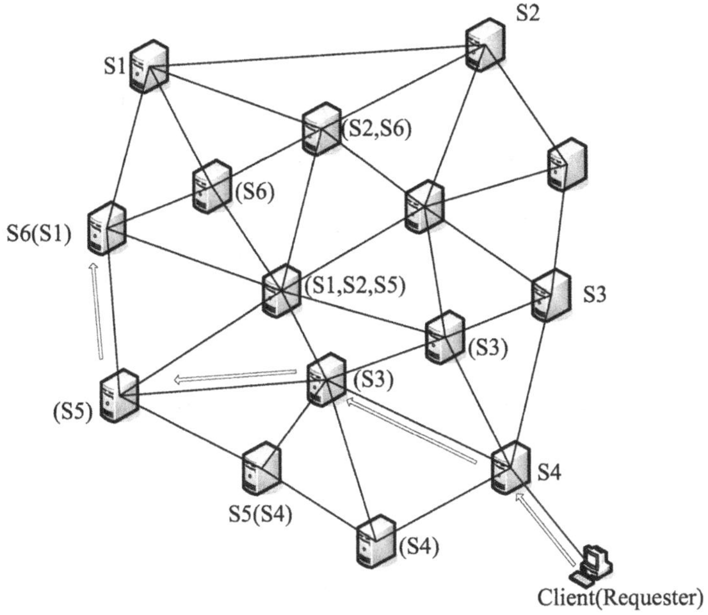

Fig. 3. A running example of service selection

The servers which have received the query will carry out the routine given in section 3. The processes of service discovery and selection which are executed on every server are noted below.

1. The service list on the first server only contains one service which satisfies the functional requirement of end users. This service, S4, also meets the QoS constraints in the query. Thus, S4 is the best candidate on the first server, and its description is attached to the end of the query. Then the query is sent to the next server. Now the query is just the same as depicted in Figure 1.

2. On the second server, S3 satisfies the functional and QoS requirements of users. So S3 and S4 should be compared to select the better one as the candidate. According to the AWT method which is introduces in section 2, the overall QoS score of S3 is 0.35 while that of S4 is 0.4. The quality of S4 is better than S3, so S4 is still the best candidate in this hop.

3. On the third server, the process is just like that on the second server. The overall QoS score of S5 is 0.25 while that of S4 is 0.4. The query is transmitted to the last server without modification.

4. On the fourth and last node, there are three services which satisfy the functional requirement in the query. But the success rate of S6 is 90%, so S6 does not meet the constraints on QoS criteria. The better scored service between S1 and S4 is the best service of this query. The scores of S1 and S4 are 0.4 and 0.25 respectively. As a result, the description of S1 is sent to the client as the best candidate and the process of service discovery and selection comes to an end.

Finally, the client receives the response of the query and then takes the service to finish the payment.

5 Conclusions and Future Work

This paper introduces and illustrates a QoS-Aware service selection approach on P2P network. This approach adopts a two-layered unstructured P2P network as its infrastructure and takes a method to quantify the overall quality of a service. In this approach, the processes of service discovery and selection are combined together and carried out in a decentralized way on every server. When the service discovery is finished, the selection has also been made at the same time. We also give an example to illustrate the execution process of the approach.

Acknowledgements. This paper is partly supported by the National Science Foundation of China under Grant No.60721002, No.60673017 and No.60736015, program for New Century Excellent Talents in University under Grant NCET-06-0440, and Jiangsu Provincial NSF Project under Grants No.BK2007137.

References

1. Gnutella Protocol draft v0.6, `http://rfc-gnutella.sourceforge.net/src/rfc-0_6-draft.html`
2. Leibowitz, N., Ripeanu, M., Wierzbichi, A.: Deconstructing the KaZaA network. In: Proceedings of The Third IEEE Workshop on Internet Applications, pp. 112–120. IEEE Press, Los Alamitos (2003)
3. Stoica, I., Morris, R., Karger, D.R., Kaashoek, M.F., Balkrishnan, H.: Chord: A Scalable Peer-to-Peer Lookup Service for Internet Application. In: Proceedings of the 2001 conference on Applications, technologies, architectures, and protocols for computer communications, pp. 149–160. ACM, San Diego (2001)
4. Balakrishnan, H., Kaashoek, M.F., Karger, D., Morris, R., Stoica, I.: Looking up Data in P2P Systems. Communications of the ACM 46, 43–48 (2003)
5. Kashani, F.B., Chen, C.C., Shahabi, C.: WSPDS: Web servies Peer-to-Peer discovery service. In: Proceedings of International Conference on Internet Computing (2004)
6. Mastroianni, C., Talia, D., Verta, O.: A super-peer model for resource discovery services in large-scale Grids. Future Generation Computer Systems 21, 1235–1248 (2005)
7. Zeng, L., Benatallah, B., Ngu, A.H.H., Dumas, M., Kalagnanam, J., Chang, H.: QoS-Aware Middleware for Web Services Composition. IEEE Transactions on Software Engineering 30(5), 311–327 (2004)
8. Hwang, C.L., Yoon, K.: Multiple Criteria Decision Making. Lecture Notes in Ecnomics and Mathematical Systems. Springer, Heidelberg (1981)

Semantic Model Driven Architecture Based Method for Enterprise Application Development

Minghui Wu[1,2], Jing Ying[1,2], and Hui Yan[2]

[1] College of Computer Science, Zhejiang University, Hangzhou, P.R. China
{minghuiwu,yingj}@cs.zju.edu.cn
[2] Dept. of Computer, Zhejiang University City College, Hangzhou, P.R. China
yanh@zucc.edu.cn

Abstract. Enterprise applications have the requirements of meeting dynamic businesses processes and adopting lasted technologies flexibly, with to solve the problems caused by the nature of heterogeneous characteristic. Service-Oriented Architecture (SOA) is becoming a leading paradigm for business process integration. This research work focuses on business process modeling, proposes a semantic model-driven development method named SMDA combined with the Ontology and Model-Driven Architecture (MDA) technologies. The architecture of SMDA is presented in three orthogonal perspectives. (1) Vertical axis is the MDA 4 layers, the focus is UML profiles in M2 (meta-model layer) for ontology modeling, and three abstract levels: CIM, PIM and PSM modeling respectively. (2) Horizontal axis is different concerns involved in the development: Process, Application, Information, Organization, and Technology. (3) Traversal Axis is referred to aspects that have influence on other models of the cross-cutting axis: Architecture, Semantics, Aspect, and Pattern. The paper also introduces the modeling and transformation process in SMDA, and describes dynamic service composition supports briefly.

Keywords: Ontology, Model-Driven Architecture, Service Oriented Architecture, Enterprise Application Modeling, UML.

1 Introduction

Nowadays, as a result of continuing evolution of technologies and the market keeps changing, enterprises have to quickly adapting their business processes to the new dynamic environments continuously if they want to stay competitive. Enterprise application system development is difficult because of the nature of heterogeneous of the cooperation processes among multiple organizations. There are following obvious problems:

(1) Various implement technologies

A lot of technologies, such as Web Services, CORBA, DCOM, .Net, J2EE, etc., have been developed and are actively used as service enabling technologies. The vast diversity of implementation increases the complexity of development process of service-based systems. Each of these technologies has its own advantages and disadvantages, and different organizations may choose different technologies.

W. Liu et al. (Eds.): WISM 2009, LNCS 5854, pp. 299–308, 2009.

(2) Common understanding

A variety of terms are often used in cooperative enterprise processes. People involved in developing application systems do not often agree on the terms used between two different organizations. Furthermore, even in a single organization, when the same terms are used, the meanings associated with them may differ, i.e. the semantics may be different.

(3) Automated discovery and integration services

There have large numbers of business services and the number is keeping increasing quickly. How to discovery and integrate necessary functions automated is a key issues in enterprise application development.

(4) Business processes verification

Only when the process can be verified formally, the integration process can be automatically.

Due to the problems mentioned above, the development of collaborative enterprise application systems is more difficult. So, corresponding actions should be taken to tackle these problems. Consequently, new methodologies and implementation techniques should be applied. To achieve this, we present a Semantic Model-Driven Architecture (SMDA) method for enterprise application system development with support of Service-Oriented Architecture (SOA). The SMDA method combines the advantages of three main-trend technologies, supports whole lifecycle of enterprise application development.

The remainder of this paper is organized as follows. Section 2 introduces the technological spaces of SMDA and Section 3 describes the architecture of SMDA. Section 4 presents main features of SMDA modeling process and related works are reviewed in Section 5. Finally, Section 6 concludes the research.

2 Technological Spaces of SMDA

2.1 Service-Oriented Architecture

SOA systems play an important role in enabling business application integration across organizations, since most enterprises are involved in cooperation processes and are also technology dependent. SOA is defined as "a set of components which can be invoked, and whose interface descriptions can be published and discovered" in [1]. In SOA, it is advocated that developers create distributed software systems whose functionality is provided by services. A service is a software entity that encapsulates business logic and provides the functionality to others through well-defined and published interfaces [2], and it is the units of modeling, design and implementation.

SOA provides a great level of flexibility and higher level of abstraction in the following ways [3, 4]:

- Services are network-enabled components with well-defined interfaces that are implementation-independent.
- Such services are consumed by message exchange and clients that are not concerned with how these services will execute their requests.
- Services are self-contained to offer functionality as defined in interfaces.
- Services are loosely coupled, and thus more independent.
- Services can be dynamically discovered.

- Services and their compositions can be independently developed and evolved.
- Composite services can be built from aggregates of other services.

In SOA, service interoperability is paramount, and it is the basis of SOA application development by services orchestration. Nowadays, although researchers have proposed various middleware technologies to achieve SOA, Web services standards better satisfy the universal interoperability needs [4]. Services will be invoked using SOAP typically over HTTP and will have interfaces described by the Web Services Description Language (WSDL). Increasingly, developers are using the Business Process Execution Language (BPEL) for modeling business processes within the Web services architecture [4]. BPEL is an XML-based standard which provides a language to describe the behavior of business processes and business interaction with their partners. Although BPEL has many advantages, it has following obvious shortcomings: (1) lack formality (2) can not provide well-defined semantics, and (3) tight coupled with Web services. All these will led the result of poor support of automated services composition and limit to a single technology platform.

2.2 Ontology

Ontology is a philosophical concept which was introduced to describe data models which were conceptually independent of specific applications. Ontology has been widely accepted as the most advanced knowledge representation model. Ontology is defined by Gruber in [5]:"Ontology is a formal, explicit conceptualization stands for an abstract model", explicit means that the elements are clearly defined, and lastly, formal means that the ontology should be machine processable.

With the advent of the Semantic Web movement and the development of XML-based ontology languages such as Web Ontology Language (OWL), ontology is becoming increasingly common in broader communities. OWL-S is an ontology specifying to services, i.e. it is a language that provides a specific vocabulary for describing properties and capabilities of Web services. Formality in the Semantic Web framework facilitates machine understanding and automated reasoning. OWL-S is composited by three parts: The Service Profile describes the capabilities of the service. The Service Model describes how the service works internally. Finally, the Service Grounding describes how to access the service. As such, the OWL-S ontology provides a uniform mechanism for describing the semantics of a web service.

2.3 Model-Driven Architecture

Model-Driven Development (MDD) has been advocated by academia and industry for many years. It can be defined as "an approach to software development where extensive models are created before source code is written" [6]. Model-Driven Architecture (MDA) proposed by the Object Management Group (OMG) is a primary example of MDD.

MDA conceives models as first class elements during system design and implementation and it separates specification of the fundamental business logic from specifications of various abstract-level models. MDA establishes a separation of the development process in three abstraction levels. Computation Independent Model (CIM) is a view of a system that does not show the details of a system structure. In software engineering it is also known as a domain model, which focuses on the domain rather than on showing details of the system structure. Platform Independent Model (PIM) is computation dependent, but it is designed independently of any technical considerations of the underlying platform, and Platform Specific Model (PSM) usually consists of an implementation-specific model geared towards the concrete implementation technique. Such an approach allows fast and effective development of systems that make use of new technologies based on existing verified business models, and makes it possible to preserve investments made into development of business models even if the technological platform is changed.

From the view of architecture, MDA is based on the four-layer meta-modeling architecture including meta-meta-model (M3) layer, meta-model (M2) layer, model (M1) layer and instance (M0) layer. In M3 layer, MOF defines an abstract language and framework for specifying, constructing and managing technology neutral meta-models. It is the foundation for defining any modeling language. All meta-models defined by MOF are positioned on the M2 layer. One of these is UML, a most accepted graphical modeling language for specifying, visualizing and documenting software systems. The models of the real world, represented by concepts defined in meta-model are on M1 layer. Finally, at M0 layer, are things from the real world.

Current most CASE tools supporting MDA is primarily based on UML. The representations provided by the UML diagrams are semi-formal, so the representation by such diagram and textual descriptions is not suitable for machine processing and validation. In this respect, ontology and languages developed for the semantic web are proving to be a good tool for formalizing the information description in a machine-processable manner and separating domain knowledge from operational concerns.

2.4 Combining by UML Extensions

To support semantic model-driven development for service oriented enterprise application, the key is modeling language. Modeling in SMDA will involve various aspects of system development and have the ability of to deal with distributed environment and heterogeneous knowledge, realize automated (semi-automated) transformation and verification. So the following requirements should be satisfied: (1) Expressiveness; (2) Independence; (3) Readability; (4) Common Understanding; and (5) Formalization.

UML 2.0 can be used as an integration platform for SMDA modeling and development. In [7] the authors indicated UML can satisfy the first three requirements mentioned above. Combining ontology technology to support formal semantics, the last two requirements will be met. Ontology allows knowledge to be shared and to be validated automatically. In the ontology language OWL, Description Logic supports formal reasoning. There are some researches on modeling ontology by UML [8, 9]. Hereby, UML is selected as modeling language in SMDA.

UML offers three extension mechanisms to extend meta-model: (1) stereotypes, which introduce new elements by extending a base element; (2) tagged values, which introduce new properties for the new element, and (3) constraints, which are restrictions on the new element with respect to its base element.

A UML profile defines standard UML extensions that combine and refine existing UML constructs to create a dialect that can be used to describe artifacts in a design or implementation model.

3 SMDA Method for SOA Application

Semantic Model-Driven Architecture (SMDA) is a method for SOA oriented enterprise application system development. In this section, we propose a framework for semantic-based model-driven architecture of service-oriented software systems, i.e. ontology as the semantic modeling technique, model-driven architecture as the development approach, and SOA system as architecture of the target application. The SMDA method combines the advantages of three main-trend technologies, and supports whole lifecycle of enterprise application development. The architecture of SMDA can be shown by figure 1, which has three orthogonal dimensions.

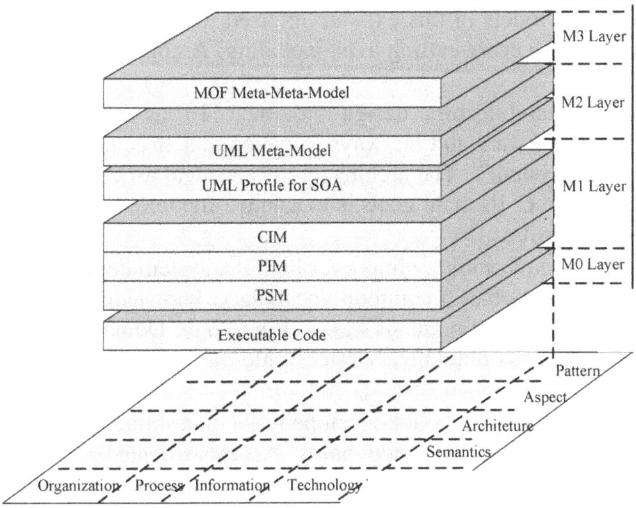

Fig. 1. Architecture of SMDA

1. Vertical Axis: Models are separated in following different abstraction levels corresponding to the four layer of MDA:

- M3 (meta-meta-model) layer is the top layer in the MDA architecture, it is MOF based.
- M2 (meta-model) layer where the modeling languages are defined to describe each model. Because the UML extension mechanism of the UML profile appears to be very useful to defining a suite of meta-model language. In SMDA, two kinds UML profiles are proposed: UML Profile for ontology, and UML Profile for SOA which includes three parts specified for CIM, PIM, and PSM respectively.
- M1 (model) layer where the CIM, the PIM and the PSM laying.

- M0 (code) layer is the goal of MDA model transformation. In SMDA, for example, BPEL is one of the target code styles.

2. Horizontal Axis: Models are separated in the different concerns involved in the enterprise application development, including: Information, Process, Organization, and Technology.

- Information: focusing on the information of resources and entities used in business processes and applications.
- Process: focusing on the business processes, which are the core of enterprise application. Process plays an important role not only in modeling domain activities, but also in modeling interaction processes in software architecture. Process interaction and composition in distributed architectures are central for service-based software systems.
- Organization: focusing on the participants responsible for the support and execution of the business processes.
- Technology: focusing on the technological environment and infrastructures supporting the enterprise applications.

3. Traversal Axis: Models in this axis are referred to aspects that have influence on other models of the cross-cutting axis, including: Architecture, Semantics, Aspect, and Pattern.

- Architecture: An architectural design is defined [10] as the design of "the overall structure of the software and the ways in which that structure provides conceptual integrity for the system". The architecture is a critical aspect in SOA-based developments since one of its main goals is to achieve flexible implementations of interacting software entities.
- Semantics: the use of ontology improves both the system description and the enterprise modeling. It enables a common vocabulary, knowledge to be shared and the reasoning. In SMDA, there are six kinds of ontology: Domain ontology, Organization ontology, Process ontology, Services ontology, Goal ontology and Policy ontology.
- Aspect: Complex software is not developed as a monolithic unit but is decomposed into modules on the basis of functionality. Aspects-oriented software development aims at providing software developers techniques and tools to better manage crosscutting concerns. AOP provides mechanisms for separating crosscutting concerns into single modules called aspects and enabling dynamic weaving aspects into the target application at runtime.
- Pattern: Patterns capture basic abstractions of software design and offer architectural solutions for reuse. In SMDA, adopted the idea from [11], patterns can be divided into four related categories: Business Patterns, Integration Patterns, Application Patterns, and Runtime Patterns.

4 Features of SMDA Modeling Process

4.1 Modeling and Transformation with Ontology

The process starts at the definition of the CIM (as figure 2 shown). In this stage, the users usually set their goals, the relationships within goals and the related constraints

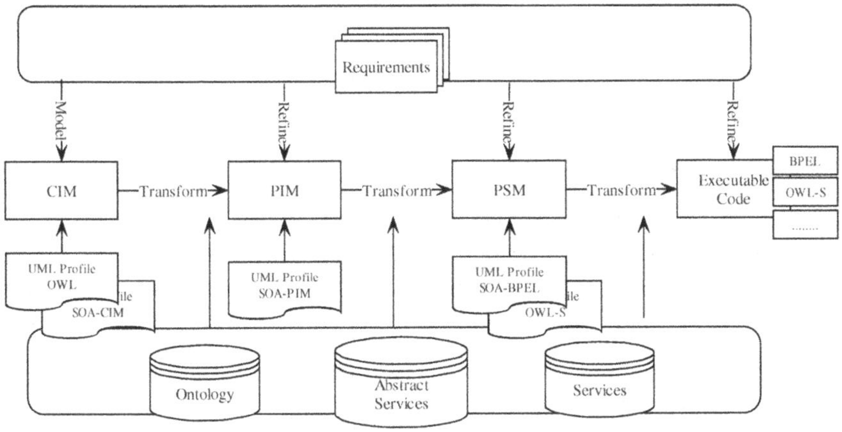

Fig. 2. Process of SMDA

firstly to represent their requirements, both functional and non-functional. The goals will be realized through tasks which operation semantics matching with requirements. An atomic unit of task is an action. Its event-driven execution may connect to explicit information. A role is a class of actors for a set of related tasks to be carried out. An actor is a user, program, or entity with certain capabilities that can be responsible for some actions. A rule is a limitation or constraint imposed by the system. Policy is external, dynamically modifiable rules and parameters that are used as input to a system. The intended use of policies is to specify actions to be performed following event detections and context changes [12]. An organization consists of a set of related roles, thus role establishes a connection between organization and specific tasks. Finally, an enterprise application system consists of all of above entities. The CIM modeling is usually a manual task. In order to facilitate it, we defined six kinds of ontology mentioned in section 3. Each of these ontologies emphasizes an aspect of enterprise application respectively, so the modeler can focus on an aspect once time without influence of others. There are two levels of ontologies in CIM: Domain level and Application level. We provide an UML Profile for OWL to model general ontology in SMDA, and an UML Profile for domain ontology of SOA application named SOA-CIM. Modeler can model application ontology by instantiating the domain ontology.

After CIM modeled, the first transformation takes place from CIM to PIM. Usually, the source model needs refining in order to drive the transformation. The PIM model does not contain details concerning specific technological platform. The PIM model is represented with aid of an UML Profile of SOA-PIM. Next, a transformation from a PIM into a PSM takes place. This transformation plays a special role in MDA. In an ideal way, the same PIM could be transformed in several PSMs, for instance, a BPEL-PSM, OWL-S-PSM or any other.

Finally, the last transformation is PSM to Code. According to the kind of PSM, the code will be corresponding executable language. A very detailed PSM can describe the whole service composition logic. But, just like the previous transformations, fine adjustments can be necessary before actually running it.

In the whole modeling process, modeler can search and retrieve reusable components in knowledge repository. Due to coupling with the formalization and DL reasoning capability, the automated services discovery, selection, composition and model verification is possible.

4.2 Enhance Flexibility by Dynamic Services Composition

The use of ontology and semantic description plays a special role in dynamic services composition. The business process described in a CIM is a description of a sequence of tasks and their control structures, those tasks will be implemented by some services. At this stage, a task is only a service description, i.e., it is an abstract service which will be binding to a concrete service later. In our previous work, we presented an extensible ontology framework named OWL-QSP for dynamic services composition [13]. In the ontology framework, Service Type was imported to improve service abstract level. The concepts of QoS, Situation and Context are also been adopted. QoS and Situation-aware service-based systems are more dynamic and flexible since services discovery, selection and execution can adapt to situation changes dynamically, thus can fit requirement of enterprise application much better. Due to the limited space, the process and support environment of dynamic services composition please refer to [13] for details.

5 Related Works

In current, web service is commonly described by WSDL, invocated through SOAP and orchestrated by BPEL. Due to lack of proper semantics, it is hard to search for appropriate services and composite them into a new service which can perform the desired task automatically and dynamically.

In order to enhance the semantics of the description of web service, ontology technology is adopted. OWL-S and WSMO are the two major examples that have been developed. Modeling language needs proper tools to support. Now there are some tools supporting ontology modeling, for instance, protégé, but the learning curve for such a tool can be steep [14]. Some researcher proposed to model ontology with UML, which are most accepted and popular in software design community. Some researches have started exploiting the connection between OWL and MDA. In [8], an MDA based ontology architecture is defined, which includes an ontology metamodel, an UML profile for ontology and a transformation of the UML ontology to OWL. OMG also issued Ontology Definition Metamodel (ODM) as adopted specification [9]. Those work is foundational, it doesn't address the service-based architectures.

In [15], UML modeling was directly connected with BPEL code generation, without the explicit ontology framework. However, combing with ontologies, which will enhance the semantic modeling and reasoning capabilities, is important in the context of service oriented software development. There are some researcher [7, 14] introduced modeling services with UML profiles. These service models are then translated into OWL-S. Their work mainly focuses the mapping between UML profiles and OWL-S, lack systematic methodology to support whole enterprise application development. The same problem is also occurred in [16]. Garrido et al [17] presents a method of groupware development process. The work focused on two specific models: a conceptual domain model formalized through domain ontology, and a system model built using a UML-based notation. Their work demonstrated the advantages of combing ontology and MDA technologies for service-oriented system development, but their work only focused on the stage of defining CIM.

6 Conclusions

In order to facilitate the development of service oriented software system of enterprise application, we propose a semantic model driven development method named SMDA. To achieve this, we study the characteristics of three main technologies in the method: SOA, MDA and Ontology. SOA is benefited in particular from semantic ontology-based modeling and development. Ontology allows a common vocabulary for knowledge sharing to be established, facilitates machine understanding and automated reasoning in distributed and heterogeneous environment. MDA proposes a separation of the development process in different abstraction levels, where emphasizes model-based transformation and automated code generation. This makes MDA suitable to develop SOA-based enterprise application.

SMDA method can be presented in a three orthogonal dimensional architecture, which includes layered abstract levels, various concerns involved in the development. We also introduce the modeling and transformation process in SMDA, and describe dynamic service composition supports briefly.

Acknowledgments

This work was partly supported by the National High-Tech Foundation (863), China (Grant No.2007AA01Z187).

References

1. World Wide Web Consortium (W3C), Web Services Glossary, Technical report (2004), `http://www.w3.org/TR/ws-gloss/`
2. Wang, G., Chen, A., Wang, C., et al.: Integrated quality of service (QoS) management in service-oriented enterprise architectures. In: The 8th IEEE Intl Enterprise Distributed Object Computing (2004)
3. Yang, Z., Gay, R., Miao, C., et al.: Automating integration of manufacturing systems and services a semantic Web services approach (2005)
4. Pasley, J.: How BPEL and SOA are changing Web services development. IEEE Internet Computing, 60–67 (May-June 2005)
5. Gruber, T.: A translation approach to portable ontology specifications. Knowledge Acquisition 5(2), 199–220 (1993)
6. Bendraou, R., Desfray, P., Gervais, M.-P.: MDA components: A flexible way for implementing the MDA approach. In: Hartman, A., Kreische, D. (eds.) ECMDA-FA 2005. LNCS, vol. 3748, pp. 59–73. Springer, Heidelberg (2005)

7. Grønmo, R., Jaeger, M.C., Hoff, H.: Transformations between UML and OWL-S. In: The European Conference on Model Driven Architecture - Foundations and Applications (ECMDA-FA) (2005)
8. Djurić, D., Gašević, D., Devedžić, V.: Ontology Modeling and MDA. Journal of Object Technology 4(1), 109–128 (2005)
9. Object Management Group. Ontology Definition Metamodel (September 2008), http://www.omg.org/docs/ptc/08-09-07.pdf
10. Shaw, M., Garlan, D.: Formulations and formalisms in software architecture. In: van Leeuwen, J. (ed.) Computer Science Today. LNCS, vol. 1000. Springer, Heidelberg (1995)
11. Zhao, L., Macaulay, L., Adams, J., et al.: A pattern language for designing e-business architecture. The Journal of Systems and Software 81, 1272–1287 (2008)
12. Maamar, Z., et al.: Towards a context-based multi-type policy approach for Web services composition. Data & Knowledge Engineering 62, 327–351 (2007)
13. Wu, M.H., Jin, C.H., Yu, C.Y.: QoS and situation aware ontology framework for dynamic web services composition. In: 12th International Conference on Computer Supported Cooperative Work in Design, CSCWD 2008, pp. 488–493 (2008)
14. Timm, J.T.E., Gannod, G.C.: A model-driven approach for specifying semantic Web services. In: The IEEE International Conference on Web Services (2005)
15. Gardner, T.: UML Modelling of Automated Business Processes with a Mapping to BPEL4WS. In: Proceedings of the First European Workshop on Object Orientation and Web Services at ECOOP (2003)
16. Marcos, L.S.: Modelling of Service-Oriented Architectures with UML. Electronic Notes in Theoretical Computer Science (194), 23–37 (2008)
17. Garrido, J.L.: Definition and use of Computation Independent Models in an MDA-based groupware development process. Science of Computer Programming (2007)

Modeling and Analyzing Web Service Behavior with Regular Flow Nets

JinGang Xie, QingPing Tan, and GuoRong Cao

Computer College of National University of Defense Technology, Changsha, Hunan, P.R. China
jgxie@nudt.edu.cn

Abstract. Web services are emerging as a promising technology for the development of next generation distributed heterogeneous software systems. To support automated service composition and adaptation, there should be a formal approach for modeling Web service behavior. In this paper we present a novel methodology of modeling and analyzing based on regular flow nets—extended from Petri nets and YAWL. Firstly, we motivate the formal definition of regular flow nets. Secondly, the formalism for dealing with symbolic marking is developed and it is used to define symbolic coverability tree. Finally, an algorithm for generating symbolic coverability tree is presented. Using symbolic coverability tree we can analyze the properties of regular flow nets we concerned. The technology of modeling and analyzing we proposed allows us to deal with cyclic services and data dependence among services.

Keywords: regular flow net, symbolic coverability tree, Web service behavior.

1 Introduction

With the recent advances in networking, computation grids and WWW, Service-oriented computing is emerging as an active area of research in both academia and industry[1]. Service-oriented computing advocates the idea of developing complex application by composing less-complex ones. However, in the most of situations, the task of composition is very difficult, error-prone and time consumed because of the heterogeneous, dynamic, distributed, and evolving nature of Web services. And because the services are developed independently, especially when we encountered a large application, the limit appears more obvious. It has been very clear that the process of service composition requires a mechanism for bridging various incompatibilities between services. Such mechanisms are called adaptation. Consequently, we have to study the method for automated services adaptation based on the formal description for Web services.

Currently, WSDL[2] is employed as standard for services description. However, WSDL descriptions do not include any semantic and behavior information of Web service. So it prominently inhibits the success of services match and automated composition. During the last years, various proposals have been put forward to feature more expressive service descriptions that include both semantic and behavior information. But all these proposals reduced the actual services behavior to a certain extent, e.g. some of them supposed that the Web services are acyclic.

W. Liu et al. (Eds.): WISM 2009, LNCS 5854, pp. 309–319, 2009.

Against this background, to pave the way for the fully automated adaptation of services and to minimize the users' efforts required to develop complex services, we propose a new formal method for modeling Web service behavior. Our method allows developers to deal with cyclic services and data dependence among services. Furthermore, this formal method provides the needed information for the a priori analysis and verification of service compositions and adaptations. Based on this formal method we will study the adaptation of Web services. Our long-term objective is to develop a formal approach for automated service adaptation capable of suitably overcoming the mismatches appearing during the composition between services.

The main contributions of this paper include:

- We give the definition of regular flow net based on Petri nets and YAWL.
- We present a technology for modeling Web service behavior using regular flow net, which add data perspective into the regular flow nets.
- To deal with cyclic services, we propose an general algorithm for the construction of symbolic coverability tree.

The rest of the paper is organized as follows. We begin by the definition of regular flow net in section 2. In the next section, we model Web service behavior using regular net. In section 4, we present the symbolic marking definition of Web service behavior and develop an algorithm to generate the symbolic coverability tree. In section 5, we discuss related work. Finally, we conclude with some directions for future research.

2 Definition of Regular Flow Net

In this section we introduce the regular flow net as the basis of service model step by step. A regular flow net is based on ordinary Petri nets[4][5] and YAWL (Yet Another Workflow Language)[6][7]. Regrettably, space limitations do not allow us to introduce Petri nets and YAWL.

The reasons that we choose a model technique based on Petri nets are twofold. On one hand, their terse graphic representation and their sound mathematical semantics allow a clear understanding of complex behavior phenomena such as concurrency, conflict, synchronization, etc. On the other hand, there are abundance of analysis techniques and tools for Petri nets. So based on the regular flow net, we can model Web service behavior with a lot of benefits.

Definition 2.1 (Flow Net, FN). A flow net N is a tuple <C, T, F, split, join, i, o> such that:

[1]C is a set of finite control places,

[2]i, $o \in C$ represent the input place and the output place respectively,

[3]T is a set of finite task nodes,

[4]$F \subseteq ((C \backslash \{o\}) \times T) \cup (T \times (C \backslash \{i\}))$ is a flow relation,

[5]every node in the graph ($C \cup T$, F) is in a directed path from i to o,

[6]split: $T \rightarrow \{AND, OR, XOR\}$ specify the split behavior of each task, and

[7]join: $T \rightarrow \{AND, OR, XOR\}$ specify the join behavior of each task.

The tuple (C, T, F) corresponds to a classical Petri Net where C corresponds to places, T to transitions, and F to the flow relation. However, there are two special places in FN: i and o. Moreover, the behavior of T is different from transitions. We use C_N, T_N, F_N, $split_N$, $join_N$, i_N, o_N to denote the components of FN N when necessary.

Definition 2.2 (pre- and post-set, Linear Structure). Given a FN N, we define for each $u \in C_N \cup T_N$ the sets

$$\bullet u = \{ v \mid \langle v, u \rangle \in F_N \} \quad u\bullet = \{ v \mid \langle u, v \rangle \in F_N \}$$

which denote respectively pre-set and post-set of u. A linear structure is a FN N satisfying:

$$\forall u \in (C_N \cup T_N).\ (u \neq i_N \rightarrow |\bullet u| = 1) \wedge (u \neq o_N \rightarrow |u\bullet| = 1).$$

In this paper, we also use $|A|$ to denote the cardinality of a finite set A, and $|p|$ to denote the length of a finite path p.

Definition 2.3 (Split / Join Node). Given a FN N, $\forall t \in T_N$, if $|t\bullet|>1$, t is called a split node; if $|\bullet t|>1$, t is called a join node.

All of the elements of FN have already been defined above. The symbols in the graphical representation of FN are depicted in Figure 1.

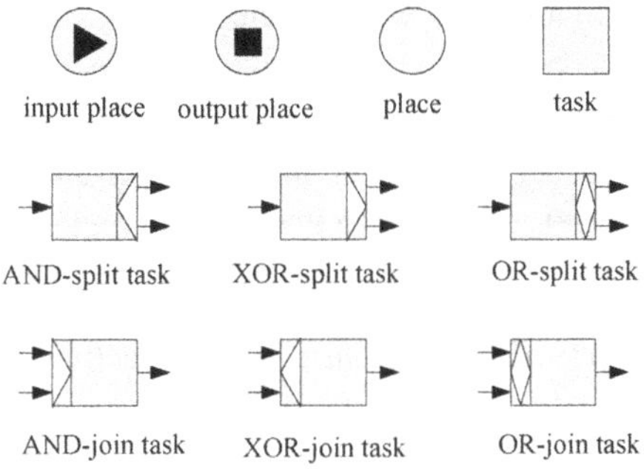

Fig. 1. Symbols used in FN

To define the regular flow net, we firstly need to define the subnet and some basic structures.

Definition 2.4 (Subnet of FN). Given a FN N,

[1] A tuple <C, T, F, split, join> is called a subnet of N where $C \subseteq C_N$, $T \subseteq T_N$, $F = F_N \cap ((C \times T) \cup (T \times C))$, and split = $split_N \restriction_T$, join = $join_N \restriction_T$.

[2] For any non-empty subset $U \subseteq C_N \cup T_N$, the subnet of N w.r.t. U, denoted as $N \restriction_U$, is defined as a tuple <C, T, F, split, join> where $C = U \cap C_N$, $T = U \cap T_N$, and F, split and join are the same as above.

Definition 2.5 (Basic Split-Join Structure). A basic split-join structure is a FN N satisfying:

[1]there is a unique split node $t_s \in T_N$, and $\langle i_N, t_s \rangle \in F_N$,

[2]there is a unique join node $t_j \in T_N$, and $\langle t_j, o_N \rangle \in F_N$,

[3]$split_N(t_s) = join_N(t_j)$,

[4]$\forall u \in (C_N \cup T_N).((u \neq i_N \wedge u \neq t_j) \rightarrow |\bullet u| = 1) \wedge ((u \neq o_N \wedge u \neq t_s) \rightarrow |u\bullet| = 1)$.

t_j is called the dual node of t_s, in this context, t_j is denoted by $dual(t_s)$ and t_s by $dual^{-1}(t_j)$. As shows in Figure 2, the dashed lines represent that some nodes are omitted, and the cardinality of post-set and pre-set for every node omitted is 1. That the split of t_s and the join of t_j are filled with black represents $split_N(t_s) = join_N(t_j)$.

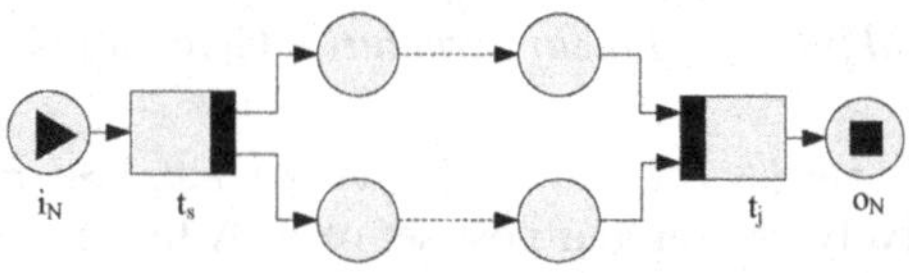

Fig. 2. Basic Split-Join Structure

Before this moment, the modeling techniques restricted the services behavior to acyclic services. However, actually, most of the services are cyclic ones. Next, we define the loop structure in FN such that we can finally model loop behavior of Web services.

Definition 2.6 (Basic Cyclic Path). Given a FN N, a cyclic path p in N is called basic if there is no cyclic path p' in N such that $|p'| < |p|$ and $Nodes(p') \subset Nodes(p)$, where Nodes(p) denotes the set of all nodes which occur in p.

Definition 2.7 (Basic Loop Structure). A basic loop structure is a FN N satisfying:

[1]the subnet $N \restriction_{C_N \backslash \{i_N, o_N\} \cup T_N}$ is a basic cyclic path,

[2]there is a unique node t_{entry} in T_N such that $<i_N, t_{entry}> \in F_N$, and $join_N(t_{entry})$=XOR,

[3]there is a unique node t_{exit} in T_N such that $<t_{exit}, o_N> \in F_N$, and $split_N(t_{exit})$=XOR.

t_{entry} is called the entry node of the loop, t_{exit} the exit node of the loop. The basic loop structure can be depicted as Figure 3(a), and normal while-do loop structure as Figure 3(b), and repeat-until loop structure as Figure 3(c), respectively.

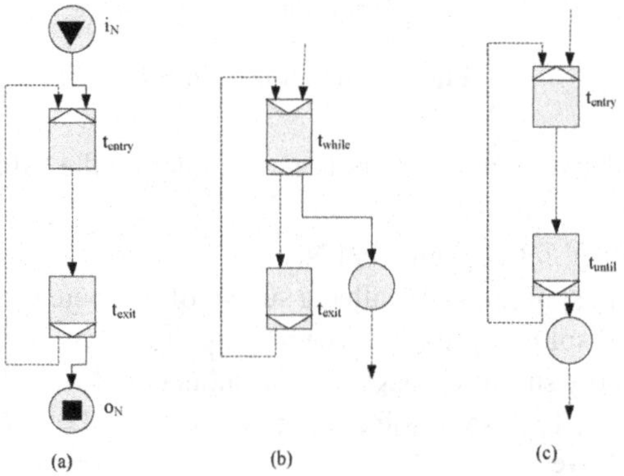

Fig. 3. Basic Loop Structures

A basic structure is a linear, a basic split-join or a basic loop structure. We have defined all the basic structures of FN, next we will define the relation between two flow nets or between a flow net and a basic structure.

Definition 2.8 (Connection of Flow Nets). Given two FN N1 and N2, the connection of N1 and N2, denoted by N1⊕N2, is a new FN N such that:

[1]$C_N = C_{N1} \cup C_{N2} \backslash \{o_{N1}\}$,

[2]$T_N = T_{N1} \cup T_{N2}$,

[3]$F_N = F_{N1} \cup (F_{N2} \backslash \{<i_{N2}, t> \mid <i_{N2}, t> \in F_{N2}\} \cup \{<o_{N1}, t> \mid <i_{N2}, t> \in F_{N2}\})$,

[4]$split_N = split_{N1} \cup split_{N2}$,
[5]$join_N = join_{N1} \cup join_{N2}$,
[6]$i_N = i_{N1}$ and $o_N = o_{N2}$.

Definition 2.9 (Embedding). Given a basic split-join or loop structure N1 and a FN N2, suppose there is a task node t in T_{N1} satisfying:

[1]if N1 is a basic split-join structure, t is different from the split or the join node of N1,
[2]if N1 is a basic loop structure, t is different from the entry or the exit node of N1.

then the embedding of N2 into N1, denoted by $N1\Theta_t N2$, is a new FN N such that:

[3]$C_N = C_{N1} \cup C_{N2} \backslash \{i_{N2}, o_{N2}\}$,
[4]$T_N = T_{N1} \cup T_{N2} \backslash \{t\}$,
[5]$F_N = (F_{N1} \backslash \{<c, t> \mid <c, t> \in F_{N1}\} \backslash \{<t, c> \mid <t, c> \in F_{N1}\}) \cup (F_{N2} \backslash \{<i_{N2}, u> \mid <i_{N2}, u> \in F_{N2}\} \backslash \{<u, o_{N2}> \mid <u, o_{N2}> \in F_{N2}\}) \cup \{<c, u> \mid <c, t> \in F_{N1} \text{ and } <i_{N2}, u> \in F_{N2}\} \cup \{<u, c> \mid <u, o_{N2}> \in F_{N2} \text{ and } <t, c> \in F_{N1}\}$,
[6]$split_N = (split_{N1} \upharpoonright_{T_{N1} \backslash \{t\}}) \cup split_{N2}$,
[7]$join_N = (join_{N1} \upharpoonright_{T_{N1} \backslash \{t\}}) \cup join_{N2}$,
[8]$i_N = i_{N1}$ and $o_N = o_{N1}$.

Sometimes we use $N1\Theta N2$ for short when t is obvious or not important.

Definition 2.10 (Regular Flow Net, RFN). A regular flow net is defined recursively:

[1]a basic structure is a regular flow net,
[2]if N1 and N2 are regular, then $N1\oplus N2$ is regular,
[3]if N1 is a basic structure and N2 is regular, then $N1\Theta N2$ is regular,
[4]no other flow net is regular.

Coming to this point, the definitions of regular flow net are finished. Based on the definition of RFN, we can model control flow perspective of services. But if we want to describe services perfectly, we have to take the data perspective into consideration. Next we add the data perspective into Regular Flow Net and then using the augmented net to model service behavior.

3 Modeling Web Service Behavior Using RFN

This section introduces how to model Web service behavior using RFN, which is a slight extension of regular flow net. The RFN is equipped with the data places, and also some essential mechanisms bound together with the data places.

Definition 3.1 (Web Service Behavior, WSB). A Web service behavior S is a tuple$<N, D, H, type, dom, predicate>$ such that:

[1]N is a RFN, which is called the base net of S,
[2]D is a set of data places, which contains two subsets ID and OD, representing the input data places and the output data places respectively, and $ID \cap OD = \varnothing$,
[3]$H \subseteq ((D \backslash OD) \times T_N) \cup (T_N \times (D \backslash ID))$ is the data flow relation satisfying: $\forall <t_1, d>, <d, t_2> \in H$, t_1 occurs in every path from i_N to t_2 in the graph $(C_N \cup T_N, F_N)$.
[4]type: $D \rightarrow TYPE$ specifies the type of each data place, where TYPE is the set of all data types,

[5]dom: TYPE$\rightarrow 2^{DOM}$ specifies the all possible values for each data type, where DOM stands for a universal individual space,

[6]predicate: $F_N \rightarrow EXP_{BOOL}$ specifies the routing condition for each OR split and XOR split arc in N, where EXP_{BOOL} is the set of all Boolean expressions which contain only the variables in D, i.e. vars(e) $\subseteq$ D for each e$\in$ EXP_{BOOL}.$\forall$<t, c>$\in$ F_N. if $split_N(t) \in$ {OR, XOR} and $|t\bullet| > 1$, then predicate(t, c)$\in EXP_{BOOL}$; otherwise, predicate(t, c) is undefined, denoted as predicate(t, c)$\uparrow$.

The input data places and output data places have no intersection. H is the data flow relation directing from a non-output data place to a task or from a task to a non-input data place. Type is a mapping from data places to data type, type(d) represents the data type of data place d. Dom is a mapping from data type to the domain that the type can denote. Predicate is a mapping from a flow relation to a Boolean expression which represents the predicate on the OR-split or XOR-split arcs of the base net. If and only if the predicate on the arc comes out to true, the arc can be enabled.

From now on we will use N_S, D_S, H_S, $type_S$, dom_S, $predicate_S$ to denote the components of WSB S when necessary. Furthermore, C_S, T_S, F_S, $split_S$, $join_S$, i_S, o_S will be used to denote C_N, T_N, F_N, $split_N$, $join_N$, i_N, o_N when N is the base net of S.

Similar with ordinary Petri nets, we define marking for Web service behavior model.

Definition 3.2 (Marking of WSB). A marking of S is a tuple $M = <M_C, M_D, M_R>$, where:

[1]M_C: $C_S \rightarrow \{0, 1\}$. $M_C(c) = 1$ means that the condition node c has a token, otherwise c does not hold any token,

[2]M_D: $D_S \rightarrow DOM_S$, and $\forall d \in D_S$, if d is inhabited, denoted as $M_D(d)\downarrow$, then $M_D(d) \in dom_S(type_S(d))$. $M_D(d)$ represents the current data hold in the data place d. $M_D(d)\uparrow$ represents d is not inhabited.

[3]M_R: $C_S \times C_S \rightarrow \{0, 1\}$. $M_R(c_1, c_2) = 1$ means that, there is a OR-split node t_{os} in T_S such that $<t_{os}, c_1> \in F_S$, $<c_2, dual(t_{os})> \in F_S$, both c_1 and c_2 occur in the same path from t_{os} to $dual(t_{os})$, and this path is split after executing the task t_{os}. Furthermore, $M_R(c_1, c_2) = 1$ also means that, to execute $dual(t_{os})$ in some future time, c_2 must have been inhabited.

Actually, the model here is also a procedure passing the tokens, and more complex, it takes data places and the split and join structure into consideration relative to ordinary Petri nets. The tokens here are the same as the tokens in ordinary Petri nets. $M_D(d)$ represents whether data place d hold data or is not inhibited. M_R is defined to deal with the OR-split.

Definition 3.3 (Enabled Task Node). Given a marking M of S, a task node t in S is enabled w.r.t. M if$\forall d \in D_S.(<d, t> \in H_S \Rightarrow M_D(d)\downarrow)$, and:

[1]if $|\bullet t| = 1$ then M(c) = 1 for the unique c such that $<c, t> \in F_S$, otherwise:

(a) if $join_S(t)$ = AND then$\forall c \in \bullet t$. $M_C(c)=1$,

(b) if $join_S(t)$ = XOR then $\exists c \in \bullet t$. $M_C(c)=1$,

(c) if $join_S(t)$ = OR then $\forall c \in \bullet t$. $((\exists c_{os} \in t_{os}\bullet .\ M_R(c_{os}, c) = 1) \Rightarrow M_C(c) = 1)$, where $t_{os} = dual^{-1}(t)$,

[2]if $split_S(t) \in$ {XOR, OR} then $\forall c \in t\bullet$. $\forall x \in vars(predicate_S(t, c))$. $M_D(x)\downarrow$.

Firstly, the data in the data places and in the predicate on the arc must hold. Then we consider the behavior of task into the situations below: For normal behavior, the input place need to have a token, for AND-join behavior, all input places need to have tokens, for OR-join behavior, at least one input places needs to have tokens, for XOR-join behavior, only one input place should have a token, and in addition, this input place should not have more than one token.

Definition 3.4 (Firing Rule). Given a marking M of S and an enabled task node t in S, we use $M \rightarrow^{t} M'$ to state that M' is the result marking after importing some input values from the input places of t, firing t at M, then generating outputs to the output places of t. $M \rightarrow^{t} M'$ holds iff:

[1] $\forall c \in \bullet t.\ M'_C(c) = 0$, and:

(a) if $split_S(t) = AND$ then $\forall c \in t\bullet.\ M'(c) = 1$,

(b) if $split_S(t) = XOR$ then $\exists c \in t\bullet.\ (predicate_S(t, c)[\theta] \wedge M'_C(c) = 1 \wedge (\forall c' \in t\bullet \setminus \{c\}.\ M'_C(c') = 0))$,

(c) if $split_S(t) = OR$ then $\forall c \in t\bullet.\ (predicate_S(t, c)[\theta] \Leftrightarrow M'_C(c) = 1)$,

where θ is the substitution which replace each variable $x \in D_S$ occurs in $predicate_S(t, c)$ with the value $M_D(x)$,

[2] $\forall c \in C_S \setminus (\bullet t \cup t\bullet).\ M'_C(c) = M_C(c)$,

[3] $\forall d \in D_S$. if $<t, d> \in H_S$ then $M'_D(d)\downarrow$ and $M'_D(d)$ is the output data of executing t, otherwise $M'_D(d) = M_D(d)$,

[4] if $join_S(t) \neq OR$ and $split_S(t) \neq OR$, $M'_R(c_1, c_2) = M_R(c_1, c_2)$ for all $<c_1, c_2> \in C_S \times C_S$, otherwise:

(a) if $join_S(t) = OR$, then $M'_R(c_1, c_2) = 0$ for all $c_1 \in t_s\bullet$ and $c_2 \in \bullet t$ where $t_s = dual^{-1}(t)$,

(b) if $split_S(t) = OR$, then $\forall c \in t\bullet.\ (predicate_S(t, c)\theta \Rightarrow M'_R(c, c') = 1)$ where $c' \in \bullet dual(t)$ and c, c' occur in the same path from t to dual(t).

For AND-split behavior, tokens are produced for all output places of the task involved, for OR-split behavior, tokens are produced for some of the output places of the task involved, for XOR-split, a token is produced for exactly one of the output place of the task involved. And for OR-split or XOR-split behavior, the arc is enabled if and only if the predicate on it comes out to true.

After transiting from marking M to M', every element in the pre-set of the task t firing this transition has no token. Whether the elements in the post-set have tokens depends on the split type of task t. If data place d is the output place of task t, after executing t to transit to M', the place d holds a data which is the output of t.

We use $RM_S(M_0)$ to denote the set of all markings which can be reached from M_0, that is, $RM_S(M_0) = \{ M \mid$ there are $t_1, \ldots t_k \in T_S$ such that $M_0 \rightarrow^{t1} M_1 \rightarrow^{t2} \ldots \rightarrow^{tk} M\}$.

4 Symbolic Coverability Tree for RFN

The construction of coverability tree is one of the most useful techniques to analyze the properties of concurrent systems modeled by Petri nets. Such a tree describes all the possible behaviors of the system, and its construction is straightforward. But there may be explosion in space and time when we want to analyze larger problems, what's more, we want to deal with cyclic services, so we propose a method to construct the symbolic coverability tree which can prominently reduce the size of the coverability tree and represent loops of the services in a elegant way.

4.1 Symbolic Marking and Fire Rule

Coverability tree generation consists of generating the full state space of a system which in the case of Petri nets means generating all reachability markings. Aiming at reducing the size of the state space, we propose to group some markings into equivalence classes, i.e. symbolic marking. Here is the definition form of the symbolic marking which is similar with the ordinary marking of RFN.

Definition 4.1 (Symbolic Marking of WSB). A symbolic marking of S is a tuple $\widehat{M} = <\widehat{M}_C, \widehat{M}_D, \widehat{M}_R>$, where:

[1] $\widehat{M}_C = M_C$,

[2] $\widehat{M}_D: D_S \rightarrow EXP$, where EXP is the set of all expressions. $\forall d \in D_S$, if $\widehat{M}_D(d)\downarrow$, then $type(\widehat{M}_D(d)) = type_S(d)$ and $vars(\widehat{M}_D(d)) \subseteq D$.

[3] $\widehat{M}_R \subseteq C_S \times C_S$.

In fact, A symbolic marking $\widehat{M}$ stands for a set of normal markings, that is:

$$\widehat{M} = \{ <M_C, M_D, M_R> \mid \forall c \in C_S. \ (\widehat{M}_C(c)=1 \Leftrightarrow M_C(c)=1) \text{ and } \forall d \in D_S. (\widehat{M}_D(d)\downarrow \Rightarrow \widehat{M}_D(d)[\theta] = M_D(d)) \text{ and } \forall c_1, c_2 \in C_S. (<c_1, c_2> \in \widehat{M}_R \Rightarrow M_R(c_1, c_2) = 1) \}$$

where θ is the substitution which replace each variable $x \in D_S$ occurs in $\widehat{M}_D(d)$ with the value $M_D(x)$.

In other words, a symbolic marking represents for a set of normal markings, these normal markings have the same distribution of tokens in the control places but different data places.

When constructing a coverability tree with ordinary markings, a method for generating new markings from old ones by transition occurrences is needed. Therefore, to develop a corresponding method for symbolic markings, the firing rule has to be modified to cope with symbolic markings.

Definition 4.2 (Symbolic Enabled Task Node). Given a symbolic marking $\widehat{M}$ of S, a task node t in S is enabled w.r.t. $\widehat{M}$ if $\forall d \in D_S. (<d, t> \in H_S \Rightarrow \widehat{M}_D(d)\downarrow)$, and:

[1] if $|\bullet t| = 1$ then $\widehat{M}_C(c)=1$ for the unique c such that $<c, t> \in F_S$, otherwise:

(a) if $join_S(t) = AND$ then $\forall c \in \bullet t. \ \widehat{M}_C(c)=1$,

(b) if $join_S(t) = XOR$ then $\exists c \in \bullet t. \ \widehat{M}_C(c)=1$,

(c) if $join_S(t) = OR$ then $\forall c \in \bullet t. \ ((\exists c_{os} \in t_{os}\bullet. \ (c_{os}, c) \in \widehat{M}_R \Rightarrow \widehat{M}_C(c)=1)$, where $t_{os} = dual^{-1}(t)$,

[2] if $split_S(t) \in \{XOR, OR\}$ then $\forall c \in t\bullet. \ \forall x \in vars(predicate_S(t, c)) \ \widehat{M}_D(x)\downarrow$.

Definition 4.3 (Symbolic Firing Rule). Given a symbolic marking $\widehat{M}$ of S, an symbolic enabled task node t in S and a Boolean expression cond. $\widehat{M} \rightarrow^{[cond]t} \widehat{M}'$ holds iff:

[1] $\forall c \in \bullet t. \ \widehat{M}'_C(c)=0$, and:

(a) if $split_S(t)=AND$ then $\forall c \in t\bullet. \ \widehat{M}'_C(c)=1$,

(b) if $split_S(t)=XOR$ then $\exists c \in t\bullet. \ (cond=predicate_S(t, c)[\theta] \wedge \widehat{M}'_C(c)=1 \wedge \forall c' \in t\bullet \backslash \{c\}. \ \widehat{M}'_C(c')=1)$,

(c) if $split_S(t)=OR$ then $\forall c \in t\bullet.(cond=predicate_S(t, c)[\theta] \Leftrightarrow \widehat{M}'_C(c)=1)$, where θ is the substitution which replace each variable $x \in D_S$ occurs in $predicate_S(t, c)$ with the expression $\widehat{M}_D(x)$,

[2] $\forall c \in C_S \backslash (\bullet t \cup t\bullet). \ \widehat{M}'_C(c)=1 \Leftrightarrow \widehat{M}_C(c)=1$,

[3] $\forall d \in D_S$. if <t, d>$\in H_S$ then $\widehat{M}'_D(d)\downarrow$ and $\widehat{M}'_D(d)$ is the output of executing t, otherwise $\widehat{M}'_D(d)=\widehat{M}_D(d)$,

[4] if $join_S(t) \neq OR$ and $split_S(t) \neq OR$, $(c_1, c_2) \in \widehat{M}'_R \Leftrightarrow (c_1, c_2) \in \widehat{M}_R$ for all $<c_1, c_2> \in C_S \times C_S$, otherwise:

(a) if $join_S(t) = OR$, then $(c_1, c_2) \notin \widehat{M}'_R$ for all $c_1 \in t_s\bullet$ and $c_2 \in \bullet t$ where $t_s = dual^{-1}(t)$;

(b) if $split_S(t) = OR$, then $\forall c \in t\bullet$. (cond = $predicate_S(t, c)\theta \Rightarrow (c, c') \in \widehat{M}'_R$) where $c' \in \bullet dual(t)$ and c, c' occur in the same path from t to dual(t).

4.2 General Algorithm for the Construction of Symbolic Coverability Tree

The symbolic firing rule allows us to construct a reduced coverability tree automatically containing a minimal number of arcs and nodes. We outline here an algorithm for construction of the SCT. We deal the loops of services in this algorithm.

Input: WSB and the initial marking

Output: a symbolic coverability tree

The initial symbolic marking $\widehat{M}_0$ contains only the ordinary initial marking M_0.

Step1: select the initial symbolic marking $\widehat{M}_0$ as the root, and mark it with a mark NEW;

Step2: if every marking is marked with a mark OLD or DEAD, then stop;

Step3: if there exists a marking marked with NEW, repeat:

[1] Select a node $\widehat{M}$ with a mark NEW;

[2] If there exists a node $\widehat{M}'$ in the path from $\widehat{M}_0$ to $\widehat{M}$, that $\widehat{M}'_C = \widehat{M}_C$, then mark M with OLD and add $\widehat{M}$ into $\widehat{M}'$, jump to [1];

[3] If there is no task can be fired w.r.t. symbolic marking $\widehat{M}$, mark $\widehat{M}$ with DEAD, jump to step2;

[4] For every task t that can fire under symbolic marking $\widehat{M}$, repeat:

a) According to Definition 4.3, generate follower $\widehat{M}'$ of $\widehat{M}$;
b) Reduce the representation of $\widehat{M}'$ if possible;
c) Add an arc from $\widehat{M}$ to $\widehat{M}'$, label the arc by t, predicate(t, c). then mark $\widehat{M}'$ with NEW;
d) If in the path from $\widehat{M}$ to $\widehat{M}'$, there is a marking M that can be group into symbolic marking $\widehat{M}'$, add M into $\widehat{M}'$.

[5] Mark $\widehat{M}$ with OLD.

Note that the marking $\widehat{M}$ stands for only one normal marking. The means of "add" in [2] is that representing the normal marking $\widehat{M}$ using $\widehat{M}'$. In a word, we use a directed path without cycle to represent loop of the services. The markings in the path are symbolic markings. It can be proved that this algorithm can terminate. We now have an algorithm to produce a SCT of a Web service behavior.

5 Related Work

A number of approaches have been proposed in the literature to model Web service behavior[1]. In [8], they modeled Web services using nondeterministic automata, however they restricted services to finite and acyclic automata. In [9], they modeled

Web services using open net, which is a extension of Petri nets. They modeled asynchronous communication of services with each other, and the services also acyclic services. In [10], they modeled Web services as automata extended with a queue, and communicate by exchanging sequence of asynchronous messages, which are used to synthesize a composition for a given specification. Their approach is extended in [11] which models services as labeled transition systems with composition semantics defined via message passing, where the problem of determining a feasible composition is reduced to satisfiability of a deterministic propositional dynamic logic formula. In [12], they represented Web services using non-deterministic state transition systems, which communicate through message passing.

In contrast, services in our approach are represented using RFN augmented with data places, which can be used to deal with the cyclic services with data flow. Our approach has been inspired by Symbolic Rechability Graph for colored Petri nets[13], and builds on insights from Parametrized Rechability Tree for predicate/transition Petri nets[14].

6 Summary and Discussion

In this paper, we propose a novel approach to model the service behavior and a symbolic technique to analyze the properties of the Web service behavior. So we can deal with more complex services.

There are a few more studies about RFN which we will present in our future work. The problems remain to study are twofold: how to check whether a flow net is a regular one, how to prove that the properties of the SCT are equivalent to the ordinary CT. Therefore, we want to develop an algorithm to check the Regularity of FN and to prove that the reachability property is equivalent for the ordinary and the symbolic coverability tree.

References

1. Pathak, J., Basu, S., Honavar, V.G.: Modeling web services by iterative reformulation of functional and non-functional requirements. In: Dan, A., Lamersdorf, W. (eds.) ICSOC 2006. LNCS, vol. 4294, pp. 314–326. Springer, Heidelberg (2006)
2. W3C. WSDL 1.1 (2001), http://www.w3.org/TR/wsdl
3. Brogi, A., Popescu, R.: Service adaptation through trace inspection. International Journal of Business Process Integration and Management 2(1), 9–16 (2007)
4. Murata, T.: Petri nets: properties, analysis, and applications. Proc. IEEE 77(4), 541–580 (1989)
5. Reisig, W.: Petri Nets: an Introduction. Springer, Berlin (1985)
6. van der Aalst, W.M.P., ter Hofstede, A.H.M.: YAWL: Yet Another Workflow Language. Inf. Syst. 30(4), 245–275 (2005)
7. van der Aalst, W.M.P., Aldred, L., Dumas, M., ter Hofstede, A.H.M.: Design and implementation of the YAWL system
8. Massuthe, P., Wolf, K.: An algorithm for matching mondeterministic services with operating guidelines. International Journal of Business Process Integration and Management 2(2), 81–90 (2007)

9. Gierds, C., Mooij, A.J., Wolf, K.: Specifying and generating behavioral service apapters based on transformation rules. Preprint CS-02-08, Universitat Rostock, Germany (August 2008)
10. Fu, X., Bultan, T., Su, J.: Analysis of Interacting BPEL Web Services. In: 13th Intl. conference on World Wide Web, pp. 621–630. ACM Press, New York (2004)
11. Berardi, D., Calvanese, D., Giuseppe, D.G., Hull, R., Mecella, M.: Automatic Composition of Transition-based Semantic Web Services with Messaging. In: 31st Intl. Conference on Very Large Databases, pp. 613–624 (2005)
12. Pistore, M., Traverso, P., Bertoli, P.: Automated Composition of Web Services by Planning in Asynchronous Domains. In: 15th Intl. Conference on Automated Planning and Scheduling, pp. 2–11 (2005)
13. Chiola, G., Dutheillet, C., Franceschinis, G., Haddad, S.: On Well-Formed coloured Nets and their Symbolic Reachability Graph. In: Proceedings of the 11th Internat. Conf. on Application and Theory of Petri Nets, Paris, June 1990, pp. 387–410 (1990)
14. Lindqvist, M.: Parameterized Reachability Trees for Predicate/Transition Nets. In: Jensen, K., Rozenberg, G. (eds.) High-level Petri Nets. Theory and Application. Springer, Berlin (1991)

Research on Passive Optical Network Based on Ant Colony Algorithms for Bandwidth Distribution in Uplink Direction

Yanping Zhu, Yongsheng Ma, Dezhong Zheng, Lulu Zhao, and Xu Han

Key Lab of Power Electronics for Energy Conservation and
Motor Drive of Hebei Province, Institute of Electrical Engineering,
Yanshan University, Qinghuangdao 066004, China
zhshq-yd@163.com

Abstract. This article design PON with working vacation mechanism about bandwidth distribution in uplink direction, and optimize the serving rate of vacation and roving by ant colony algorithms (ACA), giving the objective function. The convergence speed can be improved by setting the threshold of objectives. More and more ants concentrate towards the optimal solution space in the result of the change of hormones with the objective function about the cost of system, and the optimal solution is found. The numerical experiments show that this method can allocate rational severing rate for every ONU with high speed of convergence.

Keywords: Ant colony algorithm, Passive optical network, working vacation queuing; serving rate optimize.

1 Introduction

Ethernet-based Passive Optical Network (EPON) is the latest development of the current trends in access network. The bandwidth allocation in uplink direction is also the focus.[1-4] In order to make the queue length shorter and the average waiting time less, we use a kind of EPON with working vacation mechanism which make the system provide better service[5]. However, the problem of parameter optimization by this method hasn't been resolved well. As a high performance parallel, simulated evolutionary optimization algorithm, an ant colony algorithm has been shown very strong stability and superiority in finding global optimum. This article design a project to optimize parameter used in bandwidth distribution of passive optical network based ant colony algorithm and give the objective function. By setting threshold of the objective function, algorithm speed of convergence is improved. The simulation results show that this method is effective and reasonable.

2 System Model and Working Principle

The typical EPON system consists of OLT in exchange office, a number of ONU, and passive splitter, as shown in figure 1.

W. Liu et al. (Eds.): WISM 2009, LNCS 5854, pp. 320–327, 2009.

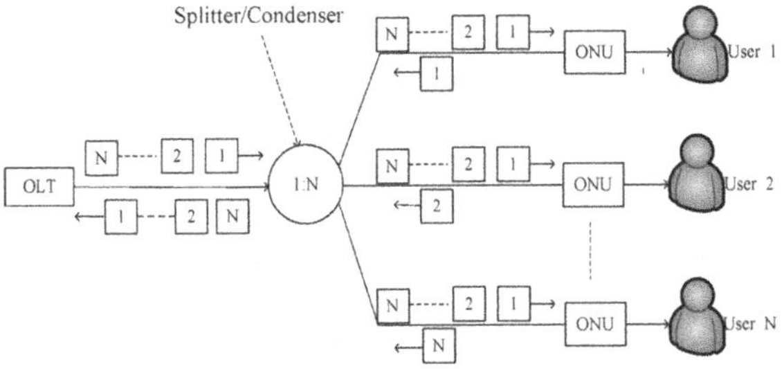

Fig. 1. Typical EPON system schematic

EPON, which provide two-way data transmission and medium shared, is multi-point-to-point structure. In downlink direction (from OLT to ONU direction), the job function by broadcasting way. However, in the uplink direction, it is necessary to prevent data collision through dispatching solution because of fiber channel shared by all of ONU. In EPON, the Multi-Point control protocol (MPCP) is in charge of bandwidth allocation and management by REPORT and GATE message in uplink. In this paper, a centralized bandwidth allocation program is adopted. When the OLT receive the arrival rate of all ONU from REPORT messages, it allocates bandwidth for them intensively by ant colony algorithm. Here uses a service mechanism with working vacation. In this system, allocating permanent slot time for queue i, the corresponding rate is recorded as μ_v^i; a part of time slot is also used as polling service ,the corresponding rate is recorded as μ^+,the total time slot of this queue is recorded as $\mu_b^i = \mu_v^i + \mu^+$. Assuming the queue is working at the rate of μ_b^i, when a time slot is in the end and the queue is empty, the polling time slot will be re-allocated to the queue $i+1$ (module M), and the queue i is allocated time slot, which will reduce to reserved time slot, the service rate is reduced to μ_v^i correspondingly. After an allocation time Δ, the service rate of queue $i+1$ (module m) will increase to $\mu_b^i = \mu_v^i + \mu^+$. The process is repeated circularly. Assuming the total service rate is $\mu^+ + \sum_{i=1}^n \mu_b^i = 1$.

3 Ant Colony Algorithm

The main title (on the first page) should begin 1-3/8 inches (3.49 cm) from the top edge of the page, centered.

Ant colony algorithm[6-10] is born from simulating ant's action of looking for food. Ant colony algorithm makes use of the capacity that ants seek optimal routes when they research for food. The problem is mainly that on the basis of the objective function of system cost is minimum, how to allocate total system cost 1 to all queues.

For the sake of simplicity, the reserved rates μ_v^i and the polling rate μ^+ in the system of 5 queues will be optimized by ant colony algorithm here. That is:

$$u^+ + \Sigma_{i=1}^{5} \mu_v^i = 1 \tag{1}$$

3.1 The Establishment of the Objective Function

Each ONU can be decomposed to a queue model in the EPON network, as shown in figure 2.

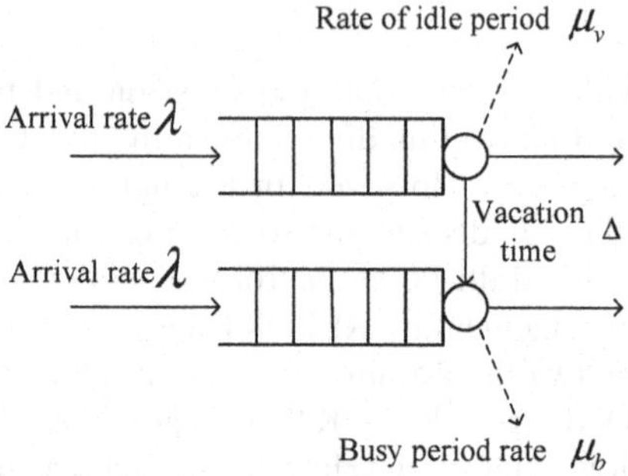

Fig. 2. Queue model

Assuming the arrival rates of data packets of queues $i, (i = 1, 2, 3, 4, 5)$ is P_i respectively, the service rate in the busy period (high-speed run-time) is $\mu_{b(i)}$ and the service rate in the working vacation period is $\mu_{v(i)}$, the working vacation rate is θ_i. Assuming the arrival rate, service rate and vacation rate is independent each other, and the service order of data packets is first-come first service.

Suppose the number of data packets in queue is $Q_v(t)$ at time t.

$$J(t) = \begin{cases} 0, \textit{the state of queue is woking vaction} \\ 1, \textit{the state of queue is busy} \end{cases} \tag{2}$$

$\{Q_v(t), J(t)\}$ is the state space of birth-death process.

$$\Omega = \{(0,0)\} \cup \{(k, j) : k \geq 1, j = 0,1\} \tag{3}$$

Suppose $\alpha = p\overline{\mu}_b / \overline{p}\, \mu_b$ and $\alpha < 1$, the joint probability density of (Q, J) [11] is :

$$\begin{cases} \pi_{00} = K[\theta + \overline{\theta}\,\overline{p}\mu_v(1-r)], \\ \pi_{k0} = Kp\overline{\theta}(1-r)r^{k-1}, \quad k \geq 1 \\ \pi_{k1} = K(p\theta/\overline{p}\mu_b) \sum_{j=0}^{k-1} r^j \alpha^{k-1-j}, \quad k \geq 1 \end{cases} \tag{4}$$

Here,

$$K = \frac{\overline{p}\mu_b(1-\alpha)(1-r)}{\overline{p}\mu_b(1-\alpha)(1-r)[\theta+\overline{\theta}\,\overline{p}\mu_v(1-r)+\overline{\theta}\,p]+p\theta}$$

$$r = \frac{1}{\overline{p}\mu_v}\left(\beta + p\overline{\mu}_v + \overline{p}\mu_v - \sqrt{(\beta + p\overline{\mu}_v + \overline{p}\mu_v)^2 - 4p\overline{\mu}_v\overline{p}\mu_v}\right) \quad \beta = \theta\overline{\theta}^{-1}$$

Then establish the cost model of queue $i, (i = 1,2,3,4,5)$. We order:

$C_v \equiv$ the cost of unit time in state of vacation

$C_s \equiv$ the cost of unit time in state of busy

$C_w \equiv$ the cost of customer stay in unit time

$L_i \equiv$ the captain of the average queue i.

E(I),E(B),E(C)is the mathematical expectation of vacation, busy period and busy loop respectively.

The cost of queue i in unit time is:

$$C(N)_i = C_s\frac{E(B)}{E(C)} + C_v\frac{E(I)}{E(C)} + C_wL = \\ C_sP\{J=1\} + C_vP\{J=0\} + C_wL_i \tag{5}$$

The system cost in unit time is:

$$C(N) = \Sigma_{i=1}^{5} C(N)_i \tag{6}$$

3.2 Algorithm Parameters and Running

Assuming that each queue has the same priority, so the largest service rate is 0.2 for each queue. Assuming that the arrival rate is p_1, p_2, p_3, p_4, p_5 respectively, and the service rate of vacation period must be greater than the arrival rate for every queue, otherwise the system will be unstable. $p_i \to 0.2$ is divided into 100 here to form the state of u_v^i. Then transform the objective function to a function about u_v^i. The problem can be described as:

$$\min C(\mu_v^i)$$
$$s.t.\begin{cases} P_i \le \mu_v^i \le 0.2, i = 1,2,3,4,5 \\ \mu_v^i 100\, state\ spaces \end{cases} \tag{7}$$

Feasible solution space is shown in Figure 3. The trail of ant is the solution space that consists of nodes from every variable. We select 5 variables to construct a 5 order decision-making problem, and each order has 100 nodes.

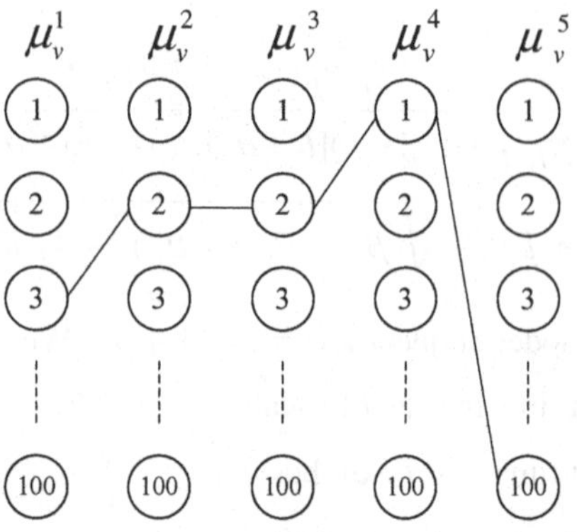

Fig. 3. Feasible solution space

First take all ants on the first level, and the initial pheromone matrix is the whole 1 matrix. The probability which ant chose the ith node at the jth order is:

$$p_{ij} = \frac{\tau_{ij}}{\sum_{i=1}^{l_i} \tau_{ij}} \tag{8}$$

Here τ_{ij} is attractive strength the ith node at the jth order.

Update equation for the trajectory is:

$$\tau_{ij}^{new} = \rho \cdot \tau_{ij}^{old} + \frac{C(u_v^i)}{f} \tag{9}$$

$C\left(u_v^i\right)$ is the objective function.

Specific algorithm steps are as follows:

Step 1: Suppose the largest number of cycle is 200, the number of ants is 180, the volatile coefficient is 0.5 and ant releasing pheromones is 100. Initialize pheromone matrix, avoiding table and feasible solution space.

Step 2: Take ants on the first level, the node of next level is chosen according to transfer probability until each ant go across every level.

Step 3: Calculate the objective function $C\left(u_v^i\right)$ through chosen value from chosen node and give the threshold value of the objective function. The pheromone of the route which is less than threshold value will be modified according to modify equation. $nc \leftarrow nc+1$.

Step 4: If $nc > Nc\_\max$, end the program, otherwise go to step 2.

4 Numerical Expriment

Assume $C_s = 17, C_v = 5, C_w = 3$, give the arrival rate of every queue in five times experiments, as shown in table 1, in which p_i is arrival rate. The corresponding experimental results are shown in table 2, in which μ_v^i is the service rate in vacation period.

Table 1. The arrival rate of 5 times experiments

Number of tests	P_1	P_2	P_3	P_4	P_5
1	0.09	0.12	0.15	0.14	0.18
2	0.07	0.15	0.05	0.10	0.12
3	0.12	0.16	0.03	0.12	0.11
4	0.05	0.14	0.17	0.06	0.09
5	0.12	0.14	0.02	0.07	0.17

Table 2. The result of 5 times experiments

Number of tests	μ_v^1	μ_v^2	μ_v^3	μ_v^4	μ_v^5	μ^+	$C(\mu_v^i)$
1	0.0900	0.1208	0.1722	0.1588	0.1908	0.2602	66.2658
2	0.0897	0.1970	0.0561	0.1414	0.1822	0.3336	48.1651
3	0.1596	0.1996	0.0300	0.1588	0.1491	0.3029	52.0063
4	0.0500	0.1958	0.1994	0.0699	0.1289	0.3560	49.7061
5	0.1838	0.1970	0.0218	0.0753	0.2000	0.3221	50.0051

From table 1 and table 2, we can draw a conclusion that if the arrival rate can be got, the algorithm can give a reasonable service rate, and service rate is proportional to arrival rate. When the total arrival rate increase, the system cost will also increase accordingly.

In order to research the relation between arrival rate and service rate, for simplicity, let the arrival time of queue 1 rang from 0.01 to 0.19 and get 19 numbers according to an interval of 0.01, and the others queue's arrival rate are constant, that is:

$$P_2 = 0.12, P_3 = 0.15, P_4 = 0.14, P_5 = 0.18$$

We observe the corresponding curve of $\mu_v^1, \mu_v^2, \mu_v^3, \mu_v^4, \mu_v^5$, shown in figure 4.

We can see from the figure 4 that the service rate of queue 1 increases with the increase of the arrival rate P_1, the relation of them is direct proportion. For other four queues, their service rates reduce with the increase of the arrival rate of queue 1. When the arrival rate of queue 1 begin to increase, the service rate of queue 2 decline

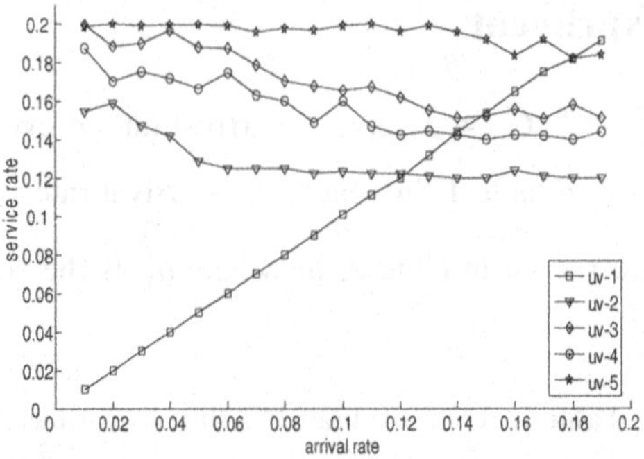

Fig. 4. Curve of service rate change with arrival rate

fastest; the descending speed of queue 3 and queue 4 is basically the same due to their same arrival rates; the service rate curve of queue 5 is steady because of the arrival rate near 0.2.

Then let the arrival time of queue 1 rang from 0.01 to 0.19 and get 19 numbers according to an interval of 0.01, the arrival rate of queue 2 0.08, 0.10, 0.12, observe the curve of the service rate of queue 1 ,shown in figure 5.

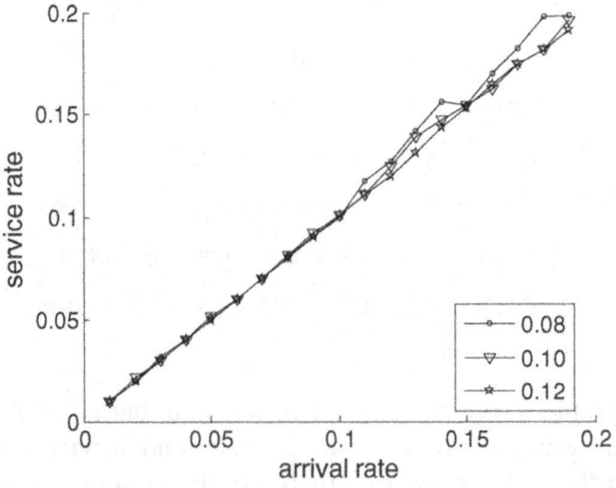

Fig. 5. The serve rate of queue 1 affected by queue 2

From the figures above, when the service rate of queue 1 rang from 0.01 to 0.10, three curves are coincident basically, little affected from queue 2. However, if the service rate of queue 1 rang from 0.10 to 0.20 and the service rate of queue 2 is 0.08, the service rate of queue 1 is greater than the others and the state of 0.10 is greater than the state of 0.12. We can get the conclusion: if the arrival rate of queue 1 is large enough, the smaller the arrival rate of queue 2, the greater the service rate of queue 1 from total service rate.

5 Conclusion

In EPON system with working vacation mechanism, assuming that the priority of each ONU is the same, when OLT get the arrival rate of all queues, it can allocate service rate to them by ant colony algorithm. Experiments show that this algorithm can allocate reasonable service rate on vacation and the polling service rate to each ONU, making system cost minimum.

References

1. Chadim, A., Yinghua, Y., Sudhir, D.: Dynamic Bandwidth Allocation for Quality-of-Service Over Ethernernet PONs. J. IEEE Journal on selected Areas in Communications 21(9), 1467–1477 (2003)
2. Dhaini, A.R., Assi, C.M., Maier, M.: Dynamic wavelength and bandwidth allocation in hybrid TDM/WDM EPON network. J. Journal of Lightwave Technology 25(1), 227–286 (2007)
3. Hsueh, Y.-L., Rogge, M.S., Yamamoto, S.: A highly flexible and efficient passive optical employing dynamic wavelength allocation. J. Journal of Lightwave Technology 23(1), 277–285 (2005)
4. McGarry, M.P., Maier, M.: WDM Ethernet passive optical network. J. IEEE Optical Communications 44(2), S18–S25 (2006)
5. Servi, L., Finn, S.: M/M/1 queue with working vacation MM/1 WV. J. Perfor. Eval. 50, 41 (2007)
6. Duan Haibin, C.: Ant colony algorithm and its application. Science Press, Beijing (2005)
7. Shiyong, L., Yongqiang, C., Yan, L., et al.: Ant colony algorithm and its application. Harbin Institute of Technolgy Press, Harbin (2004)
8. Dartgo, M., Maniezzo, V.: Colony Ant System: Optimization by a Colony Cooperating Agent. J. IEEE Transactions on Systems, Man and Cybemetics-bart B:Cybernetics 26(1), 29–41 (1996)
9. Zheng, W., Crowcroft, J.: Quality of Service Routing for supporting Multimedia Applications. J. IEEE Journal on selected Areas in Communications 14(7), 1228–1234 (1996)
10. Ying, W., Jianying, X.: An adaptive ant colony algorithm and its simulation. J. Journal of system simulation 14(1), 31–33 (2004)
11. Shunli, Y., Naishuo, T.: N-strategy and M/P/1 working vacation queue. Qinhuangdao: YanShan university master's thesis (2007)

Multifactor-Driven Hierarchical Routing on Enterprise Service Bus

Xueqiang Mi[1], Xinhuai Tang[1], Xiaozhou Yuan[2], Delai Chen[3], and Xiangfeng Luo[4]

[1] School of Software, Shanghai Jiao Tong University, 200240 Shanghai, China
alloyer@sjtu.edu.cn, tang-xh@cs.sjtu.edu.cn
[2] School of Mechanical and Dynamics Engineering, Shanghai Jiao Tong University, China
colin_yuan@hotmail.com
[3] Shanghai Key Lab of Advanced Manufacturing Environment, China Telecom Shanghai Branch, China
dlchen.2005@hotmail.com
[4] School of Computer Science, Shanghai University, China
lxf@shu.edu.cn

Abstract. Message Routing is the foremost functionality on Enterprise Service Bus (ESB), but current ESB products don't provide an expected solution for it, especially in the aspects of runtime route change mechanism and service orchestration model. In order to solve the above drawbacks, this paper proposes a multifactor-driven hierarchical routing (MDHR) model. MDHR defines three layers for message routing on ESB. Message layer gives the original support for message delivery. Application layer can integration or encapsulate some legacy applications or un-standard services. Business layer introduces business model to supplies developers with a business rule configuration, which supports enterprise integration patterns and simplifies the service orchestration on ESB.

Keywords: Enterprise Service Bus, Enterprise Integration, Message Routing.

1 Introduction

In order to accommodate reusable and flexible business processes and overcome many distributed and heterogeneous computing challenges in modern enterprises, service-oriented architecture (SOA) was designed and leaded in. An SOA provides a architecture that unifies business functionalities by modularizing applications or resources into services, which are well defined, self-contained modules and provide standard and reusable functionalities As the infrastructure of SOA, Enterprise service bus (ESB) takes on many characteristics for distributed computing in enterprise application solutions, such as standards-based and highly distributed integration, event driven architecture, process flow, and so on [1] [2].

ESB is an integration platform that combines messaging, services, data transformation and intelligent routing, and provides reliable connection and coordination between diverse coarse-grained services [1] [4]. On ESB, message routing is the

W. Liu et al. (Eds.): WISM 2009, LNCS 5854, pp. 328–336, 2009.

foremost functionality for service process as it provides the fundamental capacity to support intelligent interconnectivity between service requestors (or consumers) and service providers, and by using it as the communication backbone, the interactive topology between services can be simply point to point, and also multi-point to multi-point in complex scenarios. Since the invocations are independent of the service location and the involved protocols, they can be dynamically located and invoked [5] [6], which is important for business orchestration in an SOA.

When receiving a message, ESB route it through a set of services, which can be a fixed configuration or a dynamic composition. On ESB, services are presented as abstract service names, which can dynamically bind to different service implementations at runtime. Dynamic routing enables the routing can be changed as needed and it can be designed in two ways in general. One is dynamic service binding which can determine an optimal service provider and connect it by examining the load condition, route length, external environment and so on, and the other is convertible service routing path that can composite services and define the execution order as a result of some changes on business processes, to lead to a reconfiguration of routing path [13].

Content-based routing is an instance of convertible service routing mechanism since it can find the path depending on the message content. A pattern-based routing is proposed in our previous work to supply an integrative dynamic routing solution, which can manage abstract services at runtime, composite the cooperative services by referring to application or business requirements [3].

The rest of this paper is organized as follows: Section 2 presents the related work on dynamic routing. The proposed MDHR framework is discussed in Section 3. Section 4 gives a sample built by using MDHR. Conclusion is made in Section 5.

2 Related Work

2.1 Content-Based Routing

Content-based routing (CBR) is a message driven dynamic routing mechanism and considered as a necessary feature on ESB. In CBR, the routing path is determined by analyzing the message and applying a set of predictions to its content and it usually depends on a number of criteria, such as the message type, specific field value, existence of fields, and so on.

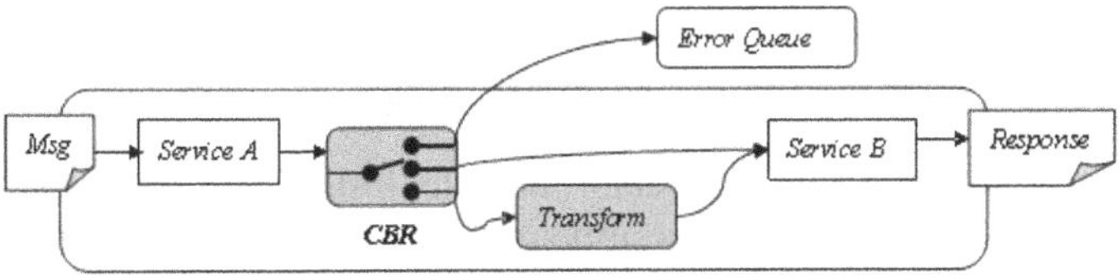

Fig. 1. Content-based Routing on ESB [3]

Fig.1. gives a basic CBR use case. After *Service A*, *Service B* is the expected process partner, but when *Service A* generates some messages which can't be accepted by *Service B*, a *Transform* module will be given for pre-processing, which will translate the messages into the expected type. An *Error Queue* is prepared for maintaining the error messages or events in this module.

2.2 Pattern-Based Dynamic Routing [3]

In conventional routing models, a service is usually maintained by single service container or unknown external server and their composition capacity usually depend on Web Services or BPEL, which are based on XML technology that has well interoperability but well-known performance drawbacks in high volume or intricate applications. Pattern-based dynamic routing (PBDR) proposed in our previous work [3] focuses on this issue and introduces application pattern (AP) concept to improve current solution.

AP is the key to support the multi-service container responsible for service orchestration in PBDR since it defines a set of interoperable services and the relationship among them, so it can helps to obtain a routing path for message delivery. Generally, an AP instance maintains a steady composition application for PBDR module. Then by analyzing a received message, PBDR will choose an appropriate one and invoke the services involved in to process the message.

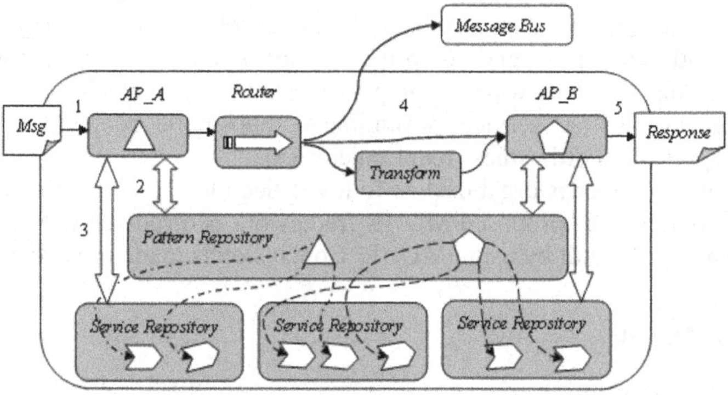

Fig. 2. Pattern-based Routing on ESB

Compared to CBR, PBDR (shown in Fig.2.) replaces the service node with AP and it has a pattern repository and service repository for AP and service registration. PBDR process a message in five steps:

1. When receiving a message, PBDR validate the message header and deliver it to an appropriate AP container (here is AP_A).

2. AP_A will select an AP instance form Pattern Repository by applying a matching predication to the message. The Pattern Repository maintains all the AP instances defined on an ESB.

3. After selection, a service executor module instantiates the related services through Service Repository, which is responsible for service abstraction, storage and reconfiguration, and invokes them to process the request.

4. A message may cross several AP containers that are linked by some router and transformation components on ESB. This step is similar with the conventional message process procedures.

5. Finally, a response is sent out to reply the request message.

To provide relatively independent services for workflow layer, the ESB should manage some composition services with mediation flows [11]. Moreover, ESB provides service containers that can host multiple services in a black box [1] [12]. PBDR follows the above two principles and provide such a more controllable and flexible routing solution.

2.3 Multifactor-Driven Hierarchical Routing

This paper proposes a hierarchical routing model to synthesize the above routing solution and introduce a domain specific language (DSL) [14] on top of PBDR layer to support enterprise integration patterns (EIPs) [15]. As shown in Fig.3., a *Message Filter* is configured in MDHR to do a message sorting for the routing modules that have three routing levels: basic level; application level and business level.

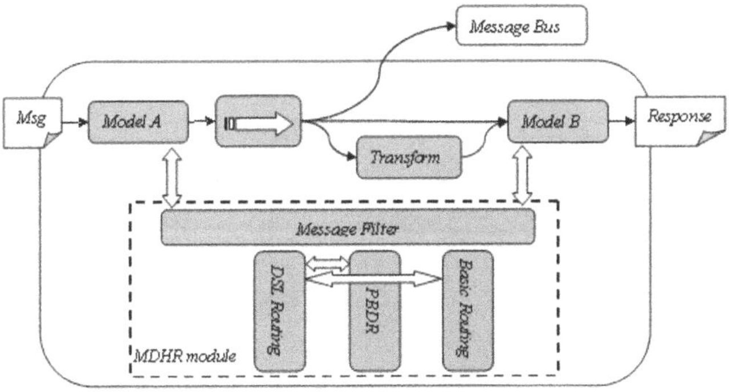

Fig. 3. The MDHR module

Basic layer provides several widely used routing mechanisms, like content-based routing, itinerary-based routing, which can determine or generate routing path by using factors closely relevant to messages and don't directly support applications or business models. If developers or deployment managers want to add some model on routing layer, they must turn to hard-code or static configuration that will make the ESB too complex to maintain.

Application layer uses PBDR to implement the application-related routing or service composition. An important functionality of PBDR is the wide-ranging technologies support. Developers can integrate the legacy systems based on TCP, FTP and many other protocols or messages, providing an independent service for external systems or top layers.

Business layer introduces DSL to allow developers to define business processes. More than twenty EIPs have been implemented in DSL which provides a powerful and flexible way to help developers build the business blocks in complex application scenarios.

The multiple layers structure allows the MDHR can be driven in manifolds, including a business rule, an application stamp or the basic message content. Multifactor-driven architecture separates the routing configuration from the application arrangement and enables a more adaptable solution when building the service mediation on ESB.

3 The MDHR Framework

3.1 MDHR ESB Solution

Current ESB products support several routing mechanisms, like content-based routing, itinerary-based routing, which are tight coupled to message or static routing configuration, but on MDHR ESB solution, some rules and patterns are defined and separated from the message details. MDHR can automatically differentiate and dispatch the message into different message process layers by analyzing the message header.

To achieve the above features, *Service Management* and *Service Mediation* are inserted into the plain ESB and as the core to process the messages (in Fig.4.). *Service Management* supplies the functionalities of service discovery and configuration, message transformation and service access interfaces. On *Service Mediation* part, MDHR Module is delegated to provide the routing service.

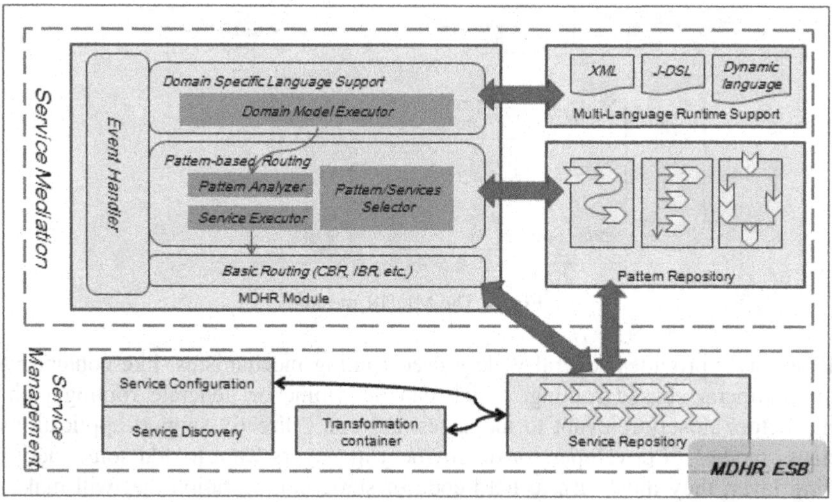

Fig. 4. The MDHR ESB Overview

Event handler is the entrance of the request messages and it dispatches the messages to different message process layers. The basic routing, pattern-based routing and domain specific language support components are set separately for different layers and they can cooperate with each other to handle a business process. *Multi-Language Runtime* provides the process engines for DSL execution. *Pattern Repository* maintains the defined application patterns for business model or external service consumers and *Service Repository* provides the functionalities of service registry and management.

Most of the current ESB products support a message-driven routing mechanism, but in MDHR ESB, multiple factors are set as drive sources to start a business and application related service. On top of CBR, which is still the basis for hierarchical routing, pattern-driven and business rule-driven routing are supported by inserting AP and BM definition on MDHR. These definitions allows developers or deploy managers can configure the routing rules at different levels and gives ESB a better adaptability when facing complex and multivariate scenarios.

3.2 MDHR Definition and Specification

Each layer in MDHR has a routing concept to specify the expected business, application or service definition. Following gives the specifications of the concepts.

Definition 1 Business Model (BM). Defines a business model, through which the developers can bind a business rule with the routing mechanism.

BM := <business_rule, {ap | where ap is an AP instance},
{service | where service is a Service instance}>, where,

- business_rule specifies the configuration file or class path of business rule;
- {ap},{service} list the application patterns and services used in this model.

The business rule part is the core of this model and the request message will be handler using the AP and Service instance referred to by this rule.

Definition 2 Application Pattern (AP). defines a set of abstract services and the running order in this pattern [3].

AP := <ap_stamp, service_order, {service | where service is a Service instance>, where,

- ap_stamp presents the accepted message type of this pattern;
- service_order gives the invoking order of the services, using service id and several symbols to express it, e.g. (0, 1), ((0, 1) -> 2);
- {service} lists the services used in this pattern.

Definition 3 Service. defines an abstract service, which maintains all the information of a service, such as the message type of input and output, service address and the involved transformers [3].

SN := <input, output, spec, {transformer | where transformer is a Transformer instance}>, where,

- input specifies the request message type of this service;
- output specifies the response message type;
- spec contains the other information of the service, such as the service name and address, and it can be extended in specific applications;
- {transformer} lists the transformers that may be used by this service.

In one business model, several application patterns and services may be configured to cooperate on this functionality. Every application pattern may also contain a bundle of services, which provides the atomic function in this model, though the service may involve some complicated external processes. Every BM, AP or Service instance in MDHR has a globally unique identifier, which relates to the version and the type of the instance and is registered at the repositories. When we have to reconfigure or update an instance, we only have to send a message to correlative repository, which is responsible to uninstall the old instance and set up the new one.

The XML schema of a BM is shown in Fig.5.

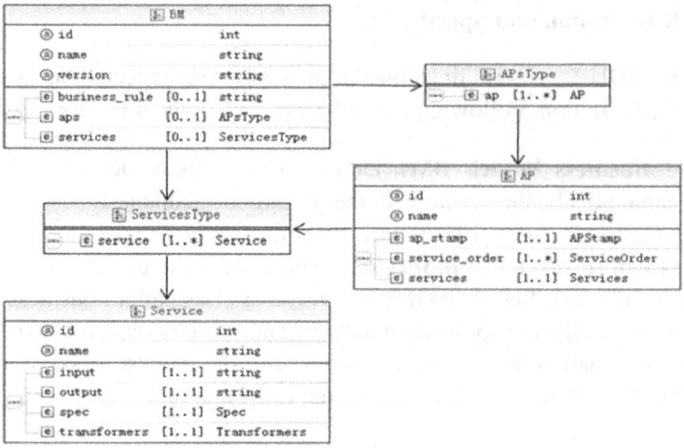

Fig. 5. Business Model Schema

4 A Sample - Travel Agency Application

Now we use MDHR model to describe the classic travel agency business scenario [16], and one of the models, travel reservation model, is shown in Fig.6. The travel

```
<?xml version="1.0" encoding="UTF-8"?>
<bm:business-model id="121301" name="travel_reservation" version="1.0" xmlns:app="http:
  <bm:business_rule>121301_business_rule.XML</bm:business_rule>
  <bm:aps>
    <bm:ap id="321864" name="AirlineApp">
      <app:ap_stamp>
        <app:expression>header/process='AirlineApp'</app:expression>
      </app:ap_stamp>
      <app:service_order><app:order>(643082->843038)</app:order></app:service_order>
      <app:services>
        <app:service id="643082" name="ReserveAirline">
            <svc:input>643082_in.XSD</svc:input>
            <svc:output>643082_out.XSD</svc:output>
            <svc:spec>
                <svc:name>ReserveAirline</svc:name>
                <svc:protocol>SOAP</svc:protocol>
                <svc:address>http://192.168.1.156:6430/</svc:address>
            </svc:spec>
            <svc:transformers>
                <svc:transformer>
                    <svc:name>JMS2XML</svc:name>
                    <svc:from>JMS</svc:from>
                    <svc:to>643082_in.XSD</svc:to>
                    <svc:path>bean:jms2xml</svc:path>
                </svc:transformer>
            </svc:transformers>
        </app:service>
        <app:service id="743001" name="CancelAirline"><!-- omission --></app:service>
        <app:service id="843038" name="ReservedAirline"><!-- omission --></app:service>
      </app:services>
    </bm:ap>
    <bm:ap id="321874" name="VehicleApp"><!-- omission --></bm:ap>
    <bm:ap id="321884" name="HotelApp"><!-- omission --></bm:ap>
  </bm:aps>
</bm:business-model>
```

Fig. 6. Travel Reservation described in MDHR

reservation business model contains three applications: *AirlineApp* (in line 5), *VehicleApp* (in line 33, omitted) and *HotelApp* (in line 34, omitted).

- Line.3 specifies the business rule configuration, which orchestrate the applications and services for this business process. Here an XML file is used to maintain the business rule.
- Line.5, 32 and 33 give the three application pattern instances used in this BM.
- Line.6-8 provides an XML expression to specify the header of the expected messages for *AirlineApp.*
- Line.9 provides the service invoking order.
- Line.10-29 presents three services: *ReserveAirline, CancelAirline* and *ReservedAirline* for reserving, cancelling and listing services.

Through XML files, ESB defines and maintains business processes at runtime. The business rule is implemented by using the DSL, which simplifies the work flow orchestration.

5 Conclusion

This paper proposes a hierarchical routing mechanism with three layers to enhance the routing functionality for ESB. Basic routing layer provides several mature routing implementations. Application routing layer supports multiple protocols or messaging systems and can integrate many kinds of underlying applications. Business layer introduces the domain specific language which supports enterprise integration patterns and enables developers can orchestrate services in a flexible and simple way. Through this hierarchical model, ESB can undertake much more complex scenarios without importing some low performance solutions, like BPEL related system.

References

1. Chappell, D.: Enterprise Service Bus. O'Reilly Media, Inc., Sebastopol (2004)
2. Papazoglou, M.P., Heuvel, W.V.: Service oriented architectures: approaches, technologies and research issues. The VLDB journal 16(3), 389–415 (2007)
3. Tang, X., Mi, X., Yuan, X., Chen, D.: PBDR: Pattern-Based Dynamic Routing on Enterprise Service Bus. In: WorldCOMP 2009: Proceedings of the World Congress in Computer Science, Computer Engineering and Applied Computing (July 2009)
4. Schulte, R.: Predicts 2003: Enterprise Service Buses Emerge. Gartner Research (December 2002)
5. Liu, Y., Gorton, I., Zhu, L.: Performance Prediction of Service-Oriented Applications based on an Enterprise Service Bus. In: COMPSAC 2007: Proceedings of the 31st Annual International Computer Software and Application Conference, pp. 327–334. IEEE Computer Society, Los Alamitos (2007)
6. Flurry, G.: Dynamic routing at runtime in WebSphere Enterprise Service Bus. IBM developerWorks (2006)
7. Keen, M., Acharya, A., et al.: Patterns: Implementing an SOA Using an Enterprise Service Bus. IBM Redbooks (2004)

8. Bai, X., Xie, J., Chen, B., Xiao, S.: DRESR: Dynamic Routing in Enterprise Service Bus. In: ICEBE 2007: Proceedings of the IEEE International Conference on e-Business Engineering, pp. 528–531. IEEE Computer Society, Los Alamitos (2007)
9. Ziyaeva, G., Choi, E., Min, D.: Content-Based Intelligent Routing and Message Processing in Enterprise Service Bus. In: ICCIT 2008: Proceedings of the International Conference on Convergence and Hybrid Information Technology, pp. 245–249. IEEE Computer Society, Los Alamitos (2008)
10. XPath: W3C (2009), `http://www.w3schools.com/xpath/`
11. Fasbinder, M.: BPEL or ESB: Which should you use? IBM developerWorks (2008)
12. Chappell, D.: ESB Integration Patterns. SYS-CON Media, Inc. (August 2004)
13. Xie, J., Bai, X., Chen, B., Xiao, S.: A Survey on Enterprise Service Bus. Computer Science 34(11), 13–18 (2007)
14. Camel DSL: The Apache Software Foundation (2009), `http://camel.apache.org/dsl.html`
15. Hohpe, G., Woolf, B.: Enterprise Integration Patterns: Designing, Building, and Deploying Messaging Solutions. Addison Wesley, Reading (2003)
16. Travel Reservation Sample: Understanding the Travel Reservation Service, SUN (2007), `http://www.netbeans.org/kb/60/soa/understand-trs.html`

A Mechanism of Modeling and Verification for SaaS Customization Based on TLA*

Shuai Luan, Yuliang Shi, and Haiyang Wang**

Shunhua Road, High-tech Development Zone, Jinan, Shandong Province, P.R. China,
Shangdong University Software College
lscat@qlsc.sdu.edu.cn, liangyus@sdu.edu.cn, why@sdu.edu.cn

Abstract. With the gradually mature of SOA and the rapid development of Internet, SaaS has become a popular software service mode. The customized action of SaaS is usually subject to internal and external dependency relationships. This paper first introduces a method for modeling customization process based on Temporal Logic of Actions, and then proposes a verification algorithm to assure that each step in customization will not cause unpredictable influence on system and follow the related rules defined by SaaS provider.

Keywords: SaaS, TLA, Customization, Verification.

1 Introduction

Software as a service [1] (SaaS) is an emerging software delivery mode according to the development of Internet and software. SaaS eliminates need for enterprises to purchase, construct and maintain infrastructure and software. Client can acquire services through Internet, and pay according to the service usage amount and time. From technical perspective, the typical character of SaaS application is 'single-instance multi-tenant', that is multiple tenants share a single software instance. Tenants' services as well as data shared physically and isolated logically [2]. However, the uniform service mode cannot satisfy different tenants who have different business demands. So a successful SaaS application must support tenants' customization to make up their own business applications. The customization includes datafield, service, process and interface. Datafield customization means tenant can add new datafields or configure existing one in application. Service customization enables tenant to configure services provided by SaaS provider and add them to his own application. Process customization allows tenant to add and configure processes. Customizing the page display of application is called interface customization.

Though customization of SaaS facilitates tenants to develop their own applications according to business demands, it brings some challenges at the same time: Firstly, there are several dependency relationships between customized actions. Tenant usually

* The research is supported by the National Natural Science Foundation of China under Grant No.60673130 and No.90818001, the Natural Science Foundation of Shandong Province of China under Grant No.Y2006G29, No.Y2007G24 and No.Y2007G38.

** Corresponding author.

W. Liu et al. (Eds.): WISM 2009, LNCS 5854, pp. 337–344, 2009.

doesn't care them during customization and may do some customized actions which cause unpredictable influence on system; secondly, SaaS provider usually defines some rules to restrict tenant's customization. Tenant sometimes ignores the rules during customization, which may lead to incorrect result.

To guarantee the customized result correct, in this paper we develop a customized behavior model based on temporal logic for the problems. In model we describe the atomic customized action in TLA and guarantee the dependency relationship through valid logic. After that, we propose an algorithm to check tenant's customization and prevent the invalid action which may cause unpredictable influence on system.

The paper is organized as follows: Section 2 is related works. In Section 3, a brief review of Temporal Logic of Actions is provided. Section 4 presents how to model customized behavior base on TLA. Verification algorithm is described in section 5. Finally, we draw a conclusion and discuss future works.

2 Related Work

In present research, solutions proposed for personalized customization are mostly concerned with dynamic selection, on demand customization, dynamic alteration of process and integration with rules. Kuo ZHANG [3] proposes a SaaS-oriented service customization approach. It allows service providers to publish customization policies along with services. If tenant's customization requirement is in agreement with policy, providers will update service. This approach will greatly burden service providers. [4] describes SaaS from perspective of configuration and customization. It depicts the difference and their competency model under multi-tenant, giving the skeleton diagram of customized dependence relationship between different components in SaaS and presenting configuration policy execution framework.

Rules are increasingly used in process to achieve the goal of constraint and verification. In [5], a framework has been proposed to support the semi-automatic generation of business processes. User can choose an appropriate template, and then enrich it to generate BPEL schema, and then embeds business rules to generate an executable BPEL. This framework is able to help user orchestrate correct process, and dynamically choose services at runtime. Pietro Mazzoleni [6] introduces a business driven model to determine overall impact of a new requirement in a complex SOA environment. It uses fact rules to describe relationship between elements in business Layer and IT layer, and provides impact propagation rules. When a new requirement comes, it is convenient to determine which elements are impacted. However, this method lacks flexibility, and does not allow users to define rules on their own.

In recent years, various kinds of temporal logic emerge for modeling and analyzing real world systems. Temporal logic now has become not only an important tool of modeling but also a tool of model analysis because of its ability of reasoning. [7] uses TLA to model and analyze workflow, and validates the implication relationship between two formulas to find the deadlock. Afterward, [8] does some expanding work. It uses TLA to model the business process which is described by BPEL or OWL-S, and then uses the model checking technology to verify some key properties.

3 Temporal Logic of Action

This section presents a brief reviewed of Temporal Logic of Actions, abbreviated as TLA. For completed specification of Temporal Logic of Actions, please refer to [9]. Customization in SaaS involves process, service and datafield, which is discrete and ordered. Aimed on these, we use TLA to model and analyze tenant's behavior.

TLA is an important temporal logic to depict dynamic system. All TLA formulas can be expressed in terms of familiar mathematical operators (such as $\wedge$, $\vee$) plus four new ones:'(prime), $\Box$, $\Diamond$, $\exists$.'(prime) means reach the next state of prime state, $\Box$ represents *always*, $\Diamond$ represents *eventually*, and $\exists$ means *exist*. The semantics of TLA is defined in terms of *behavior*, *state*, and *action*. *Behavior* is an infinite sequence of states. *State* is an assignment of values to variables. *Action* is a boolean-valued expression formed from variables, primed variables, and constant symbols. We describe the syntax, formal semantics and axioms which may be used in this paper.

Syntax of TLA

$\langle formula \rangle \triangleq \langle predicate \rangle \mid \Box[\langle action \rangle]_{\langle state\ function \rangle} \mid \neg \langle formula \rangle \mid \langle formula \rangle \wedge \langle formula \rangle \mid \Box \langle formula \rangle$

$\langle action \rangle \triangleq$ boolean-valued expression containing constant symbols,variables,and primed variables

$\langle state\ function \rangle \triangleq$ nonboolean expression containing constant symbols and variables

Semantics of TLA

$s[\![f]\!] \triangleq f(\forall 'v' : s[\![v]\!]/v)$ $\langle s_0, s_1, \ldots \rangle$ $s[\![A]\!]t \triangleq A(\forall 'v' : s[\![v]\!]/v, t[\![v]\!]/v')$

Additional notation

$p' \triangleq p(\forall 'v' : v'/v)$ $\Diamond F \triangleq \neg \Box \neg F$ F *Leads to* $G \triangleq F \mapsto G \triangleq \Box(F \Rightarrow G)$

$[A]_f \triangleq A \vee (f' = f)$ $\langle A \rangle_f \triangleq A \wedge (f' \neq f)$ $SF_f(A) \triangleq \Box \Diamond \langle A \rangle_f \vee \Diamond \Box \neg Enabled \langle A \rangle_f$

p is a $\langle state\ function \rangle$ or $\langle predicate \rangle$ σ is a behavior f is a $\langle state\ function \rangle$ $s, s_0, s_1, \ldots$ are states

F and G are $\langle formula \rangle$s A is an $\langle action \rangle$ $(\forall 'v' : \ldots / v, \ldots / v')$ denotes substitution for all variables v

4 Modeling of Customization Based on Temporal Logic

In this section, we first propose a figure, which elaborates atomic customized action and dependence in each level, and then model each atomic customized action in TLA, and generate the complete customized behavior model at last.

4.1 Hierarchical Customized Behavior and Dependence Figure

Figure1 has listed all the customized action and dependency in process, service and datafield layer. Customized action in each layer not only influences other actions in this layer, but affects some ones in other layers. In addition, SaaS provider usually defines some extra rules in a special layer (we call it rule layer) to restrict some customized actions. Therefore, in the figure we use solid lines to represent internal dependency and broken lines to represent external constraint defined by provider, and then string discrete customized actions into an ordered behavior.

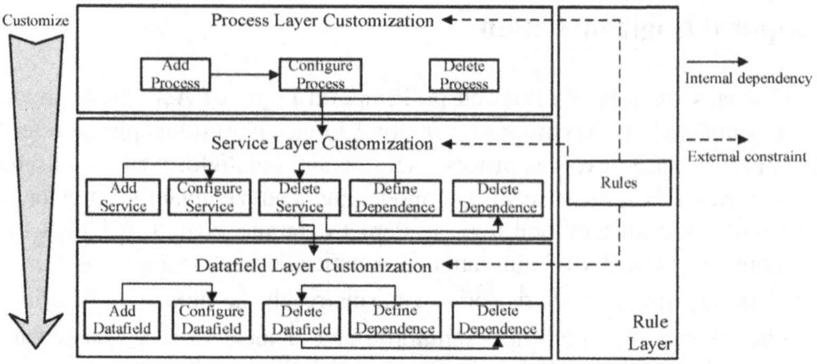

Fig. 1. Hierarchical customized behavior and dependence figure

In this paper, we consider each datafield can be seen as a special service (stored information) while each service can be seen as a special process (only one service), so whether datafield, service or process customization can be thought of beginning with process. As shown in figure 1, customized behavior is a top-down mode. Adding process leads to configure it, and configuring process means to customize each service and relationship between services. Configuration of service refers to datafields and relationship between them. The ordered behavior can also guarantee rationality of customization. In allusion to this, we model customization process based on temporal logic to assure that each step in customized behavior will not cause unpredictable influence on system and follow the related rules defined by SaaS provider.

4.2 Modeling of Customization Based on Temporal Logic of Action

To model customization process, we must describe semantic of customization process using temporal logic. We use the *behavior* to represent a whole customization process. The *states* describe state of process, service and datafield. We use *action* to represent atomic customized action. We define several sets based variables in TLA:

Definition 1

P, S, D represent the initial process set, service set and datafield set;

$P_{CIN}, S_{CIN}, D_{CIN}$ represent the process, service and datafield set being customized;

$P_{CED}, S_{CED}, D_{CED}$ represent the sets which have been customized;

S_{DEP}, D_{DEP} represent the service dependency set and datafield dependency set.

Now according to the actions, we give the corresponding description in TLA.

Process Layer

- Add process: Tenant can customize a new process. The important thing is that adding a process is bound to configure it. The action of adding process is based on the configuration of new process. The action is denoted by Add_P:

$$Add\_P(p_i) \triangleq (p_i \in P) \wedge (P'_{CED} = P_{CED}) \wedge (P'_{CIN} = P_{CIN} \cup \{p_i\})$$

Where p_i is the process which will be added, $\triangleq$ means *equals by definition.*

- Configure process: The action exactly means to add, configure, delete services and dependency relationships. Configuration of process is over iff (if and only if) all services in process have been customized. The action is denoted by $Config\_P$:

$$Config\_P(p_i) \triangleq \Box(\bigwedge_{k=1}^{m} SA(s_k) \mapsto CP(p_i))$$
$$CP(p_i) \triangleq (P'_{CIN} = P_{CIN} - \{p_i\}) \wedge (P'_{CED} = P_{CED} \cup \{p_i\})$$

Where SA is customized action of service layer, $s_1, s_2, \ldots s_m$ are services in p_i.

- Delete process: Deleting process is to delete all the services and service dependency relationships in process. The action is denoted by $Delete\_P$:

$$Delete\_P(p_i) \triangleq (p_i \in P_{CED}) \wedge (P'_{CIN} = P_{CIN}) \wedge (P'_{CED} = P_{CED} - \{p_i\})$$
$$\bigwedge_{k=1}^{n} Delete\_SDep(Depend\_S_k) \bigwedge_{k=1}^{m} Delete\_S(s_k)$$

$s_1, s_2, \ldots s_m$ are all the services in process p_i, $Delete\_SDep$ represents deleting service dependency relationships in process, $Depend\_S_k$ represent the service dependency relationships in process, $Delete\_S$ means to delete service.

The customized action of entire process layer can be expressed as PA:

Definition 2

$$PA \triangleq (Add\_P \mapsto Config\_P) \vee Config\_P \vee Delete\_P$$

Service Layer

- Add service: Tenant can select a service to his application. However, adding a service is bound to configure it. The action is denoted by Add_S:

$$Add\_S(s_i) \triangleq (s_i \in S) \wedge (S'_{CED} = S_{CED}) \wedge (S'_{CIN} = S_{CIN} \cup \{s_i\})$$

s_i represents the service to be added.

- Configure service: The action exactly means to add, configure, delete datafields and dependency relationships. The action is denoted by $Config\_S$:

$$Config\_S(s_i) \triangleq \Box(\bigwedge_{k=1}^{m} DA(d_k) \mapsto CS(s_i))$$
$$CS(s_i) \triangleq (S_{CIN} = S_{CIN} - \{s_i\}) \wedge (S_{CED} = S_{CED} \cup \{s_i\})$$

Where DA is customized action of datafield layer, $d_1, d_2, \ldots d_m$ are datafields in s_i.

- Define dependence: Tenant can define dependency relationships between services. It determines execution order and has transitivity. It is denoted by $Depend\_S$:

$$Depend\_S(s_i, s_j) \triangleq (s_i, s_j \in S_{CED}) \wedge (S'_{CIN} = S_{CIN}) \wedge (S'_{CED} = S_{CED})$$
$$\wedge (S'_{DEP} = S_{DEP} \cup \{s_i \mapsto s_j\})$$

s_i, s_j represent two services, and s_i should be executed before s_j.

$Depend\_SAL \triangleq \bigwedge_{i} Depend\_S_i$ represents all dependency relationships in process.

- Delete dependence: Tenant can delete the customized dependency relationship between services. The action is denoted by $Delete\_SDep$:

$$Delete\_SDep(Depend\_S(s_i, s_j)) \triangleq (S'_{CIN} = S_{CIN}) \wedge (S'_{CED} = S_{CED}) \wedge (S'_{DEP} = S_{DEP} - \{s_i \mapsto s_j\})$$

- Delete service: Deleting service is to delete all the datafields and datafield dependency relationships in service, and keep all service dependency relationships in process at the same time. The action is denoted by $Delete\_S$:

$$Delete\_S(s_i) \triangleq (S_i \in S_{CED}) \wedge (S'_{CIN} = S_{CIN}) \bigwedge_{k=1}^{n} Delete\_DDep(Depend\_D_k) \bigwedge_{k=1}^{m} Delete\_D(d_k) \wedge Depend\_SAL \wedge (S'_{CED} = S_{CED} - \{s_i\})$$

$d_1, d_2, \ldots d_m$ are all the datafields in service s_i , $Delete\_DDep$ represents deleting datafield dependency relationships in the service, $Depend\_D_k$ represent datafield dependency relationship in the service , $Delete\_D$ means to delete datefield.

The customized action of entire service layer can be expressed as SA :

Definition 3

$$SA \triangleq (Add\_S \mapsto Config\_S) \vee Config\_S \vee Delete\_S \vee Depend\_SAL \vee Delete\_SDep$$

Datafield Layer

- Add datafield: Tenant can add a datefield in service. It is denoted by Add_D:

$$Add\_D(d_i, d_{Ai}) \triangleq (d_{Ai} \in D_A) \wedge (D' \in D \cup \{d_i\}) \wedge (D'_{CIN} = D_{CIN} \cup \{d_i\})$$

d_i is the datafield to beadded, d_{Ai} represents data type, D_A represents all data types.

- Configure datafield: Tenant can configure the existing or customized datafield in service. The action is denoted by $Config\_D$:

$$Config\_D(d_i, d_{Ai}) \triangleq (d_{Ai} \in D_A) \wedge (D' = D) \wedge (D'_{CIN} = D_{CIN} - \{d_i\}) \wedge (D'_{CED} = D_{CED} \cup \{d_i\})$$

- Define dependence: Tenant can define dependency relationships between datafields. It is denoted by $Depend\_D$:

$$Depend\_D(d_i, d_j) \triangleq (d_i, d_j \in D_{CED}) \wedge (D'_{CIN} = D_{CIN}) \wedge (D'_{CED} = D_{CED}) \wedge (D'_{DEP} = D_{DEP} \cup \{d_i \mapsto d_j\})$$

d_i, d_j represent two datafields, and d_i should be handled before d_j.

$Depend\_DAL \triangleq \bigwedge_i Depend\_D_i$ represents all dependency relationships in service.

- Delete dependence: Tenant can delete the customized dependency relationship between datafields. The action is denoted by $Delete\_DDep$:

$$Delete\_DDep(Depend\_D(d_i, d_j)) \triangleq (D_{CIN} = D_{CIN}) \wedge (D'_{CED} = D_{CED}) \wedge (D'_{DEP} = D_{DEP} - \{d_i \mapsto d_j\})$$

- Delete datafield: Tenant can delete customized datafield in service, but should keep all datafield dependency relationships in service. It is denoted by $Delete\_D$:

$$Delete\_D(d_i) \triangleq (d_i \in D_{CED}) \wedge (D'_{CIN} = D_{CIN}) \wedge (D'_{CED} = D_{CED} - \{d_i\}) \wedge (D' = D - \{d_i\}) \wedge Depend\_DAL$$

The customized action of entire datafield layer can be expressed as DA:

Definition 4

$$DA \triangleq (Add\_D \mapsto Config\_D) \vee Config\_D \vee Delete\_D \vee Depend\_DAL \vee Delete\_DDep$$

To definition a customization behavior model, besides above actions, we need to describe another semantic which represents the initial state of customization.

Definition 5

$Init_\sigma \triangleq (P_{CIN} = P_{CED} = S_{CIN} = S_{CED} = S_{DEP} = D_{CIN} = D_{CED} = D_{DEP} = \Phi) \wedge D \wedge S \wedge (P = \Phi)$

Now we can define the customization behavior model as flow: A customization behavior model is described as a formula of TLA. The formula is denoted by σ:

$$\sigma = Init_\sigma \wedge \Box[A]_f \wedge SF_f(A)$$

$Init_\sigma$ represents the initial state of customization , $A \triangleq PA \wedge SA \wedge DA$ represents customized action which is composed of several atomic actions. $f \triangleq \langle P_{CIN}, P_{CED}, S_{CIN}, S_{CED}, S_{DEP}, D, D_{CIN}, D_{CED}, D_{DEP} \rangle$ represent state function which is to map states to values. We use state function f_{Rule} to describe rules provided by SaaS provider. If there are several rules, the customization must satisfy all of them.

5 Verification Algorithm

After modeling customized actions, we proposed a verification algorithm to verify tenant's customization. We first define the following theorem as foundation.

Theorem: For current system state set s (s is true) and customized action A, if existing state set t ($t \triangleq s[\![A]\!] \wedge Rules$) is true, the customized action A is valid. Where $Rules$ is rule set defined by SaaS provider, $Rules \triangleq \{Rule_1 \wedge Rule_2 \wedge ... \wedge Rule_n\}$. If the provider doesn't define any interrelated rules, $Rules \equiv true$.

```
For each p_m in A//Activity A customizes m processes
    t_p ≜ Init_σ[[PA_m]]
    For each s_i in p_m //process p_m have i services
        t_s ≜ Init_σ[[SA_i]]
        For each d_k in s_i //service s_i have k datafields
            t_{d_k} ≜ Init_σ[[DA_k]]
        end For
        t_{s_i} ≜ ∧_k t_{d_k} ∧ t_s
    end For
    t_{p_m} ≜ ∧_i t_{s_i} ∧ t_p
end For
t_1 ≜ ∧_m t_{p_m}
If t_1 = false            //Not meet
  then A is not valid   // Internal dependence
  else if t_1 ∧ Rules=false       //Not meet
        then A is not valid   //External dependence
     else A is valid
```

Algorithm 1. The verification algorithm

To prove the correctness of Theorem, we define a new state set $t_1 (t_1 \triangleq s[\![A]\!])$ which can be got by replacing variable states in s with states in A. If t_1 is true, A will not have friction with former system state set. That is, action A satisfies all the internal dependency relationships. Well, if $t_1 \wedge Rules$ is true, it means that A also satisfies external constraints defined by provider. Therefore, if $t \triangleq s[\![A]\!] \wedge Rules$ is true, A satisfies both internal dependency and external constraint, so action A is valid. According to Theorem, we use Algorithm1 to verify tenant's customization.

Complexity Analysis: Each atomic customized action is to change system state set, where state can be expressed as: $State = \{P_{CIN}, P_{CED}, S_{CIN}, S_{CED}, S_{DEP}, D, D_{CIN}, D_{CED}, D_{DEP}\}$.So each atomic action takes $O(n)$ time. Tenant's customized action A is usually composed of several atomic customized actions in process, service and datafield layer, so A takes $O(n^3)$ time. The provider defines finite rules to restrict system state set, so $Rules$ also takes $O(n)$ time. Therefore, the algorithm takes $O(n^3)$ time.

6 Summary

With gradually mature of SOA, SaaS has gradually become a popular software delivery mode. In this paper, we model tenant's customization process based on temporal logic and propose an algorithm. Through analyzing and computing the customized behavior model, the algorithm can judge whether tenant's customized action has satisfied all the dependency and constraint, and prevent the invalid action. The future work focuses on expanding the customized behavior model, and put the verification mechanism into a SaaS platform to generate a complete SaaS application.

References

1. Microsoft, Architecture Strategies for Catching the Long Tail (April 2006)
2. Stefan, A., Torsten, G.: Multi-Tenant Databases for Software as a Service: Schema-Mapping Techniques. In: ACM SIGMOD/PODS Conference, pp. 1195–1206 (2008)
3. Kuo, Z., Xin, Z.: A Policy-Driven Approach for Software-as-Services Customization. In: CEC/EEE 2007, pp. 123–130. IEEE Press, Tokyo (2007)
4. Wei, S., Kuo, Z.: Software as a Service: An Integration Perspective. In: Krämer, B.J., Lin, K.-J., Narasimhan, P. (eds.) ICSOC 2007. LNCS, vol. 4749, pp. 558–569. Springer, Heidelberg (2007)
5. Corradini, F., Meschini, G.: A Rule-Driven Business Process Design. In: 29th International Conference on Information Technology Interfaces, Cavtat, pp. 401–406 (2007)
6. Mazzoleni, P., Srivastava, B.: Business Driven SOA customization. In: Bouguettaya, A., Krueger, I., Margaria, T. (eds.) ICSOC 2008. LNCS, vol. 5364, pp. 286–301. Springer, Heidelberg (2008)
7. Yang, B., Wang, H.: Management as a service for IT service management. In: Bouguettaya, A., Krueger, I., Margaria, T. (eds.) ICSOC 2008. LNCS, vol. 5364, pp. 664–677. Springer, Heidelberg (2008)
8. Yuan, W., Yushun, F.: Using Temporal Logics for Modeling and Analysis of Workflows. In: CEC-East 2004, pp. 169–174. IEEE Press, Beijing (2004)
9. Lamport, L.: The temporal logic of actions. DEC Systems Research Center, Palo Alto, CA: Technical Report 79 (1991)

EPN-Based Web Service Composition Approach*

Min Gao and Zhongfu Wu

College of Computer Science, Chongqing University, Chongqing 400030, China
gaomin@cqu.edu.cn, wzf@cqu.edu.cn

Abstract. Web service composition is highly significant for the efficiency of software development in service-oriented architecture (SOA). Till now, most PN-based composition approaches are based on HPN (High Level Petri Net), but the resources of HPN are not suitable for web services. It leads to the inefficiency of those composition approaches. This paper presents a novel approach based on EPN (Elementary Petri Net) to the issue. The research has adapted EPN to device a method to model web services in order to facilitate the composition by an automatic process which is controlled by transition rules and conditions. Finally, an example is given to validate the applicability of the composition approach.

Keywords: SOA; Web Service Composition; Petri Net; Web Services.

1 Introduction

In recent years, many organizations have engaged their core business competencies with a collection of web service over the Internet. Web services provide a language-neutral, loosely-coupled, and platform independent way for linking applications within organizations or enterprises across the Internet [2]. Developers can then solve complex problems by combining available services. Lots of publications have presented the benefits of web service composition, particularly for heterogeneous applications integration and SOA.

A number of Web Service-oriented Architecture Description Languages (WSADLs), such as Web Services Flow Language (WSFL), Business Process Execution Language for Web Services (BPEL4WS), Web Service Choreography Interface (WSCI), and XLANG have been proposed and applied in both academe and industry. However, less of these WSADLs support automatic composition.

Petri Net (PN) is well-founded process modeling technique and has a useful formalism in the information technology industry, which is suitable for presenting complex services composition processes and analyses and verification. Many researchers have proposed high-level Petri Net-based approaches to the composition of web

* This work is supported by "National Social Sciences Foundation" of China under Grant No. ACA07004-08, Supported by "Science and technology" and "11th 5-year Plan" Programs of Chongqing Municipal Education Commission under Grant No. KJ071601 and 2008-ZJ-064. This research is an extended version of the paper presented at IEEE International Conference on Networking, Architecture, and Storage (IEEE NAS'08) [1].

W. Liu et al. (Eds.): WISM 2009, LNCS 5854, pp. 345–354, 2009.

services. GUO proposed a Colored Petri Net (CPN) model for web service composition [3], Hamadi proposed a Petri Net-based algebra for modeling web services control flows [4], and Petri Net patterns are developed for enhancing service composition modeling by Chi [5]. All these approaches are helpful for services composition. However the kind of resource which they described is different with web services. It leads to low efficiency. Consequently, a proper mechanism is crucial for services integration. To address the issues, this paper describes web services and dynamic composition based on EPN.

The rest of the paper is structured as follows. Section 2 discusses Petri Net-based services composition methods and their limitations. Section 3 describes an improved EPN model and a services model. Then Section 4 proposes a composition framework, composition algorithm, and a case study. Finally the last section draws the conclusions.

2 Background

Web services are software components that are developed using specific technologies, such as an XML-based description format, an application messaging protocol, and a collection or transport protocol. Each of the software components are treated as self-contained, self-describing, marked-up software resources that can be published and retrieved across the Internet [7]. A web service is normally defined to perform a specific task. Service composition is a process of selecting relevant web services, relating purposely the web services, and packaging the web services to meet the business needs. This process consists of complex procedures, conditions and actions driven by users' requirements. Petri Net possesses capabilities to handle such complexity of the composition process [2~6].

2.1 Petri Net

Petri net [6] is a mathematical representation of a workflow model which captures behavior composed of a finite number of *Place (P)* and *Transition (T)* between those places and actions. A *P* is associated with a *T* by an arc *F*. A set of *P*, *T*, and *F* will form a net. A message passing between *P*s is conducted by tokens in the input *P* firing condition and actions through *T* to the output *P*. All kinds of Petri net are based on a Net. A Net depicts the components of a distributed system [7]. A Net is composed by a group of tuple *(P, T, F)*. The set F is subject to the constraint of $P \cap T = \varnothing$, $P \cup T \neq \varnothing$ and $F \subseteq P \times T \cup T \times P$. Therefore

$$Net(P, T, F) \Leftrightarrow P \cap T = \varnothing \wedge P \cup T \neq \varnothing \wedge F \subseteq P \times T \cup T \times P \wedge dom(F) \cup cod(F) = P \cup T \quad (1)$$

Where $dom(F) \cup cod(F) = P \cup T \mid dom(F) = \{x \mid \exists y : (x, y) \in F\}$, $cod(F) = \{y \mid \exists x : (x, y) \in F\}$.

There are three kinds of Petri nets: Petri Net (PN), HPN and EPN.

Petri Net depicts dynamic information changing process in a distributed system. A 6-tuple $\sum = (P, T, F, K, W, M_0)$ is a Petri Net. K is a capability function of *N*, *W* is a weighing function of N, and M_0 is initial mark of N. In formalization, the definition of Petri Net is: $Petri\ Net\ (P, T, F, K, W, M_0) \Leftrightarrow Net(P, T, F) \wedge K{:}P \rightarrow N \cup \{\infty\} \wedge W{:}F \rightarrow N \wedge M_0{:}P \rightarrow N_0 \wedge \forall p \in P{:}\ M_0\ (p) \leq K\ (p)$. The net transition rules are complicated, details in [6].

HPN is even more complicated, which is some Petri net formalisms that extend the Petri Net. Maturescent high-level Petri Net includes Pr/T_ (predication and transition system), CPN (Colored Petri Net), hierarchical Petri net, TPN (Timed Petri Net). Most of them use a complicated P-element to express more than one kind of resource, and a complicated T-element to express more than one kind of transition. For example, Pr/T_ is a 9-tuple $\sum = (P, T, F, D, V, A_P, A_T, A_F, M_0)$, where: (P, T, F) is a *Net*; D is a finite set of all elements in $\sum$; V is a set of elements in D; A_p: $P \rightarrow \pi$, π is a variable set of predications; A_T: $T \rightarrow f_D$, f_D is a set of formulae in D; A_F: $F \rightarrow f_S$, and f_S is symbolic sum of D. M_0: $P \rightarrow f_S$. They are more complex than the common Petri Net is.

The advantage of EPN is the way by which the basic characters of distributed systems are described both conceptually and mathematically.

$$EPN = (B, E, F, C) \tag{2}$$

Where B is a set of behavior as P in (1), E is a set of events as T in (1), F is a set of flow relation (arcs), C is a set of condition, $c \subseteq B$ is a condition, and c_{in} represents an initial condition. The relation of EPN elements (2) can be expressed through

$^{\cdot}e = \{x | (x, e) \in F\}$ or $^{\cdot}e = F^{-1} \cap (B \times E)$

$e^{\cdot} = \{y | (e, y) \in F\}$ or $e^{\cdot} = F \cap (B \times E)$,

as illustrated in fig 1. An event $e \in E$ will be fired at condition set c iff $^{\cdot}e \subseteq c$ & $e^{\cdot} \cap c = \varnothing$. It is marked by $c[e>$, and means that c makes e fire. In another word, an event e is fired if and only if the inputs of e are *contained* by condition set c and the outputs of e are *disjoint* with c.

2.2 Petri Net-Based Approaches for Web Service Composition and Challenges

GUO [3] proposed a CP-net (Colored Petri Net) model for web service composition. For each service, a CP-net model is constructed to describe the logical relation of components graphically. Dynamic behaviors of services can be simulated and analyzed by the CP-net model.

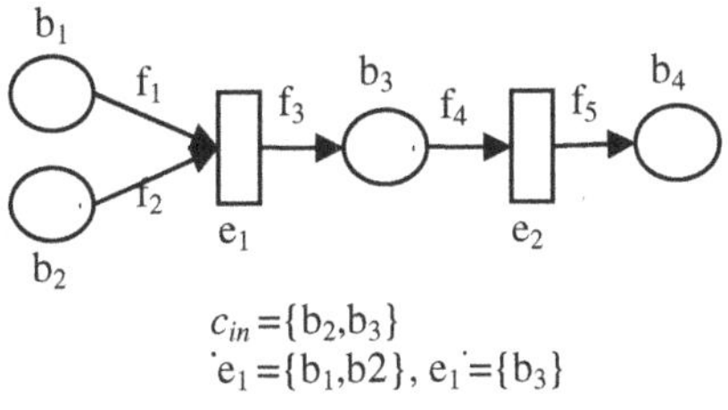

$c_{in} = \{b_2, b_3\}$

$^{\cdot}e_1 = \{b_1, b2\}$, $e_1^{\cdot} = \{b_3\}$

Fig. 1. An EPN Model

Hamadi [4] proposed Petri Net-based algebra for modeling web services control flows. The model is used to capture the semantic meaning of complex service combination. And the algebra is defined for the creation of dynamic and transient relationships among services.

Chi [5] proposed a composition platform based on a formal modeling language. Petri Net patterns are developed for enhancing service composition modeling, workflow quality measurement and scripting productivity.

All these approaches are very helpful for the services composition research. They are mostly based on high-level Petri Nets. Resources that Petri Net or High-level Petri Net described are material resources. Every place represents a kind of physical resources, and K(s) is a space resource (a kind of physical resources) which expresses capability of containing s resource. W (s, t) and W (t, s) represent the s resource numbers transition t consumed and produced. Hence, W indicates the relationship between transitions and substantial resources.

However, the resources in service composition are information resources instead of substantial resources. Hence, K and W functions are useless for services composition. They make the transition rules complex and lead to the low efficiency in composition. So an efficient model method is urgent needed by web service composition.

Every information resource has only two states that are "having" and "lack". Every place at most has a token in EPN. If a place does not have a token, it means "false" state, otherwise it means "true" state. So the places in EPN are appropriate to describe web services.

3 An EPN-Based Approach for Web Service Modeling

Information resources will not decrease or disappear when they are used. It is different to transition rules of EPN. So we redefine the transition rules and events' relationship calculation methods.

3.1 Improved EPN Transition Rule

Definition 3.1. Improved transition rules (TR)

$$\text{An event } e \in E \text{ is fired at condition set } c \textit{ iff } {}^{\bullet}e \subseteq c. \tag{3}$$

It is marked by $c[e>$. It is also called that c makes e fire. It means an event e is fired if and only if the inputs of e are *contained* by condition set c.

If $c[e>$ then the input states of e will not be changed but the output states of e become true and the condition set c becomes subsequence condition set c', marked by $c[e>c'$, $c' = c \cup e^{\bullet}$.

It means two things will happen when an event is fired. Firstly, all tokens will be kept. Secondly, new tokens will be added in conditions indicated by arrows that originate from the transition.

Let $e_1, e_2 \in E$, there are only two relationships between e_1 and e_2 at c: sequence and concurrency. Because there is not competition, if ${}^{\bullet}e \subseteq c$, ${}^{\bullet}e$ will always have token, token will not disappear when it is used. If $c[e_1> \wedge c[e_2>$ then $c[\{e_1, e_2\}>$.

Definition 3.2. Relationships between events
Let e_1, $e_2 \in E$, e_1 and e_2 can be related at c in three ways: sequence, conflict, and concurrency.

Sequence means e_1 can occur at c but not e_2, however, after e_1 has occurred e_2 can occur. In formalization, e_1 and e_2 are in sequence at c *iff* $c[e_1> \wedge \neg (c[e_2>) \wedge c' [e_2>$ where $c[e_1>c'$.

Conflict means e_1 and e_2 can occur individually at c but they cannot occur together at C. In formalization, e_1 and e_2 are in conflict at c *iff* $c[e_1> \wedge c[e_2> \wedge \neg (c[\{e_1, e_2\}>)$.

Concurrency means e_1 and e_2 can occur at C without interfering with each other. Moreover no order is specified over their occurrences. In formalization, e_1 and e_2 can occur concurrently at c *iff* ${}^{\bullet}e_1 \cup {}^{\bullet}e_2 \subseteq c$.

3.2 EPN-Based Web Service Model

A web service behavior is basically a partially ordered set of operations. Therefore, it is directly mapped into an improved EPN-based web service composition model (WSC). Services operations are modeled by events and the states of services are modeled by behaviors. The arrows between behaviors and events are used to specify relationships. The formal representation of WSC is (4)

$$WSC = \{B, E, F, C\} + \mathrm{TR} \tag{4}$$

Where $\{B, E, F\}$ is a directional net which is composed of behaviors, services, and directional arcs;

C is a set of condition as C in (2).

B: $b \rightarrow \{true, false\}$ is a finite set of behaviors that represents inputs and outputs of a service.

$E = \{s_1, s_2 \ldots s_n\}$ is a finite set of services operations.

$F \subseteq \{<{}^{\bullet}s, s>, <s, s^{\bullet}>\}$ includes only two kinds of arcs: One is service inputs to service, and another is service to service outputs.

TR determines which services can be enabled in the condition set c as c in (3):

An event $s \in E$ is fired at condition set c *iff* ${}^{\bullet}s \subseteq c$, marked by $c[s>$.

If $c[s>$ then the inputs states of s will be kept and the outputs states of s will become true, and the condition set c becomes subsequence condition set c', marked by $c[s>c'$, where $c' = c \cup s^{\bullet}$.

Definition 3.3. Sequence and parallel relationships
Given s_1, $s_2 \in E$ is two different service operations s_1 and s_2, the sequence relationship represents s_1 followed by s_2 and parallel relationship represents s_1 and s_2 can be executed in parallel.

Iff $c[s_1> \wedge \neg c[s_2> \wedge c'[s_2> \wedge c[s_1>c'$ then s_1 and s_2 have sequence relationship in the condition c, marked by $s_1 \cdot s_2$. s_1 and s_2 have parallel relationship *iff* ${}^{\bullet}s_1 \cup {}^{\bullet}s_2 \subseteq c$, marked by $s_1 \,||\, s_2$.

Definition 3.4. Liveness of WSC
The property of liveness of WSC is used to check whether the system will reach a deadlock. Let $r= \{(c, c') \mid c, c' \in C \wedge \exists s \in E: c[s>c'\}$ then r* means reach-ability relation where $r^* = r^0 \cup r^1 \cup r^2 \cup \ldots = \bigcup_{i=0}^{\infty} r^i$. If $\forall c \in C, \forall s \in E, \exists c' \in C: c\ r^* c \wedge c'\ [s>$ then *WSC* is *live* (no deadlock).

4 A WSC-Based Composition Algorithm and a Case Study

4.1 WSC Model

In our composition framework, service providers and requesters do not communicate with each other. Services are required, matched, composed, and executed by an agent Mediator. Mediator models services and requirements and then eliminates different names for the same semantic meanings. Then it uses a heuristic forward chain algorithm of Artificial Intelligent to compose services.

Suppose that every service has only an operation. When Mediator models service, a service is modeled by an event *s*, and its inputs and outputs are represented by $^{\bullet}s$ and $s^{\bullet}$. There are only two kinds of arcs in a single WSC model: one is from input to service, and another is from service to output. So a single service can be simply represented by formal description $s\ (^{\bullet}s;\ s^{\bullet})$.

For example, a given weather service named WS_1, its inputs are a book's name and an author's name, and its output is the introduction of the book or books. The service can be represented by *WS_1 (Book_name, Author_name; Book_Info)*. Fig.2 illuminates its *WSC* model.

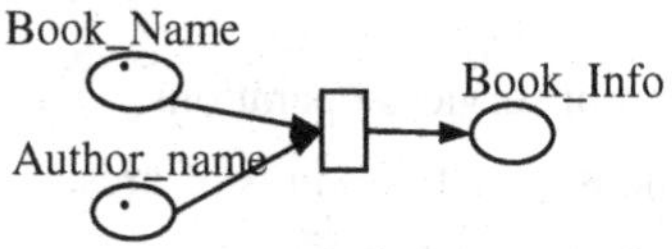

Fig. 2. WSC Model

Finally, mediator deduces which services will be used and the usage sequence according to users' inputs. If the inputs of a model are matched with the user's inputs, then the event can be fired. Then Mediator adds its outputs to user input set. These steps will be looped until all user requirements are in the input set.

4.2 Heuristic Forward Chain Composition Algorithm

The mediator applies forward reasoning heuristic linearity search algorithm. The detailed steps are as following.

(1) To model every candidate service to a *WSC* model.
(2) To put user's input data into initial condition set c_{in}; if $b \in c_{in,}$ add a token to place b; let *c* equal $c_{in,}$.
(3) To match initial condition set $c_{in,}$ with $^{\bullet}s$. If no $^{\bullet}s \subseteq c$, then no service can be fired at condition set c ($\neg\exists\ c[s>$), composition failed and the algorithm is over; else

some services can be execute, add their *WSC* model to composed *WSC* model, add $s^\bullet$ to condition set c.

(4) To match condition set c to user's goal set. If the goal is reached then reasoning is successful, and exit the algorithm; else execute step (5).

(5) To match condition set c with $^\bullet s$. If there is not $^\bullet s \subseteq c$, then no service can be fired at condition set c ($c[s>$ does not exist), composition is failing and the algorithm is over; else some services can be executed, go to step (6).

(6) If exist $c[s>$, according to avoiding backdating policy, add the WSC model to the composed WSC model; connect them by $^\bullet s$; add $s^\bullet$ to c. Go to step (4).

Composition Algorithm: *NonbacktraceWebServicesComposition*

UserInputs: a user's input;
UserOutputs: the user's goal;
C: known data set;
c_{in}: initial data set;
WebServicesSet($^\bullet s;s^\bullet$): all service set;
UsedWebServices: composition services set;
WSC NonbacktraceWSComposition(*UserInputs*, *UserOutputs*, *WebServicesSet*)
{Model every candidate service to a *WSC* model, form *WebServicesSet*($^\bullet s;s^\bullet$) to a *WSC* set;
Initial c_{in} = *UserInputs*, *UsedWebServices*= Null;
Choose a service s satisfied $^\bullet s \subseteq c_{in}$ from *WSC* set;
if (s does not exist) then { Output: Composing is failed; break};
else if ($s^\bullet \subseteq c$) then delete s from *WSC* set;
else { Add the *WSC* model of s to the composed *WSC* model;
Add all $s^\bullet$ to C, $C=C+s^\bullet$;
Add s into *UsedWebServices*;
Delete s from *WSC* set; }
Repeat
{Choose a service s satisfied $^\bullet s \subseteq c_{in}$ from *WSC* set;
if (s does not exist) then { Output: Composing is failed; break};
else if ($s^\bullet \subseteq c$) then delete s from *WSC* set;
else
{ Add the *WSC* model of s to the composed *WSC* model, connected them by $^\bullet s$;
Add all $s^\bullet$ to C, $C=C+s^\bullet$;
Add s into *UsedWebServices*;
Delete s from *WSC* set;
}
} Until *UserOutputs*$\subseteq C$;
Output: Composition is successful;
Output: *UsedWebServices* is the service set that satisfied the user's requirement;
}

4.3 A Case Study of Service Composition

Suppose that a user inputted a company name and time, and he wanted to know the company's location, the company's detail information, and the company stock's value (Chinese currency) no matter where stock market is. There are eight available web services.

S_1 National Information Service: *INS(CountryName;CountryCode, Info)*
S_2 Country Information Service: *CIF(CountryCode; CountryName, Info)*
S_3 Pounds for Dollars : *UStoUK(USprice, DateTime;UKprice)*
S_4 RMB for Dollars: *UStoRMB(USprice, DateTime;RMBprice)*
S_5 RMB for Pounds: *UKtoRMB(UKprice, GivenDateTime;RMBprice)*
S_6 London Stock Query: *LondonStock(CompanyID, SomeDatetime;UKprice)*
S_7 NewYork Stock Query: *NewYorkStock(CompanyID, GivenDatetime; USprice)*
S_8 Company Yellow Pages: *YellowPages(CompanyName; CompanyID, Address, CountryID, Phone,Info)*

According to the user's requirement, the *UserInputs* is {*CompanyName*, *DateTime*} and the *UserOutputs* is {*CountryName*, *CountryInfo*, *RMBprice*}. The initial Condition Set is {*CompanyName*, *DateTime*}. Dynamic composing steps are as following:

(1) Because the *UserInputs* {CompanyName,DateTime} only can fire S_8, add S_8 to *WebServicesUsed* set.
(2) After executing S_8, there are tokens in CompanyID, Address, CountryID, Phone and Info. It means we get these values. Add them to *InputsSet* and remove S_8 from *WebServicesSet*.
(3) The *InputsSet* cannot achieve *UserOutputs*:{CountryName, CountryInfo, RMBprice}. Continue to discover which transitions can be fired, S_2, S_6, and S_7 can be fired, add them to *WebServicesUsed*.
(4) After executing S_2, S_6, and S_7, there are tokens in CountryInfo, Ukprice, and Usprice. Add them to *InputsSet*, and *InputsSet* becomes {CompanyName, DateTime, CompanyID, Address, CountryID, Phone, CompanyInfo, CountryInfo, Ukprice, Usprice}. Remove S_2, S_6, and S_7 from *WebServicesSet*.
(5) The *InputsSet* cannot achieve *UserOutputs*. Continue to find which transitions can be fired, S_1, S_3, S_4, and S_5 can be fired. The *Outputs* of S_1 and S_3 are CountryCode and CountryInfo which already exist in InputsSet. So directly remove the services from *WebServicesSet*. And then add S_4 and S_5 to WebServicesUsed.
(6) Now *InputsSet* becomes {CompanyName, DateTime, CompanyID, Address, CountryID, Phone, CompanyInfo, CountryInfo, Ukprice, Usprice, RMBprice} which have satisfied *UserOutputs*, the composing is successful. We get the composition services and service executing order.
(7) Finally, *WebServicesUsed*={S_8,S_2,S_6,S_7,S_4,S_5}. Fig.3 is the composition result.

Finally, we get the graphic web service composition model. The user requirement can be achieved by S_8, S_2, S_6, S_7, S_4, and S_5. We analyzed this model using the analysis methods mentioned in section 3. The composition model is live and no deadlock which means every service can be executed.

- The accessibility property justifies that the goal condition set {*CountryName*, *CountryInfo*, *RMBprice*} is realizable.

- There are $c_{in}[S_8>$and $\neg c_{in}[S_2>\wedge\neg c_{in}[S_6>\wedge\neg c_{in}[S_7>$ and $(c[S_2>\wedge c[S_6>\wedge c[S_7>)$. So c is subsequence set of c_{in} marked $c_{in}[S_8>c$. It means there are *sequence* relationship between S_8 and $\{S_2,S_6,S_7\}$. $\dot{S}_6\cup\dot{S}_7\subseteq c$ and $\dot{S}_4\cup\dot{S}_5\subseteq c'$, where c' is subsequence set of c. This means there are *sequence* relationship between $\{S_6, S_7\}$ and $\{S_4,S_5\}$. $\dot{S}_2\cup\dot{S}_6\cup\dot{S}_7\subseteq c$ means there are *parallel* relationships between S_2, S_6, and S_7. $\dot{S}_4\cup\dot{S}_5\subseteq c'$ means there are *parallel* relationships between S_4 and S_5. Finally the composition service mathematical expression is: $S_8\cdot(S_2||S_6||S_7)\cdot(S_4||S_5)$.

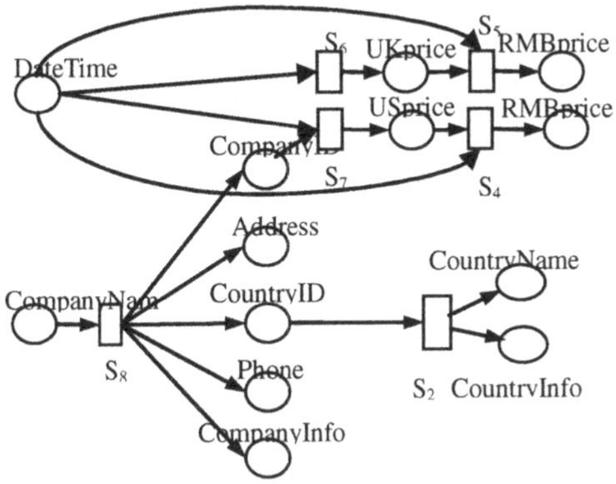

Fig. 3. A Composed Model

5 Conclusions and Future Work

In this paper, we proposed an EPN-based approach to characterise services and their composition with a new transition rule and algorithm. After modelling every service, we dynamically composed services according to the match level of places. The composition was transformed to a typical AI problem and solved by a heuristic forward chain reasoning algorithm. In the end, an example was given to illuminate the composition approach.

The WSC model facilitates the automatic composition of web services, but it is at the outset. There are still some challenging issues, such as quantitatively performance analysis and comparison for the EPN-based composition algorithms, importing "time" to EPN transition rule for the temporal characteristic of the composition, and introducing the precondition and effect, and taking the quality of web services into consideration.

References

[1] Gao, M., Sun, L., Wu, Z.F., Jiang, F.: An EPN-based method for web service composition. In: Proceedings of the IEEE International Conference on Networking, Architecture, and Storage, pp. 163–164 (2008)

[2] Fu, Y., Dong, Z., He, X.: An Approach to Web Services Oriented Modeling and Validation. In: IW-SOSE, Shanghai, China (2006)
[3] Guo, Y.-B., Du, Y.-Y., Xi, J.-Q.: A CP-Net Model and Operation Properties for Web Service Composition. Chinese Journal of Computers 29, 1067–1074 (2006)
[4] Hamadi, R., Benatallah, B.: A Petri Net-based Model for Web Service Composition. In: Conferences in Research and Practice in Information Technology, Adelaide, Australia. Australian Computer Society (2003)
[5] Chi, Y.-L., Lee, H.-M.: A formal modeling platform for composing web services. Expert Systems with Applications (2007)
[6] Petri, C.A.: Kommunikation mit Automaten. University of Bonn (1962)
[7] Petri, C.A.: Nets, time and space. Theoretical Computer Science 153, 3–48 (1996)
[8] Aalst, W.V.D.: Don't go with the flow: Web service composition standards exposed. IEEE Intelligent System 15, 72–85 (2003)
[9] Charif, Y., Sabouret, N.: An Overview of Semantic Web service composition Approaches. Electronic Notes in Theoretical Computer Science 146, 33–41 (2006)
[10] Koehler, J., Srivastava, B.: Web service composition: Current solutions and open problems. In: Proceedings of the 13th International Conference on Automated Planning & Scheduling, Trento, Italy (2003)
[11] Lee, J., Wu, C.-L.: A Possibilistic Petri-Nets-Based Service Discovery. In: Proceedings of the 2004 IEEE International Conference on Networking, Sensing & Control, Taipei, Taiwan (2004)
[12] Milanovic, N., Malek, M.: Current solutions for Web service composition. IEEE Internet Computing 8, 51–59 (2004)
[13] Murata, T.: Petri Nets: Properties, Analysis and Applications. Proceedings of the IEEE 77, 541–580 (1989)
[14] Paolucci, M., Kawamura, T., Payne, T.R., Sycara, K.: Importing the semantic Web in UDDI. In: Proceedings of Web services, E-business and semantic Web workshop, Toronto, Canada (2002)
[15] Qian, Z.-Z., Lu, S., Li, X.: Automatic Composition of Petri Net Based Web Services. Chinese Journal of Computers 29, 1057–1066 (2006)
[16] Zhu-Zhong, Q., Sang-Lu, L., Li, X.: Automatic Composition of Petri Net Based Web Services. Chinese Journal of Computers 29(7), 1057–1066 (2006)
[17] Jensen, K.: An Introduction to the Theoretical Aspects of Coloured Petri Nets. In: A Decade of Concurrency: Reflections and Perspectives: REX School/symposium, Noordwijkerhout, the Netherlands, June 1-4 (1993)
[18] Rao, J., Su, X.: A Survey of Automated Web Service Composition Methods. In: Cardoso, J., Sheth, A.P. (eds.) SWSWPC 2004. LNCS, vol. 3387, pp. 43–54. Springer, Heidelberg (2005)
[19] Ezpeleta, J., Colom, J.M., et al.: A Petri net based deadlock prevention policy for flexiblemanufacturing systems. IEEE Transactions on Robotics and Automation 11(2), 173–184 (1995)
[20] Majithia, S., Shields, M., et al.: Triana: a graphical Web service composition and execution toolkit. In: IEEE International Conference on Web Services, 2004. Proceedings, pp. 514–521 (2004)
[21] Dustdar, S., Schreiner, W.: A survey on web services composition. International Journal of Web and Grid Services 1(1), 1–30 (2005)

Quantum CSMA/CD Synchronous Communication Protocol with Entanglement

Nanrun Zhou, Binyang Zeng, and Lihua Gong

Department of Electronics Information Engineering, Nanchang University,
Nanchang, 330031, China
znr21@163.com

Abstract. By utilizing the characteristics of quantum entanglement, a quantum synchronous communication protocol for Carrier Sense Multiple Access with Collision Detection (CSMA/CD) is presented. The proposed protocol divides the link into the busy time and leisure one, where the data frames are sent via classical channels and the distribution of quantum entanglement is supposed to be completed at leisure time and the quantum acknowledge frames are sent via quantum entanglement channels. The time span between two successfully delivered messages can be significantly reduced in this proposed protocol. It is shown that the performance of the CSMA/CD protocol can be improved significantly since the collision can be reduced to a certain extent. The proposed protocol has great significance in quantum communication.

Keywords: Quantum Synchronous Communication, Quantum Entanglement, EPR Pairs, CSMA/CD.

1 Introduction

The classical computer networks have witnessed impressive progress in the past decades. The advent of multimedia data applications and the growing use of the Internet, the telephone, and other network services have brought an increasing transmission rates demand. The need for higher transmission rates has led to the progressive implantation of quantum networks. The developments of the current communication networks and the advanced state of current photonic technology have permitted the experimental realization of quantum cryptography, the transmission of information in superdense codes, the teleportation, and the quantum network communication. Quantum networks play a key role in quantum information processing [1], [2]. In such quantum networks, quantum states can be prepared initially and shared between neighboring stations, i.e., entanglement can be generated, and their resource is then to be used for quantum communication [3], or distributed quantum computation involving arbitrary stations of the network. The works on quantum networks attract much attention, and a lot of schemes for quantum networks communication have been proposed by exploiting quantum entanglement, quantum entanglement swapping, single photons, and continuous variable quantum state. One of the main tasks is then to design protocols which use

W. Liu et al. (Eds.): WISM 2009, LNCS 5854, pp. 355–362, 2009.

the available quantum correlations to entangle two stations of the network, and to realize these protocols in current and future quantum networks.

By exploiting quantum entanglement, many tough issues which could not be solved in classical ways can be well resolved with quantum methods. In 1993, Bennett et al. proposed the first quantum teleportation scheme with an unknown single-particle state [4], which has attracted wide attention in recent yeas. In 1997, Bouwmeester et al. first realized quantum teleportation experimentally [5] which will be a critical ingredient for quantum computation networks. Gisin et al. proposed two quantum methods to resolve Byzantine problem effectively which can not be solved in classical ways [6], [7]. By employing the non-locality of Einstein-Podolsky-Rosen (EPR) correlation pairs and quantum teleportation, a novel quantum synchronous communication protocol to resolve the two-army problem effectively was proposed in Ref. [8]. Based on quantum entanglement correlation, Zhou et al. [9] presented a quantum communication protocol for data link layer to enhance the maximum throughput and the performance of the stop-and-wait protocol, which has great significance in quantum communication. There will be inevitable collisions in the process of message transmission with the classical CSMA/CD protocol.

It takes time and energy to deal with collisions in the classical CSMA/CD protocol, which influences the efficiency of the classical one to a certain extent. Moreover, there are few references about how to improve the performance of the CSMA/CD protocol with quantum methods. In this paper, we will first review the classical CSMA/CD protocol and then propose a CSMA/CD quantum synchronous communication protocol by utilizing quantum entanglement. From the simplified workflow of quantum CSMA/CD protocol, we can find that the quantum version one has great improvement. Finally, we will give an analysis of the performance of the classical CSMA/CD protocol and the quantum version one.

2 Classical CSMA/CD Protocol

CSMA is an improved protocol of ALOHA system with additional hardware interface. When a sending station has messages to send, it first listens to the bus to see whether anyone else is transmitting at the moment. If the bus is busy, the sending station waits until it becomes idle. While the sending station detects an idle bus, it transmits messages. If a collision occurs, the sending station waits a random amount of time and starts all over again.

CSMA/CD is a developed protocol of CSMA with the additional function of transmitting jamming signal to other sending stations immediately. Since it is possible for two sending stations to listen to the bus at the same time and to detect an idle bus, the two sending stations could then transmit message at the same time. When this occurs, a collision will take place and then a jamming signal is sent throughout the bus in order to notify all sending stations of the collision. The sending stations will wait for a random time which has been worked out with back-off algorithm before retransmitting their messages. In this way, the bus can be fully used and the performance of the CSMA can be improved. The workflow of the CSMA/CD protocol is shown in Fig.1.

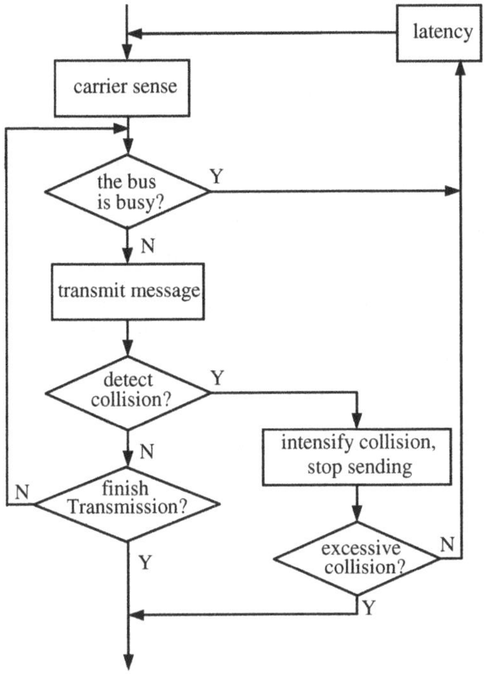

Fig. 1. Workflow of CSMA/CD protocol

3 Quantum CSMA/CD Synchronous Communication Protocol

The proposed protocol divides the link into the leisure time and the busy one, where the message is sent via classical channel and the preparation and distribution of EPR correlation pairs are supposed to be completed at the leisure time. The transmission of quantum information can be instantaneous according to Refs. [9] and [10]. Suppose quantum acknowledge (QACK) frames are believable, each sending station and the bus share quantum ACK frames in different quantum states $|\varepsilon\rangle_m = \alpha_m|0\rangle + \beta_m|1\rangle$ (m represents different stations, $|\alpha_m|^2 + |\beta_m|^2 = 1$). The EPR entanglement pairs shared by the sending station and the bus at the leisure time are in the fixed state $|\phi^-\rangle = \frac{1}{\sqrt{2}}(|01\rangle - |10\rangle)$. When the bus has a quantum ACK frame to send, the bus utilizes the equipment which can distinguish Bell states $|\psi^\pm\rangle = \frac{1}{\sqrt{2}}(|01\rangle \pm |10\rangle)$ and $|\varphi^\pm\rangle = \frac{1}{\sqrt{2}}(|00\rangle \pm |11\rangle)$ to make a joint measurement on the system composed of the particles $|\varepsilon\rangle_m$ and $|\phi^-\rangle$. One can obtain

$$|\varepsilon\rangle_m \otimes |\phi^-\rangle = \frac{1}{2}\big[|\psi^-\rangle(-\alpha_m|0\rangle - \beta_m|1\rangle) + |\psi^+\rangle(-\alpha_m|0\rangle + \beta_m|1\rangle) \\ + |\varphi^-\rangle(\beta_m|0\rangle + \alpha_m|1\rangle) + |\varphi^+\rangle(-\beta_m|0\rangle + \alpha_m|1\rangle)\big] \quad (1)$$

The measurement result of the bus is in one of the four Bell states with equal probability $1/4$. When it comes to the quantum ACK frame measurement time, each sending station measures the particle entangled with the bus, if the bus has sent out the quantum ACK frame, the measurement result must be in one of the four states $-\alpha_m|0\rangle \pm \beta_m|1\rangle$ and $\alpha_m|1\rangle \pm \beta_m|0\rangle$ with equal probability $1/4$. If it is, the bus allows the sending station to send message; if not, the sending station will wait until the bus becomes idle. If the bus hasn't sent out the quantum ACK frame, the measurement result is $\frac{1}{\sqrt{2}}\begin{bmatrix}1\\0\end{bmatrix}$ or $\frac{1}{\sqrt{2}}\begin{bmatrix}0\\1\end{bmatrix}$. If it is, the sending station will wait and resend a request frame for communication. The six states $-\alpha_m|0\rangle \pm \beta_m|1\rangle$, $\alpha_m|1\rangle \pm \beta_m|0\rangle$, $\frac{1}{\sqrt{2}}\begin{bmatrix}1\\0\end{bmatrix}$ and $\frac{1}{\sqrt{2}}\begin{bmatrix}0\\1\end{bmatrix}$ can be distinguished under appropriate conditions shown in Refs. [10] and [11]. Therefore, the time span between two successfully delivered date frames can be significantly reduced since the collision can be reduced to a certain extent.

3.1 Protocol Description

The simplified workflow of quantum version of CSMA/CD is shown in Fig.2 and the details of quantum version one are as follows:

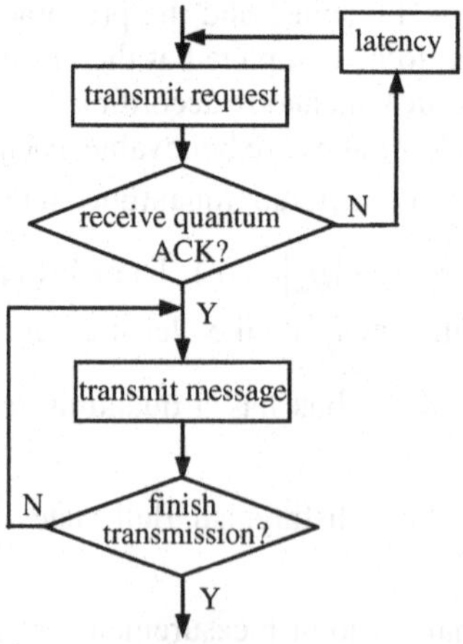

Fig. 2. Workflow of CSMA/CD synchronous communication protocol

S1) At leisure time, each sending station sets the QACK measurement time γ, the random delay time δ and sends a request frame to the bus for communication.

S2) After receiving the request, the bus notes down all request frames order sent by each sending station and decides if it is currently available. If available, the bus feeds back the quantum ACK frame to the sending station whose request frame received by the bus at first and permit the sending station to send message; if not, the bus defers the attempt until the end of the current carrier event.

S3) When it comes to the measurement time γ, each sending station carries out quantum ACK frame measurement to see whether the bus has sent out the quantum ACK frame or not. If it is, the sending station sends message; if not, the sending station will wait and resend a request frame after the random delay time δ.

S4) After accomplishing message transmission, the sending station notifies the bus at the end of its transmission. Then the bus feeds back the relevant quantum ACK frame to the next sending station according to the request frames order and turns to S3).

S5) The receivers prepare to receive message momentarily.

By repeating the above steps, messages can be sent to receivers efficiently.

3.2 Performance Analysis

By transmitting quantum ACK frames to inform each sending station of the information whether the bus is occupying or not, collision can be reduced effectively. Therefore, the workflow in Fig.2 is simpler than in Fig.1.

As for the classical CSMA/CD protocol, suppose there are N sending stations on the bus, each sending station with the probability p to send messages, the end-to-end transmission time is τ, the contention period is defined as two times of τ, the transmitting time of a message is T_f. The average time of sending a data frame, including contention period time, message transmitting time and interval time, is noted as T_{av}. The total average time of sending a message successfully is shown in Fig.3.

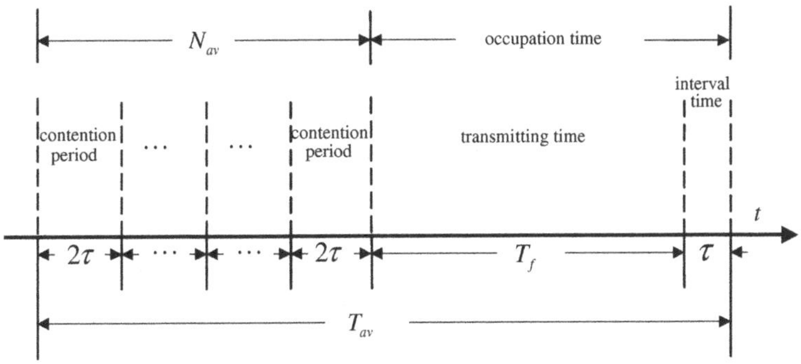

Fig. 3. Average time T_{av} of sending a message successfully

It is shown in Fig.3 that the average time T_{av} of sending a message successfully is composed of the average contention periods N_{av} and the occupation time, in which the occupation time contains transmitting time and interval time. The interval time is determined by the end-to-end transmission time τ. Suppose A is the probability of sending a message successfully, then one can obtain

$$A = Np(1-p)^{N-1}. \tag{2}$$

The contention period appears when a message fails to send. In fact, there are a number of contention periods before sending a message successfully and the probability of sending a message after n contention periods is $(1-A)^n A$. The times of resending a message N_r are equal to the average contention period N_{av}. According to the definition of mathematical expectation, the average contention period on the channel is the expectation N_{av} of A with the probability $(1-A)^n A$, one can obtain

$$N_r = N_{av} = \sum_{n=0}^{\infty} n(1-A)^n A = \frac{1-A}{A}. \tag{3}$$

The normalized throughput S [12] on the bus with the classical CSMA/CD protocol is

$$S = \frac{T_f}{T_{av}} = \frac{T_f}{2\tau N_r + T_f + \tau} = \frac{1}{1+\alpha(2A^{-1}-1)}. \tag{4}$$

where the parameter α is the ratio of the end-to-end transmission time τ to the transmitting time of a message T_f, i.e., $\alpha = \tau / T_f$. From Eq. (4) one can learn that if one wants to get the maximum normalized throughput, the value of A must be on the maximum. From Eq. (2), one can know that only the parameter $p = 1/N$ can get the maximum value of A, noted as A_{max}

$$A_{max} = \left(1 - \frac{1}{N}\right)^{N-1}. \tag{5}$$

When $N \to \infty$, $\lim_{N\to\infty} A_{max} = 1/e \approx 0.368$. In fact, when $N = 20$, the value of A approaches to A_{max}. So the maximum normalized throughput value S_{max} under ideal conditions is

$$S_{max} = \frac{1}{1+4.44\alpha}. \tag{6}$$

Eq. (6) indicates that S_{max} increases while α decreases.

As for the CSMA/CD quantum synchronous communication protocol, by employing quantum ACK frame, the bus can notify each station whether the bus is occupying or not, while collision may occur with the classical CSMA/CD protocol due to the propagation delay of ACK frames. Let N, p, T_f and τ be the same to the classical CSMA/CD protocol. Because collision can be reduced effectively in the proposed quantum CSMA/CD protocol, the average time of sending a message $T_{av}^{'}$ reduces by $2\tau N_r$. So the normalized throughput $S^{'}$ in our proposed CSMA/CD protocol is

$$S^{'} = \frac{T_f}{T_{av}^{'}} = \frac{T_f}{T_f + \tau}. \tag{7}$$

Compared Eq. (7) with Eq. (4), one can get

$$\lambda = \frac{S^{'}}{S} = 1 + \frac{2\tau N_r}{T_f + \tau} > 1. \tag{8}$$

From Eq. (8), it is seen clearly that the normalized maximum throughput of the proposed protocol was enhanced effectively when the communication process is under non-ideal conditions, i.e., $p \neq 1/N$ and N is limited.

For the ideal communication process, i.e., $p = 1/N$ and $N \to \infty$, $S_{\max}$ is determined by the parameter α with the classical CSMA/CD protocol. The greater of α means the smaller of the normalized throughput $S_{\max}$. With the distance increasing, the time τ grows up as well as the parameter α, which results in a smaller maximum normalized throughput. Therefore, it is not suitable for telecommunication with the classical CSMA/CD protocol. Suppose message transmitting time T_f is invariant, the time τ can be ignored in the proposed protocol. For the transmission of quantum ACK frame is instantaneous, τ is close to 0, i.e., $\alpha \to 0$. Then the maximum normalized throughput $S_{\max}^{'}$ in the proposed protocol can approach to 100% in theory. In fact, it is impossible to achieve the upper limit because it needs time to carry out measurement on the entangled particles, while ways can be done to reduce the measurement time with modern methods. It is obvious that the proposed protocol is more efficient and feasible than the classical CSMA/CD protocol.

4 Conclusion

We introduced a quantum synchronous communication protocol for CSMA/CD by utilizing quantum entanglement. The proposed protocol divides the link into the leisure time and the busy one, where the message is sent via classical channel and the preparation and distribution of EPR pairs are completed at leisure time. The transmission of quantum information can be instantaneous; therefore, the collision can be reduced to a certain extent. It is shown that the time span between two successfully delivered messages can be significantly reduced and the performance of CSMA/CD protocol can be improved effectively. So our protocol is a more efficient and feasible one with current techniques. In fact, we get the maximum throughput only under the ideal conditions. Therefore, the actual maximum throughput of the proposed protocol should be associated with the practical channels.

Acknowledgement. This work was supported by the National Natural Science Foundation of China (10647133), the Natural Science Foundation of Jiangxi Province (2007GQS1906), the Research Foundation of the Education Department of Jiangxi Province ([2007]22), the Key Project in the 11th Five-Year Plan on Education Science of Jiangxi Province (07ZD017), the Innovation Project of Jiangxi Graduate Education (YC08A026).

References

1. Cirac, J.I., Zoller, P., Kimble, H.J., Mabuchi, H.: Quantum State Transfer and Entanglement Distribution among Distant Nodes in a Quantum Network. Phys. Rev. Lett. 78, 3221–3224 (1997)
2. Boozer, A.D., Boca, A., Miller, R., Northup, T.E., Kimble, H.J.: Cooling to the Ground State of Axial Motion for one Atom Strongly Coupled to an Optical Cavity. Phys. Rev. Lett. 97, 083602-1-4 (2006)
3. Braunstein, S.L., Van, L.P.: Quantum Information with Continuous Variables. Rev. Mod. Phys. 77, 513–577 (2005)
4. Bennett, C.H., Brassard, G., Crepeau, C., Jozsa, R., Peres, A., Wootters, W.K.: Teleporting an Unknown Quantum State via Dual Classical and Einstein-Podolsky-Rosen Channels. Phys. Rev. Lett. 70, 1895–1899 (1993)
5. Bouwmeester, D., Pan, J.W., Mattle, K., Eibl, M., Weinfurter, H., Zeilinger, A.: Experimental Quantum Teleportation. Nature 390, 575–579 (1997)
6. Fitzi, M., Gisin, N., Maurer, U.: Quantum Solution to the Byzantine Agreement Problem. Phys. Rev. Lett. 87, 217901-1-4 (2001)
7. Fitzi, M., Gisin, N., Maure, U., Von, R.O.: Unconditional Byzantine Agreement and Multiparty Computation Secure against Dishonest Minorities from Scratch. In: Knudsen, L.R. (ed.) EUROCRYPT 2002. LNCS, vol. 2332, pp. 482–501. Springer, Heidelberg (2002)
8. Zhou, N.R., Zeng, G.H., Zhu, F.C., Liu, S.Q.: The Quantum Synchronous Communication Protocol for Two-Army Problem. Journal of Shanghai Jiaotong University 40, 1885–1889 (2006)
9. Zhou, N.R., Zeng, G.H., Gong, L.H., Liu, S.Q.: Quantum Communication Protocol for Data Link Layer Based on Entanglement. Acta Phys. Sin. 56, 5066–5070 (2007) (in Chinese)
10. Zhang, Y.D.: Principles of Quantum Information Physics. Science Press, Beijing (2006)
11. Peres, A.: How to Differentiate between Non-Orthogonal States. Phys. Lett. A. 128, 19 (1988)
12. Xie, X.R.: Computer Network. Publishing House of Electronics Industry, Beijing (2008)

Development of EPA Protocol Information Enquiry Service System Based on Embedded ARM Linux

Daogang Peng, Hao Zhang, Jiannian Weng, Hui Li, and Fei Xia

College of Electric Power and Automation Engineering
Shanghai University of Electric Power
Shanghai 200090, China
jypdg@163.com, hzhangk@yahoo.com.cn, xmulollipop@gmail.com,
elmerlee@163.com

Abstract. Industrial Ethernet is a new technology for industrial network communications developed in recent years. In the field of industrial automation in China, EPA is the first standard accepted and published by ISO, and has been included in the fourth edition IEC61158 Fieldbus of NO.14 type. According to EPA standard, Field devices such as industrial field controller, actuator and other instruments are all able to realize communication based on the Ethernet standard. The Atmel AT91RM9200 embedded development board and open source embedded Linux are used to develop an information inquiry service system of EPA protocol based on embedded ARM Linux in this paper. The system is capable of designing an EPA Server program for EPA data acquisition procedures, the EPA information inquiry service is available for programs in local or remote host through Socket interface. The EPA client can access data and information of other EPA equipments on the EPA network when it establishes connection with the monitoring port of the server.

Keywords: EPA standard; Industrial Ethernet; Information inquiry service; Embedded ARM; Embedded Linux.

1 Introduction

Industrial Ethernet is a newly-developed technology for industrial network communications. Since the international Fieldbus standard of IEC 61158(Third Edition) was published in 2000, big multinationals of automation companies took an active part in developing their own technology of industrial Ethernet and gave out a series of techniques and standards, such as ProfiNET, Ethernet/IP, MODBUS/TPRS, EPA, EtherCAT, PowerLink, Vnet, etc. In the field of industrial automation in China, EPA is the first standard accepted and published by ISO, and has been included in the fourth edition IEC61158 Fieldbus of NO.14 type. EPA is abbreviated to "Ethernet for Plant Automation". It was proposed in order to realize the seamless integration of information at all network levels of integrated automation system in industry based on Ethernet and wireless technology, which includes management layer, monitor layer, field device layer etc. According to EPA standard, field devices such as industrial

W. Liu et al. (Eds.): WISM 2009, LNCS 5854, pp. 363–372, 2009.

field controller, actuator and other instruments are all able to realize communication based on Ethernet standard.

Embedded system is an application-oriented system, which can adapt to the strict requirements of functions, reliability, cost, size and power consumption etc. Embedded Linux is applied widely nowadays as one of the most promising directions Linux technology is supposed to step into. It has unique advantages on its open source code, wide range of technique support, excellent stability and reliability, efficient operation, scalability and compatibility for many embedded hardware. In this sense, ARM Linux is outstanding enough for ARM platform. Being a target system, ARM Linux is used to transplant Linux kernel into an ARM-based processor. In this paper, an information inquiry service system of EPA standard based on the embedded ARM Linux is designed with Atmel AT91RM9200 development board as its target board. Local or remote users are able to access data and information of other EPA equipments with the help of this system.

2 EPA Standard Industrial Ethernet Technology

2.1 EPA Standard Summary

EPA is short for "Ethernet for Plant Automation". "EPA system architecture and communication specification for use in industrial control and measurement systems" (Hereinafter referred to as "EPA standard") funded by China National HI-TECH Project for priority subjects, has been the first national fieldbus standard GB/T20171-2006 with independent intellectual property. It was formulated by a group consisting of Zhejian University and other corporations, academes, colleges and universities as the standard of real-time Ethernet communication for plant automation.

In March, 2005, EPA was published as the standardization document of Publicly Available Specification IEC/PAS 62409 with 95.8% of approval among IEC/SC65C member countries. It is the first industrial automation standard formulated by China with its own intellectual property.According to the voting result released by International Electrotechnical Commission on October 5th, 2008, international real-time Ethernet standard application rules of IEC61784-2 and international fieldbus standard IEC61158-300/400/500/600 have passed the FDIS ballot with 100% of approval, which means that they have been qualified to be published as international standards. EPA standard protocol was listed as TYPE14 in IEC61158-300/400/500/600(Data Link Layer Services/ Data link layer protocol Application Layer Services/Application layer protocol), while the applied technology of EPA was include in the international real-time Ethernet standard IEC61784-2 as CPF14.

With the standardization of EPA, resources involved in developing EPA and its standards will be constantly increasing. Concerns from users in the mean time are experiencing a big rise to promote the development of EPA standard.

2.2 EPA System Architecture

With the reference to interconnection model of ISO/OSI open system, EPA standard adopts the first, second, third, fourth and seventh layer based on which the eighth

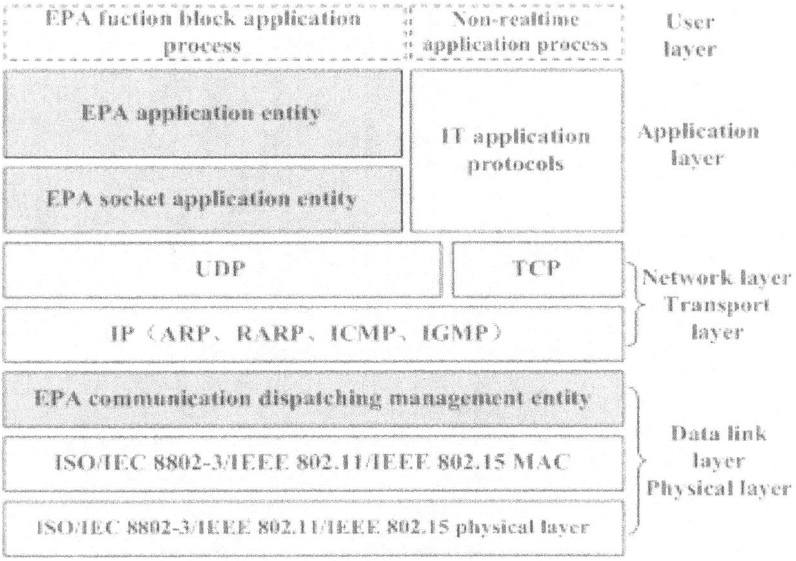

Fig. 1. EPA communication protocol model

layer(user layer) is added, so that the communication model is made up of six layers in all, as shown in Figure 1.

Apart from protocol components such as ISO/IEC 8802-3/IEEE 802.11/IEEE 802.15、TCP (UDP) /IP、SNMP、SNTP、DHCP、HTTP、FTP, the architecture of EPA system is composed of other parts as follows:

1) Application process, which includes application process of EPA function block and Non-real-time application process.

2) EPA application access entity.

3) EPA communications dispatching management entity.

EPA Communications dispatching management entity at the data link layer can help to avoid collision on the network, reducing the uncertainty in the process of Ethernet communication, which makes it to meet requirements of field-level communication on the industrial network through its real-time property.

Each EPA equipment is made up of at least one function block entity, EPA application access entity, EPA system management entity, EPA Sockets mapping entity, EPA link object, EPA communication dispatching management entity and UDP / IP protocol etc.

EPA function block is based on the basic model of IEC61499 (standard of industrial process measurement and control system function block). Each equipment on the network functions by coordinating with each other. An application process can go with one or more equipment. And a device may includes one completed application process or parts of it.

EPA system architecture provides a system framework to demonstrate how many different devices communicate with each other, how they exchange data and how to configurate. Figure 2 presents the process of communication between two EPA devices.

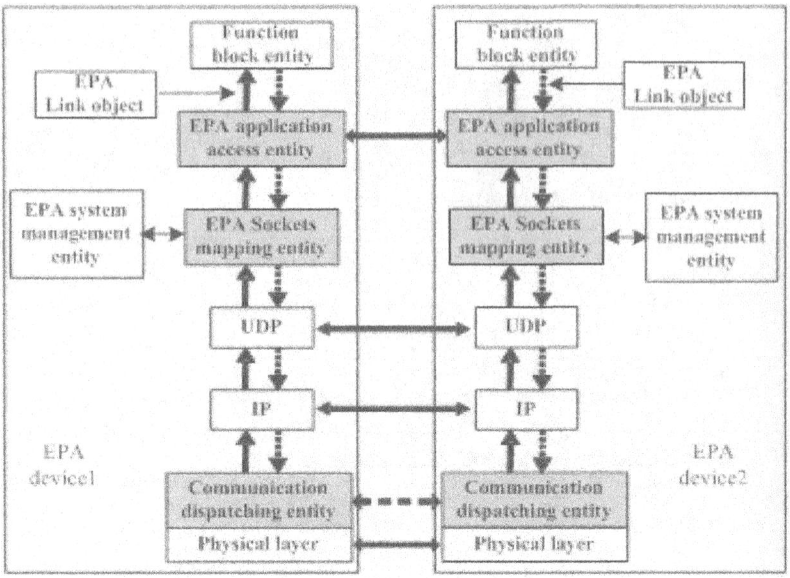

Fig. 2. Communication between EPA devices

2.3 Architecture of Distributed Network Control System Based on EPA Standard

EPA is a distributed system making use of networks of ISO/IEC8802-3、IEEE802.11 、IEEE802.15 protocol in order to connect devices, small systems and control and monitoring devices in the field. Through such a system, devices are likely to operate all together, implementing measurement and controlling in the process of industrial production. It boasts of openness, consistency, and easy integration.

The design of the architecture of distributed network control system based on EPA standard follows the principle of being "dispersed overall but centralized in particular parts". The design divides industrial control network into two levels: process monitoring network and industrial field device network, as shown in Figure 3. Process monitoring network as backbone of the system is used to connect control devices such as engineer station, operator station, data server, device for manufacturing execution systems etc. It is constituted by redundant high-speed Ethernet.

EPA control system is consisted of EPA main device, EPA field device, EPA network bridge, EPA proxy, EPA wireless access device, and so on. EPA main device is a device used in process monitoring layer L2 network segment. It has EPA communication port and doesn't need to have control module or some other function module. EPA main device is generally means configuration, monitoring device or man-machine interface of EPA control system. EPA field device is a device used in the industrial field, such as transmitter, actuator, switch, data acquisition device and field control device, and so on. EPA field device needs EPA communication entity and includes one function module at least. EPA network bridge is a device used to connect

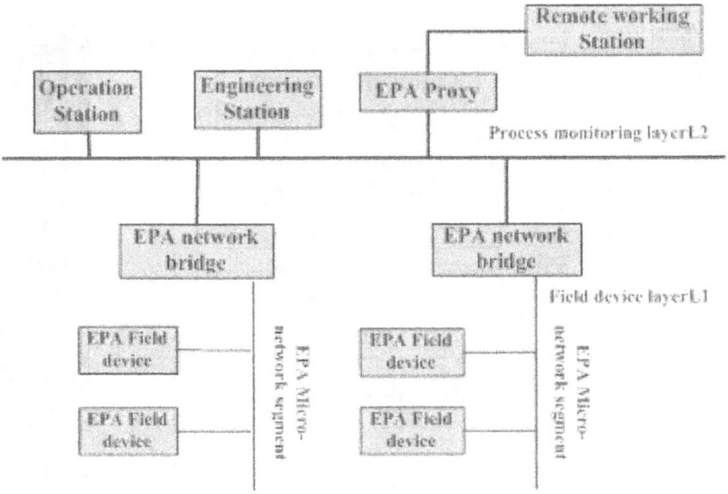

Fig. 3. Topology graph of EPA system

micro-network segment with other micro-network segment. One EPA network bridge must at least have two communication interfaces, separately connected to two micro-network segment. EPA network bridge can be configured by users and it has functions of communication isolation, message transponder, and control.

3 EPA Query Service System Based on ARM and Linux

3.1 The General Design of System

In order to make users access EPA device data easier, this paper designs a EPA server service program using EPA data acquisition program based on embedded ARM and Linux. Local program or some other remote host can call the function of EPA information inquiry service through Socket. This makes the program have more performance of reusability and scalability. Once EPA client connected to the monitoring port of server, it can call the function of EPA information inquiry service. Figure4 is the system structure of EPA Server.

EPA Server program uses Socket(SOCK_STREAM) to guarantee the reliability of connection. Server program is made up of four parts: main program, EPA module, communication module and command parsing module.

1) Main program is used to initialize the address information of Server, creates socket, binds it to server address, and then converses it into monitoring socket, waiting for receiving. When the server receives a new connection, it creates a sub-process, which serves for this new connection. Parent process still waits on monitoring socket. This executing mode is called "concurrent server". It means allocating one server process for every client process.

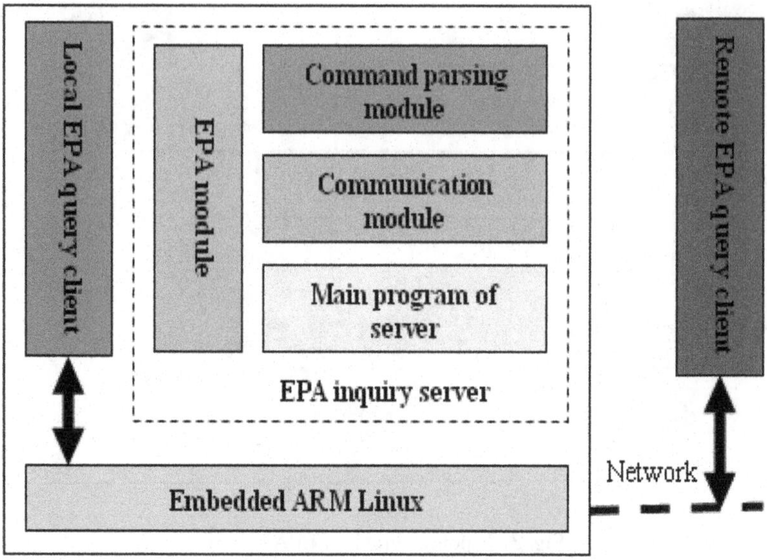

Fig. 4. Diagram of system structure of EPA Server

2) EPA module is used to monitor EPA protocol on the network. It creates a data format called Socket to monitor EPA message of the net, and saves the receiving data into buffer waiting for user query. EPA module mainly provides three functions:

epa_list_ip function: List all IP address of EPA device monitored by EPA module.

epa_get_type function: Return the type of EPA device with certain IP. It has a parameter used to query IP address of device.

epa_get_io function: Return the input and output value of EPA device with certain IP. It has two parameters. One is for IP address and the other is for device type.

3) Communication module is used to write and read socket data which provides two functions:

read_line function: Read one line from socket buffer. This paper uses "\n" as the end of data record. Reading one line means receiving a data record.

send_line function: Write a line into socket buffer. Also uses "\n" as the end of data record here. Sending a line means sending a data record.

4) Command parsing module is used to parse data taken out from socket, and then call the corresponding EPA module function. Only providing a function here:

Query function: Parsing out function names and parameters from the input string and calling corresponding EPA module functions.

3.2 EPA Server Main Program Design

EPA Server main program needs to create a Socket for external service. Some related data definitions of main program are as following:

#define EPASEVPORT 9999 //monitoring port of the Server□used when client connected

#define MAXUSER 5 //the maximal connection requests at the same time

The definition of variables:

```
int svr_lis_sock,svr_new_sock; //Define monitoring Socket(lis_sock) and Socket
  (new_sock)used in data transmition
struct sockaddr_in svr_local_add;         //Local address
struct sockaddr_in svr_remote_add;        //Client address
socklen_t svr_remote_addrlen;
int epa_lis_sock;   //Define EPA monitoring Socket
int epa_con_count;                        //device amount counting
data_node * epa_conlist_head=NULL; //define the head of device information
struct sockaddr_in epa_local_add;         //epa local address
struct sockaddr_in epa_remote_add; //epa client address
```

After the initialization of variables by main program, a Socket to monitor EPASEVPORT port is created according to general steps. If the connection is initiated by client, a new Socket will be generated to deal with the requests of client. Close this Socket when EPA module processing result is returned.

3.3 The Program Design of EPA Module

According to EPA standard, EPA devices use UDP/IP transmit its data unit of application layer protocol, so, the programming this module creates a data format called Socket to monitoring the EPA information frame on the network. EPA module should realize three functions: epa_list_ip, epa_get_type and epa_get_io. EPA module also needs to create a device information table to save the latest device data. The data structure of the node is as following:

```
typedef struct m_data_node{
struct in_addr ip;          //IP address of the device
u_int8_t        type;       //the types of the device
u_int32_t       time;       // update time of the device data
u_int8_t data[MAX_DATA]; //latest data of the device
struct m_data_node *next;
}data_node,*lp_data_node;
```

The related definition of some data in EPA module is as following:

Predefinition:

```
#define EPAPORT 35004          //Port number of  EPA device message
#define MAX_DATA 200           //the size of EPA device message
#define MAX_CON 50   //the amount of EPA device
```

The definition of function:

```
u_int16_t epa_list_ip(data_node *,u_int8_t *);
u_int16_t epa_get_type(data_node *,const struct in_addr);
void epa_get_io(data_node *,const struct in_addr,u_int8_t *,u_int16_t);
```

EPA module creates a data format called Socket to monitor the EPA message of EPAPORT port. And then, according to EPA message format, they are parsed into EPA device data, and later saved into device information table. When receiving the command of calling command parsing module, take the corresponding data of EPA module information table and return them to command parsing module. Figure 5 shows this process.

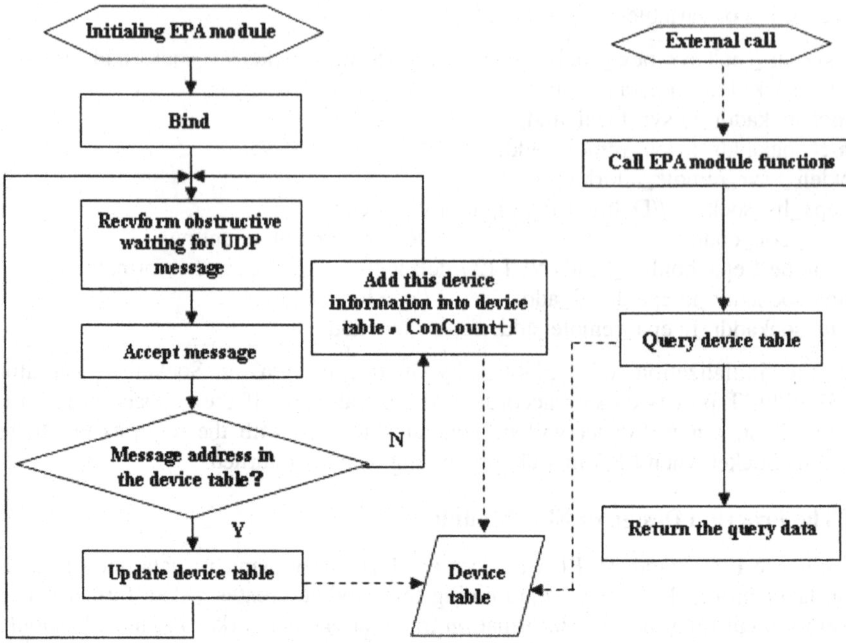

Fig. 5. EPA module running process

3.4 The Program Designing of EPA Commands Parsing

The command parsing module of EPA parses the input commands. The definition of function is as following:

```
int query(char * query_str, char * res_buff);
```

The return value of the function is a pointer to character string. Parameter query_str is the query character string and res_buff is the return data. The function returns the length of return data and a form for querying character string can be used as follows：

COMMAND(@IP ADDRESS)\n

For example : QVC@127.0.0.1\n or ALLIP\n

ALLIP, QIPTYPE, QVC, QDA can be used as COMMAND，which with the optional parameter in brackets and IP ADDRESS as the IP address of EPA device.

ALLIP: res_buff returns the IP list of all the online device and the keyword "#" is used for spacing every IP address.

QIPTYPE: res_buff returns the type of EPA device, and the return value is VC(VC module)or DA(DA module) etc.

QVC: res_buff returns the analog input value of EPA device.

QDA: res_buff returns the analog output value of EPA device.

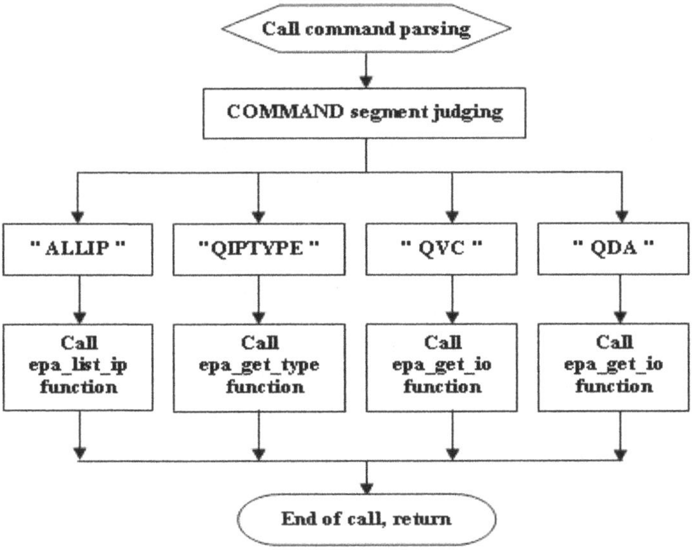

Fig. 6. EPA command parsing process

4 Conclusions

Industrial Ethernet is a new technology for industrial network communications developed in recent years. In the field of industrial automation in China, EPA is the first standard accepted and published by ISO, and has been included in the fourth edition IEC61158 Fieldbus of NO.14 type. According to EPA standard, Field Devices such as industrial field controller, actuator and other instruments are all able to realize communication. Embedded system is an application-oriented system, which can adapt to the strict requirements of functions, reliability, cost, size and power consumption etc. Embedded Linux is applied widely nowadays as one of the most promising directions. It has unique advantages on its open source code, wide range of technique support, excellent stability and reliability, efficient operation, scalability and compatibility for many embedded hardware. An information inquiry service system of EPA standard based on the embedded ARM Linux is designed with Atmel AT91RM9200 development board as its target board in this paper. Local or remote users are able to access data and information of other EPA equipments with the help of this system.

Acknowledgments

This work was supported by High Technology Research and Development Program of China (No. 2007AA041301), Program of Shanghai Subject Chief Scientist (No.09XD1401900) and Natural Science Foundation of Shanghai (No. 09ZR1413300).

References

1. Feng, D.Q., Yu, H.B., Jin, J.X., et al.: EPA Rea-time Ethernet and Its Standardization. Process Automation Instrumentation 26(9), 1–3 (2005)
2. Gao, L., Yu, H.B., Wang, H., et al.: Architecture of EPA Network. Computer Engineering 30(17), 81–83 (2004)
3. Li, H., Zhang, H., Peng, D.G.: Design and Research of Embedded Data Processing Platform on Power Equipments. Journal of Information & Computational Science 5(4), 1875–1883 (2008)
4. Peng, D.G., Zhang, H., Yang, L., et al.: Design and Realization of Modbus Protocol Based on Embedded Linux System. In: The 2008 International Conference on Embedded Software and Systems Symposia, Chengdu, Sichuan, China, July 29-31, pp. 275–280 (2008)
5. Zhou, K., Zhang, H., Wang, X.P.: A design of embedded-system in data acquisition and data monitoring based on PowerPC. Mechatronics 13(3), 30–33 (2007)
6. Peng, D.G., Zhang, H., Zhang, K., et al.: Research and Development of the Remote I/O Data Acquisition System Based on Embedded ARM Platform. In: 2009 International Conference on Electronic Computer Technology, Macau, China, February 20-22, pp. 341–344 (2009)
7. Wang, J.F., Zhang, H., Peng, D.G.: Design of ARM-based embedded remote monitoring systems. East China Electric Power 36(2), 139–142 (2008)
8. Peng, D.G., Zhang, H., Jiang, J.N.: Design and Realization of Embedded Web Server Based on ARM and Linux. Mechatronics 14(10), 37–40 (2008)

EasyKSORD: A Platform of Keyword Search Over Relational Databases

Zhaohui Peng[1], Jing Li[1], and Shan Wang[2,3]

[1] School of Computer Science and Technology, Shandong University, 250101 Jinan, China
[2] Key Laboratory of Data Engineering and Knowledge Engineering, Ministry of Education, 100872 Beijing, China
[3] School of Information, Renmin University of China, 100872 Beijing, China
pzh@sdu.edu.cn, soumer@mail.sdu.edu.cn, swang@ruc.edu.cn

Abstract. Keyword Search Over Relational Databases (KSORD) enables casual users to use keyword queries (a set of keywords) to search relational databases just like searching the Web, without any knowledge of the database schema or any need of writing SQL queries. Based on our previous work, we design and implement a novel KSORD platform named EasyKSORD for users and system administrators to use and manage different KSORD systems in a novel and simple manner. EasyKSORD supports advanced queries, efficient data-graph-based search engines, multiform result presentations, and system logging and analysis. Through EasyKSORD, users can search relational databases easily and read search results conveniently, and system administrators can easily monitor and analyze the operations of KSORD and manage KSORD systems much better.

Keywords: Relational Database, Keyword Search, Information Retrieval.

1 Introduction

Keyword Search Over Relational Databases (KSORD) enables casual users to use keyword queries (a set of keywords) to search relational databases just like searching the Web, without any knowledge of the database schema or any need of writing SQL queries[1, 2]. There are many studies on KSORD, which can be categorized into two types according to the search mechanism, schema-graph-based and data-graph-based. The former includes BANKS[3,4], ObjectRank[5], NUITS[6,7], and etc., while the latter includes DBXplore[8], DISCOVER[9], IR-Style[10], SEEKER[11], SPARK[12], and etc.

Based on our previous work, we design and implement a novel KSORD platform named EasyKSORD for users and system administrators to use and manage different KSORD systems in a novel and simple manner. Compared with the previous single KSORD systems, EasyKSORD has improvements in input, output, search engine and system management. EasyKSORD supports advanced queries, efficient data-graph-based search engines, multiform result presentations, and system logging and analysis. Through EasyKSORD, users can search relational databases easily and read search results conveniently, and system administrators can easily monitor and analyze the operations of KSORD and manage KSORD systems much better.

W. Liu et al. (Eds.): WISM 2009, LNCS 5854, pp. 373–382, 2009.

2 Architecture

The architecture of EasyKSORD is shown in figure 1. In input, EasyKSORD supports simple keyword queries as well as advanced keyword queries with conditions; in output, EasyKSORD provides three kinds of result presentation: List, TreeCluster, and S-CBR. Furthermore it supports users to use a relevance feedback method named VSM-RF to reformulate a new query and acquire result lists containing more relevance results. As of the search engine, EasyKSORD integrates backward expansion algorithm and two-way expansion algorithm in BANKS system and DPBF algorithm in DETECTOR system to provide choices for users. In system management, EasyKSORD can log users' operations and system executions, and provide batch processing and analysis for system administrators.

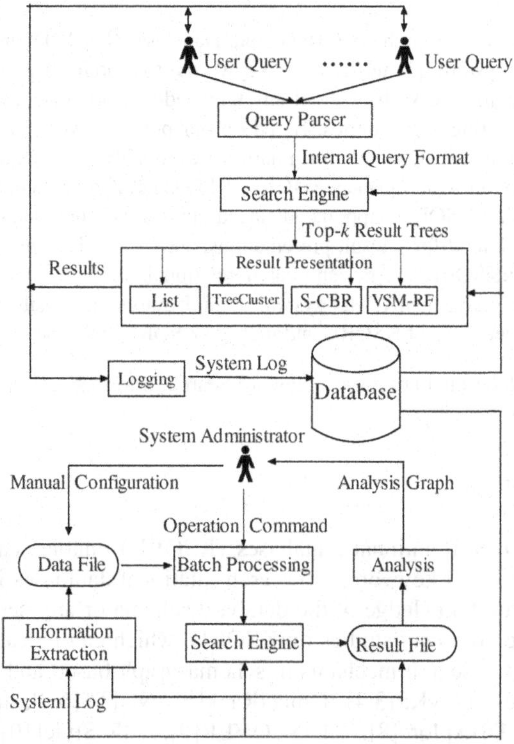

Fig. 1. The EasyKSORD architecture

EasyKSORD adopts browser/server architecture. Users submit keyword queries and view the search results in the browser-side. EasyKSORD concurrently executes queries issued by different users. Keyword queries submitted by users are first parsed by the query parser module, which transforms various queries (simple queries and advanced queries) into an internal format, and then the search engine processes the internal query to generate top-*k* result trees. EasyKSORD shows these result trees according to the users' choice on result presentation.

The execution of each query and the mouse clicks of users' browsing and reading query results will be automatically logged and form the system logs to be stored in the database.

EasyKSORD provides an approach to batch processing queries. It parses keyword queries in the data file and sends them to search engines circulating, and after executing queries, it writes search results to the result files. According to the system administrators' need, EasyKSORD processes and analyzes the result files, drawing various graphics to provide a reference to the administrators.

Data flies contain a series of query requests, each of which specifies query keywords, search engine, dataset, top-*k* value, the method of result presentation, and etc. Data files can be manually configured by the administrator, and also can be constructed by extracting the relevant information from the system logs stored in the database. In addition, we can directly extract the relevant information from the system logs to construct the result files, which will be processed by the analysis module.

3 Advanced Queries

Most of the existing KSORD systems' query language is a set of free keywords, each of which can be a word or a combination of multiple words enclosed by quotation marks, for example "*Jim Gray*". In addition to supporting such simple keyword queries, EasyKSORD extends the query language to support the advanced queries, which can assist users to express query requests more exactly.

3.1 Keyword Specification

EasyKSORD supports three types of keywords: simple keyword, metadata keyword and conditional keyword. The syntax format is as follows:

Keyword ::= simple keyword|metadata keyword| conditional keyword

- Simple keyword

A simple keyword refers to a single semantic word, in form of either a word or a multi-words sequence enclosed in quotation marks. For example, *database*, or "*Jim Gray*". The syntax of simple keyword is:

Simple Keyword ::= word| "word word …."

- Metadata keyword

Users usually don't know the relational database schema (It is the advantage of KSORD that users do not need to understand the database schema). However, sometimes users need to use metadata. A situation is that a keyword may have many meanings and appear in many attributes and relations, so the search results may be large and include many ones that users do not need. At this time, we can use metadata to restrict search space. Another situation is that users expect to search the entire relations, where metadata is also needed. The syntax format of metadata keyword is:

Metadata Keyword ::= relation name:keyword| relation alias:keyword| attribute name:keyword| attribute alias:keyword| relation name:*| relation alias:*

For example, *Author:Gray* is a metadata keyword, which indicates that the keyword *Gray* appears in *Author* relation, and it will be only searched in *Author* relation.

A wildcard "*" represents any possible keyword, so at this time all tuples of the relation satisfy conditions. For example, if a user is interested in any authors who wrote or cited a relevant paper on *database*, he/she can issue a keyword query with *Author:* database*. Thus, all tuples in the relation *Author* are related to the first keyword *Author:**.

Generally, users don't know the relation or attribute name, so metadata keyword supports aliases. The above example can also be written as: *Writer:* database*, as long as the alias "*Writer*" has been configured in advance by system administrators.

- Conditional keyword

EasyKSORD allows users to specify conditions associated with a keyword. The conditions are typically concerned with some attributes of relations the keyword belongings to, and these attributes must be in numerical or text types. For example, if users want to search a paper on *database* and published after the year of *2005*, then they can use such a conditional keyword: *database year>2005*. The condition means that, if a tuple containing the keyword *database* has an attribute named *year*, its value must be greater than *2005*. If satisfying the condition, the tuple is related to *database year>2005*. The syntax format of conditional keyword is:

Conditional Keyword ::= keyword condition condition …

Condition ::= numeric attribute name op1 value| numeric attribute alias op1 value| text attribute name op2 value | text attribute alias op2 value

Op1 ::= = |< |<= |>= |> |!=|~

Op2 ::= =|!=

From the grammar, we can see that a conditional keyword can contain multiple conditions.

In addition to commonly used arithmetic operators, with respect to numeric attributes, a special operator "~" can be used to represent approximation. For example, if users search the paper published closer to the year of *2000* on *database*, they can use a conditional keyword: *database year~2000*, which means that the related tuples must have an attribute *year* and the value of *year* is closer to 2000, the relevance between the tuple and *database year~2000* is higher.

Only these two operators "=" and "!=" can be used for text attributes. For example, *Jim sex=M* represents the related tuples containing *Jim* must have an attribute *sex*, and the value of *sex* is *M*.

3.2 Keyword Query Specification

EasyKSORD supports two types of queries: simple queries and advanced queries. Simple queries consist of a series of simple keywords, and the syntax is as follows:

Simple Queries ::= simple keyword simple keyword …

For the system supporting *AND* semantics, such as BANKS, the above format is equivalent to:

Simple Queries ::= simple keyword *AND* simple keyword *AND* …

For the system supporting *OR* semantics, such as IR-Style, the above format is equivalent to:

Simple Queries ::= simple keyword *OR* simple keyword *OR* …

Advanced queries of EasyKSORD contain simple keywords as well as metadata keywords and conditional keywords, and also support *AND*, *OR*, *NOT*. Users can achieve complex Boolean queries by the combinations of *AND*, *OR*, *NOT*. For example, we can search all papers on *database* or *algorithm* written or cited by *Ullman*. The query is written as: *Ullman AND* (*database OR algorithm*). The syntax of advanced queries is as follows:

Advanced Queries ::= keyword|(advanced queries)| advanced queries *AND* advanced queries| advanced queries *OR* advanced queries| *NOT* advanced queries.

As described in the syntax format, advanced queries are expressions linking keywords with Boolean operators *AND*, *OR*, *NOT* and brackets representing the highest priority.

3.3 Implementation

Here, we briefly introduce the approach to implement advanced queries for the data-graph-based search engine.

Keyword queries are analyzed by the query parser, which transforms queries into an internal format, and then sends it to the search engine. In fact the query parser firstly parses queries to the format of disjunctive normal form. Assume the result disjunctive normal form is V, and V can be defined as follows:

V ::= V OR M

M ::= M AND R

R ::= Keyword| NOT Keyword

For example, the query *(database OR AI) AND (Jim OR Kate)* is parsed to the *OR* operation of the following four queries: *database AND Jim*, *database AND Kate*, *AI AND Jim*, *AI AND Kate*.

With respect to metadata keywords, the query parser will find the real table names or column names in the database and send them as search conditions to the search engine. Likewise, as to conditional keywords, the query parser sends the conditions to search engine as well. When the search engine searches for the related tuples on the data graph, it uses those conditions to filter and only gets tuples satisfying search conditions.

When implementing the *OR* operation of multiple expressions, we use a Boolean array of 2^n elements as a checker, and here n is the number of keywords. For example, there are 3 keywords k_1, k_2, k_3, and the *OR* operation composed by them is: k_1 *AND* k_2 *OR* k_2 *AND* k_3 *OR NOT* k_3 *AND* k_1, thus the length of Boolean array b is 8, ranging from 000 to 111. We can conclude that b[110]=true from k_1 *AND* k_2 and b[011]=true from k_2 *AND* k_3 and b[111]=true can be drawn from the above two expressions. b[100]=true and b[110]=true can be drawn from *NOT* k_3 *AND* k_1. The remaining b[000], b [001], b[010], and b[101] are false, because they don't satisfy any expressions.

4 Search Engine

EasyKSORD integrates three kinds of data-graph-based search algorithms, which includes backward extension algorithm and bidirectional extension algorithm of BANKS[3,4] and DPBF algorithm[7], constructing a multi-KSORD platform. In this section, we first introduce DPBF algorithm briefly, which is brought forward by our research group, and then we compare its performance to other algorithms.

Like two other algorithms of BANKS, DPBF considers relational database as a data-graph, where each node represents a tuple, and each edge represents the primary-foreign key relationship between tuples. Each node and edge is given a cost. When a node or an edge has a bigger weight, it will have a smaller cost. For example, when the relationship between two tuples is closer, the edge representing this relationship will have a smaller cost. The cost of a result tree equals to the summation of the cost of all nodes and edges. The bigger the cost of a result tree is, the smaller the relevance score of it is.

Suppose a graph $G(V,E)$, given a keyword search $k_1,k_2,\ldots,k_l$, whose length is L. Suppose V_i is a subset of V, and V_i is the set containing the nodes related to the keyword k_i. The result of keyword search is a rooted cost tree, which contains at least one node in each V_i. So, the problem DPBF faced is how to find the top-k minimum cost trees. When k is 1, this is the famous problem of minimum cost group Steiner tree (GST), which is a NP complete problem. The problem of finding top-k can be described as GST-k.

DPBF is an efficient search algorithm. It first uses a dynamic programming approach to find GST-1. It can be proved that GST-1 is the optimal result with the time complexity of $O(3^L N + 2^L((L + \log N) N + M))$. N and M are the number of nodes and edges in graph G. Because the number of keywords, L, is small in most situations, this solution can handle graphs with a large number of nodes efficiently.

Furthermore, this method can be extended easily to support GST-k. In brief, DPBF computes top-k minimum cost Steiner tree by steps, and do not need to compute and sort all the results.

[7] presents the implementation details of DPBF and gives the comparison on time efficiency, space efficiency, average cost of top-k result trees among DPBF, BANKS-I (backward extension algorithm) and BANKS-II (bidirectional extension algorithm). It proves that DPBF has the least average cost in finding result trees and the time efficiency of DPBF surpasses that of BANKS-I and BANKS-II. It should be noted that the top-1 of DPBF is in global optimum, and other top-k results of the three algorithms are all in local optimum.

Based on experiment conclusions, when user chooses algorithms provided by EasyKSORD, if he attaches more importance to time efficiency, DPBF is the most appropriate. If he pays more attention to space efficiency, BANKS-II is the best choice. Although the average cost of result trees produced by DPBF is the minimum, the ranking mechanism of KSORD may not satisfy the needs of current user query, so the result quality still depends on the user experiences.

5 Result Presentation

When keyword queries are issued, users can specify which approach to be used to present the results among the result presentation methods of list (just ranking and listing results), TreeCluster (clustering results based on tree isomorphs)[13], S-CBR(Schema-based Classification, Browsing and Retrieving)[14] and VSM-RF(Vector-Space-Model-based Relevance Feedback)[15].

List is usually used when k is small in top-k search. When using list, users can choose to use scoring strategy based on vector space model or that provided by

KSORD system itself to rank the results. According to current need, users can employee user feedback or pseudo feedback to reformulate the keyword query and issued the new query to get more related results.

When the value of *k* is bigger, TreeCluster can cluster results into two levels of clusters, which helps to find needed results quickly and is also a way to information discovery.

If users are not certain about the value of *k*, they can use S-CBR, which assists users to know all the possible results (including null results), and guides users a further search on some specific class. Also, S-CBR may recommend common result patterns to users and users can choose the pattern they are interested in and search for data matching the pattern directly.

6 Logging and Analysis

EasyKSORD provides logging and analysis on KSORD systems, which can record system logs, and provide batch processing mode for KSORD systems. EasyKSORD can automatically analyze the results of system logs and batch operations and present the analysis results in graph for system administrators.

6.1 Logging

System logs mainly consist of two aspects of contents, in which one is about users' queries, and another is the operation trails when users are browsing the results. These logs can be recorded by EasyKSORD and then stored in the database.

The information of users' queries includes keywords entered by users, operating parameters chosen by users, system retrieval time, and etc. Operating parameters consist of search engines chosen by users, data sets, the value of top-*k*, the result presentation methods and other runtime parameters.

After the system returns results to the browser, EasyKSORD must capture the user's mouse operations, collect information operated on results, including which result trees and nodes clicked by the user.

It is helpful to improve the system to make summaries and statistics on system logs, analyze and compare various algorithms, data sets, the value of top-*k* and query effects under various result presentation methods. On the other hand, we can grasp users' preferences and querying habits so that we can improve the systems' personalization. Researchers can use these data to analyze users' behavior to realize personalized search and implicit feedback.

In implementation, we create four tables in database to record the user information, the query information, the result tree information, as well as information on the result tree nodes.

When the user logins the server through the browser, EasyKSORD will write information including login time, IP address and port number to the user information table. After the user chooses parameters, inputs keywords and submits the query, EasyKSORD writes information including user ID, current time, query keywords, operating parameters and query running time to the query information table. When the user views results, clicks on a certain result tree, EasyKSORD will write current time, query ID, feature strings (strings which identify structures of the result tree), as well as scores of the result tree to the result tree information table. When the user clicks on a node of the result tree, EasyKSORD will write current time, query ID, table name the node belonging to, the rowid and score of the node to the node information table.

6.2 Batch Processing

EasyKSORD provides batch processing mode, in which the system reads data files, sends it to search engines circularly after the query requests of data files are handled by the query parser, and writes retrieval results to the result files. Each query request specifies the search engine, data sets, the value of top-*k*, result presentation method, query keywords, and etc.

Data files can be manually configured by the administrator, and also can be formed by extracting relevant information from system logs stored in the database.

Batch processing can be used to estimate and analyze performance and effectiveness of KSORD systems.

6.3 Analysis

EasyKSORD can read result files according to parameter files and draw various graphics for system administrators. The parameter files must be pre-configured by system administrators specifying data to be drawn, and result files the data coming from, and etc.

According to the purpose of analysis, result files can be generated by a batch processing, or directly formed by extracting the relevant information from the system logs.

7 Related Work

EasyKSORD is a platform of keyword search over relational databases, whose purpose is to provide a good interface for users and administrators to use KSORD systems. It has done a lot of improvements in input, output, and system management. On the contrary, current other KSORD systems are all single systems, which always focus on retrieval algorithms rather than the usability of KSORD.

In users' input, most KSORD systems only provide simple keyword queries for text attributes of relational databases. SEEKER[11] can search metadata and numerical attributes in databases. The query language of SEEKER includes three kinds of keywords: (1) *Keyword*; (2) *Keyword:Keyword*; (3) *Keyword:<op>Value*. The second and third types of keywords of SEEKER are familiar to metadata keywords and conditional keywords of EasyKSORD, but the search capability of EasyKSORD is more powerful. For example, the metadata keywords of EasyKSORD support the wildcard, and the conditional keywords support "/=" operation for numerical attributes and "=", "/=" operations for text attributes. Some KSORD systems like BANKS[3,4] and DISCOVER[9] only support the semantic of *AND* between keywords, while IR-Style[10] and SEEKER support the semantic of *OR*. None of them supports the semantic of *NOT* and Boolean queries, while EasyKSORD implements these advanced queries.

The result presentation methods mentioned in this paper, such as TreeCluster[13], S-CBR[14] and VSM-RF[15] provided by EasyKSORD are creative work and not equipped by current other KSORD system.

EasyKSORD tries to give users, administrators, researchers a platform to use, manage and study KSORD. Towards this goal, EasyKSORD integrates multi search algorithms, and provides logging and analysis, which are all new features compared to other KSORD systems.

8 Conclusion

EasyKSORD is a platform of keyword search over databases whose goal is to improve usability and practicality of KSORD. It supports advanced keyword queries, and integrates the novel result presentation methods that can be used in different situations, such as TreeCluster, S-CBR and VSM-RF. It also implements the integration of several KSORD search engines, and provides the function of logging and analysis, improving usability and practicality of KSORD systems to the best. The platform makes users and administrators easier to use and manage KSORD, and also makes it convenient for researchers to study KSORD. Based on a good interface, the problems of KSORD are easier to find and promote researchers to do further research.

Acknowledgments. This work was supported by the National Natural Science Foundation of China (No.60473069 and No.60496325) and the Special Fund for Postdoctoral Innovative Project of Shandong Province (No. 200703084).

References

1. Wang, S., Zhang, K.L.: Searching Databases with Keywords. J. Comput. Sci. Tech. 20(1), 55–62 (2005)
2. Wang, S., Zhang, J., Peng, Z.H., Zhan, J., Wang, Q.Y.: Study on Efficiency and Effectiveness of KSORD. In: Dong, G., Lin, X., Wang, W., Yang, Y., Yu, J.X. (eds.) APWeb/WAIM 2007. LNCS, vol. 4505, pp. 6–17. Springer, Heidelberg (2007)
3. Bhalotia, G., Hulgery, A., Nakhe, C., Chakrabarti, S., Sudarshan, S.: Keyword Searching and Browsing in Databases using BANKS. In: Agrawal, R., et al. (eds.) Proc. of the 18th Int'l. Conf. on Data Engineering, pp. 431–440. IEEE Press, Los Alamitos (2002)
4. Kacholia, V., Pandit, S., Chakrabarti, S., Sudarshan, S., Desai, R., Karambelkar, H.: Bidirectional Expansion for Keyword Search on Graph Databases. In: Böhm, K., et al. (eds.) Proc. of the 31st Int'l. Conf. on Very Large Data Bases, pp. 505–516. ACM, New York (2005)
5. Balmin, A., Hristidis, V., Papakonstantinou, Y.: ObjectRank: Authority-Based Keyword Search in Databases. In: Nascimento, M.A., et al. (eds.) Proc. of the 30th Int'l. Conf. on Very Large Data Bases, pp. 564–575. Morgan Kaufmann Publishers, San Francisco (2004)
6. Wang, S., Peng, Z.H., Zhang, J., Qin, L., Wang, S., Yu, J., Ding, B.L.: NUITS: A Novel User Interface for Efficient Keyword Search over Databases. In: Dayal, U., et al. (eds.) Proc. of the 32nd Int'l. Conf. on Very Large Data Bases, pp. 1143–1146. ACM, New York (2006)

7. Ding, B.L., Yu, J., Wang, S., Qin, L., Zhang, X., Lin, X.M.: Finding Top-k Min-Cost Connected Trees in Databases. In: Proc. of the 23rd Int'l. Conf. on Data Engineering, pp. 836–845. IEEE Press, Los Alamitos (2007)
8. Agrawal, S., Chaudhuri, S., Das, G.: DBXplorer: A system for keyword-based search over relational databases. In: Agrawal, R., et al. (eds.) Proc. of the 18th Int'l. Conf. on Data Engineering, pp. 5–16. IEEE Press, Los Alamitos (2002)
9. Hristidis, V., Papakonstantinou, Y.: DISCOVER: Keyword search in relational databases. In: Bernstein, P.A., et al. (eds.) Proc. of the 28th Int'l. Conf. on Very Large Data Bases, pp. 670–681. Morgan Kaufmann Publishers, San Francisco (2002)
10. Hristidis, V., Gravano, L., Papakonstantinou, Y.: Efficient IR-style keyword search over relational databases. In: Freytag, J.C., et al. (eds.) Proc. of the 29th Int'l. Conf. on Very Large Data Bases, pp. 850–861. Morgan Kaufmann Publishers, San Francisco (2003)
11. Wen, J.J., Wang, S.: SEEKER: Keyword-based Information Retrieval Over Relational Databases. Journal of Software 16(7), 1270–1281 (2005)
12. Luo, Y., Lin, X.M., Wang, W., Zhou, X.F.: SPARK: Top-k Keyword Query in Relational Databases. In: Chan, C.Y., et al. (eds.) Proc. of the ACM SIGMOD International Conference on Management of Data, pp. 115–126. ACM, New York (2007)
13. Peng, Z.H., Zhang, J., Wang, S., Qin, L.: TreeCluster: Clustering Results of Keyword Search over Databases. In: Yu, J.X., Kitsuregawa, M., Leong, H.-V. (eds.) WAIM 2006. LNCS, vol. 4016, pp. 385–396. Springer, Heidelberg (2006)
14. Peng, Z.H., Zhang, J., Wang, S.: S-CBR: Presenting Results of Keyword Search over Databases Based on Database Schema. Journal of Software 19(2), 323–337 (2008)
15. Peng, Z.H.: A Study on New Result Presentation Technology of Keyword Search over Databases. Doctoral Thesis, Renmin University of China (2007)

Real-Time and Self-adaptive Method for Abnormal Traffic Detection Based on Self-similarity

Zhengmin Xia[1], Songnian Lu[1,2], Jianhua Li[1,2], and Jin Ma[2]

[1] Department of Electronic Engineering, Key Lab of Information Security Integrated Management Research, Shanghai Jiao Tong University, 200240, Shanghai, P.R. China
[2] School of Information Security Engineering, Key Lab of Information Security Integrated Management Research, Shanghai Jiao Tong University, 200240, Shanghai, P.R. China
{miaomiaoxzm,snlu,lijh888,majin}@sjtu.edu.cn

Abstract. Abnormal traffic detection is a difficult problem in network management and network security. This paper proposes an abnormal traffic detection method based on a continuous LoSS (loss of self-similarity) through comparing the difference of Hurst parameter distribution under the network normal and abnormal traffic time series conditions. Due to the needs of fast and high accuracy for abnormal traffic detection, the on-line version of the Abry-Veitch wavelet-based estimator of the Hurst parameter in large time-scale is proposed, and the detection threshold could self-adjusted according to the extent of network traffic self-similarity under normal conditions. This work also investigates the effect of the parameters adjustment on the performance of abnormal traffic detection. The test results on data set from Lincoln lab of MIT demonstrate that the new abnormal traffic detection method has the characteristics of dynamic self-adaptive and higher detection rate, and can be implemented in a real-time way.

Keywords: Network traffic, Anomaly detection, Hurst parameter, Time series, Self-similarity.

1 Introduction

Along with the development and popularization of network, more and more serious attacks have been presented and led the Quality of Service (QoS) performance of network degrade, so the security of network becomes more and more important. It can be seen that network traffic related to attacks, especially distributed denial-of-service (DDoS) attacks, is "pulse" from the perspective of dynamical aspects for limited time interval in physics[1], so we call these attack-contained traffic abnormal traffic and call these attack-free traffic normal traffic. There are mainly two kinds of abnormal traffic detection: misuse detection and anomaly detection. Misuse detection is a kind of passive method that is based on a library of known signatures to match against network traffic. Hence, unknown signatures from new variants of an attack mean 100% miss. Anomaly detection does not suffer from this problem, but it is usually found to have a lower probability of detection, and time consuming. Hence, a self-adaptive and real-time method for abnormal traffic detection was proposed.

W. Liu et al. (Eds.): WISM 2009, LNCS 5854, pp. 383–392, 2009.

As shown by Leland, et al.[2], and supported by a number of later papers including [3][4], traffic captured from Local Area Networks and the Internet exhibits self-similar behavior. Self-similar is the property that associated with the objects whose structure is unchanged at different time scales. The work presented in [4] shown that the self-similarity of internet traffic distributions could often be accounted for a mixture of the actions of a number of individual users, hardware and software behaviors at their originating hosts, multiplexed through an interconnection network.

Li[5] quantitatively described the abnormal traffic statistics, and found that the averaged Hurst parameter of abnormal traffic usually tended to be significantly smaller than that of normal traffic. The works in [6][7][8][9] presented new methods of detecting the possible presence of abnormal traffic without a template of normal traffic. These methods used LoSS definition with the self-similarity or Hurst parameter beyond the fixed threshold. But the Hurst parameter estimation takes huge time, and the real-world network traffic self-similarity extent is changing over time, so these detection methods could not be realized in real-time, and had low detection rate and high false alarm rate.

In this paper, we present a real-time and self-adaptive abnormal traffic detection method based on the self-similar nature, and the detection threshold can change with the network traffic self-similarity extent automatically. The method mainly includes three steps: (i) estimates the Hurst parameter of incoming network traffic use on-line Abry-Veitch wavelet-based estimator, and (ii) decides whether the incoming traffic is normal or not by comparing the Hurst parameter with the detection threshold, then (iii) updates the detection threshold if the incoming traffic is normal, otherwise informs the network administrators. This method has the characteristics of dynamic self-adaptive and real-time detection.

The remainder of this paper is organized as follows. Section 2 briefly introduces the theoretical background of self-similarity and on-line Abry-Veitch wavelet-based Hurst parameter estimation. Section 3 describes the detection principle and explains the abnormal traffic detection process in detail. The discussions of parameters adjustment on detection's performance using data set from Lincoln Lab of MIT are presented in Section 4. Finally, a brief summary of our work and future research directions are provided in section 5.

2 Self-similarity and Hurst Parameter Estimation

2.1 A Brief Review of Self-similarity

Self-similarity means that the sample paths of the process $X(t)$ and those of rescaled version $c^H X(t/c)$, obtained by simultaneously dilating the time axis by a factor $c > 0$, and the amplitude axis by a factor c^H, cannot be statistically distinguished from each other. H is called the self-similarity or Hurst parameter. Equivalently, it implies that an affine dilated sunset of one sample path cannot be distinguished from its whole.

Let $X = \{X_i, i \in \mathbb{Z}_+\}$ be a wide-sense stationary discrete stochastic process with constant mean μ, finite variance σ^2, and autocorrelation function $r(k), (k \in \mathbb{Z}_+)$. Let $X^{(m)}$ be a m-order aggregate process of X,

$$X_i^{(m)} = (X_{mi-m+1} + \cdots + X_{mi})/m \quad i, m \in \mathbb{Z}_+ . \tag{1}$$

For each m, $X^{(m)}$ defines a wide-sense stationary stochastic process with autocorrelation function $r^{(m)}(k), \ (k \in \mathbb{Z}_+)$.

Definition 1. A second-order stationary process X is called exactly second-order self-similar (ESOSS) with Hurst parameter H=1-$\beta/2$, $0 < \beta < 1$, if the autocorrelation function satisfies

$$r^{(m)}(k) = r(k), \quad (k, m \in \mathbb{Z}_+) . \tag{2}$$

Definition 2. A second-order stationary process X is called asymptotical second-order self-similar (ASOSS) with Hurst parameter H=1-$\beta/2$, $0 < \beta < 1$, if the autocorrelation function satisfies

$$\lim_{m \to \infty} r^{(m)}(k) = r(k), \quad (k \in \mathbb{Z}_+) . \tag{3}$$

where $r(k) = [(k+1)^{2-\beta} - 2k^{2-\beta} + (k-1)^{2-\beta}]/2$.

2.2 On-Line Hurst Parameter Estimation

Many methods had been developed to estimate the Hurst parameter, such as aggregated variance[10], local whittle[11], and the wavelet-based methods[12]. These estimators have been used as a batch estimator, that is, the data set is collected and analyzed off-line. So it is unrealistic to use these estimators directly for abnormal traffic detection, because the detection emphasized on not only the accuracy, but also the speed.

By far, the wavelet-based estimator of the Hurst parameter stands out as one of the most reliable estimators in practice for it is more robust with respect to smooth polynomial trends and noise[13]. Wavelet-based Hurst parameter estimation mainly includes three methods: wavelet variance analysis, wavelet power spectra analysis, and wavelet energy analysis. These methods are consistent in essence.

For the purpose of real-time estimation of the Hurst parameter, we revise the wavelet-based Hurst parameter estimation using the multiple resolutions feature of wavelet analysis, and realize an on-line Abry-Veitch wavelet-based Hurst parameter estimation[14]. The on-line estimation progress is summarized as follows:

- Phase 1, for a given traffic trace time series X, compute the wavelet coefficients $d_X(j,k)$ using a pyramidal filter bank in an online fashion as shown in figure 1 for each scale j and position k. At each level in the recursive structure, the bandpass (BP) output $d_X(j,\cdot)$, and the lowpass (LP) output $a_X(j,\cdot)$ occur at half the rate of the input $a_X(j-1,\cdot)$.

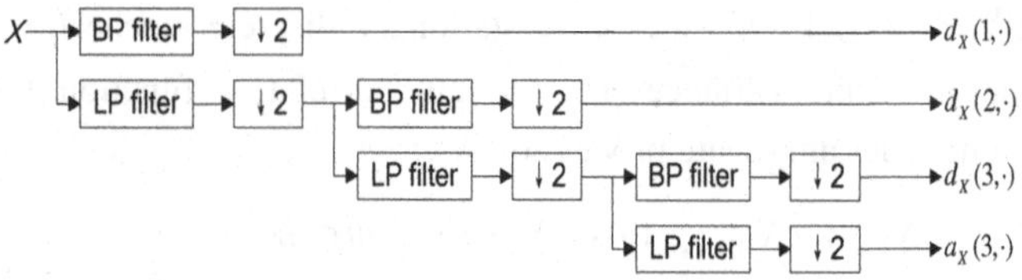

Fig. 1. Filter bank

- Phase 2, let the current stored sum of squares at scale j calculated from the first n_j values be

$$S_j = \sum_{k=1}^{n_j} d_X^{\ 2}(j,k) \ . \tag{4}$$

Assume that the arrival of the new data $x(n)$ results in a new coefficient $d_X(j,n_j+1)$ at scale j from the filter bank. The sum is then updated as follows:

$$\begin{aligned} n_j &\leftarrow n_j + 1 \\ S_j &\leftarrow S_j + d_X^{\ 2}(j,k) \end{aligned} \ . \tag{5}$$

When the variance estimation at scale j is required for the Phase 3, it can be calculated as $\mu_j = S_j / n_j$.

- Phase 3, make a plot of $log_2(\mu_j)$ versus scale j and apply linear regression over the curve region that looks linear, and compute the slope α . There is no need to compute the Logscale Diagram every time a new data is acquired, and they may be recalculated only as needed.
- Phase 4, estimate the Hurst parameter as:

$$H=(\alpha+1)/2 \cdot \tag{6}$$

3 Abnormal Traffic Detection

3.1 Detection Principle

For given discrete time series $X = \{X_i, i \in \mathbb{Z}_+\}$, $Y = \{Y_i, i \in \mathbb{Z}_+\}$ and $Z = \{Z_i, i \in \mathbb{Z}_+\}$, let X and Y represent normal traffic and abnormal traffic respectively, and Z be the attack traffic during transition process of attacking. X and Z are uncorrelated[5], so Y can be abstractly expressed by $Y = X + Z$.

Figure 2 illustrates the components of traffic the target received at time i . Let $x_k(i)$ represents the number of bytes send out by node k at time i for normal network services, and $z_q(i)$ represents the number of bytes send out by node q at time i for network attacks. X_i , Z_i and Y_i be the normal traffic, attack traffic and total traffic the target received at time i , respectively.

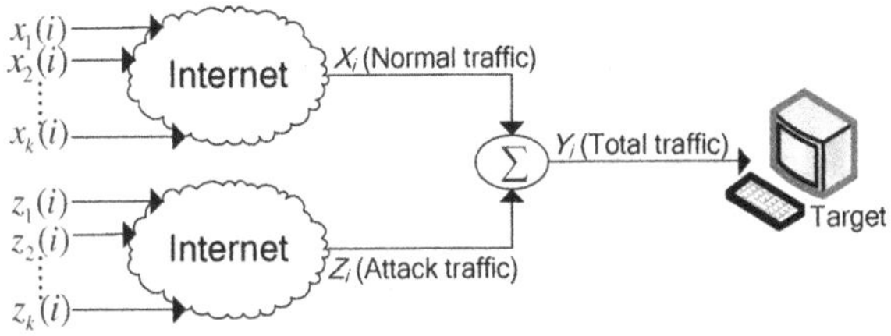

Fig. 2. Composition of network traffic

Let r_{XX}, r_{ZZ} and r_{YY} be autocorrelations of X, Z and Y respectively. During the transition process of attacking, $\|r_{YY} - r_{XX}\|$ is noteworthy[5], and $r_{YY}=r_{XX}+r_{ZZ}$. For each value of $H \in (0.5,\ 1]$, there is exactly one autocorrelation function with self-similarity as it can be seen from Beran (1994, pp. 55). Thus, a consequence is that $\|H_Y - H_X\|$ is considerable, where H_Y and H_X are average Hurst parameters of Y and X, respectively. Hence, H is a parameter that can yet be used to describe the abnormality of network traffic.

3.2 Detection Process

In general, the network traffic of one host or one LAN is almost equal and remains steady as a whole under normal conditions for a period of time. But for longer time, the traffic is unstable, and it changes with the network circumstance (network load, and number of users, et al.), so the detection threshold also should change accordingly. Figure 3 is the flowchart of self-adaptive abnormal traffic detection.

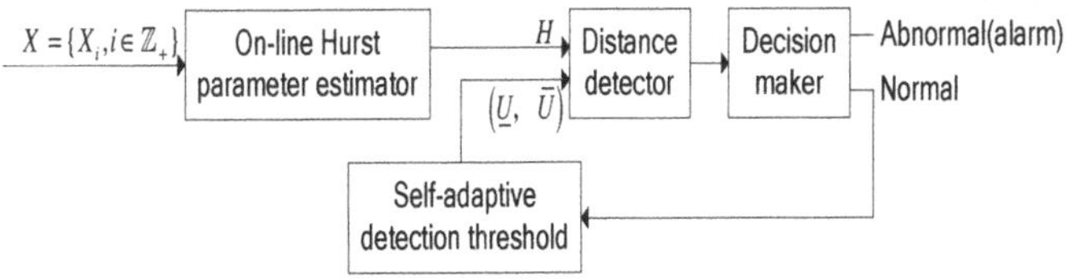

Fig. 3. Self-adaptive and real-time abnormal traffic detection process

The detailed detection process is as follows:

Step one: detection threshold initialization

Before the detection, the detection threshold should be initialized. According to [8], the typical Hurst parameter of network traffic is 0.75, and the traffic will be considered as abnormal if the change of Hurst parameter is more than 0.15. So, let the down threshold $\underline{U}$ and up threshold $\overline{U}$ be 0.6 and 0.9, respectively.

Step two: abnormal traffic detection

Let there are two registers, named l and m, and initialize l and m with 0 and 1 separately. For every incoming of the new traffic data at time segment n, we update the l as $l = l+1$, and estimate S_j according to phase 1 and phase 2 if $l < L$, and then wait for the next traffic data. When l reaches to L, we estimate the Hurst parameter according to phase 1 to phase 4, and record it as $H_{(m)}$, and then reset the register l, and update the m as $m = m+1$. Averaging $H_{(m)}$ in terms of index m yields

$$H_n = \frac{1}{M}\sum_{m=1}^{M} H_{(m)} . \tag{7}$$

H_n represents the Hurst parameter of network traffic at time segment n. The traffic is considered normal if $H_n \in (\underline{U}, \overline{U})$, and then resets the registers l, m and goes to step three. Otherwise the traffic is considered abnormal, and then goes to step one, and at the same time reports the abnormal to the network administrators.

Step three: detection threshold updating

When the incoming traffic is considered normal, the detection threshold should be updated before the next detection. Considering the Hurst parameters H_n (n=1,2,⋯, N) of network traffic at time segment N and $N-1$ consecutive time segments before time segment N, a normality assumption for H_n is quite accurate in most cases for $M \geq 10$ [15].

Let $\overline{H}$ and σ_H^2 be the mean and variance estimates of H_n separately,

$$\overline{H} = E[H_n] = \frac{1}{N}\sum_{n=1}^{N} H_n$$

$$\sigma_H^2 = \frac{1}{N}\sum_{n=1}^{N} H_n^2 - \overline{H}^2$$

Then, the probability density function of H_n can be expressed as:

$$f\left(H_n; \overline{H}, \sigma_H^2\right) = \frac{1}{\sqrt{2\pi}\sigma_H} e^{-\frac{\left[H_n - \overline{H}\right]^2}{2\sigma_H^2}} . \tag{8}$$

The confidence interval of H_n with $(1-\alpha)$ confidence coefficient is given by $\left(\overline{H} - \sigma_H z_{\alpha/2}, \overline{H} + \sigma_H z_{\alpha/2}\right)$. Considering the self-similar network traffic's Hurst parameter is located between $(0.5,\ 1]$, so the down threshold $\underline{U}$ and up threshold $\overline{U}$ could be set as:

$$\begin{aligned} \underline{U} &= \max\,(0.5,\ \overline{H} - \sigma_H z_{\alpha/2}) \\ \overline{U} &= \min\,(\overline{H} + \sigma_H z_{\alpha/2},\ 1] \end{aligned} . \tag{9}$$

4 Experiment

4.1 Data Preparation

To testing the proposed method, we used the traffic data set from Lincoln Lab of MIT, named DARPA 2000 LL_DDoS_2.0.2. It collected from U.S. Air Force base over a span of approximately 1 hour 45 minutes on April 16, 2000. This data set includes five phases of DDoS attack: probe a host; break in-to the host; upload DDoS software and attack script; initiate attack; launch the DDoS. Some of the traffic is displayed in figure 4, and the merge time scale is 100ms.

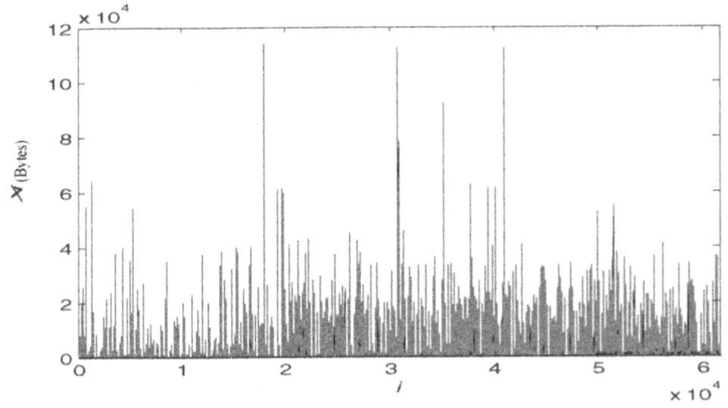

Fig. 4. Traffic of LL_DDoS_2.0.2

4.2 Experimental Results and Analysis

Let $M=N=10$, $L=128$, and the change trend of H_n, $\underline{U}$ and $\overline{U}$ are displayed in figure 5.

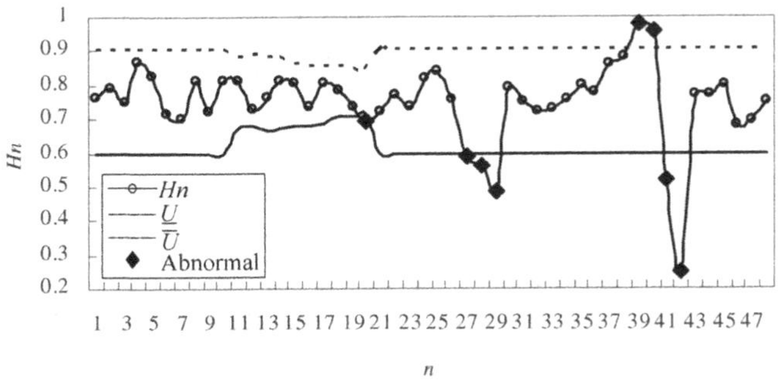

Fig. 5. Change trend of H_n, $\underline{U}$ and $\overline{U}$ ($L=128$, $\alpha=0.05$)

In figure 5, time segments 1~10 are the initialization of detection threshold, so the down threshold $\underline{U}$ and up threshold $\overline{U}$ could be set as 0.6 and 0.9, respectively. At time segment 11, the traffic of 10 consecutive time segments before time segment 11 is normal, so the $\underline{U}$ and $\overline{U}$ could be updated using equation (9) with H_n ($n=1,2,\cdots,10$) under the confidence level of 95% ($\alpha=0.05$). The detection result is $H_{11} \in (\underline{U}, \overline{U})$, so the traffic of time segment 11 is considered normal. At time segment 20, the H_{20} is out of the range $(\underline{U}, \overline{U})$, so the traffic of time segment 20 is considered abnormal. Then, report this abnormal to network administrators, and reinitialize the detection threshold for the next ten consecutive time segments after the time segment 20. Using this detection method, the traffic is considered abnormal at time segments 20, 27-29, 39-42. Change the confidence level to 99% ($\alpha=0.01$), and the H_n, down threshold $\underline{U}$ and up threshold $\overline{U}$ of each time segment are displayed in figure 6. From figure 6, we can see that the traffic is considered abnormal at time segments 27-29, 39-42.

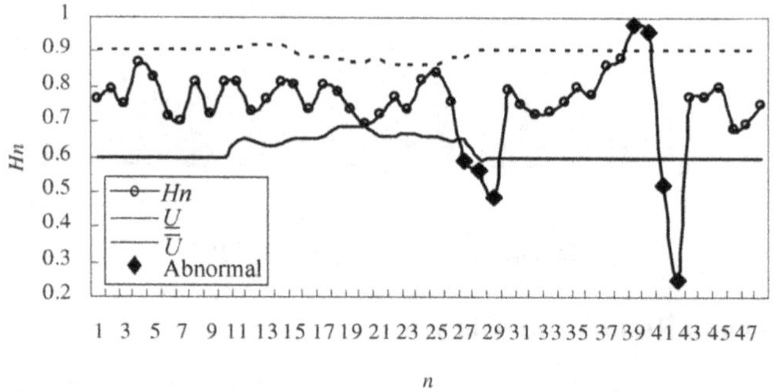

Fig. 6. Change trend of H_n, $\underline{U}$ and $\overline{U}$ ($L=128$, $\alpha=0.01$)

Comparing the figure 5 and figure 6, we can find that the wider the confidence level is, the less abnormal traffic will be detected; the narrower the confidence level is, the more abnormal traffic will be detected. This is because under the condition of wider confidence level, some low-rate abnormal traffic will be missed, and under the condition of narrower confidence level, some normal change of traffic will be taken as abnormal.

Let $L=256$, and the change trend of H_n, $\underline{U}$ and $\overline{U}$ are displayed in figure 7.

By detection, the traffic is considered abnormal at time segments 14-15, 20-21, corresponding to time segments 28-30, 40-42 when $L=128$.

Comparing the figure 5 and figure 7, we can find that the longer the L is, the less abnormal traffic will be detected; the shorter the L is, the more abnormal traffic will be detected. This is because when L is long, some short-time abnormal traffic will be missed, and when L is short, the data used to estimate the Hurst parameter is less, so the accuracy of Hurst parameter estimation will be degrade and lead to the false detection.

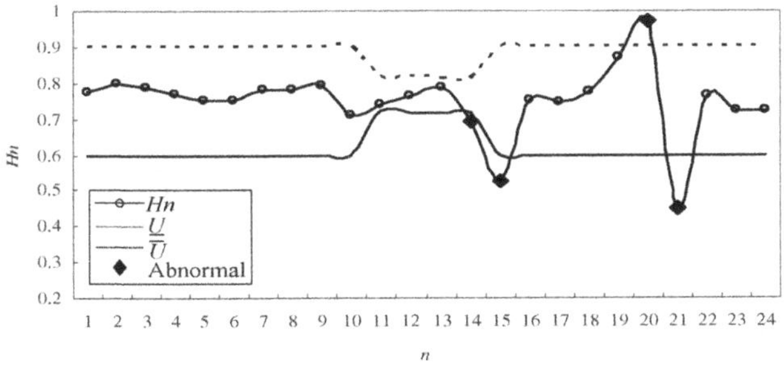

Fig. 7. Change trend of H_n, $\underline{U}$ and $\overline{U}$ (L=256, α=0.05)

5 Conclusion

This paper compared the difference of Hurst parameter distribution under network normal and abnormal traffic time series conditions, and designed an abnormal traffic detection method using self-adaptive detection threshold. The detection results on data set DARPA 2000 LL_DDoS_2.0.2 demonstrated that the new detection method had the characters of self-adaptive and higher detection rate, and can be implemented in a real-time way. These merits will bring a great benefit to network administrators for early finding out abnormal traffic and taking effective measures to prevent more damages to the network. Recent research found that wireless network traffic also had the character of self-similarity, so our future research direction will focus on abnormal traffic detection of wireless network.

Acknowledgments. This work is sponsored by the National High Technology Research and Development Plan of China (No.2005AA145110 and No.2006AA01-Z436), and Pudong New Area Technology Innovation Public Service Platform of China (No.PDPT2005-04).

References

[1] Toma, G.: Practical test functions generated by computer algorithms. In: Gervasi, O., Gavrilova, M.L., Kumar, V., Laganá, A., Lee, H.P., Mun, Y., Taniar, D., Tan, C.J.K. (eds.) ICCSA 2005. LNCS, vol. 3482, pp. 576–584. Springer, Heidelberg (2005)
[2] Leland, W.E., Taqqu, M.S., Willinger, W., et al.: On the self-similar nature of ethernet traffic (extended version). IEEE/ACM Transactions on Networking 2(1), 1–15 (1994)
[3] Paxson, V., Floyd, S.: Wide area traffic: The failure of poisson modeling. IEEE/ACM Transactions on Networking 3(3), 226–244 (1995)
[4] Mark, E., Azer, B.: Self-similarity in World Wide Web traffic: Evidence and possible causes. IEEE/ACM Transactions on Networking 5(6), 835–846 (1997)
[5] Li, M.: Change trend of averaged Hurst parameter of traffic under DDOS flood attacks. Computers & security 25(3), 213–220 (2006)
[6] Schleifer, W., Mannle, M.: Online error detection through observation of traffic self-similarity. IEE Proceedings on Communications 148(1), 38–42 (2001)

[7] William, H., Gerald, A.: The loss technique for detecting new denial of service attacks. In: IEEE Proceedings on SoutheastCon, pp. 302–309. IEEE Press, Los Alamitos (2004)
[8] Ren, X., Wang, R., Wang, H.: Wavelet analysis method for detection of DDOS attack on the basis of self-similarity. Frontiers of Electrical and Electronic Engineering in China 2(1), 73–77 (2007)
[9] Mohd, F., Mohd, A., Ali, S., et al.: Uncovering anomaly traffic based on loss of self-similarity behavior using second order statistical model. International Journal of Computer Science and Network Security 7(9), 116–122 (2007)
[10] Taqqu, M., Teverovsky, V., Willinger, W.: Estimators for long-range dependence: An empirical study. Fractals 3(4), 785–798 (1995)
[11] Taqqu, M.S., Teverovsky, V.: Robustness of Whittle-type estimates for time series with long-range dependence. Stochastic Models 13, 723–757 (1997)
[12] Patrice, A., Darryl, V.: Wavelet analysis of Long-Range-Dependence Traffic. IEEE Transactions on Information Theory 44(1), 2–15 (1998)
[13] Stoev, S., Taqqu, M., Park, C., et al.: On the wavelet spectrum diagnostic for Hurst parameter estimation in the analysis of Internet traffic. Computer Networks 48(3), 423–445 (2005)
[14] Roughan, M., Darryl, V., Patrice, A.: Real-time estimation of the parameters of long-range dependence. IEEE/ACM Transactions on Networking 8(4), 467–478 (2000)
[15] Bendat, J.S., Piersol, A.G.: Random Data: Analysis and Measurement Procedure. John Wiley & Sons, Chichester (2000)

Trust-Based Fuzzy Access Control Model Research

Rina Su, Yongping Zhang, Zhongkun He, and Shaojing Fan

School of Electronic and Information Engineering
Ningbo University of Technology
315016, Ningbo, China
srn2009@126.com, zhangyp1963@yahoo.com, hzk@nbut.cn, fsj@nbut.cn

Abstract. Trust-based access control model uses the value of users' static trust corresponding to the static attributes to get security of access control. However, facing the large number of users in the open system, the calculation of trust not only includes the static properties but also the context information with the ever-changing environment in the system. Using the fixed value to express the trust of the user to grant permissions without considering the dynamic context information may reduce the effectively of access control. Therefore, this paper proposes to use the fuzzy theory into the trust-based access control to solve the problem of dynamic access control by constructing reasonable structure with the fuzzy membership interval of the context information, along with establishing fuzzy rules for fuzzy inference.

Keywords: Trust-based access control, fuzzy inference, dynamic access, context information.

1 Introduction

With the rapid development of the computer technology and internet applications, facing the large number of users and organizations in the open system, how to make the access control and permission management in the system safety and effective becomes an important issue. The traditional security mechanism for access control includes Discretionary Access Control (DAC) and Mandatory Access Control (MAC). The common drawback of them is to bind the permission operation to users directly, resulting in the user can access the resources in the system directly according to the assigned permissions. Thus, with the changing of the number of users and permissions, the cost of maintenance permissions management will be greatly increased. Sandhu et al. proposed the technology of Role-Based Access Control (RBAC) [1] to solve the problem of access control, in which they introduced the concept of role in the access control model, and converted the relationship of traditional user-permission directly to user-role-permission indirectly. This model separated the assignment of users and permissions, simplified the management of users and permissions. However, with the number and the status changing in the open system, it is difficult to determine which role the user belongs to before the access. The assignment of the user to role is dynamic, so Sudip Chakraborty et al. built confidence-mechanism in

W. Liu et al. (Eds.): WISM 2009, LNCS 5854, pp. 393–399, 2009.

the assignment of user to role by introducing the concept of trust in the model based on the RBAC [2]. According to certain attributes of each user, the trust was computed, and then mapped to the role for access control, and also the permissions incentive rules were set. So whether the users get the permissions successfully depended on its trust and assigned role. The assignment from the user to role was dynamic. Trust-based access control (TrustBAC) can solve the problem of dynamic allocation of role effectively, but the role assigned to the permission is still static. Calculation of trust usually included the static attributes about the level of users' experience, knowledge, recommendation. However, the trust, determined by the context information (for example: time, users' location, state of the environment, interpersonal relationships, etc.), is a fuzzy value. This paper expands the TrustBAC model, introducing dynamic trust mechanism in the role assigned to permission. The fuzzy set theory is applied to the calculation of trust. We establish fuzzy rules, carry out fuzzy inference, and realize the dynamic assignment from user to permission eventually. The establishment of trust-based fuzzy access control model (TBFAC) can solve the problem of inaccurate access control effectively.

2 Trust-Based Fuzzy Access Control

2.1 System Model

Trust expresses the confidence degree of the subject in the system, as it is complex and ambiguous essentially, so it is actually fit to use fuzzy set theory to measure and forecast. Trust usually can be categorized as static and dynamic in access control technology. Both of them can be expressed by fuzzy numbers.

Static trust of the users usually expresses the level of experience confidence, knowledge confidence, and recommendation confidence. Dynamic trust in the system is to calculate the context information, including information space and physical characteristic space. Usually, it can be divided into four categories [3], that is, the calculating context (such as network connectivity, communication costs, communication bandwidth and the nearby resources, etc.), the user context (such as the user's status, location, current interpersonal relationships, the activities, etc.), the physical context (such as light, the degree of noise, conditions and temperature, etc.), and the history context. Trust-based fuzzy access control technology calculates the roles' dynamic trust by context-aware fuzzy membership interval and fuzzy rule in the process of role requesting permissions to achieve dynamic authorization.

Trust-based fuzzy access control model is shown in Figure 1.

The model divides the authorization process into two parts, *UA* part and *PA* part. *UA* part is in charge of the assignment from users to roles, and its principle is based on the calculation of users' static trust. *PA* part is in charge of the assignment from roles to permissions, taking into account the actual situation of access control, we introduce the Fuzzy Rule module to carry out fuzzy reasoning, by the calculation of dynamic trust to determine whether the permission is assigned to the role finally.

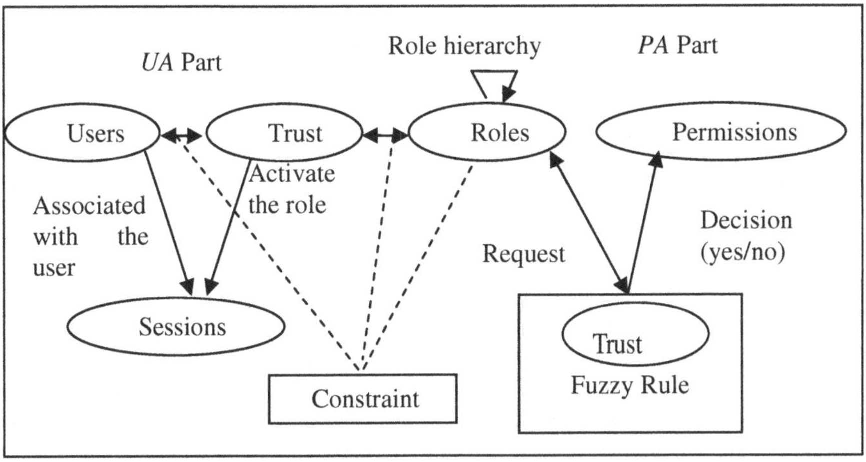

Fig. 1. Trust-based Fuzzy Access Control

This model contains five main entities: Users, Trust, Roles, Permissions and Sessions, and two parts of *UA* and *PA*.

Users: represent the subject of using the system's data or other resources. Mainly refer to the visitors of the system.

Trust: represent the extent of confidence by truster. It is divided into static trust (*TS*) and dynamic trust (*DS*). *TS* is used to indicate users' degree of confidence to the role according to the users' experience, knowledge, recommendation. It is a fuzzy logic value. *DS* is used in fuzzy evaluation of the assignment from roles to permissions, representing the degree of confidence to assign permissions. It is calculated by the context-aware fuzzy membership interval and fuzzy inference rules.

Roles: represent the abstract description of a certain type users based on departments, functions, powers and so on. Roles are inheritance and responsibility separated. In addition, RBAC also provides the concept of static mutex, dynamic mutex and capacity constraints, etc. [4].

Permissions: represent the powers to access resources in system. It is divided into two types of data access control and object access control.

Sessions: represent the conversation to activate the role. It is responsible for the mapping each session to a single user and the mapping each session to a set of roles.

Session s_i has the permission of all roles r' junior to roles activated in the session, i.e.

$$\{p \mid ((p, r) \in PA \vee (p, r') \in PA) \wedge r \in roles(s_i) \wedge r \geq r'\}.$$

UA part: the operation of assignment from the users to the roles, $UA_{TS} \subseteq (U \times R)$. Establish many-to-many relationship between users to roles by the static trust.

PA part: the operation of assignment from the roles to the permissions, $PA_{DS} \subseteq (P \times R)$. Establish many-to-many relationship between roles and permissions by the dynamic trust.

Fuzzy evaluation: when the trust of role with dynamic mechanism is equal to or greater than the required threshold of trust through fuzzy evaluation of competence mechanism, permissions are assigned to the role.

Rule: represent the set of fuzzy safety policy rules with the permissions assignment by dynamic trust. It can be expressed by the logical reasoning as follows.

$$Rule_i: \; F_i = (C_1, w_1, [x_1^-, x_1^+]) \wedge (C_2, w_2, [x_2^-, x_2^+]) \wedge \ldots \wedge (C_n, w_n, [x_n^-, x_n^+])$$
$$F_i \rightarrow (P_i, DS_i) \,. \tag{1}$$

F_i represents the logical rule conditions in the reasoning. There, C_i represents the context information; w_i represents the weight of the impact by the context information C_i; $[x_i^-, x_i^+]$ represents the satisfaction interval of the context information. (P_i, DS_i) represents the logical rule reasoning result. "→" represents logical derivation.

The rule means that when role R_i requests permission P_i, if its membership interval in the case of context-aware C_i by weight of w_i is in range $[x_i^-, x_i^+]$, then the dynamic trust of assignment permission P_i needed is DS_i. If the meeting interval is near $[x_i^-, x_i^+]$, then we will carry out the fuzzy inference to calculate the practical dynamic trust.

2.2 Fuzzy Inference Mechanism

Fuzzy reasoning process can be carried out by fuzzy relations synthesized. Assuming that there is a rule:

Rule: if x is $\tilde{A}$ then y is $\tilde{B}$

Its reasoning model is:

$$\tilde{B} = \tilde{A} \circ \tilde{R} = \begin{bmatrix} a_{11} & a_{12} & \cdots & a_{1m} \\ a_{21} & a_{22} & \cdots & a_{2m} \\ \cdots & \cdots & \cdots & \cdots \\ a_{n1} & a_{n2} & \cdots & a_{nm} \end{bmatrix} \circ \begin{bmatrix} r_{11} & r_{12} & \cdots & r_{1m} \\ r_{21} & r_{22} & \cdots & r_{2m} \\ \cdots & \cdots & \cdots & \cdots \\ r_{n1} & r_{n2} & \cdots & r_{nm} \end{bmatrix} = (b_{ij})_{n \times m} \,. \tag{2}$$

$$b_{ij} = \bigvee_{l=1}^{k} a_{il} \wedge r_{lj} \quad (i = 1,2,\ldots,n;\; j = 1,2,\ldots,m) \,. \tag{3}$$

Usually, n = 1. In this way, $\tilde{R}$ is the new fuzzy set derived from the above *Rule*, and $\tilde{B}$ is the new conclusion.

Fuzzy inference process is to draw reliable control conclusions from incomplete and fuzzy information. According to the above way, the TBFAC reasoning process that we set in this paper is as follows:

Denote by F_i the logical rule conditions by the context information known in the system. Its fuzzy rule is shown as follows:

$$F_i = (C_1, w_1, [x_1^-, x_1^+]) \wedge (C_2, w_2, [x_2^-, x_2^+]) \wedge \ldots \wedge (C_n, w_n, [x_n^-, x_n^+]) \rightarrow (P_i, DS_i) \,. \tag{4}$$

Let L_i represents the matching facts of given role R_i during the process of request permission P_i :

$$L_i = (D_1, [y_1^-, y_1^+]) \wedge (D_2, [y_2^-, y_2^+]) \wedge \ldots \wedge (D_n, [y_n^-, y_n^+]) \,. \tag{5}$$

Where D_i is the actual context information, interval $[y_i^-, y_i^+]$ represents the membership range of R_i satisfied to D_i.

We use q to express the degree of the logical facts L_i matching to logical rule condition known F_i. The matching degree q can be calculated using the mathematical operation of matrix, because L_i and F_i are all matrices.

In the present, through a large number of tests, we introduce the following method to calculate the value of q:

$$q=Exp\,(L_i, F_i)/Var\,(L_i, F_i)\,. \tag{6}$$

Where *Exp* represents the mathematical expectation of L_i *and* F_i, *Var* represents the variance of them.

Finally, dynamic trust of the role R_i to request the permission P_i can be calculated by $DS_i' = DS_i \times q$. It is obvious that if $L_i = F_i$, that is, the degree of role R_i matching the facts is equal to its degree of matching the known context information, also we can draw a conclusion that $DS_i' = DS_i$.

So, when the role R_i requests permission P_i, first, we will judge the degree of R_i meet with the context information C_i. If the membership interval is in the range of $[x_i^-, x_i^+]$ known, then the request is successful. If the value is near the dynamic trust, the level of approaching can be calculated by carrying out fuzzy inference. Second, calculate the new dynamic trust of R_i named DS_i' according to $Rule_i$ and the real-time membership range with the context information through fuzzy reasoning process. Last, judge whether DS_i' is more than P_i's dynamic trust threshold, named θ_D, and if DS_i' reaches or exceeds the threshold set by the system, the request is successful, otherwise the request of the permissions is refused [5].

3 An Application Example

This example simulates users download remote resources. If there are two resources named Resource1 and Resource2 on the remote server, it provides remote web-services for users to download resources. Users can log in the system remotely to download resources at the same time.

Assume that the set of users is:

Users = {Peter, Mike, Rose}.

We assume that the initial static trust thresholds of Resource1 and Resource2 are θ_{S1} and θ_{S2}, the dynamic trust thresholds of Resource1 and Resource2 are θ_{D1} and θ_{D2} separately:

Resource1 trust set :{ $\theta_{S1} = 0.4$, $\theta_{D1} = 0.5$}

Resource2 trust set :{ $\theta_{S2} = 0.5$, $\theta_{D2} = 0.6$}

First of all, calculate the static trust *TS* based on users' experience, knowledge, recommendation (specific method of calculation is not discussed in this paper; please refer to the corresponding literature). When user accesses the system, if its *TS* is over the static threshold θ_S, it means the user is legitimate, so assign the role to the user to access the system, otherwise the request is denied. The results of computed users' static trust are as shown in table 1.

Table 1. List of users' static trust

Users	Resources	Static trust
Peter	Resource2	0.9
Mike	Resource1	0.6
Rose	Resource2	0.4
Peter	Resource1	0.8
Mike	Resource2	0.6

Secondly, we need to calculate the dynamic trust to determine whether to assign the permissions to the role.

We select three context information expressed by C_1, C_2, C_3: C_1 means the time scope of the request; C_2 means the flow range of the resources; C_3 means resources vacancy or not. All of these values of C_1, C_2 and C_3 need to be transformed to fuzzy membership interval initially.

Establish fuzzy rules of dynamic trust as shown in table 2.

Table 2. Fuzzy rules of dynamic trust

$Rule_1$: F_1=(C_1,0.1,[0.4,0.5]) ∧ (C_2,0.2,[0.4,0.8]) ∧ (C_3,0.7,[0.6,0.7]) →(Resource1,0.7)
$Rule_2$: F_2=(C_1,0.2,[0.6,0.7]) ∧ (C_2,0.3,[0.3,0.5]) ∧ (C_3,0.4,[0.4,0.8]) →(Resource2,0.8)

As an example, we assume that the user Roes requests Resource2, according to table 1, Roes' static trust value is 0.4, and it is less than Resource2 static trust threshold $\theta_{S2} = 0.5$, so the request is refused.

As another example, we assume that the user Peter requests Resource1. Suppose the matching facts with current context information and fuzzy interval of Peter expressed by L:

$$L= (D_1, [0.5, 0.6]) \wedge (D_2, [0.4, 0.5]) \wedge \ldots \wedge (D_n, [0.6, 0.7]) . \tag{7}$$

According to table 1, Peter' static trust value is 0.8, and it is greater than Resource1 static trust threshold value 0.4, then the role of legitimate users is assigned to Peter.

Next we will calculate the Peter's dynamic trust. According to Peter's current state of context information L and fuzzy rules $Rule_1$ from table 2:

$$F_1=(C_1,0.1,[0.4,0.5]) \wedge (C_2,0.2,[0.4,0.8]) \wedge (C_3,0.7,[0.6,0.7]) \rightarrow (\text{Resource1},0.7) \tag{8}$$

Using the fuzzy inference mechanism, we can calculate the actual dynamic trust value of Peter is 0.625, and it is greater than the Resource1 dynamic trust threshold θ_{D1}= 0.5, so the system will assign the requested permission Resource1 to Peter.

If Peter requests Resource2, according to matching facts L and fuzzy rules $Rule_2$ from table 2, we can calculate the dynamic trust value of Peter is 0.439, it is less than the Resource2 dynamic trust threshold θ_{D2} = 0.6, so the request is refused.

Thus through the trust-based fuzzy access control model, we realize the dynamic authorization from users to roles and roles to permissions.

4 Summary

The paper proposed a novel fuzzy access control model based on the TrustBAC. The model extended the concept of trust and used the fuzzy set theory to calculate the dynamic trust of role to determine whether to assign the permissions to the roles. The assignment process was accomplished in the basis of established fuzzy rules. It provides the technology of calculation the roles' dynamic trust by context-aware fuzzy membership interval and fuzzy rule for dynamic authorization in system. The next step is to research how to establish a reasonable membership relationship and improve the fuzzy evaluation mechanism in access control.

References

[1] Sandhu, R., Coyne, E., Feinstein, H.: Role-based Access Control Models. IEEE Computer 29(2), 38–47 (1996)

[2] Ray, I., Chakraborty, S.: TrustBAC-Integrating Trust Relationships into the RBAC Model for Access Control in Open Sytems. In: SACMAT 2006, Lake Tahoe, California, USA, pp. 7–9 (2006)

[3] Weiser, M.: The computer for the twenty-first century. Scientific American 265(3), 94–104 (1991)

[4] Sandhu, S.: The ARBAC 1997 Model for Role-Based Administration of Roles: Preliminary Description and Outline. In: Proceedings of the second ACM workshop on Role-based access control, November 1997, pp. 41–50 (1997)

[5] Zhang, L., Wang, X.: Trustworthiness-based fuzzy adaptive access control model for pervasive computing. Application Research of Computers (1), 311–316 (2009)

[6] Hilary, H.H.: Security is fuzzy: applying the fuzzy logic paradigm to the multipolicy paradigm. In: Proc. of the ACM Workshop on New Security Paradigms, pp. 175–184. ACM Press, New York (1993)

[7] Janczewski, L.J., Portougal, V.: Need-to-know principle and fuzzy security clearances modeling. Information Management & Computer Security 8(5), 210–217 (2000)

[8] Chang, E., Thomson, P., Dillon, T., et al.: The fuzzy and dynamic nature of trust. In: Katsikas, S.K., López, J., Pernul, G. (eds.) TrustBus 2005. LNCS, vol. 3592, pp. 161–174. Springer, Heidelberg (2005)

[9] Charalampos, P., Karamolegkos, P., Voulodimos, A., et al.: Security and privacy in pervasive computing. IEEE Pervasive Computing 6(4), 73–75 (2007)

An Evaluation Model of CNO Intelligence Information Confidence*

Yunyun Sun[2,1], Shan Yao[2], Xiaojian Li[1,2,3], Chunhe Xia[1,2], and Songmei Zhang[2]

[1] State Key Laboratory of Virtual Reality Technology and Systems, Beihang University, Beijing, China
[2] Key Laboratory of Beijing Network Technology, Beihang University, Beijing, China
[3] College of Computer Science & Information Technology, Guangxi Normal University, Guilin, China
{syy,yaoshan,amei}@cse.buaa.edu.cn, {xiaojian,xch}@buaa.edu.cn

Abstract. Intelligence activity is one of the most important activities during Computer Network Operation (CNO) command and decision-making process. Specially, Evaluation of Intelligence Information Confidence is the basic element of intelligence activity and process. As intelligence is essential foundation when forming Course Of Action (COA), confidence evaluation, with the main function of determining facticity and reliability of intelligence, will effect the quality and efficiency of CNO command and decision-making. In this paper, an evaluation model of CNO intelligence information confidence IICEM was described with respect to the reliability of collectors and the credibility of the information content through analyzing the intelligence information evaluation role model IIERM. The results of experiments on the prototype based on IICEM show that different confidence information could be distinguished by IICEM, which affect the following analysis and production activities.

Keywords: CNO intelligence; confidence; evaluation model; IICEM.

1 Introduction

CNO is defined as the activities and behaviors taken to enhance and defend own information capabilities, prevent and weaken adversary's information capabilities in computer network for the purpose to gain the information superiority [1]. Intelligence Process is the process by which information is converted into intelligence and made available to users. The process consists of six interrelated intelligence operations: planning and direction, collection, processing and exploitation, analysis and production, dissemination and integration, and evaluation and feedback [2] [3]. In this paper, we focus on the evaluation of intelligence information confidence in analysis and

* This work is supported by three projects: the National 863 Project "Research on high level description of network survivability model and its validation simulation platform" under Grant No.2007AA01Z407; The Co-Funding Project of Beijing Municipal education Commission under Grant No.JD100060630 and National Foundation Research Project.

W. Liu et al. (Eds.): WISM 2009, LNCS 5854, pp. 400–412, 2009.

production operation. Signal value in intelligence information is object of our evaluation model, which include targets' running status and platform features reported by collectors. Information confidence is the degree to which this information is confirmed as true, is subject's cognition of information. Evaluation of information confidence is the process of calculating degree of confidence assigned to each piece of information with respect to the reliability of the source (in CNO, source refers to collector) and the credibility of the information content. Until now, research on information evaluation mainly focuses on theory, which reaches consensus on two evaluating factors. However, problems exist when introducing the theory into CNO intelligence process, so we need to analyze roles in CNO intelligence information evaluation process and their relationships, so we built IICEM, CNO intelligence information confidence evaluation model.

In this paper, the roles of CNO intelligence process will be firstly analyzed to find out the relationships between the reliability of collectors and the credibility of the information content, based on which, the evaluation model of CNO intelligence information confidence will be constructed.

In section 2, current research status of information evaluation is introduced. In section 3, the evaluation model of CNO intelligence information confidence are discussed, including the construction of IIERM, the analysis of concepts and process of intelligence information evaluation, and establishment of the automaton of IICEM. In section 4, a prototype is designed. In section 5, a scenario specification is introduced. Finally, we give our conclusion and acknowledgements.

2 Related Works

Input information of CNO intelligence system mainly come from state about target collected by CNE(Computer Network Exploitation), while in the real world, it often happens that there is undiscovered or uncertain threats, inadequate information or adversary cheat behaviors, which conduces to the uncertainty of information from different sources [4]. Information evaluation has drawn more and more researchers' attentions. JP2-01 [2] talks that Each new item of information is evaluated by the appropriate analysis and production element with respect to the reliability of the source and the confidence of the information, which is the same as STANAG [5] [6] from NATO. This paper is based on the similar idea, in which we systematically analyze roles of CNO intelligence process and extract evaluation notions. L. Cholvy [7] [8] propose a model of information evaluation, which takes into account three notions, with total ignorance about the reliability of an unknown source. Lisa Krizan [9] analyzes three aspects of source and information itself contributing to an initial assessment of the value of a particular piece of information to the intelligence production process. The theory and model about information evaluation above mainly focus on the general modeling that seems to be applicable to all situations, while don't cost much energy on analyzing cause of information uncertainty and correlation among information. In this paper, cause of CNO intelligence information uncertainty is firstly studied to support modeling, and algorithm of evaluation is mainly based on correlation among information.

3 CNO Intelligence Information Confidence Evaluation Model

In this section, we specify intelligence information evaluation role model (IIERM) and the concepts of the entities involved in information confidence evaluation and provide the formal analysis of evaluation activities and the interdependence among them. Then, based on these works, we present the information confidence evaluation model (IICEM) at last.

3.1 Intelligence Information Evaluation Rule Model

The concepts of roles used in the intelligence information evaluation process and their relationships are shown in figure 1. Information reported by collectors according to collection intention, transferred to intelligence production system, integrated and grouped by integration activity, will be delivered to evaluators to evaluate its degree of confidence. There are three important roles in intelligence information evaluation process: target, collector, and evaluator. In CNO, target is the identification of objects that friend and adversary operate attack, defense or exploit on [1]. Collector is the operator that carries out activities and behaviors of data obtaining. Collection activities of CNE are just data obtaining activities, including three basic activities, such as probing, grabbing and intercepting.

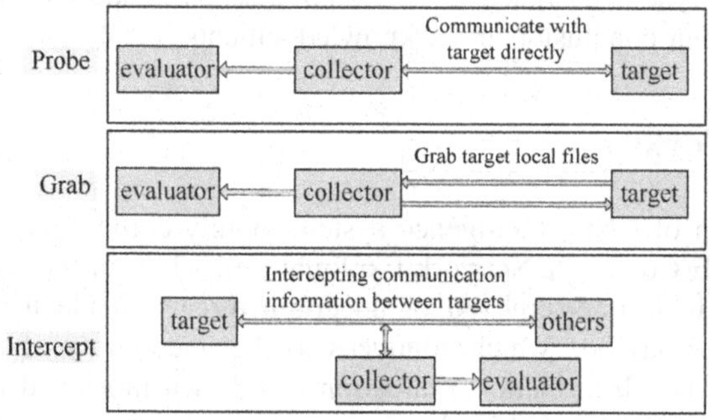

Fig. 1. Intelligence information evaluation role model

3.2 Cause Analysis of Uncertainty of Intelligence Information

Cause of uncertainty of information divided into two aspects:

1. The distance between the real truth and information which target appears, because targets don't show its real state under some conditions;
2. The distance between information that target appears and information reported by collectors, because reliability of collectors is influenced by its location and currently network environment.

It is difficult to evaluate behavior of target directly, but behavior of collectors and information content is what we can contact and evaluate. According to factors that should be considered in information evaluating [2][5] and IIERM, evaluating reliability of collectors, credibility of information content and integration of their evaluating results are extracted as three aspects composing the evaluating process of CNO intelligence information confidence.

3.3 Analysis of Collectors' Behavior

From the aspects of influencing information confidence evaluation, collectors' behavior was analyzed as followed.

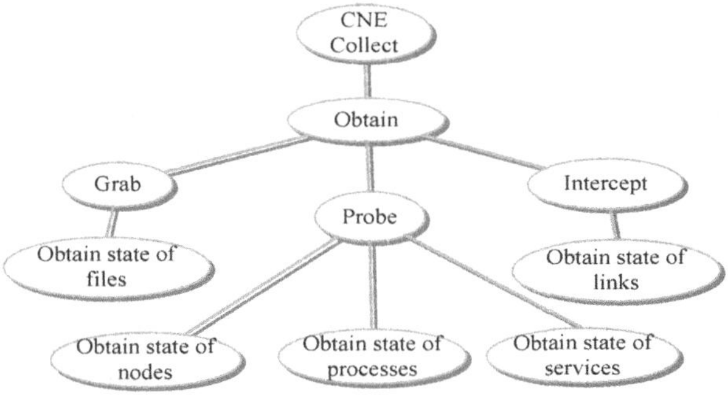

Fig. 2. Activities of collectors

Collectors are divided into three basic activities according to obtaining means. In CNE, probing obtains data of target's (including network, node, process, service) state through "signal intelligence" direct-finding[10]; intercepting listens to the communication among targets, like nodes, processes or services, to get data of state transferring in connections of communication; grabbing fetches targets' data like its authorization and configuration through reading targets' local files.

Reliability of three basic activities of collectors needs to be evaluated separately in real network environment, because even the same network environment may have different effects on reliability of these three basic collection activities. The result of probing and intercepting has close relationships to current network environment, while data intercepting are being affected by location of collection process[11]. When reporting information about targets' platform features like OS type and service, collectors will append precision of this information based on its own judge.

Definition 3.1. Let precision be the degree to which collected platform information tally with tuples of features. This value could be used in the following evaluation process of information confidence.

While evaluating reliability of collectors, evaluator might need to communicate with other evaluators. The reasons are followed:

1. There are one more evaluators in intelligence system. They subscribe information from collectors according to missions or roles with publish/subscribe mechanism based on P2P network, so one evaluator is not possible to subscribe all information reported by one collectors;
2. As analyzed above, reliability of one collector is not same when it carries out different activity. However, evaluator only has chances to communicate with some activity of one collector, which result in partial knowledge of this

collector. So evaluator needs to communicate with others to know the overall knowledge, in case evaluating information from unknown collectors without any communicating history.

3.4 Analysis of Intelligence Information Content

Besides evaluation of reliability of collectors, information itself also has a bearing on its degree of confidence. Object of information confidence evaluation is CNO intelligence information, which is recorded in intelligence database, including status of CNO target, identification of collectors, and precision (not necessary).

Definition 3.2. Let CNO TARGET be a set of 2-tuple (IEF, LOCATION) where IEF is a set of identification of friend or adversary and LOCATION is a set of identification of target objects [1].

$$\text{TARGET}::=\{(\omega,l)|\omega\in\text{IEF}, l\in\text{LOCATION}\} \tag{1}$$

$$\text{LOCATION}::=\{\text{lnet},\text{lnode},\text{lprocess},\text{lservice},\text{lfile},\text{ln-connect},\text{lp-connect},\text{ls-connect}\} \tag{2}$$

State of a CNO target includes: status varying anytime and platform feature stable comparably, which all imply some possible development tendency. We introduce correlation analysis into the evaluation of information content. Information content can be evaluated according to its kind.

1. Vertical correlation: designed for running level information. Correlation of running level of the same target from the collector with the same obtaining means;
2. Horizontal correlation: designed for signal platform feature information from different collectors.

Other intelligence information like configuration files are not in the area of my study.

3.5 CNO Intelligence Information Confidence Evaluation Concepts

Definition 3.3. Let the evaluation process of intelligence information confidence be a set of 4-tuple which includes CNO target information, degree of confidence of target information, evaluation activities and their order. Evaluation activities include evaluation of reliability of collectors, evaluation of credibility of information content and integration evaluation.

$$\Psi::=(\text{TINFO},\text{COFINDECE}, \Phi_{\text{EVALUATION}}, \xi_{\text{EVALUATION}}). \tag{3}$$

$$\begin{aligned} &\psi^{\text{EVALUATION}}: \text{TINFO}\rightarrow\text{CONFIDENCE};\\ &\Phi_{\text{EVALUATION}}:\{\varphi^{\text{EofCR}}, \varphi^{\text{EofCC}}, \varphi^{\text{Integration}}\};\\ &\xi_{\text{EVALUATION}}: \varphi^{\text{EofCR}} \prec \varphi^{\text{EofCC}} \prec \varphi^{\text{Integration}}. \end{aligned} \tag{4}$$

There are three actions in the evaluation process of CNO intelligence information confidence: ξ_{EofCR}, ξ_{EofCC} and $\xi_{\text{integration}}$, shown as followed.

Fig. 3. Activities of evaluation process of CNO intelligence information confidence

Definition 3.4. Let evaluation process of reliability of collectors be a set of 4-tuple, as followed, including CNO targets' information confidence, reliability of collectors, a set of reliability evaluation actions Φ_{EofCR}, and their orders ξ_{EofCR}.

$$\psi^{EofCR} ::= \left(\text{TINFO,CONFIDENCE, RELIABILITY}, \Phi_{EofCR}, \xi_{EofCR}\right); \quad \xi_{EofCR} = \delta^{feedback} \prec \delta^{communication} \quad (5)$$

$\delta^{feedback}$ is the confirming action according to information feedback, which based on new information's confidence from this evaluator and confidence changes happened among old information.

$$\delta^{feedback}: \text{CONFIDENCE} \rightarrow \text{RELIABILITY}. \quad (6)$$

$\delta^{communication}$ is update action of collector's reliability through communicating with other ones when evaluator encounters with some collector or obtain means that he never met.

$$\delta^{communication}: \text{RELIABILITY} \rightarrow \text{RELIABILITY'}. \quad (7)$$

Definition 3.5. Let evaluation process of credibility of information content be a set of 5-tuple, as followed, including CNO targets' information, credibility of targets' information, reliability of collectors, set of information content credibility evaluation actions Φ_{EofCC}, and their orders ξ_{EofCC}.

$$\psi^{EofCC} ::= \left(\text{TINFO,CREDIBILITY, RELIABILITY}, \Phi_{EofCC}, \xi_{EofCC}\right); \quad \xi_{EofCC} = \delta^{vertical} \prec \delta^{horizontal} \quad (8)$$

$\delta^{vertical}$ is vertical correlation action. Firstly, we need to conduct experiment to establish priori distributions of running levels under similar environment as targets. Then credibility of level information is calculated based on the possibility it hold in priori distributions. Finally, this new information participates in the update of priori distribution with Bayesian statistics theory to reduce the distance between priori distribution and target's practical situation.

$$\delta^{vertical}: \text{TINFO(level)} \rightarrow \text{CREDIBILITY}. \quad (9)$$

$\delta^{horizontal}$ is horizontal correlation action, including assigning credibility among the same kind of information of the same target from different collectors using statistical analysis theory.

$$\delta^{horizontal}\text{: TINFO(signature)}\rightarrow\text{CREDIBILITY.} \quad (10)$$

3.6 CNO Intelligence Information Confidence Evaluation Model

Based on the analysis above, we propose an evaluation model of CNO intelligence information confidence as follows.

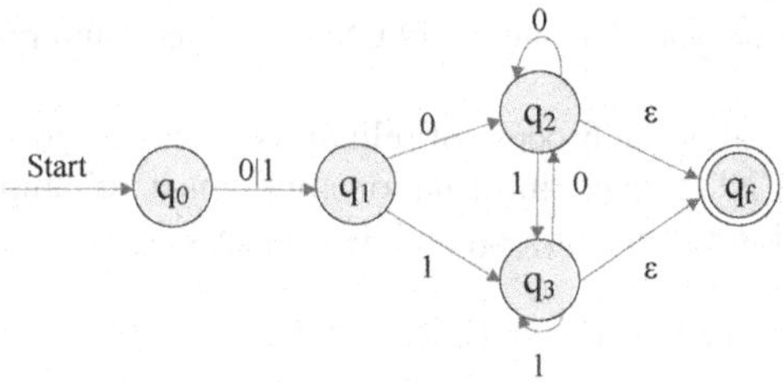

Fig. 4. the Automaton Model of IIERM

The automata model of IIERM: $M = \{Q,\Sigma,\delta,q_0,F\}$.
In the equation,
$Q = \{q_0, q_1, q_2, q_3\}$, $\Sigma = \{0,1\}$, $F = \{q_f\}$.
The meanings of the states are as follows:
q_0: Initial State (0: when not existing communication history of this collectors, communicate with other evaluators; 1: when existing, calculate reliability of this collector directly); q_f: End State; q_1: State of determining reliability of collectors (0: when evaluation objects are running state of targets, evaluate credibility of information vertically; 1: when evaluation objects are platform features of targets, evaluate credibility of information horizontally); q_2: State of Determining credibility of vertical evaluation (0: when the next evaluation information is running state about targets, return to the same state; 1: when the next evaluation information is platform features about targets, evaluate the credibility of information vertically; ε: when the next evaluation information is none, integrate and calculate the degree of confidence placed on the information, enter end state); q_3: State of Determining credibility of horizontal evaluation (opposition of q_2). Transition matrix of IICEM automaton is followed:

Table 1. Transition matrix of IICEM automaton

	0	1	ε
q_0	q_1	q_1	
q_1	q_2	q_3	
q_2	q_2	q_3	q_f
q_3	q_2	q_3	q_f
q_f			

Proposition 1. The automaton of IICEM accepts character string $(0|1)(0|1)^n$. If input character string is accepted, means reachable. First '1/0' in input string denotes that whether there is knowledge of collector's reliability or not. Other '0/1's represent whether information being evaluated is running status or platform features.

Proof: Let n=1,ω=(0|1)(0|1), according to transfer function:
$(q_0,00) \vdash (q_1,0) \vdash (q_2,\varepsilon) \vdash q$. Same as when inputs are '01', '10', '11'.
Let n=k, $\omega=(0|1)(0|1)^k$, according to transfer function:
$(q_0,\omega) \vdash^* (q_2,\varepsilon) | (q_3,\varepsilon) \vdash q$. Then let $n = k + 1$, $\omega=(0|1)(0|1)^{k+1}$, we have:
$(q_0,\omega) \vdash (q_1,(0|1)^{k+1}) \vdash (q_2,(0|1)^{k+1}) | (q_3,(0|1)^{k+1}) \vdash (q_2,(0|1)^{k-1}) | (q_3,(0|1)^{k-1}) \vdash (q_2, \varepsilon) | (q_3, \varepsilon) \vdash q$□.

4 Intelligence Information Confidence Evaluation Prototype

Based on IICEM, we designed and implemented a CNO intelligence information confidence evaluation prototype as follows.

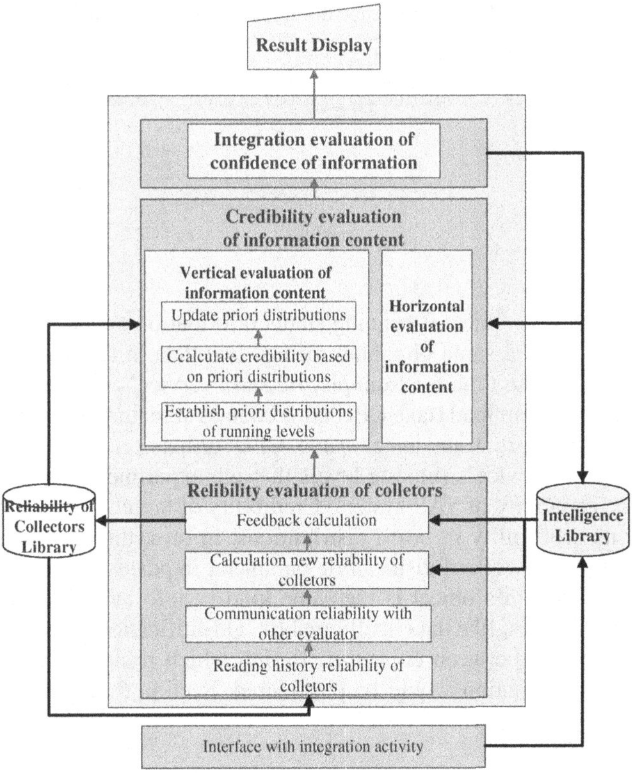

Fig. 5. Intelligence information confidence evaluation model

4.1 Reliability Evaluation of Collectors

This part is evaluation of reliability of collectors, history knowledge of which is stored in reliability collectors' library. Once confidence of new information reported are evaluated or old information's confidence are updated, reliability of collectors need to be calculated again. The reliability of collectors is defined in reference to its use in the past [7], which can be measured, as the average of the confidence of information. When encountering some collector that has never met, evaluator needs to get reliability from other evaluators and write back to the reliability library. The item in the reliability library is recorded like (CollectorId, ObtainId, Reliability), the same as the information communicated among evaluators.

4.2 Credibility Evaluation of Information Content

Evaluation of information content credibility is mainly based on analysis of information in the intelligence database. The item recorded in intelligence database is as followed:

Table 2. Intelligence library schematic diagram

Intelligence Library[table]

ID	CollectorID	ObtainID	target	status	feauture	confidence	precision ...
1.	p_1	o_1		t_1	s_1	f_1	c_1
2.	p_1	o_1		t_1	s_2	f_2	c_2
3.	p_1	o_1		t_1	s_3	f_3	c_3
4.	p_2	o_1		t_1	s_4	f_4	c_4
5.	p_3	o_1		t_2	s_5	f_5	c_5
⋮	⋮						

Vertical evaluation's object is running status information about one target from the same collector using the same obtaining means, like data in the quadrate box. Take target service's response time for example. Because service's response time has close relationships with system load (take CPU utilization as signature of system load), priori distributions of response time under same CPU utilization are established before evaluation of target service's running levels through experiments under similar situation using proc capability of SAS system. Credibility of target service's response time is defined as its probability in priori distributions. In turn, this new response time is involved in the update work of distribution parameter in posterior distribution.

Horizontal evaluation's object is the same kind of information about same target from different collectors, like data in elliptic box. Quantification of credibility refers to support or confliction between collectors, during which reliability of collectors and precision of the information would be considered. Section 5 scenario will introduce application of it specifically.

4.3 Integration Evaluation of Confidence of Information

Based on two parts above, the reliability of collectors (RC) and credibility of information content (CC) have been quantized, and then the integrated formula is shown:

$$\text{Confidence of Information (Cf)} = \text{P} \times \text{Reliability of Collectors (RC)} + \text{Q} \times \text{Credibility of Information Content (CC)} . \quad (11)$$

$P \in [0,1]$, $Q = 1 - P$, $RC, CC \in [0,1]$. P and Q in this formula, which indicate the acquaintance of effects of reliability of collectors and credibility of information content on the degree of confidence of information, are appointed by the evaluator or commander. This rate should be written into confidence item of intelligence library.

5 Scenario Specification

A simple scenario is used to illuminate the application of our prototype. Figure 7 presents the topology of the scenario. CNO intention is to destroy fileserver in target network.

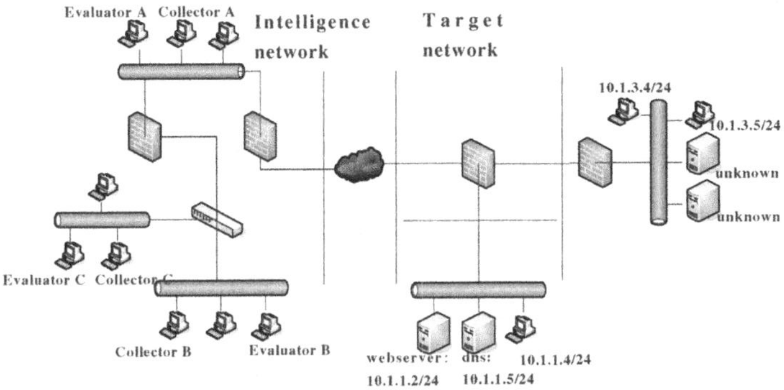

Fig. 6. Experiment topology

In the following part, collector A is called C_A for short, the same as C_B and C_C. Information about target network reported by collectors will be transferred to intelligence system to evaluate the degree of confidence, to analyze and produce vulnerability and attack means which can be used to implement the attack operation. Through analysis of variation of information's confidence and result of collectors' reliability evaluation, the application value of IICEM will be shown. Table 3 shows segment of intelligence library about webserver in target network before attack operation.

Table 3. Webserver's information in intelligence library before attack operation

ID	ReportTime	CollectorID	ObtainID	precision	address	mask	port	Info	level	S-Version
1	2009-4-23 21: 12: 00	C_A	g	0.9	10.1.1.2	255.255.255.248	80	webserver	9	4.0.5.0
2	2009-4-23 21: 12: 00	C_B	p	0.8	10.1.1.2	255.255.255.248	80	webserver	7	4.0.4.0
3	2009-4-23 21: 12: 00	C_C	g	0.8	10.1.1.2	255.255.255.248	80	webserver	8	4.0.5.0

CollectorID(C_i): identification of collectors; ObtainID: identification of obtaining means by collectors; Level: grade of this service's response time; S-Version: version of this service.(g : grab; p : probe; i : intercept).

Table 4. Segment of reliability of collectors library

Collectors	C_A	C_A	C_A	C_B	C_B	C_B	C_C	C_C	C_C
ObtainId	g	p	i	g	p	i	g	p	i
Reliability	0.9	0.8	0.7	0.8	0.7	0.6	0.7	0.6	0.5

Reliability: reliability of collectors according to its past appearances evaluated by evaluators. Calculate reliability of collectors based on the history knowledge in table 4 and precision of information in table 3.

$rc(db_1)=rc(C_A,g)*pre(db_1)=0.9*0.9=0.81$; $rc(db_2)=rc(db_3)=0.56$;.

After calculating reliability of these three collectors, we evaluate the content of information horizontally from analysis of consistency among different collectors' information. (Because the evaluation objects are not running status, we just show how to evaluate credibility horizontally.)

1. C_A report S-Version = 4.0.5.0 (a)
2. C_B report S-Version = 4.0.0.0 (b)
3. C_C report S-Version = 4.0.5.0 (a)

For expressing more distinctly, we denote 4.0.5.0 S-Version by "a", while denoting 4.0.0.0 by "b".

$$\mathbf{cc(db_a)} = \frac{\Sigma_1^n (\mathbf{rc}(\mathbf{db_i}))}{\Sigma_1^m (\mathbf{rc}(\mathbf{db_j}))} \times 100\% . \tag{12}$$

n equals the number of information include a, and db_i equals reliability of collectors report this information. m equals the number of information include S-version of target, and db_j equals reliability of collectors report this information.

$$\begin{gathered} \mathbf{cc(db_a)} = \frac{\Sigma_1^2 (\mathbf{rc}(\mathbf{db_i}))}{\Sigma_1^3 (\mathbf{rc}(\mathbf{db_j}))} \times 100\% \approx 71\%; \mathbf{cc(db_b)} \approx 29\%; \\ \mathbf{cc(db_1)} = \mathbf{cc(db_3)} = \mathbf{cc(db_a)} = 71\%; \mathbf{cc(db_2)} = \mathbf{cc(db_b)} = 29\% \end{gathered} \tag{13}$$

We suppose that P and Q in (8) are 0.5. Integration calculations of three information's confidence are followed:

$Cf(db_1)=P*RC(db_1)+Q*CC(db_1)=76\%$; $Cf(db_2)= Cf(db_3)63.5\%$.

We suppose that the degree of confidence of information can be accepted by decision-maker is 50%, so db_1 and db_2 would be used in the analysis and produce activity, while S-Version in db_3 may be discarded. Now evaluation of information finished a filtration process of untrusted information.

Using information in db_1 and db_2 that webserver's service version (S-Version) in target network is 4.0.5.0, intelligence reasoning activity derives that there is vulnerability CVE-2002-0364 on webserver, which can be exploited to get its root authority, then using webserver to launch denial of service attack to the target fileserver.

Table 5. Webserver's information in intelligence library after attack operation

ID	ReportTime	CollectorID	address	port	Info	level
1	2009-4-23 21: 12: 00	CA	10.1.3.8	21	fileserver	9
2	2009-4-23 21: 13: 00	CA	10.1.3.8	21	fileserver	7
3	2009-4-23 21: 14: 00	CA	10.1.3.8	21	fileserver	5
4	2009-4-23 21: 15: 00	CA	10.1.3.8	21	fileserver	3
5	2009-4-23 21: 16: 00	CA	10.1.3.8	21	fileserver	1
6	2009-4-23 21: 17: 00	CA	10.1.3.8	21	fileserver	0

Table 5 shows fileserver's running level after attack, which is presenting a downtrend. We can see the success of attack means which prove the effect of evaluation model of intelligence information confidence. The running level information need to be evaluated too, with vertical algorithm not mentioned in this example.

6 Conclusions

In this paper, we propose an evaluation model of intelligence information confidence. This model includes all the basic elements to be considered when evaluation information proposed by Joint Publication and STANAG. Meanwhile, compared to their information evaluation theory, this model provide not only framework of evaluation, also relevant methods to quantify reliability of sources and credibility of information, which can be applied to CNO intelligence information evaluation directly, and easy to be abstracted to support other kind of information evaluation process.

In this model, we focus on the evaluation of information confidence about CNO targets through analysis of collectors' reliability and information content. From analyzing information's confidence about one target, we can figure out this target's behavior, which can assist decision in computer operations.

Acknowledgments. I am deeply grateful to Ms. Qi Li and Mr. Jie Feng for many substantial suggestions. Detailed comments by all of them greatly improve the paper.

References

1. Xiaojian, L.: Research on Computer Network Self-Organizing Operations (in Chinese). PhD Thesis, Beihang University (2007)
2. Joint Chiefs of Staff: Joint Publication 2-01: Joint and National Intelligence Support to Military Operations. Washington DC: Joint Publications, as amended through, October 7 (2004)
3. Pace, P.: Joint Publication 2-0: Joint Intelligence. Washington DC: Joint Publication, June 22 (2007)
4. Shan, Y.: Research and Implementation of the Tactical Intelligence System of Computer Network Operation based on peer-to-peer communication (in Chinese). Master thesis, Beihang University (2007)

5. Annex to STANAG 2022 (Edition 8) North Atlantic Treaty Organization (NATO) Information Handling Services, DODSID, Issue DW9705
6. AAP-6-2006, North Atlantic Treaty Organization NATO Standardization Agency, NSA (2006), http://www.nato.int/docu/stanag/aap006/AAP-6-(2006)
7. Cholvy, L., Nimier, V.: Information Evaluation discussion about: STANAG 2022 recommandations 2003. In: NATO-IST Symposium "Military data and information fusion", Prague, rublique Tchue, October 20-22 (2003)
8. Cholvy, L.: Modelling information evaluation in fusion. In: 10th International Conference on Information Fusion 2007, July 9-12, pp. 1–6 (2007)
9. Lisa, K.: Intelligence Essentials for Everyone. Joint Military Intelligence College, USA (1999)
10. Joint Chiefs of Staff: Joint Publication 1-02: Department of Defense Dictionary of Military and Associated Terms. Washington DC: Joint Publication, as amended through, December 7 (1998)
11. Changhua, Z., Changxing, P., Jian, L.: Network measurement and its key technologies (in Chinese). Jounal of Xidian University (2002)

A Verified Group Key Agreement Protocol for Resource-Constrained Sensor Networks

Mi Wen[1], Jingsheng Lei[1], Zhong Tang[1], Xiuxia Tian[1], Kefei Chen[2], and Weidong Qiu[3]

[1] Department of Computer Science and Engineering, Shanghai University of Electric Power, Shanghai 200090, China
[2] Department of Computer Science and Engineering, Shanghai Jiaotong University of China, Shanghai 200240, China
[3] School of Information Security Engineering Shanghai Jiaotong University of China, Shanghai 200240, China
superwm_9@yahoo.com, jshlei@126.com, tangzhong64@163.com, tianxiuxia_76@sina.com, kfchen@sjtu.edu.cn, qiuwd@sjtu.edu.cn

Abstract. As a result of the growing popularity of wireless sensor networks (WSNs), secure group communication becomes an important research issue for network security because of the popularity of group-oriented applications in WSNs, such as electronic monitoring and collaborative sensing. The secure group key agreement protocol design is crucial for achieving secure group communications. As we all know, most security technologies are currently deployed in wired networks and are not fully applicable to WSNs involving mobile nodes with limited capability. In 2008, we proposed a Blom's matrix based group key management protocol (BKM) with robust continuity for WSNs. Unfortunately, in this paper we present that the BKM has a security weakness in which participants cannot confirm that their contributions were actually involved in the group key establishment. This is an important property of group key agreement. Therefore, we propose a verified group key agreement protocol (V-BKA) for resource-constrained WSNs. We show that the proposed protocol produces contributory group key agreement. We demonstrate that the proposed protocol is a perfect key management protocol and is well suited for resource-constrained WSNs.

Keywords: Wireless Sensor Networks (WSNs); Group key agreement; Continuity; Authentication; Group key.

1 Introduction

Wireless sensor networks (WSNs) have revolutionized the field of target tracking, battlefield surveillance, intruder detection and scientific exploration [1]. Multicasting in WSN, as an efficient communication mechanism for delivering information to a large group of recipients, has led to the development of a range of powerful applications in both commercial and military domains. Key management serves as the crucial foundation to enable such secure group communications. However, the large size of

W. Liu et al. (Eds.): WISM 2009, LNCS 5854, pp. 413–425, 2009.

the serving group, combined with the dynamic nature of group changes, poses a significant challenge on the scalability and efficiency on key management research [2]. In this paper, we study the problem of verified group key agreement over WSNs.

A group key establishment protocol allows group members to construct a group key encrypt/decrypt all the messages destined to the group over an open channel. There are two kinds of group key establishment protocols: group key distribution and group key agreement. One advantage of the group key agreement protocol is that no participant can predetermine the group key. A group key agreement protocol involves all participants cooperatively establishing a group key. Here, we focus on a secure group key agreement protocol design. In the past decades, many Diffie–Hellman based key agreement [3-6] schemes were proposed to improve the system performances. But, the computational costs required by these previously proposed schemes were beyond the computing capability of mobile nodes in WSNs.

In such a scenario, a good group key agreement should efficiently manage the group key when members join and leave; this is especially true in WSNs where the nodes are highly mobile and the network topology is dynamic. At the same time, given the limited capabilities on sensor nodes, the key continuity which enables all the sensors to establish and update both of the pair-wise key and group key with the same key pre-distribution information should also be considered, because it can significantly decreases the resource consuming of key management in resource-constrained WSNs. Unfortunately, although considerable research has been done on key management over WSNs, only a few them addressed the continuity between the group key and the pairwise key, and none of them efficiently solved the verified group key agreement problem simultaneously. Detailed analysis will be discussed in the next section.

Motivated by the mentioned above, in this work, we consider the scenario of WSN wherein, each of the participating nodes has some pre-loaded secrets and wish to establish pairwise key and group key with its neighbors in their group for further communication. Earlier, we proposed a key management scheme with robust continuity for WSNs, called BKM [7]. In this paper, we extend the BKM to include member verification to ensure the group key established is legal and their contributories are also included in it, and also with robust continuity, called V-BKA. The proposed protocol requires only two rounds to construct a group key. We show that the proposed protocol is a verifiable contributory group key agreement protocol. We demonstrate that the proposed protocol is still secure against node compromise attacks and keep their robust continuity when members join and leave. Additionally, the proposed protocol has absolute group confidentiality and perfect group fairness.

The rest of this paper is organized as follows. Section 2 describes the related works. Section 3 provides a modified version of Blom's matrix construction and key distribution. Section 4 reviews the BKM protocol and presents its security weakness. Section 5 presents the verified group key agreement protocol, V-BKA. Performance analysis will be presented in section 6. Finally, we conclude in section 7.

2 Related Works

Several secure group communication key management schemes have been presented in the past decade. The most well-known distributed scheme is the Group

Diffie–Hellman (GDH) method [3-5]. This approach requires a linear number of expensive public-key operations, while there have been efforts to reduce the number of public-key operations [5, 6]. However, due to the computational overhead imposed by public-key operations, each user needs to negotiate with its communication peers to maintain the communication group. In general, this class of schemes is not suitable for resource limited WSNs.

Recently, some group key distribution schemes are proposed for WSNs. In 2003, Zhu proposed a LEAP protocol [8], which can support establishment of 4 types of keys: Individual Key, paiwise key, Cluster Key and Group Key. The establishment of group key in LEAP bases on the establishment of the previous three keys. Thus, it's too complicated and resource consuming. At the same time, Liu [9] proposed an efficient self-healing group key distribution with revocation capability, which introduced the two kinds of polynomials to establish and rekeying the group key. Later, Bohio[10] and Eltoweissy [11] proposed two group key distribution methods by using subset difference and combinatorial optimization respectively. Also, some literatures use reversed hash chains to construct group key distribution schemes, such as [12].

These literatures do give some security supports for secure group communications in WSNs. But, we should notice that most of them are group key distribution schemes. That is, the group key is decided by one node. In many scenarios, however, the key distribution approach is not appropriate because the powerful node might become a single failure point for the group's security.

In 2008, we proposed Blom's matrix based key management protocol (BKM) with robust continuity for WSNs. The protocol can establish group key and paiwise key by using the same pre-loaded secrets, and it can update the group key and pairwise key at them same time, thus, the group key has good forward and backward secrecy and the pairwise key has strong compromise resistance and network connectivity. It is very suitable for resource-constrained WSNs. Unfortunately, it is still a group key distribution scheme. Therefore, in this paper we will extend the BKM to be a good group key agreement scheme.

Table 1. Basic notations

N_i	A node *i* $(i=,1,\ldots,n)$, where N_n is the trusted dealer and it is a powerful node, i.e. the cluster head in a cluster.
s_i	Row seed of the matrix *D*; the row seed used in each row *i* of matrix *D* should not bigger than s_i.
g^i	Column seed of matrix *B*
K_{ij}	The pairwise key between node N_i and N_j
K_G	The group key of the initial set $N=\{ N_1, N_2, \ldots, N_n \}$

3 Modified Version of Blom's Key Distribution Approach

Blom proposed a key distribution approach [13], which allows any pair of nodes in a network to be able to find a pairwise secret key. As long as no more than λ nodes are compromised, the network is perfectly secure. Du [14] gave an extended version to construct the multiple key spaces. For the sake of key updating, we modify the Blom's symmetric matrix construction in [16]. We briefly describe how our modified version of Blom's key distribution approach works as follows. Some used notations in this paper are given in Table 1.

The base station (BS) in WSNs (acting as a trusted server) first computes a $n*n$ matrix B over a finite field GF(q), B is considered as the public information, q is a prime, and $q<n$. One example of such a matrix is a Vandermonde matrix whose element $b_{ij}=(g^j)^i \bmod q$, where g is the primitive nonzero element of $GF(q)$ and g^j is the jth column seed. That means:

$$B=\begin{bmatrix} 1 & 1 & 1 & & 1 \\ g & g^2 & g^3 & & g^n \\ & & & & \vdots \\ g^{n-1} & (g^2)^{n-1} & (g^3)^{n-1} & \cdots & (g^n)^{n-1} \end{bmatrix}$$

This construction requires that $n^2<\phi(q)$ i.e., $n^2<q-1$. Since B is a Vandermonde matrix, it can be proved that the n columns are linearly independent when $g, g^2, g^3 \cdots, g^n$ are all distinct.

Next, the BS generates n row seeds $s_1, \cdots s_n$, where $s_i (i=1,\dots,n)$ is the random prime number of $GF(q)$ and it is only known to the powerful node (N_n) in the network, e.g., the cluster head of WSNs. And then BS creates a random $n*n$ symmetric matrix D over $GF(q)$. Each row of the D is composed of hash values of the row seeds. Differing from the construction of matrix B, the elements in symmetric matrix D are generated as follows:

$$for(i=1; i\le n; i++)$$
$$for(j=1; j\le n; j++)$$
$$\{if\,(i>j)\ d_{ij}=H^i(s_j); else\ d_{ij}=H^j(s_i);\}$$

Where d_{ij} is the element in matrix D. An example of matrix D with size $3*3$ is shown as follows:

$$\begin{bmatrix} H^1(s_1) & H^2(s_1) & H^3(s_1) \\ H^2(s_1) & H^2(s_2) & H^3(s_2) \\ H^3(s_1) & H^3(s_2) & H^3(s_3) \end{bmatrix}$$

At last, the BS computes a $n*n$ matrix $A=(DB)^T$, where T indicates a transposition of the matrix. The elements in matrix A denote as a_{ij}, where $a_{ij}=\sum_{\beta=1}^{n} d_{j\beta} b_{\beta i}$. The matrix B is public while the matrix D is kept secret by the base station. Since D is symmetric, the key matrix $K=AB$ can be written as:

$$\mathrm{K}=(\mathrm{DB})^{\mathrm{T}}\mathrm{B}=\mathrm{B}^{\mathrm{T}}\mathrm{D}^{\mathrm{T}}\mathrm{B}=\mathrm{B}^{\mathrm{T}}\mathrm{DB}=(\mathrm{AB})^{\mathrm{T}}=\mathrm{K}^{\mathrm{T}}$$

Thus K is also a symmetric matrix and $K_{ij}=K_{ji}$, where K_{ij} is the element of K at ith row and jth column. We take K_{ij} (or K_{ji}) as the pairwise key between node N_i and node N_j. To carry out the above computation, nodes N_i and N_j should be able to compute K_{ij} and K_{ji} respectively. This can be easily achieved using the following key predistribution procedure, for node N_i:

(1) Store the ith row of matrix A at node N_i, denoted as $r_i(A)$, i.e., $r_i(A) = [a_{ij}]$.

(2) Store the ith column seed g^i of matrix B at N_i.

4 Analysis of the BKM Protocol

In this section, we briefly review the BKM [7] protocol. We present that the protocol is not a contributory group key agreement protocol because the participants cannot confirm that their contributions were involved in establishing the group key.

4.1 Pairwise Key Establishment

After deployment, each node has a piece of secret information as described in section 3. When nodes N_i and N_j need to find the pairwise key between them, they first exchange their column seeds of matrix B (since B is the public information, it can be sent in plaintext). Then, by using the preloaded secrets, they can compute K_{ij} (or K_{ji}) respectively as: $K_{ij} = \sum_{\beta=1}^{n} a_{i\beta} b_{\beta j}$. It can be proved that the above scheme is n-secure because all the rows in D are linearly independent. And this property guarantees the uniqueness of the pairwise keys in the cluster.

4.2 Group Key Establishment

Without loss of generality, let $N=\{N_1, N_2, ..., N_n\}$ be the initial set of participants in each cluster group that want to generate a group key. Assume that there are n-1 mobile nodes and a powerful node N_n. The group distribution protocol is depicted in Fig.1, and the detailed steps are presented as follows.

Step 1: Initially, each node $N_i (1 \le i \le n-1)$ is preloaded a row $r_i(A)$ from matrix A and column seed g^i as described in section 2. Then, after deployment, each node precomputes K_{in} and K_{ii} (i.e, $K_{ii} = \sum_{\beta=1}^{n} a_{i\beta} b_{\beta i}$). N_i sends the enciphered message (N_i, $C_i = E_{K_{in}}(K_{ii})$) to the powerful node N_n. K_{in} is the pairwise key between node N_i and N_n. || stands for message concatenation.

Step 2: The powerful node N_n computes K_{nn} as above. Upon receiving each $(N_i, C_i)(1 \le i \le n-1)$, the powerful node N_n deciphers them and computes $K_G = \Pi_{i=1}^{n} K_{ii}$. Next, powerful node N_n computes $y_i = K_G / K_{ii}$. Finally, the powerful node N_n broadcasts (N_n, $y_1 ... y_{n-1}$) to other nodes.

Step 3: On receiving the broadcast, each node $N_i (1 \le i \le n-1)$ computes the common group key $K_G = (y_i . K_{ii})$.

Until now, the participating nodes can establish an identical group key by using the same pre-distributed secrets as the pairwise key establishment.

4.3 Analysis of BKM

Although the participating nodes can successfully establish a group key, this protocol is not a real group key agreement protocol. As we all know, one advantage of a key agreement protocol over a key distribution protocol is that no participant can predetermine the common key. In many scenarios, however, the key distribution approach is not appropriate because the powerful node might become a single failure point for the group's security. Additionally, in some cases it is not possible for the powerful node N_n to be strongly trusted by all clients.

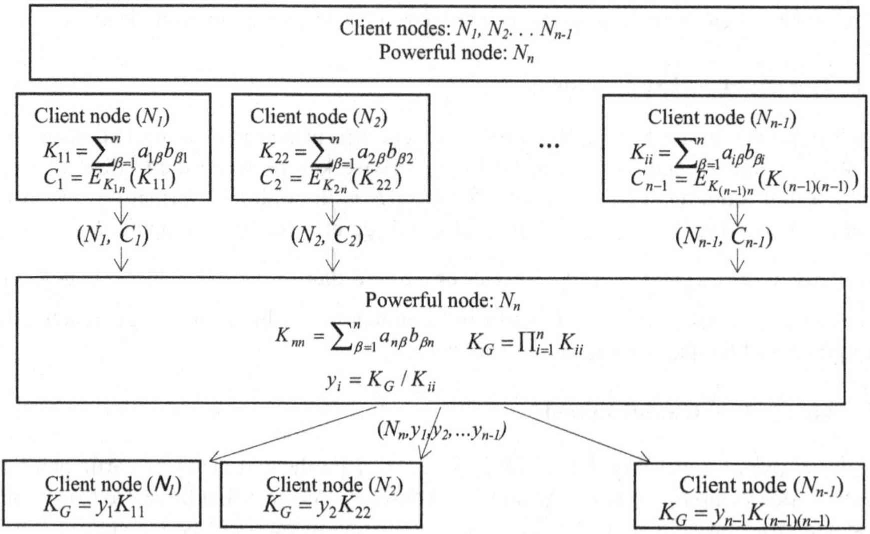

Fig. 1. The process of group key establishment in BKM [7]

For example, we will show that there exists another kind of attack in BKM protocol [7], even very low probability. We call it the inner cheating attack. The attack is possible due to the fact that the group key is computed by the powerful node N_n and then distributed to all the other members. If the powerful node N_n chooses an illegal key K' instead of K_G and uses K' to compute the broadcasting message y_i, the other members in the group also can deduce an common group key K' and can not know K' is a substitutor of K_G and their contributions are not included in it. If K' is advantaged to the malicious attackers, it could cause harmful problems in communications.

5 V-BKA: Verified Group Key Agreement

In this section, we describe V-BKA by extending BKM [7] with verification among users. Here, we should give some explanation that, since V-BKA is the extension of BKM, and its aim is to improve its group key distribution protocol to be a group key agreement protocol, so they use the same secrets pre-distribution mechanism and the pairwise key establishment is the same. The key updating is also similar to BKM and with a little difference. We will show that, this extended one has better security and performance than BKM and it is suitable to resource-constrained WSNs.

5.1 The Group Key Agreement

Next, we will introduce the proposed V-BKA protocol in details. The system parameters are the same as ones in the BKM protocol reviewed in Section 4. Without loss of generality, also we let $N=\{N_1, N_2, ..., N_n\}$ be the initial set of participants that want to generate a group key. Assume that there are n-1 mobile nodes and a powerful node N_n. The group distribution protocol is depicted in Fig.2. The detailed steps are presented as follows.

Step 1: Initially, each node $N_i (1 \le i \le n-1)$ is pre-loaded a row $r_i(A)$ from matrix A and column seed g^i as described in section 3. Then, after deployment, each node pre-computes $K_{in,}$ K_{ii} (i.e, $K_{ii} = \sum_{\beta=1}^{n} a_{i\beta} b_{\beta i}$) and K_{ii}^{-1}. N_i sends the enciphered message (N_i, $C_i = E_{K_{in}}(K_{ii})$) to the powerful node N_n and keeps K_{ii}^{-1} in its local memory. K_{in} is the pairwise key between node N_i and N_n. || stands for message concatenation.

Step 2: The powerful node N_n computes K_{nn} as above. Upon receiving each $(N_i, C_i)(1 \le i \le n-1)$, the powerful node N_n deciphers them and computes $x_i=K_{nn}K_{ii}$. Next, powerful node N_n computes $K_G = \prod_{i=1}^{n-1} x_i$. Finally, the powerful node N_n broadcasts (N_n, x_1... x_{n-1}) to other nodes.

Step 3: On receiving the broadcast messages, each node $N_i (1 \le i \le n-1)$ computes the common group key $K_G = x_j K_{jj}^{-1} \prod_{i=1}^{n-1} x_i$.

Note that the client N_i may pre-compute K_{ii}^{-1} to reduce the computational load. Until now, storage limitation is becoming less of a concerning issue as many add-on memory cards are widely available. In the following, we show that the proposed protocol is a contributory group key agreement protocol.

Theorem 1. *By running the proposed protocol, an identical group key can be established by all nodes. Each node may then confirm that its contribution was included in the group key.*

Proof. According to the proposed protocol, the powerful node N_n broadcasts (N_n, x_1... x_{n-1}) to all nodes, and each node N_i may use its secret exponent K_{ii}^{-1} to compute an identical group key K_G. Since an identical group key K_G has been established, this means that the following equation holds.

$$K_G = x_1 K_{11}^{-1} \prod_{i=1}^{n-1} x_i = x_2 K_{22}^{-1} \prod_{i=1}^{n-1} x_i = \cdots = x_{n-1} K_{n-1n-1}^{-1} \prod_{i=1}^{n-1} x_i$$

Thus, we have a value $V= K_G (\prod_{i=1}^{n-1} x_i)^{-1}$, such that, $V=x_1 K_{11}^{-1} = x_2 K_{22}^{-1} = ...x_{n-1} K_{n-1n-1}^{-1}$.

Therefore, we have:

$$x_1 = VK_{11}$$
$$x_2 = VK_{22}$$
$$.......$$
$$X_{n-2} = VK_{(n-2)(n-2)}$$
$$X_{n-1} = VK_{(n-1)(n-1)}$$

Observing the above equations, each x_i includes the node N_i's secret information K_{ii}^{-1}. Since $K_G = x_j K_{jj}^{-1} \prod_{i=1}^{n-1} x_i$. Therefore, the group key K_G contains all nodes' secret information K_{ii}^{-1}, that is, each node may confirm that its contribution was included in the group key.

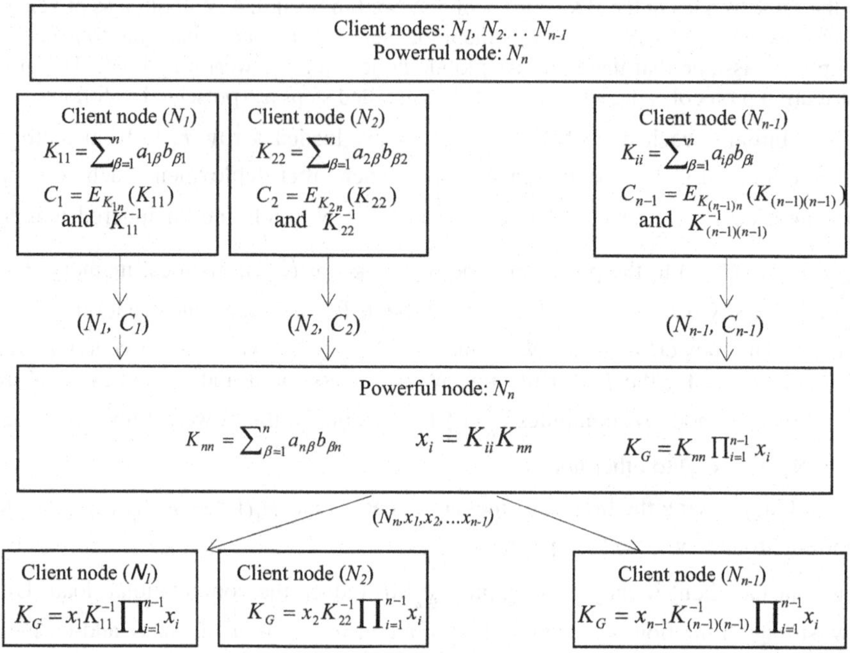

Fig. 2. Proposed group key agreement protocol

5.2 Join Operation and Key Extension

Before deploying the new node into the operating WSNs, some secret key information needs to be pre-loaded into the new node's memory and the key information extensions have to be done in existing innocent nodes, which ensures that the new node will establish a pairwise key with the existing ones. Thus, the main issue in this procedure is the key information extension problem. Here we identify a real-time extension technique to solve this problem.

***Real-time Extension (RTE)*:** In real-time extension approach, neighboring nodes can collaborate with each other to effectively protect and appropriately use the preloaded keys to extend their key information. For convenience, we denote the new node as N_{n+1}. The main tasks in this phase are the generation of the new node's key information and the extension of the existing nodes' key information.

Step 1: The base station first generates a new column seed g^{n+1} and stores it in the new node N_{n+1}. Then it generates a row seed s_{n+1} and distributes it to the powerful node N_n.

Step 2: N_n first extends the rows of the original matrix D and does the following:

(1) It computes the elements in the (n+1)th row of matrix D as:

$d_{(n+1)j}=H^{n+1}(s_j)$,($j=1,\ldots, n+1$)

(2) It distributes the elements $d_{i(n+1)}(i=1,\ldots,n)$ to the existing nodes. Note that $d_{i(n+1)}= d_{(n+1)i}$ in matrix D.

(3) It constructs $r_{n+1}(A)$ of matrix A for the new node N_{n+1} as: $b_{i(n+1)}=(g^{n+1})^{i-1}$; $a_{(n+1)i}=\sum_{j=1}^{n+1} d_{ij} b_{i(n+1)}$, (i=1,…, n+1).

Step 3: Upon receiving the elements $d_{i(n+1)}$ from new node N_{n+1}, each existing node N_i first extends column element of matrix B as: $b_{ji}= (g^i)^j$, (j=1,…, n+1). Then it extends row $r_i(A)$ as: $a_{ij}=a_{ij}+d_{j(n+1)}b_{(n+1)i}$, (j=1,…,n) and $a_{i(n+1)} = \sum_{j=1}^{n+1} d_{(n+1)j} b_{ji}$.

The drawback of real-time extension approach is that it introduces additional communication and computation overhead for each innocent node, which does not appear in reuse approach in the previous schemes [8-12]. However, the previous schemes suffer just because of this repeated use of fixed key space. The capture of each node increases the fraction of keys known by the adversary. The addition of new nodes to the network with keys from the same key pool will not help because the keys in the new nodes are already compromised. While in our approach all key spaces can be updated benefiting from the real time extension mechanism, nodes' compromise cannot affect the keys in the newly constructed network. Thus, the additional overhead is desirable.

5.3 Leave Operation and Key Information Shrink

We assume that a compromised node can eventually be detected by most of its neighbors within a certain time period. To achieve this purpose, the collaborative intruder detection and identification scheme [15] and schemes for remote code attestation like [17] may be used. After node compromise detection, the innocent nodes should update their key information to prevent the adversary from utilizing the captured keys. The main issue in this procedure is the key information shrink problem. Also, we identify two types of techniques to address this problem.

Regeneration: in this type of approach, the BS first regenerates all of the key information for the remaining nodes, including row seeds of matrix D, column seed of matrix B and row elements of matrix A. Then, the new key information will be distributed to the innocent nodes with the help of the previous secure pairwise keys. Finally, the remaining sensor nodes can establish new pairwise key without influencing the rest network security. The regenerating process is same to the description in section 3. The drawback of this approach is that it introduces substantive communication and computation overhead. Thus, this approach can be applied only at the pivotal moment, such as cluster head compromise etc.

Real-time Shrink (RTS): the real-time shrink approach needs the local collaboration of the neighboring innocent nodes to shrink the rows of matrix A. For clarity, we denote the evicted node as node k.

Step 1: The base station announces that the column seed g^k and the row seed s_k are both invalid.

Step 2: The powerful node N_n generates $d_{ki}(i<k)$ as $d_{ki}=H^k(s_i)$ and distributes it to node i (i<k) because $d_{ki}(i<k)$ is independent of the row seed s_k (s_k is public because it was announced invalid by BS) and node i (i<k) cannot generate d_{ki} by itself. In contrast, node i (i≥k) can generate $d_{ki}(i\geq k)$ by itself as $d_{ki}=H^i(s_k)$ from the row seed s_k.

Step 3: Each remaining node N_i deletes a_{ik} from its secret information $r_i(A)$ and shrinks the elements in the rest of $r_i(A)$ as $a_{ij}=a_{ij}-d_{ik}*b_{ki}$, ($i\neq j, j=1,\ldots,n$).

Though this real-time evolution also introduces additional computation overhead for each remaining node, which is much smaller than that in the regeneration approach above, and it is also smaller than that in the key revocation approach in [9-12].

5.4 Group Key Updating

Typically, group key management for WSNs should take both security and service quality into consideration.

Therefore, the group keys should be updated if a node is evicted or re-added. Consequently, after finishing the secret information updating in section 3.3 and 3.4, each innocent node N_i should update its contribution K_{ii} and send it to the powerful node N_n to update the group key. The group key rekeying process is similar to the group key establishment in the above.

Step 1: each node re-computes K_{in}, K_{ii} (i.e, $K_{ii}=\sum_{\beta=1}^{n}a_{i\beta}b_{\beta i}$) and K_{ii}^{-1} after finishing the secret information updating in section 3.2. N_i sends the enciphered message (N_i, $C_i=E_{K_{in}}(K_{ii})$) to the powerful node N_n.

Step 2: The powerful node N_n re-computes K_{nn}. Upon receiving each $(N_i, C_i)(1\leq i\leq n-1)$, N_n deciphers them and re-computes $x_i=K_{nn}K_{ii}$ and $K_G=\prod_{i=1}^{n-1}x_i$. At last, the powerful node N_n broadcasts (N_n, $x_1 \ldots x_{n-1}$) to other nodes.

Step 3: On receiving the broadcast, each node $N_i(1\leq i\leq n-1)$ re-computes the common group key $K_G=x_jK_{jj}^{-1}\prod_{i=1}^{n-1}x_i$.

Now, with the aid of the real-time key extension and shrink, the secrets in each node and their pairwise key are updated, so as the group key.

6 Performance Analysis

The BKM [7] protocol meets the requirements specified in Section 1. It has robust continuity between the group key and pairwise key establishment, it also can dynamically extend the key information in existing nodes and update the keys accordingly. But, it is not a group key agreement protocol. Though, we have proved that the BKM [7] and the modified Blom's key matrix construction and distribution method [16] is better that most of the existing key management schemes, this protocol is not perfect and requires to be improved before they were applied into the resource-constrained WSNs. Next, we will show that the V-BKA is a perfect group key agreement protocol, and it has all of the good properties as that of BKM and furthermore it has more merits, and the computation and communication overhead are not increase any more.

6.1 Security Analysis

V-BKA uses the same mechanism, modified version of Blom's matrix construction and key distribution, to agree upon the group key as BKM [7]. Thus, V-BKA also has the same merits as BKM in both pairwise key and group key, such as strong compromise resistance, strong network connectivity, perfect forward secrecy and perfect backward secrecy. Apart from these merits, the V-BKA has another two advantages over the other existing protocols [3-7]: absolute group confidentiality and perfect group fairness.

Absolute group confidentiality: In V- BKA, nodes in a group can securely communicate with each other and the information being shared inaccessible to anybody outside the group. For the message (N_i, C_i), only the powerful node N_n can decrypt the message to get the contributions of the client nodes by using its pairwise key with the client nodes. With the same reason, only the contributory nodes can verify and recover the group key by using their own contribution. This lead the proposed protocol is resistant to the inner cheating attack. Therefore, the group key is absolute group confidential.

Perfect Group fairness: One advantage of a key agreement protocol over a key distribution protocol is that no participant can predetermine the common key. In V-BKA, although there needs a powerful node N_n to collect the other nodes contributions and compute the group key. But, the powerful node N_n also is supervised by the other nodes. It can not replace the illegal group key by another one on its own. This property guarantees that all of the members in the group have the same right to join the group key establishment, and each of them has the same right to verify if its contribution is included in their currently used group key. Thus, V- BKA has perfect group fairness.

6.2 Overheads

In the following, we analyze the computational complexity and the communicational cost. We show that the proposed protocol V-KBA is almost the same as BKM and also is well suited for an imbalanced wireless network. For convenience, the following notations are used to analyze the communication cost and the computational complexity: H: hash computation; M: multiplication; Exp: exponentiation; D: Decryption; E: Encryption; I: inversion. Table 2 lists the comparisons between BKM protocol and the proposed protocol. We consider the performance comparisons in terms of the contributory property, the number of rounds, and both the message size and the computational complexity required for each client and the powerful node. It is obvious that the required computational complexity required by the powerful node in the proposed protocol is almost the same as that in BKM. Therefore, we can say that the proposed protocol V-BKA has perfect property to be applied into the resource-constrained WSNs. It not only has good key continuity between group key and pairwise key, which can greatly decreases the resource consuming of key management in resource-constrained WSNs; it also has perfect securities: strong compromise resistance, strong network connectivity, perfect forward secrecy, perfect backward secrecy, absolute group confidentiality and perfect group fairness.

Table 2. Comparisons between BKM and the new protocol AUTH- BKA

	BKM	**VS-BKA**
Contributory group key agreement	Existing a weakness	Yes
Number of rounds	2	2
Message sent by each client (via unicast)	2	2
Message sent by the powerful node (via broadcast)	n	n
Computational costs required by each client	2n M+1E	(2n+1) M+1E+1I
Computational costs required by the powerful node	(n-1)D+2nM+1Div	(n-1)D+(2n+1)M

H: hash computation; M: multiplication; Exp: exponentiation; D: Decryption; Div:division; E: Encryption; I: inversion.

7 Conclusion

We have presented that the BKM protocol has a security weakness. And it is not a real group key agreement. Therefore, a secure group key agreement protocol design for WSNs is an important issue to provide secure services among mobile nodes with limited computing capability. We proposed a new and secure group key agreement protocol,V-BKA, for resource-constrained WSNs. We have shown that the proposed protocol is a real contributory group key agreement. The analysis results show that the proposed protocol is a perfect key management protocol and is well suited for mobile nodes with limited computing capability.

Acknowledgment

This work is supported by the National Natural Science Foundation of China under Grant No.60863001 and 60903188, the National High Technology Development 863 Program of China under Grant No. 2007AA01Z456, and the National Grand Fundamental Research 973 Program of China under Grant No. 2007CB311201.

References

1. Akyildiz, I.F., Su, W., Sankarasubramaniam, Y., Cayirci, E.: A Survey on Sensor Networks. IEEE Communications Magazine 40(8), 102–116 (2002)
2. Li, J.H., Bhattacharjee, B., Yu, M., Levy, R.: A scalable key management and clustering scheme for wireless ad hoc and sensor networks. Future Generation Computer Systems 24(8), 860–869 (2008)
3. Burmester, M., Desmed, Y.: Efficient and Secure Conference key Distribution. In: Lomas, M. (ed.) Security Protocols 1996. LNCS, vol. 1189, pp. 119–129. Springer, Heidelberg (1997)

4. Ateniese, G., Steiner, M., Tsudik, G.: Authenticated group key agreement and friends. In: Proceedings of the 5th ACM conference on Computer and communications security, San Francisco, California, United States, pp. 17–26. ACM Press, New York (1998)
5. Kim, Y., Perrig, A., Tsudik, G.: Simple and fault-tolerant key agreement for dynamic collaborative groups. In: Proceedings of the ACM Conference on Computer and Communications Security, pp. 235–244 (2000)
6. Nam, J., Lee, J., Kim, S., Won, D.: DDH-based group key agreement in a mobile environment. The Journal of Systems and Software 78(1), 73–83 (2005)
7. Wen, M., Zheng, Y.F., Ye, W.J., Chen, K.F., Qiu, W.D.: A key management protocol with robust continuity for sensor networks. Journal of Computer Standards & Interfaces 31(4), 642–647 (2009)
8. Zhu, S., Setia, S., Jajodia, S.: LEAP: Efficient Security Mechanisms for Large-Scale Distributed Sensor Networks. In: Proc. of the 10th ACM Conf. on Computer and Communications Security (CCS 2003), pp. 62–72 (2003)
9. Liu, D., Ning, P., Sun, K.: Efficient Self-Healing Group Key Distribution with Revocation Capability. In: Proc. of the 10th ACM Conference on Computer and Communications Security (CCS 2003), pp. 231–240 (2003)
10. Bohio, M.J., Miri, A.: Self-healing in group key distribution using subset difference method. In: Proceeding of the Third IEEE International Symposium on Network Computing and Applications, NCA 2004 (2004)
11. Eltoweissy, M., Heydari, H., Morales, L., Sadborough, H.: Combinatorial Optimization of Key Management in Group Communications. J. Network and Systems Management 12(1), 33–50 (2004)
12. Jiang, Y., Lin, C., Shi, M., Shen, X.: Self-Healing Group Key Distribution Schemes with Time-Limited Node Revocation for Wireless Sensor Networks. Ad Hoc Networks 5(1), 14–23 (2007)
13. Blom, R.: An optimal class of symmetric key generation systems. In: Beth, T., Cot, N., Ingemarsson, I. (eds.) EUROCRYPT 1984. LNCS, vol. 209, pp. 335–338. Springer, Heidelberg (1985)
14. Du, W., Deng, J., Han, Y.S., Varshney, P.K., Katz, J., Khalili, A.: A pairwise key predistribution scheme for wireless sensor networks. ACM Transaction on Information and System Security 8(1), 228–258 (2005)
15. Wang, G., Zhang, W., Cao, G., La Porta, T.: On Supporting Distributed Collaboration in Sensor networks. In: IEEE Military Communications Conference (MILCOM 2003), vol. 2, pp. 752–757 (2003)
16. Wen, M., Chen, K.F., Zheng, Y.F., Li, H.: A reliable pairwise key-updating scheme for sensor networks. Journal of Software 18(5), 1232–1245 (2007)
17. Shaneck, M., Mahadevan, K., Kher, V., Kim, Y.: Remote Software-Based Attestation for Wireless Sensors. In: Molva, R., Tsudik, G., Westhoff, D. (eds.) ESAS 2005. LNCS, vol. 3813, pp. 27–41. Springer, Heidelberg (2005)

A Bulk Email Oriented Multi-party Non-repudiation Exchange Protocol*

Lu Bai[2,1], Chunhe Xia[1,2], Xiaojian Li[1,2,3], and Haiquan Wang[2]

[1] State Key Laboratory of Virtual Reality Technology and Systems, Beihang University, Beijing, China
[2] Key Laboratory of Beijing Network Technology, Beihang University, Beijing, China
[3] College of Computer Science & Information Technology, Guangxi Normal University, Guilin, China
{bailu,xch,whq}@buaa.edu.cn, xiaojian@mailbox.gxnu.edu.cn

Abstract. Much research has been done in the area of non-repudiation protocol. There are lots of research results on multi-party non-repudiation protocols. However, only a few of them present examples of applying the protocol to the Email-system. In this paper, a Bulk Email Oriented Fair Multi-party Non-repudiation Protocol is proposed. With the use of the online TTP to collect evidence from the sender and the receivers, the non-repudiation property among the sender and the receivers is provided, as well as the fairness among the receivers. The correctness of this protocol is proved using the formal method of SVO logic.

Keywords: Encryption; protocol; non-repudiation; fairness; SVO logic; bulk email.

1 Introduction

Email is becoming an integrated routine of people's daily life. Corresponding social responsibilities should be taken by all Email participants. But the protocols that current email-transmission based on have no ability to obtain the evidence for testifying whether or not someone has sent or received the message. Consequently, it is impossible to tell which participant should take the responsibility for the transmission [1] [2] [3]. In this condition, email user feel free to pursue phishing emails or deny having participated in a part or the whole of the email transmission.

To solve such responsibility problem, the non-repudiation protocol is one choice. Still, a good non-repudiation protocol should also provide the fairness property, which means that before the transmission finishes, either all the participants get their expected messages and the corresponding evidence or no one get any at all [4].

* This work is supported by three projects: the National 863 Project "Research on high level description of network survivability model and its validation simulation platform" under Grant No.2007AA01Z407; The Co-Funding Project of Beijing Municipal education Commission under Grant No.JD100060630 and National Foundation Research Project.

W. Liu et al. (Eds.): WISM 2009, LNCS 5854, pp. 426–438, 2009.

At present, most of the fair non-repudiation protocol involves a TTP (Trusted Third Party) to ensure the fairness property of the protocol. They put more emphasis on the fairness property between the sender and the receivers rather than that among the receivers. When bulk email transmission finishes, there may be some receivers that have not received the email and no receiver discovers it. In this situation, we consider that the receivers who have received the email are in a dominant position over those who have not. Therefore, in bulk email transmission the fair position among all the receivers should also be considered, i.e., when one receives an email, not only the sending evidence should be obtained from the sender but also the receiving evidence should be obtained from the other receivers.

Aim at solving the problems discussed above, this paper presents a Bulk Email Oriented Fair Multi-Party Non-repudiation Protocol. Section 2 introduces related works on multi-party non-repudiation protocol. Section 3 presents our protocol and it is proved with the formal method of SVO logic in section 4. Section 5 gives an overall conclusion.

2 Related Works

The multi-party non-repudiation protocol [5] proposed by Markowitch and Kremer allows one to send the same message to many receivers, in order to support fairness property, the Chinese Remainder Theorem(CRT) is introduced to construct a group signcryption scheme to protect the shared key K [6] so that only the set of the receivers included in R'(e.g., , R', a subset of R which is the set of all the receivers, contains the receivers that give the receiving evidence.) are able to get the shared key K. However, the execution of this protocol depends on the behavior of the sender, who has the right to decide the members of the set R' and whether to send the shared key K to the TTP or not.

The protocol discussed above puts the sender at a dominating position, for the sender is able to keep his shared key from the receiver who has already provided the receiving evidence. To eliminate such unfairness, He Bing addressed a fairer multi-party non-repudiation protocol which hands the right of confirming the R' set to the TTP by using a double group encryption scheme in 2005 [7]. It weakens the dominance of the sender and keeps the confidentiality. Nevertheless, in this protocol the fairness among the receivers was not taken into account, each of the receivers only knows whether himself\herself has received the message, none is clear on the receiving situation of the others.

Combining the signcryption scheme [8] proposed by Chandana Gamage with the Domain-verifiable signcryption [9] proposed by Seo and Kim, Wang Caifen gave a new signcryption scheme [10] that can be used in certified mail protocol. A multi-party certified mail protocol was proposed based on this scheme, with its fairness and non-repudiation property proved. However, this protocol has not considered the fairness among the receivers, neither.

It can be seen from the above discussion that although researches on the fair multi-party non-repudiation protocols have been conducted, only a few of them have the connection with the bulk email application. Most of them focus on the non-repudiation and fairness between the sender and the receivers. For the application of the non-repudiation protocol in bulk email system, it is necessary for each of the receivers to obtain the receiving evidence of all the other receivers. In the next section a new protocol is presented to solve this problem.

3 The Bulk Email Oriented Multi-party Non-repudiation Exchange Protocol

In this paper, the group signcryption scheme discussed above is used to presents a bulk email oriented multi-party non-repudiation exchange protocol. Involving the online TTP to collect evidence from the senders and the receivers, we successfully provide the non-repudiation property and the fairness property between the sender and the receivers, as well as the fairness among the receivers. The details of the protocol are addressed in this section.

3.1 Denotations Referred

P,Q : Entity that participates in the execution of the protocol.

A : The originator of the email, A means Alice.

B_i : Every receiver of the mail, i = 1, 2, ……, n, B means Bob.

TTP : Trusted third party.

N_A, N_A', N_{B_i} : Random number built by A or Bi.

m : Content of the mail

C : Cipher text of the mail.

K_{AB} : The symmetry key shared by A and Bi, used to encrypt the mail.

K_A : Public key of A.

K_A^{-1} : Private key of A.

ACK : TTP's confirmation of the successful finish of the protocol

smi : Simple Mail Information, the unique identifier of one time mail transmission. smi is concatenation of smi_1 and smi_2 , e.g., $\text{smi}=\text{smi}_1 \bullet \text{smi}_2$; $smi_1 = H(\text{mailinfo}, \text{ GenerateIP}, \text{ GenerateTime})$, built by the mail sender; smi_2 is a random number built by TTP.

X : The message transferred among the participants.

$\{X\}_{K_A}$: The cipher text of encrypting X with A's public key.

$\{X\}_{K_A^{-1}}$: The cipher text of encrypting X with A's private key.

$\{X\}_{K_{B_i}}$: The cipher text of encrypting X with all the public key of receivers.

$\{X\}_{K_{B_i}}^{-1}$: The cipher text of encrypting X with the private key of receivers separately.

$\{X\}_{K_{AB}}$: The cipher text of encrypting X with the key shared by A and B.

$list$: The sequence of all the email addresses of the receivers.

$P \rightarrow Q : X$: P send message X to Q

$P \leftrightarrow Q : X$:P send message X to Q forwardly, if Q did not receive the message, Q could get X from P through internet (it is similar to the situation that P provides the electrical affiche)

EOO : Evidence Of Original, signed by A

EOR : Evidence Of Receipt, signed by Bi

EOT : Evidence Of TTP, signed by TTP

In addition, the relationship of c and m is: $c = \{m\}_{K_{AB}}, m = \{c\}_{K_{AB}}$.

3.2 Content of the Protocol

M1. $A \rightarrow TTP : \{N_A, A, EOO_1\}_{K_{TTP}}$

where, $EOO_1 = \{H(N_A, A)\}_{K_A^{-1}}$

M2. $TTP \rightarrow A : \{N_A + 1, N_{TTP}, EOT_1\}_{K_A}$

where, $EOT_1 = \{H(N_A + 1, N_{TTP})\}_{K_{TTP}^{-1}}$

M3. $A \rightarrow TTP : \left\{ \begin{array}{l} N_{TTP} + 1, N_A', c, smi_1, \\ list, \{K_{AB}\}_{K_{B_i}}, EOO_2 \end{array} \right\}_{K_{TTP}}$

where, $EOO_2 = \left\{ H\left(\begin{array}{l} N_{TTP} + 1, N_A', c, \\ smi_1, list, \{K_{AB}\}_{K_{B_i}} \end{array} \right) \right\}_{K_A^{-1}}$

M4. $TTP \rightarrow A : \{N_A' + 1, smi_2, EOT_2\}_{K_A}$

where, $EOT_2 = \{H(N_A' + 1, smi_2)\}_{K_{TTP}^{-1}}$

M5. $A \rightarrow B_i : \{c, smi, A, list, EOO_3\}_{K_{B_i}}$

where, $EOO_3 = \{H(c, smi, A, list)\}_{K_A^{-1}}$

M6. $B_i \rightarrow TTP : \{c, smi, B_i, N_{B_i}, EOR_i\}_{K_{TTP}}$

where, $EOR_i = \{H(c, smi, B_i)\}_{K_{B_i}^{-1}}$

IF(within a specified timeframe, $TTP\ received \sum_{i=1}^{n}(EOR_i)$)

{

$$M7.\, TTP \leftrightarrow B_i : \left\{ \begin{matrix} N_{B_i} + 1, ACK, \{K_{AB}\}_{K_{B_i}}, \\ smi, EOR_{complete}, EOT_3 \end{matrix} \right\}_{K_{B_i}}$$

where, $$EOT_3 = \left\{ H \begin{pmatrix} N_{B_i} + 1, ACK, \{K_{AB}\}_{K_{B_i}}, \\ smi, EOR_{complete} \end{pmatrix} \right\}_{K_{TTP}^{-1}}$$

$$EOR_{complete} = \sum_{i=1}^{n} EOR_i, the\ gather\ of\ all\ the\ EOR_i$$

$$M8.\, TTP \leftrightarrow A : \{N_A' + 2, ACK, EOR_{complete}, EOT_4\}_{K_A}$$

where, $$EOT_4 = \{H(N_A' + 2, ACK, EOR_{complete})\}_{K_{TTP}^{-1}}$$

}

The blueprint of the protocol is given as Figure 1.

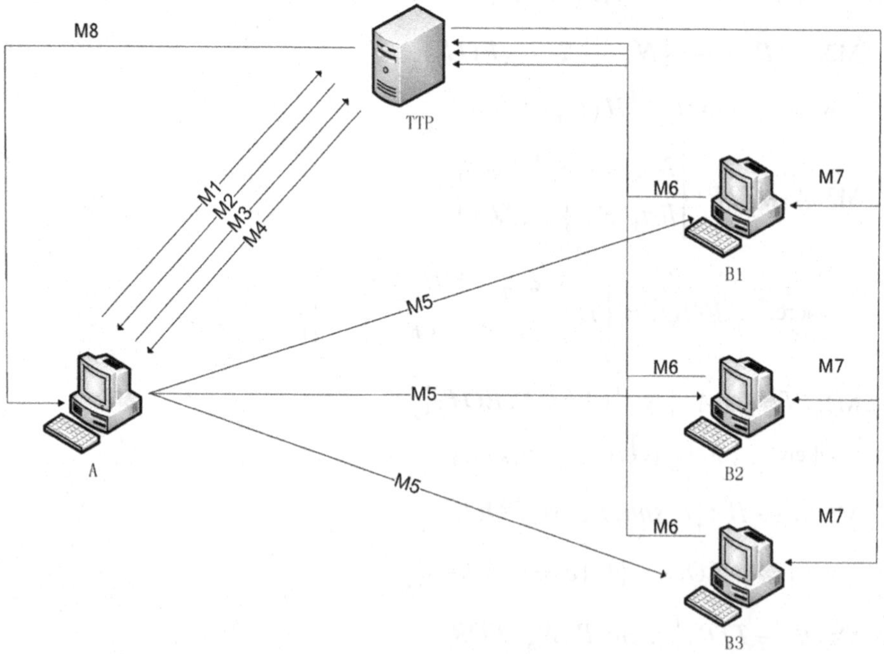

Fig. 1. The blueprint of the bulk email oriented fair multi-party non-repudiation protocol

3.3 Characteristics of the Protocol

Compared with other fair multi-party non-repudiation exchange protocols, our protocol inherits the strongpoint from He Bing's protocol which hands the right of publishing the shared key to the TTP, only when TTP has received all the receivers' valid receiving evidence, the shared key $\{K_{AB}\}_{K_{B_i}}$ would be published. Consequently, it prevents the sender from being at the dominating position in the transmission. Besides, the TTP will collect EOR_i from the receiver Bi, after TTP has gathered all the valid EOR_i, it will send the shared key along with the $EOR_{complete}$ to the receivers. Therefore, each of the receivers will know that the rest of them have received the cipher text of the email and will get the shared key ultimately. Thus the fairness among the receivers is guaranteed.

4 Analysis with SVO Logic

4.1 Basical Hypothesis

A0: The communication channel is neither safe nor reliable, failures would occur in the channel. The messages transferred in the channel can be stolen and used to launch replay attack.
A1: All the participants have published their public keys.
A2: The private key is visible only to the owner.
A3: $TTP\ believes\ K_A^{-1}$
A4: $TTP\ believes\ K_{B_i}^{-1}$
A5: $P\ believes\ K_{TTP}^{-1}$, P means any entity in the protocol.
A6: $P\ believes\ (TTP\ believes\ X) \supset P\ believes\ X$
A6 means that, with the credibility of TTP, we believe that TTP will not ally itself with the sender/receivers to cheat anyone else, so P can believe what TTP believes.
A7: $A\ said\ (c, smi) \wedge A\ said\ (K_{AB}, smi_1) \supset A\ said\ m$
A8: $B_i\ received\ (c, smi) \wedge B_i\ received\ (K_{AB}, smi) \supset B_i\ received\ m$
A7、A8 means that the smi can associate two different messages in the same round of the protocol.
A9: $TTP\ believes\left(A\ said\ (c, smi_1) \wedge \sum_{i=1}^{n}(B_i\ received\ (c, smi))\right)$
$\wedge\ TTP\ received\left(\{K_{AB}\}_{K_{B_i}}, smi_1\right) \supset TTP\ said\ (\{K_{AB}\}_{K_{B_i}}, smi)$
A10: $P\ believes\ PK_s(Q, K_Q) \wedge P\ received\ \{X\}_{K_Q^{-1}} \supset P\ believes\ (Q\ said\ X)$
A11: $A\ believes\ fresh(smi_1)$
A12: $A\ believes\ fresh(N_A)$
A13: $A\ believes\ fresh(N_A{}')$
A14: $B_i\ believes\ fresh(N_{B_i})$

A15: $TTP\ believes\ fresh(N_{TTP})$

A16: $TTP\ believes\ fresh(smi_2)$

A17: $TTP\ received \sum_{i=1}^{n} EOR_i \supset TTP\ says\ EOR_{complete}$

$$TTP\ said\ EOR_{complete} \supset TTP\ received \sum_{i=1}^{n} EOR_i$$

A17 means when TTP received $\sum_{i=1}^{n} EOR_i$, then it will say $EOR_{complete}$; if TTP said $EOR_{complete}$, it must have received $\sum_{i=1}^{n} EOR_i$.

A18: $A\ believes\ (B_i\ has\ K_{TTP})$

A19: $B_i\ believes\ (A\ has\ K_{TTP})$

A20: $TTP\ said\left(\{K_{AB}\}_{K_{B_i}}, smi\right) \supset TTP\ received\left(\{K_{AB}\}_{K_{B_i}}, smi_1\right)$

A20 means that because of the credibility of TTP, so if the TTP said ($\{K_{AB}\}_{K_{B_i}}$, smi), we can believe that it received ($\{K_{AB}\}_{K_{B_i}}$, smi_1) from the message M3.

A21: $TTP\ received\ M3 \wedge TTP\ received \sum_{i=1}^{n} EOR_i \supset TTP\ says\ ACK$

$$TTP\ said\ ACK \supset TTP\ received\ M3 \wedge TTP\ received \sum_{i=1}^{n} EOR_i$$

A21 means that only when TTP received the message M3 and all the EOR_i , will TTP say the "ACK"; and if TTP said "ACK" we believe that it has received message M3 and all the EOR_i .

4.2 The Goals of the Protocol

The general goal:

G1: $A\ believes \sum_{i=1}^{n} (B_i\ received\ m)$

G2: $\sum_{i=1}^{n} (B_i\ believes\ (A\ said\ m))$

G3: $\sum_{i=1}^{n} \left(B_i\ believes \sum_{j=1}^{n} ((B_j\ received\ m)) \right)$, $j \neq i$

The arbitration goal:

G4: $J\ believes \sum_{i=1}^{n} (B_i\ received\ m)$

G5: $J\ believes (A\ said\ m)$

4.3 Received Message Premises

Interpret every message as the assertion form of receiving that message.

[RMP.1] $TTP\ received \{\mathrm{N_A}, A, EOO_1\}_{K_{TTP}}$

[RMP.2] $A\ received \{N_A + 1, N_{TTP}, EOT_1\}_{K_A}$

[RMP.3] $TTP\ received \left\{ \begin{matrix} N_{TTP} + 1, N_A', c, smi_1, \\ list, \{K_{AB}\}_{K_{B_i}}, EOO_2 \end{matrix} \right\}_{K_{TTP}}$

[RMP.4] $A\ received \{N_A' + 1, smi_2, EOT_2\}_{K_A}$

[RMP.5] $B_i\ received \left\{ c, smi, A, list, \{EOO_3\}_{K_{AB}} \right\}_{K_{B_i}}$

[RMP.6] $TTP\ received \left\{ c, smi, B_i, N_{B_i}, EOR_i \right\}_{K_{TTP}}$

[RMP.7] $\sum_{i=1}^{n} \left(B_i\ received \left\{ \begin{matrix} N_{B_i} + 1, ACK, \{K_{AB}\}_{K_{B_i}}, \\ smi, EOR_{complete}, EOT_3 \end{matrix} \right\}_{K_{B_i}} \right)$

[RMP.8] $A\ received \left\{ N_A' + 2, ACK, EOR_{complete}, EOT_4 \right\}_{K_A}$

4.4 The Derivation of the General Goal

In the derivation below, I1 ~ I20 refer to the 20 axioms of SVO logic.

By M7, RMP.7, A1, A2, A10, A13, A14, I1, I4, I9, I7, we got:

(F1): $A\ believes \left(TTP\ received \sum_{i=1}^{n} EOR_i \right)$

By F1, M6, RMP.6, A1, A2, A10, A16, I1, I4, I7, I9, we got:

(F2): $A\ believes \left(TTP\ believes \sum_{i=1}^{n} (B_i\ said\ (c, smi)) \right)$

By F2, A6, M5, RMP.5, we got:

(F3): $A\ believes \sum_{i=1}^{n} (B_i\ reveived\ (c, smi))$

By M8, RMP.8, A1, A2, A10, I1, I4, I7, we got:

(F4): $A\ believes\left(TTP\ believes\left(A\ said\left(c, smi_1\right)\right)\right)$

By M8, RMP.8, A1, A2, A10, A13, A16, I1, I4, I7, I9, M5,RMP.5, we got:

(F5): $A\ believes\left(TTP\ believes\sum_{i=1}^{n}\left(B_i\ received\left(c, smi\right)\right)\right)$

By M8, RMP.8, A1, A2, A10, I1, I4, I7, we got:

(F6): $A\ believes\left(TTP\ received\left(\left\{K_{AB}\right\}_{K_{B_i}}, smi_1\right)\right)$

By F4, F5, F6, A9, we got:

(F7): $A\ believes\left(TTP\ said\left(\left\{K_{AB}\right\}_{K_{B_i}}, smi\right)\right)$

By F7, M7, RMP.7, A2, we got:

(F8): $A\ believes\sum_{i=1}^{n}\left(B_i\ received\left(K_{AB}, smi\right)\right)$

By F3, F8, A8, we got:

(F9): $A\ believes\sum_{i=1}^{n}\left(B_i\ received\ m\right)$

G1 is fulfilled.

By M5,RMP.5, A1, A2, A10, I4, I7, we got:

(F10): $\sum_{i=1}^{n}\left(B_i\ believes\left(A\ said\left(c, smi\right)\right)\right)$

By M7, RMP.7, A1, A2, A10, A14, A20, I1, I4, I7, I9, we got:

(F11): $\sum_{i=1}^{n}\left(B_i\ believes\left(TTP\ received\left(\left\{K_{AB}\right\}_{K_{B_i}}, smi_1\right)\right)\right)$

By F11, M3, RMP.3, A1, A2, A6, A10, A15, I1, I4, I7, I9, we got:

(F12): $\sum_{i=1}^{n}\left(B_i\ believes\left(A\ said\left(\left\{K_{AB}\right\}_{K_{B_i}}, smi_1\right)\right)\right)$

By F10, F12, A7, we got:

(F13): $\sum_{i=1}^{n}\left(B_i\ believes\left(A\ said\ m\right)\right)$

G2 is fulfilled.

By M7, RMP.7, A1, A2, A10, A14, I1, I4, I7, I9, we got:

(F14): $\sum_{i=1}^{n}\left(B_i\ believes\sum_{j=1}^{n}\left(B_j\ received\ \left(K_{AB}, smi\right)\right)\right), j \neq i$

By M7, RMP.7, A1, A2, A10, A14, I1, I4, I7, I9, we got:

(F15): $\sum_{i=1}^{n}\left(B_i\ believes\left(TTP\ received\sum_{i=1}^{n}EOR_i\right)\right)$

By F15, M6, RMP.6, A1, A2, A6, A10, A16, I1, I4, I7, I9,M5, RMP.5, we got:

(F16): $\sum_{i=1}^{n}\left(B_i\ believes\left(\sum_{j=1}^{n}\left(B_j\ received\ (c,smi)\right)\right)\right), j \neq i$

By F14, F16, A8, we got:

(F17): $\sum_{i=1}^{n}\left(B_i\ believes\left(\sum_{j=1}^{n}\left(B_j\ received\ m\right)\right)\right), j \neq i$

G3 is fulfilled.

4.5 The Derivation of the Arbitration Goal

In the case that someone denies having participated in the transmission process of the email, others might produce their relevant evidence for the judge to tell the truth. Here we present the derivation of the arbitration goal by which the judge can reproduce the truth.

We suppose that the judge has the public keys of all the participants (including the sender, the receivers and TTP) and believes their private keys.

Case 1: the receiver Bi denied having received the email m. In this case, the sender A may give (m, K_{AB}) to the judge J, and TTP gives $\left(N_{TTP}+1, N_A', c, smi, list, \{K_{AB}\}_{K_{B_i}}, EOO_2, EOR_{complete}\right)$ to J. J will prove that the receiver Bi has actually received the email m through the following steps.

(1). J checks up the (m, K_{AB}) from A and the cipher text c from TTP. If $m=\{c\}_{K_{AB}}$, then J believes that the K_{AB} from A is the key that used to encrypt the email m.

(2). J produces $\{K_{AB}\}_{K_{B_i}}'$ by encrypting the K_{AB} with the public keys of all the receivers listed by the "$list$" from TTP. If $\{K_{AB}\}_{K_{B_i}}' = \{K_{AB}\}_{K_{B_i}}$, then J believes that the K_{AB} in the $\{K_{AB}\}_{K_{B_i}}$ which is given by A is the key used to encrypt the email m.

(3). J verifies that all the receivers have received the K_{AB}.

Due to the $(c, smi, list, EOR_{complete})$ provided by TTP, we got:

(F18): $J\ Believes\left(TTP\ said\ EOR_{complete}\right)$

By F18, A1, A2, A10, A17, I1, I7, M5, RMP.5, we got:

(F19): $J\ believes\left(TTP\ believes\sum_{i=1}^{n}\left(B_i\ received\ (c,smi)\right)\right)$

Due to the $\left(N_{TTP}+1, N_A', c, smi, list, \{K_{AB}\}_{K_{B_i}}, EOO_2\right)$ provided by TTP, we got:

(F20): $J\ believes\left(TTP\ received\left(\begin{matrix}N_{TTP}+1, N_A', c, smi_1,\\ list, \{K_{AB}\}_{K_{B_i}}, EOO_2\end{matrix}\right)\right)$

By F20, A1, A2, A10, I1, I7, we got:

(F21): $J\ believes\left(TTP\ believes\left(A\ said\left(c, smi_1\right)\right)\right)$

Due to the $\{K_{AB}\}_{K_{B_i}}$ provided by TTP, we got:

(F22): $J\ believes\left(TTP\ received\{K_{AB}\}_{K_{B_i}}\right)$

By F19, F21, F22, A9, I1, I4, M7, RMP.7, we got:

(F23): $J\ believes\sum_{i=1}^{n}\{B_i\ received\ K_{AB}\}$

(4). J verifies that the EOR_i in the $EOR_{complete}$ is the production of signing (c, smi, B_i) with the private key of receiver Bi, the Bi in the (c, smi, B_i) is the email address, it can be found from the "$list$".

Due to the $(c, smi, EOR_{complete})$, we got:

(F24): $J\ received\sum_{i=1}^{n}(c, smi, B_i, EOR_{complete})$

By F24, A1, A2, A10, I7, M6, RMP.6, we got:

(F25): $J\ believes\sum_{i=1}^{n}\{B_i\ received\ (c, smi)\}$

By F23, F25, A8, I1, we got:

(F26): $J\ believes\sum_{i=1}^{n}(B_i\ received\ m)$

G4 is fulfilled.

Case 2: the sender denied having sent the email m. In this case, the receiver Bi may give $\left(\begin{matrix}c, smi, A, list, EOO_3, N_{B_i}+1, ACK,\\ \{K_{AB}\}_{K_{B_i}}, EOR_{complete}, EOT_3, m, K_{AB}\end{matrix}\right)$ to the judge J. J will prove that the sender A has actually sent the email m through the following steps.

(1). J verifies that the EOT_3 is the production of signing $\left(N_{B_i}+1, ACK, \{K_{AB}\}_{K_{B_i}}, smi, EOR_{complete}\right)$ with the private key of TTP.

Due to the $\left(N_{Bi}+1, ACK, \{K_{AB}\}_{K_{B_i}}, smi, EOR_{complete}, EOT_3\right)$ provided by Bi, we got:

(F27): $J\ received \begin{pmatrix} N_{B_i}+1, ACK, \{K_{AB}\}_{K_{B_i}}, \\ smi, EOR_{complete}, EOT_3 \end{pmatrix}$

By F27, A1, A2, A6, A10, A15, A20, I1, I4, I7, I9, M3, RMP.3, we got:

(F28): $J\ believes \left(A\ said \left(\{K_{AB}\}_{K_{B_i}}, smi_1 \right) \right)$

(2). J verifies that EOO_3 is the production of signing $(c, smi, A, list)$ with the private key of A.

Due to the $(c, smi, A, list, EOO_3)$ provided by Bi, we got:

(F29): $J\ received\ (c, smi, A, list, EOO_3)$

By F29, A1, A2, A10, I1, I7, we got:

(F30): $J\ believes\ (A\ said\ (c, smi))$

(3). J checks up the (m, c, K_{AB}) from Bi. If $m = \{c\}_{K_{AB}}$, then we got:

(F31): $J\ believes\ (A\ said\ m)$

G5 is fulfilled.

5 Conclusion

In this paper, we proposed a bulk email oriented multi-party non-repudiation exchange protocol. With the application of this protocol, it can be prevented that the email sender and the receivers deny having participated in the transmission process of the email. Meanwhile, all the participants are at a fair position during the transmitting process. When the transmission finished successfully, all the receivers receive the email, the sender gets the receiving evidence of the receivers, and the receivers get both the sending evidence of the sender and the receiving evidence of the other receivers. If the transmission failed, none of the receivers would receive the email. The sender and the receivers would not get their needed evidence. This keeps any receivers from being at the dominating position. Finally, we used the formal method of SVO logic to demonstrate its correctness.

In the protocol proposed above, the online TTP needed to guarantee the fairness can probably be the bottleneck that affects the efficiency, so how to reduce the computing weight of the TTP should be considered as the future work. Besides, in order to apply our protocol in the email system without affecting the basic SMTP protocol, the compatibility between our protocol and the SMTP protocol should also be taken into account.

Acknowledgments. I would like to express my gratitude to Mr. Jianzhong Qi for many substantial suggestions which greatly improve this paper.

References

1. Klensin, J.: Simple Mail Transfer Protocol [EB/OL]. RFC2821 (April 2001)
2. Myers, J., Rose, M.: Post Office Protocol - Version 3 [EB/OL]. RFC1939 (May 1996)
3. Crispin, M.: Internet Message Access Protocol - Version 4 rev1 [EB/OL]. RFC3501 (March 2003)
4. Zhou, J., Gollmann, D.: A fair non-repudiation protocol. In: Proceedings of the IEEE Symposium on Security and Privacy, Oakland, CA, pp. 55–61 (1996)
5. Kremer, S., Markowitch, O.: A multi-party nonrepudiation protocol. In: Proceedings of the 15th IFIP International Information Security Conference, August 2000, pp. 271–280. Beijing (2000)
6. Chiou, W.C.: Secure broadcasting using the secure lock. IEEE Transaction on Software Engineering 15(8) (1989)
7. Bing, H., Xiaojian, L., Chunhe, X., Kejian, X.: A Fair Multi-Party Non-repudiation Protocol. Computer Engineering and Applications 156, 120–122 (2005)
8. Gamage, C., Leiwo, J., Zheng, Y.: Encrpted Message Authentication By Firewalls. In: Imai, H., Zheng, Y. (eds.) PKC 1999. LNCS, vol. 1560, pp. 69–81. Springer, Heidelberg (1999)
9. Moonseog, Kim, K.: Electronic funds transfer protocol using domain-verifiable signcryption scheme. In: Song, J.S. (ed.) ICISC 1999. LNCS, vol. 1787, pp. 269–278. Springer, Heidelberg (2000)
10. Wang, C.-f., Jia, A.-k., Liu, J.-l., Yu, C.-z.: Multi-party Certified Mail Proocol Based on Signcryption. Acta Electronica Sinica 33(11), 2070–2073 (2005)

Software Fault Feature Clustering Algorithm Based on Sequence Pattern*

Jiadong Ren[1,2], Changzhen Hu[2], Kunsheng Wang[3], and Dongmei Zhang[1]

[1] College of Information Science and Engineering, Yanshan University, Qinhuangdao, 066004, China
jdren@ysu.edu.cn, zhangdongdongmei@126.com
[2] School of Computer Science and Technology, Beijing Institute of Technology, Beijing, 100081, China
chzhoo@bit.edu.cn
[3] China Aerospace Engineering Consultation Center
Beijing, 100037, China
kshwang@126.com

Abstract. Software fault feature analysis has been the important part of software security property analysis and modeling. In this paper, a software fault feature clustering algorithm based on sequence pattern (SFFCSP) is proposed. In SFFCSP, Fault feature matrix is defined to store the relation between the fault feature and the existing sequence pattern. The optimal number of clusters is determined through computing the improved silhouette of fault feature matrix row vector, which corresponds to the software fault feature. In the agglomerative hierarchical clustering phase, entropy is considered as the similarity metric. In order to improve the time complexity of the software fault feature analysis, the fault features of the software to be analyzed are matched to each centroid of clustering results. Experimental results show that SFFCSP has better clustering accuracy and lower time complexity compared with the SEQOPTICS.

Keywords: sequence pattern; software fault feature clustering; improved silhouette; entropy.

1 Introduction

Software security [1] modeling and analysis become widely concerned in the software security field, since the risk of software vulnerability becomes increasingly urgent with the complication and the wide usage of software.

Software fault can be attributed to several causes. Many research papers are concerned with the software security modeling and analysis so as to locate the vulnerabilities [2-3] of the software to improve the software security property. Software Fault Tree (SFT) [4] analysis is one of the traditional software fault analysis methods. SFT

* This work is supported by the National High Technology Research and Development Program ("863"Program) of China (No. 2009AA01Z433) and the Natural Science Foundation of Hebei Province P.R. China (No.F2008000888).

W. Liu et al. (Eds.): WISM 2009, LNCS 5854, pp. 439–447, 2009.

involves in specifying a top event to analyze, followed by identifying all of the associated elements in the system that could lead to the top event. Software Fault Tree analysis is logical structured process that can help identify potential causes of system fault. However, the time complexity of the analysis is high. Guy Helmer proposed a Software fault tree and colored Petri net–based specification [5] to structure the specification and design of the IDS. But the time complexity of this method is still high, since the matching scope is large. It becomes a significant problem to analyze the existing software fault features, which are in the form of sequences, in the fault feature database in order to search the fault pattern. There are several research papers on the subject of clustering sequences, such as CLUSEQ [6], seqPAM [7], and the two-phase hybrid method for biological sequence clustering proposed by W. B. Chen [8], which combines the strengths of the hierarchical agglomerative clustering methods and the partition clustering methods. However, most of them match the sequences to the existing ones so as to locate the source of the fault. Few of them address the problem efficiently, since the matching scale is too large. Therefore, it becomes a necessary problem to analysis the inner relation between the existing features and then the sequence matching scope can be saved to improve the performance of the fault analysis.

In this paper, we focus on clustering the software fault features in order to improve the time complexity of the software fault feature analysis. A software fault feature clustering algorithm based on sequence pattern is presented. In the proposed approach, fault feature matrix is defined to reflect the relation between the fault features and their corresponding row vectors, considering that the information whether the fault feature supports the existing sequence patterns can be stored in the corresponding row vector of the fault feature matrix. Thus, the similarity between the fault features can be obtained through computing the similarity between the corresponding row vectors. Improved silhouette is defined to determine the optimal number K of sequence clusters and entropy is introduced into the agglomerative hierarchical clustering as the similarity metric. The fault feature is assigned to the cluster corresponding to minimum incremental entropy. When the software fault features to be analyzed arrive, we match them with the centroids of the sequence clusters to locate the most similar sequence cluster, therefore, the matching scale is largely reduced when doing the software fault feature analysis and the efficiency is improved.

This paper is organized as follows: In section 2, we give the basic concepts and definitions. In section 3, we present our hierarchical clustering algorithm called SFFCSP. Section 4 shows the experimental results of the clustering algorithm. Finally we conclude the paper in section 5.

2 Problem Definitions

A software fault feature is represented by an event sequence $\{e_1, e_2, e_3 \ldots e_i \ldots e_n\}$, where e_i is the basic event resulting in the fault. The fault event sequences are stored in the software fault feature database. Since the length of the sequence is different, it is necessary to unify the format of the sequence to analyze the software fault feature. Entropy [9] is used to measure the degree how each cluster is composed of the single type objects. The more entropy is closes to 0, the better the effect of clustering will be. For cluster i, the probability p_{ij} and the members of cluster i belonging to class j are calculated, where $p_{ij}= n_{ij}/n_i$, n_{ij} is the number of objects belonging to cluster i. The entropy of cluster C_i can be calculated as follows:

$$e(c_i) = -\sum_{j=1}^{k} p_{ij} \log_2 p_{ij} \tag{1}$$

Definition 1. Fault feature matrix
Let D be a fault feature database, and S be the existing fault sequence patterns of D. Suppose that D is $\{s_1,s_2,\ldots,s_n\}$ and S is $\{p_1, p_2, p_3, \ldots, p_m\}$. The Fault feature matrix transforms the fault features into row vectors based on D and S. If the fault feature $s_i \subseteq D$ supports the fault sequence pattern $p_j \subseteq S$, 1 is set at row i column j in the fault feature matrix, otherwise 0.

Therefore, each fault feature s_i corresponds to a row vector v_i of the fault feature matrix. Thus, a n×m matrix is constructed based on D and S.

A simple example will be given below to illustrate how to construct the fault feature matrix:

Suppose that fault feature database D is composed of five fault features: $s_1=\{e_1, e_2, e_3, e_5, e_6, e_7\}$, $s_2=\{e_4, e_5, e_6, e_7\}$,$s_3=\{e_3, e_5, e_6, e_8\}$, $s_4=\{e_1, e_4, e_6, e_7\}$, $s_5=\{e_2, e_3, e_4, e_5, e_7, e_8\}$ and the existing fault sequence patterns of D are $p_1=\{e_1, e_7\}$, $p_2=\{e_2, e_3, e_5\}$, $p_3=\{e_2, e_5, e_7\}$, $p_4=\{e_5, e_6, e_7\}$, $p_5=\{e_3, e_6\}$, $p_6=\{e_3, e_8\}$, then the fault feature matrix is constructed as follows, where each fault feature sequence s_i corresponds to a row vector v_i of the fault feature matrix.

$$\begin{array}{c} \\ v_1 \\ v_2 \\ v_3 \\ v_4 \\ v_5 \end{array}\begin{array}{c} \begin{array}{cccccc} p_1 & p_2 & p_3 & p_4 & p_5 & p_6 \end{array} \\ \begin{bmatrix} 1 & 1 & 1 & 1 & 1 & 0 \\ 0 & 0 & 0 & 1 & 0 & 0 \\ 0 & 0 & 0 & 0 & 1 & 1 \\ 1 & 0 & 0 & 0 & 0 & 0 \\ 0 & 1 & 1 & 0 & 0 & 1 \end{bmatrix} \end{array}$$

Definition 2. improved silhouette
Let S be a given data set, which is composed of sequence clusters $C_1, C_2 \ldots C_k$, and C_i is composed of sequences. For each sequence s_i, the improved silhouette of sequence s_i is denoted as $isilhouette_{si}$. $Isilhouette_{si}$ is defined as the formula (2).

$$isilhouette_{si} = \frac{a_i - b_i}{\min(a_i, b_i)} \tag{2}$$

Where a_i is the similarity between the ith sequence $s_i(s_i \in C_j, j \in [1,k])$ and the centroid of sequence cluster that s_i belongs to, and b_i is the maximum of similarity between s_i and the centroids of each other k-1 clusters. The value of isilhouette is in the continuous range [-1, 1]. A positive value of isilhouette expresses a good clustering of sequence i, while a negative value expresses a bad one.

Each fault feature corresponds to a row vector of fault feature matrix. The more similar the row vectors are, the higher the similarity between the fault features will be. Thus, the similarity between the row vectors reflects the similarity of their corresponding fault features.

Consequently, the similarity between features can be obtained through calculating the corresponding row vectors of the fault feature matrix. The formula is as follows:

$$Sim(v_1, v_2) = (v_1 \wedge v_2)/(v_1 \vee v_2) \tag{3}$$

$$Sim(s_1, s_2) = Sim(v_1, v_2) \tag{4}$$

The centroids X_i of fault feature cluster C_i ($1 \le i \le n$) is represented by a m-dimensional vector. In order to determine the value of the n-dimensional vector, we need to decide the value of the n attributes p_{i1}, p_{i2}... p_{ij} ($1 \le j \le m$) to be 0 or 1. A minimum support threshold sup_j ($1 \le j \le n$) is assigned to each attribute p_{ij}. If the mean support of sequences for p_j is more than sup_j, p_j is considered as the feature attribute, and the value of p_j is set at 1, otherwise 0.

The similarity between the fault sequences can be obtained through calculating the mean isilhouette of the cluster consisting of the corresponding row vectors. The isilhouette ranges from -1 to 1, and the closer the value to 1, the higher similarity between row vectors will be.

3 Software Fault Feature Clustering Algorithm Based on Sequence Pattern

In the SFFCSP, software fault feature matrix based on the existing sequence patterns is constructed for the software fault feature database, thus each fault feature, which is a sequence, corresponds to a unique row vector in the software fault feature matrix. The similarity between the fault features can be obtained through calculating that between the corresponding row vectors. Thus, the optimal number K of feature clusters can be gotten by computing the mean improved silhouette of the corresponding row vectors. Entropy of feature cluster is introduced to SFFCSP. Entropy reflects the feature cluster purity consisting of a single category. Entropy is used as the similarity metric, and the closer the entropy is to 0, the higher the purity of the sequence cluster consisting of a single category will be. The agglomerative hierarchical clustering is carried out according to the entropy change of feature cluster; the fault feature is always assigned to the cluster whose incremental entropy is small. We continue this procedure until only K clusters remain.

The parameters of SFFCSP are as follows: D is the fault feature database; n is the number of fault features; S is the existing sequence pattern set of D,and S= {p_1, p_2, p_3... p_m}; s′ is the fault feature of the software to be analyzed; sup_j is minimum support threshold of p_i; C_{sim} represents the fault feature cluster, which is most similar to s′.

```
Algorithm SFFCSP
   Input: D, n, S, s', sup_j
   Output: C_sim
   Begin
      1: for (each software fault feature s_i in D) do
      2:     {if ( s_i supports p_j)
      3:           1 is set at row i column j in the software
      fault feature matrix M;
```

4: else
5: 0 is set at the row i column j of M;
6: } //Construct the software fault feature matrix
7: Partition the row vectors in M into K′ clusters C_1, C_2… C_{K}. based on the distribution of element in M;
8: $sup_j = \sqrt{\sup(p_j)}$; // sup (p_j) is the support of sequence p_j in the sequence database.
9: Calculate o_1, o_2… o_K respectively, where o_i is the centroid of C_i //o_i (i=1,2…K′)is a m-dimensional vector;
10: repeat
11: {for (each row vector v_j) // v_j corresponds to the software fault feature s_i, which belongs to C_i
12: calculate $is_i(v_j)$ based on formula (2); // $is_i(v_j)$ is the improved silhouette of v_j
13: calculate $\overline{is_i}$, the means improved silhouette of C_j consisting of t row vectors: $\overline{is_i} = \frac{1}{t}\sum_{i=1}^{t}\sum_{v_j \subseteq C_j} is_i(v_j)$;
14: draw the graph of $\overline{is_i}$ for K′ in the 2-D coordinate system;
15: K′=:K′+1;
16: }
17:until (the $\overline{is_i}$ in the graph reaches the maximum)
18: K:= K′;
19: repeat
20: {
21: label each fault feature sequence in D as a single cluster;
22: Combine the sequences corresponding to the most similar row vector of M to form t clusters initially (t<K);
23: Calculate the incremental entropy of each cluster supposing the sequence is joined. Assign the sequence to the cluster whose incremental entropy is least;
24: re-label the clusters obtained from step 23;
25: n:=n-1;
26: } until (n=K)
27: Compute entroid$_1$, centroid$_2$… centroid$_K$, which are obtained from step 26;
28: Calculate the similarity between s and the centroid$_i$(i=1, 2… K) based on formula (3-4), match the feature sequence s to the centroids of sequence cluster C_{mim} (d_1, d_2, d_3, …d_K), assign it to the most similar cluster;
End

In SFFCSP, how to decide the value of sup_j is critical while calculating the centroid of the feature. The appropriate sup_j can correctly measure whether the attribute p_j is significant compared with the attributes of the other sequences. In this paper, sup_i is assigned as $\sqrt{\sup(p_i)}$, where sup (p_i) is the support of the sequence p_i in the sequence database.

The SFFCSP algorithm is presented in order to facilitate the fault feature analysis. The fault feature matrix reflecting the relation between the fault feature and the row vector is constructed under the premise that the sequence pattern of the fault feature database exists. Thus, the similarity between sequences can be obtained through computing that between their corresponding row vectors. The optimal number K of sequence clusters corresponds to the maximum of the graph which is incrementally drawn based on the mean isilhouette of fault feature cluster. In the clustering phase, the fault features are agglomeratively clustered according to the entropy increment of sequence cluster until the number of clusters equals to K. Finally, each feature sequence of the software to be analyzed is matched to the centroids of the K clusters which are obtained from the above clustering operation. Consequently, the time complexity of the feature analysis is largely improved.

4 Experimental Results

Our experiments are conducted on a computer with 2.4Ghz Intel CPU and 512M main memory. The operating system of the computer is Microsoft Windows XP. SFFCSP is compared with SEQOPTICS [8] to evaluate the performance of SFFCSP. All the algorithms are implemented in Visual C++.

We perform our experiments on the synthetic sequence data sets D_1, D_2, D_3, D_4 under the condition that the corresponding sequence patterns have been found. The time complexity and the cluster accuracy are mainly compared between the algorithms SFFCSP and SEQOPTICS [10].

All of the data sets are generated through running the IBM generator. Data set D_1 contains 147 sequences from three different clusters: 75 sequences of cluster A_1, 24 sequences of A_2, 48 sequence of A_3. D_2 contains 298 sequences from two different clusters: 167 sequences of B_1, 121 sequences of B_2. D_3 contains 970 sequences from five different clusters: 182 sequences of cluster C_1, 174 sequences of C_2, 398 sequence of C_3, 216 sequences of C_4. D_4 contains 8000 sequences from four different clusters: 1750 sequences of cluster E_1, 476 sequences of E_2, 4480 sequence of E_3. 1294 sequences of E_4. Table 1 shows the parameters of four sequence data sets.

Table 1. Parameters of the testing sequence data sets

Data sets	Size	Clusters
D_1	147	A_1,A_2,A_3
D_2	298	B_1,B_2
D_3	970	C_1,C_2,C_3,C_4
D_4	8000	E_1,E_2,E_3,E_4

4.1 The Complexity Analysis

SFFCSP and SEQOPTICS are carried out on the testing sequence data sets respectively. The results of the both algorithms on time complexity are showed in Fig.1.

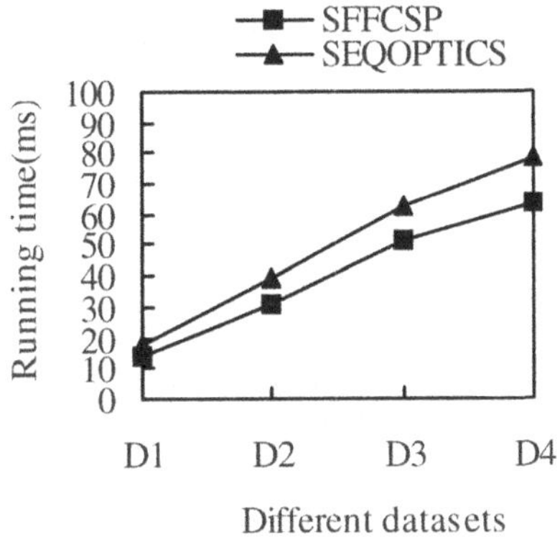

Fig. 1. Clustering time comparison between SEQOPTICS and SFFCSP

In the SFFCSP algorithm, the optimal number of each sequence cluster is determined through calculating the mean isilhouette, and then entropy is considered as the similarity metric in the hierarchical clustering stage. Since SFFCSP avoids utilizing the distance as the similarity metric, its efficiency is improved extensively. In contrast, SEQOPTICS needs to compute the pairwise distance between the sequences. From Fig. 1 we can get the conclusion that the results of the SFFCSP are better in terms of the time complexity on each data set, especially when the data set is large.

4.2 The Accuracy Analysis

The experiment is performed on the synthetic data set D4 with the varying number of existing sequence pattern. The result of the experiment is showed in Fig.2.

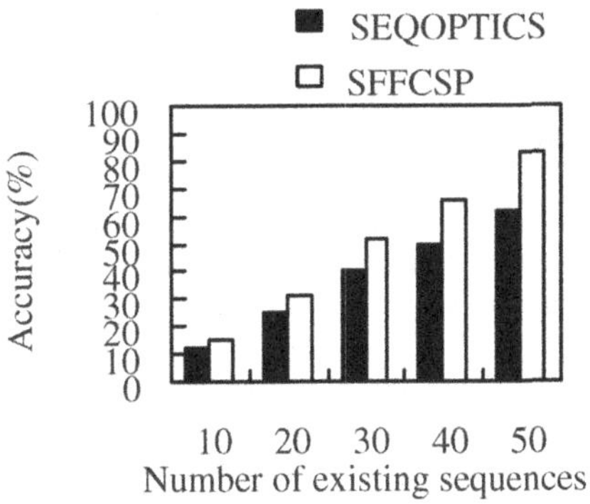

Fig. 2. Clustering accuracy with different number of existing sequence pattern

From Fig.2, we can draw the conclusion that the more number of existing sequences pattern, the better the clustering result. In the SFFCSP, the similarity between sequences can be obtained through computing the similarity between the corresponding vectors of the sequences. The more the row vectors support the same sequence patterns the more similar they are. In addition, entropy is introduced to the agglomerative hierarchical clustering as the similarity avoiding the effect of outliers to improve the clustering accuracy.

5 Conclusion

A software fault feature clustering algorithm based on sequence pattern is presented in this paper. The new approach decides the optimal number K of clusters through calculating the mean isilhouette of software fault sequence cluster. In the subsequent clustering phase, the fault feature sequences are agglomeratively clustered based on the change of the sequence cluster's entropy until the number of sequence clusters equals to K. When analyzing the software fault features, the similarity between the fault feature of the software to be analyzed and the centroids of the K clusters is calculated in order to locate the most similar sequence cluster with the fault feature to be analyzed, consequently, the matching scope of feature sequences is largely reduced and the efficiency of the software fault features analysis can be improved. The experimental results show that SFFCSP can not only reduce the time complexity but also improve the accuracy of the software fault feature analysis.

References

1. Barnum, S., McGraw, G.: Knowledge for software security. IEEE Security & Privacy 3(2), 74–78 (2005)
2. Hadavi, M.A., Sangchi, H.M., Hamishagi, V.S., Shirazi, H.: Software Security; A Vulnerability-Activity Revisit. In: Proceedings of the 2008 Third International Conference on Availability, Reliability and Security, Barcelona, Spain, pp. 866–872 (2008)
3. Nichols, E.A., Peterson, G.: A Metrics Framework to Drive Application Security Improvement. IEEE Security and Privacy 5(2), 88–91 (2007)
4. Needham, D., Jones, S.: A Software Fault Tree Metric. In: 22nd IEEE International Conference on Software Maintenance, Philadelphia, PA, United states, pp. 401–410 (2006)
5. Helmer, G., Wong, J., Slagell, M., Honavar, V., Miller, L.: Software fault tree and colored Petri net–based specification, design and implementation of agent-based intrusion detection systems. International Journal of Information and Computer Security 2(1-2), 109–142 (2007)
6. Yang, J., Wang, W.: CLUSEQ: Efficient and effective sequence clustering. In: Nineteenth International Conference on Data Engineering, Bangalore, India, pp. 101–112 (2003)
7. Kumar, P., Bapi, R.S., Krishna, P.R.: SeqPAM: A sequence clustering algorithm for web personalization. International Journal of Data Warehousing and Mining 3(1), 29–53 (2007)

8. Chen, W.B., Zhang, C.C.: A Robust Method for Biological Sequence Clustering. In: Proceedings of the 2006 IEEE International Conference on Information Reuse and Integration, Waikoloa Village, HI, United states, pp. 286–291 (2006)
9. Jenssen, R., Hild, K.E., Erdogmus, D., Principe, J.C., Eltoft, T.: Clustering using Renyi's Entropy. In: Proceedings of the International Joint Conference on Neural Networks, Portland, OR, United states, pp. 523–528 (2003)
10. Chen, Y.H., Reilly, K.D., Sprague, A.P., Guan, Z.J.: Seqoptics: A protein sequence clustering method. In: First International Multi-Symposiums on Computer and Computational Sciences, Hangzhou, Zhejiang, China, pp. 69–75 (2006)

Identification of Malicious Web Pages by Inductive Learning*

Peishun Liu and Xuefang Wang

Department of Computer Science and technology, Ocean University of China Qingdao 260071, China
Department of Mathematics, Ocean University of China Qingdao 260071, China
liups@ouc.edu.cn, ouc.wxf@163.com

Abstract. Malicious web pages are an increasing threat to current computer systems in recent years. Traditional anti-virus techniques focus typically on detection of the static signatures of Malware and are ineffective against these new threats because they cannot deal with zero-day attacks. In this paper, a novel classification method for detecting malicious web pages is presented. This method is generalization and specialization of attack pattern based on inductive learning, which can be used for updating and expanding knowledge database. The attack pattern is established from an example and generalized by inductive learning, which can be used to detect unknown attacks whose behavior is similar to the example.

Keywords: Malicious detection; drive-by downloads; inductive learning; generalization; specialization.

1 Introduction

Internet attacks that use a malicious, hacked, or infected Web server to exploit unpatched client-side vulnerabilities of visiting browsers are on the rise. As the web browser requests content from a web server, the server returns an exploit embedded in the web pages that allows the server to gain complete control of the client system. Traditional defenses are ineffective against these new threats because they cannot deal with zero-day attacks.

In this paper, we provide a detailed study of the pervasiveness of so-called drive-by downloads on the Internet. Drive-by downloads are caused by URLs that attempt to exploit their visitors and cause Malware to be installed and run automatically. Billions of URLs over a ten-month period are analyzed in [1] and it shows that a non-trivial amount of over 3 million malicious URLs initiate drive-by downloads. An even more troubling finding is that approximately 1.3% of the incoming search queries to Google's search engine returned at least one URL labeled as malicious in the results page.

* This project is supported by the National Natural Science Foundations of China (No.60703082 and No.60672102).

W. Liu et al. (Eds.): WISM 2009, LNCS 5854, pp. 448–457, 2009.

For the most part, the techniques in use today for delivering web malware can be divided into two main categories. In the first case, attackers use various social engineering techniques to entice the visitors of a website to download and run malware. The second, more devious case, involves the underhanded tactic of targeting various browser vulnerabilities to automatically download and run—i.e., unknowingly to the visitor—the binary upon visiting a website. When popular websites are exploited, the potential victim base from these so-called drive-by downloads can be far greater than other forms of exploitation because traditional defenses (e.g., firewalls, dynamic addressing, proxies) pose no barrier to infection. While social engineering may, in general, be an important malware spreading vector, in this work we restrict our focus and analysis to malware delivered via drive-by downloads.

Recently, Provos et al. [4-6] provided insights on this new phenomenon, and presented a cursory overview of web-based malware. Manual analyses of exploit sites have recently emerged. Although they often provide very useful and detailed information about which vulnerabilities are exploited and which malware programs are installed, such analysis efforts are not scalable and do not provide a comprehensive picture of the problem. Unfortunately, the high-level specifications of malicious behavior used by these advanced malware detectors are currently manually developed. Creating specifications manually is a time-consuming task that requires expert knowledge, which reduces the appeal and deployment of these new detection techniques.

Our research work in this paper is introducing a technique to automatically derive specifications of malicious behavior from a given malware sample. Such a specification can then be used by a malware detector, allowing for the creation of an end-to-end tool-chain to update malware detectors when a new malware appears.

This paper presents a novel classification method that identifies malicious web pages based on statically analyzing the initial HTTP response denoted by the URL by inductive learning. This method requires neither the entire web page to be retrieved nor a dedicated environment that client honeypots do. Application of this method leads to significant performance improvements.

2 Drive-By Download Attack

The classification method utilizes common elements of malicious web pages. In this section, a description of these elements is presented. To fully understand the dramatic shift to using the Web browser as the attack tool, it is useful to revisit the history of major Internet-based computer attacks. During the "Internet worm era," when attacks like Code Red, Blaster, Slammer and Sasser, hackers used remote exploits against the vulnerabilities of operating system. Some malicious executables, such as Melissa, were also attached to e-mail or they arrived via instant messaging or peer-to-peer applications.

Microsoft, Businesses and consumers reacted to the worm attacks in a positive way. They added a firewall, which is turned on by default in Windows XP SP2, and implemented several anti-worm mitigation mechanisms in the operating system. With automatic updates enabled on Windows, end users got some assistance with regularly applying operating system patches. These factors forced attackers to shift tactics, moving up the stack to target third-party applications and to perfect the art of social engineering. This, in turn, has lead attackers to seek other avenues of exploitation. An equally potent alternative is to simply lure web users to connect to (compromised) malicious servers that subsequently deliver exploits targeting vulnerabilities of web browsers or their plugins. Adversaries use a number of techniques to inject content under their control into benign websites. In many cases, adversaries exploit web servers via vulnerable scripting applications. Typically, these vulnerabilities (e.g., in phpBB2 or Invision Board) allow an adversary to gain direct access to the underlying operating system. In general, upon successful exploitation of a web server the adversary injects new content to the compromised website. In most cases, the injected content is a link that redirects the visitors of these websites to a URL that hosts a script crafted to exploit the browser. To avoid visual detection by website owners, adversaries normally use invisible HTML components (e.g., zero pixel IFRAMEs) to hide the injected content.

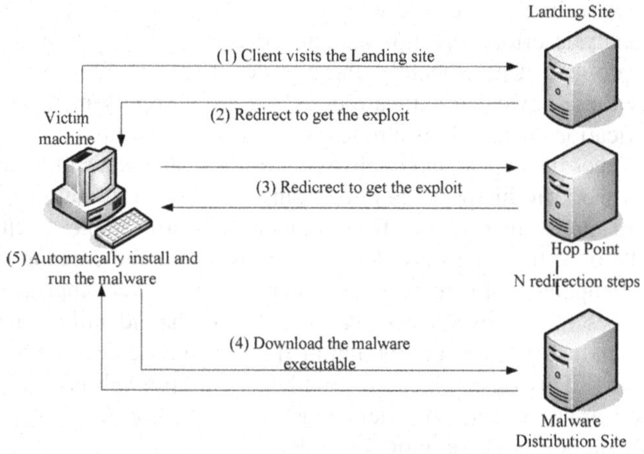

Fig. 1. A typical Interaction of drive-by download victim with a landing URL

Figure 1, from the Google Anti-Malware Team [12], illustrates the main phases in a typical interaction that takes place when a user visits a website with injected malicious content. Upon visiting this website, the browser downloads the initial exploit script (e.g., via an IFRAME). The exploit script (in most cases, javascript) targets vulnerability in the browser or one of its plugins. Successful exploitation of one of these vulnerabilities results in the automatic execution of the exploit code, thereby triggering a drive-by download. Drive-by downloads start when the exploit instructs the browser to connect to a malware distribution site to retrieve malware executable(s).

The downloaded executable is then automatically installed and started on the infected system. Finally, attackers use a number of techniques to evade detection and complicate forensic analysis. For example, the use of randomly seeded obfuscated javascript in their exploit code is not uncommon. Moreover, to complicate network based detection attackers use a number or redirection steps before the browser eventually contacts the malware distribution site.

Over a recent ten-month period, the Google Anti-Malware Team crawled billions of pages on the Web in search of malicious activity and found more than three million URLs initiating drive-by malware downloads. There are three core elements contained on a malicious web page [1]: the exploit itself, the delivery mechanism that brings the browser to the exploit, and mechanisms to hide the exploit or the delivery mechanism from detection.

Exploit: The exploit is the attack code that targets a specific vulnerability of the browser, its plug-ins, or underlying operating system. It is specific to the vulnerability of target and can make use of a variety of techniques. In Fig.1 exploit is always in distribution site.

Exploit Delivery Mechanism: While the exploit is the central part of a malicious web page, the web page might not contain the exploit directly. Exploits might be "imported" from a central exploit server to provide a mechanism for a reusable and modular design. There are two types of imports: direct includes of resources and redirects. Direct includes of resources is a feature naturally supported by HTML. The source attribute, which exists on several HTML tags, is able to import resources from local and remote web servers.

Obfuscation: Hiding the exploit or the exploit delivery mechanism through obfuscation is a common mechanism used by malicious web pages. Script code is provided in obfuscated form alongside a custom de-obfuscation function, which can convert the obfuscated code snippet into its clear form. Once converted, the code can be executed. Signature-based detection algorithms are unable to analyze the obfuscated JavaScript.

All the elements described above have their legitimate purposes, but attackers also use them. As a result, a web page cannot be classified as malicious if one merely observes these elements contained in an HTTP response. A victim is required to visit the malware hosting server or any URL linking to it in order to download the malware. This behavior puts forward a number of challenges to defense mechanisms (e.g., malware signature generation schemes) mainly due to the inadequate coverage of the malware collection system. Anti-virus engines use the signature of exploit to detect the malicious code. The research of Google Anti-Malware Team shows that the detection capability of the anti-virus engines is lacking, with an average detection rate of 70% for the best engine. These results are disturbing as they show that even the best anti-virus engines in the market (armed with their latest definitions) fail to cover a significant fraction of web malware. A more suitable mechanism is needed, which is described next.

3 Model of Malware Detection

In this section, a description on how characteristics of mechanism of exploit delivery and obfuscation can be taken advantage of to classify web pages is presented.

We describe the behavior characteristics of the malware using a set of features FV={F1,F2,...,Fn}and the relationship of the features RV=f(F1,F2,...,Fn). A feature associated with the context of an event and every event consists of three domains: action, object, result. A feature vector for malware detection can contain any number of nominal, discrete, and/or continuous features. In addition, a feature can also be as complex as a compound feature.

Definition 1 (behavior signature): A behavior signature Sig={FV, RV}is a set of features and the relationship of features. FV is a set of features FV={F_1, F_2, …, F_n}, where F_i is a feature associated with the context of a event, Fi={Action, Object, Result}, i=1,2,…,n, Action is the operation of the event, Action={add, del, …, }, Object the target of the action is the defining domain w.r.t a computing resource. Result the outcome of the action is also the defining domain. RV is a subsequent expressions, its operator is identical with operator of the set.

This model represents the form in which patterns are internally stored and matched. Externally, a language can be designed to represent signatures in a more programmer natural framework, and programs in the language compiled to this internal representation of matching.

The event is the smallest entity of a pattern and defined as the occurrence of something that might be a part of an attack, so it is identifier of worm. The basic building block of a worm pattern is a sequence of events. We use a simple language [5] (called Event Descriptive Language, shortly as EDL) that allows to specify an attack as a list of event sets. The model described by EDL consists of two parts: one is the definition of the event; the other is the relationship of events and it is a concurrent expression. The definition includes many fields of event, such as the type, role, operation, object, result, time constraint and name. Every field can be a variable or a constant. The relation is composed of 'FOLLOW', 'AND', 'OR' and 'POWER'. It can be transferred to CPN model by the method in [5].

Several thousand potentially malicious web pages were inspected using the honeypot Capturer. As the client honeypot inspected potentially malicious web pages, the network traffic was recorded. Web pages identified as malicious were marked as such and the corresponding HTTP response was saved. Table 1 shows the mainly characteristics of the mechanism of exploit delivery and obfuscation.

Table 1. Characteristics of the mechanism of exploit delivery and obfuscation

No.	Name	Example
1	iframe	<iframe src= 'www.xxx.com' width=0 height=0></iframe>
2	Java script	<script type="javascript"> document.write("<iframe width='0' height='0' src='www.xxx.com '> </iframe>"); </script>
3	Body redirect	<body onload="window.location='www.xxx.com';"></body>
4	CSS redirect	background-image: url(t:open("http://www.X.com/muma.htm", "newwindow" , "border="1" Height=0, Width=0, top=1000, center=0, toolbar=no, menubar=no, scrollbars=no, resizable=no, location=no, status=no"))

Table 1. *(continued)*

No.	Name	Example
5	img	<html> <iframe src="www.xxx.com" height=0 width=0> </iframe> <img src="www.yyy.com"></center> </html>
6	frame	<frameset rows="444,0" cols="*"> <frame src="www.xxx.com" framborder="no" scrolling="auto" noresize marginwidth="0" marging-height="0"> <frame src="www.yyy.com" framebor-der="no" scrolling="no" noresize marginwidth="0" margingheight="0"> </frameset>
7	innerHTML	top.document.body.innerHTML = top.document.body.innerHTML + '\r\n <iframe height="0" src="http://www.xxx.com/ /"></iframe>'
8	window.open	<SCRIPT language=javascript> window.open ("www.xxx.com", "","toolbar=no, lo-cation=no,directories=no,status=no,menubar=no,scrollb ars=no, width=1, height=1"); </script>

For example, the EDL and CPN models mainly of the mechanism with iframe and CSS are show in Fig.2 and Fig.3.

Attack "iframe" *[tag, src, width, hight]*
Tag= "iframe"

Src= "www.xxx.com"

Width=0

Hight=0

Host=URL <>Src

Relation=FOLLOW (AND(tag,OR(width,hight),host))

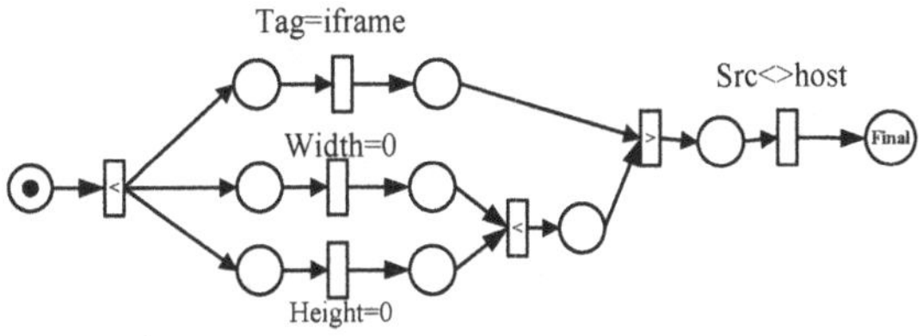

Fig. 2. Model of the mechanism with iframe

Attack "CSS" *[tag, src, width, hight]*
Tag= " background-image: url(t:open("

Src= "www.xxx.com"

Width=0

Hight=0

Host=URL <>src

Relation=FOLLOW (AND(tag,OR(width,hight),host))

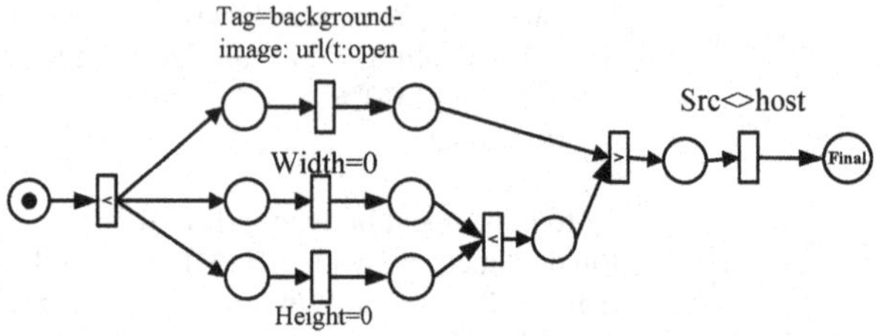

Fig. 3. Model of the mechanism with CSS redirect

4 The Model Generalized by Inductive Learning

A malicious often metamorphoses to avoid detection; fortunately its main behavior is remained. To deal with its metamorphoses, we need to generalize the model. To do this we use inductive learning and the generalization can be used to detect the attack whose behavior is similar to the examples.

The generalization algorithm based on inductive learning established three concept spaces of attack. A typical process can be divided into three steps: suppose the primal expression described by EDL be α,

(1) First step is generalizing the result field of event which includes four tasks: result generalization, relation generalization, doing with isolated value and background knowledge.

If these events are common to all the examples, they are referred to as background knowledge. As background knowledge is visible for each example, all the facts that can be derived from the background knowledge and an example are parts of the extended example.

(2) Second step is generalizing the object field of event. This generalization also includes four tasks similar to the first step. The output of this step is α_2.

(3) Last step is using dropping condition rule to generalize the result, object and constraint field of all events in α_2, and combine the events which have the same object field. The output of this step is α_3.

The model of attack instance is generalized by turning constants into variables and extending the range of variables in this generalization process. This generalization makes the model extending from one instance to a space. The background knowledge can enhance the veracity of the generalization and be strengthened during the process. The learning process is incremental.

The tree layer concept space made up of α_1, α_2, α_3 is generalized step by step. These concepts describe experience at different levels of generality. Space layer α_1 is more similar than α_2 and it can detect attacks more similar to the instance described byα. α_2 can detect more attacks than α_1, but it causes more false alarms, too. α_3 only include the type and operation of the event and it can increase efficiency of detection by eliminating useless event.

The process of detection is described in [5]. The system matches the event from monitor in the three-layer concept space of the attack model during worm detection. If the fields of event in model include the fields of event of monitor, it is a success matching. The matching is not strict but approximate.

By generalization the system can detect more attacks, but the false alarm is also increased, which is excess generalization. The concept space should cover all instances and eliminate all counter examples, which need specialization of the concept space. Fig.4, Fig. 5 and Fig.6 are the generalization models of the mechanism with iframe and CSS.

Attack "α_1" [tag, src, width, hight]

Tag= {" background-image: url(t:open(" ,"iframe"}

Src= "www.xxx.com"

Width=X <minx

Hight=Y<miny

Host=URL <>src

Relation=FOLLOW (AND(tag,OR(width,hight),host))

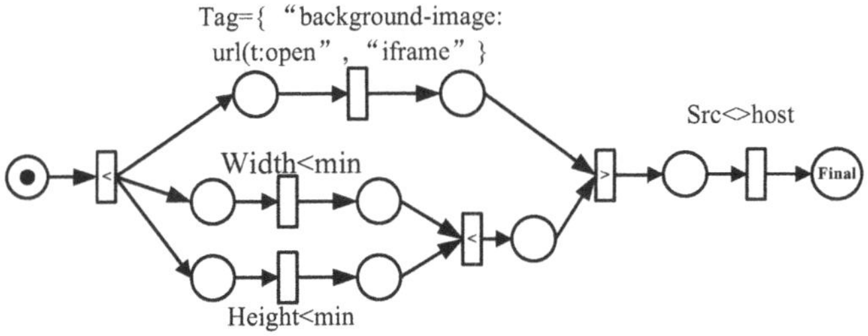

Fig. 4. The first model of the mechanism with iframe and CSS

Attack "α_2" [tag, src, width, hight]

Tag= {" background-image: url(t:open(" ,"iframe" , "frame", "window.open"}

Src= URLs

Tag ={ "" ,""}

Hight=Y<miny

Host=URLh <>Src

Relation=FOLLOW (AND(tag,OR(width,hight),host))

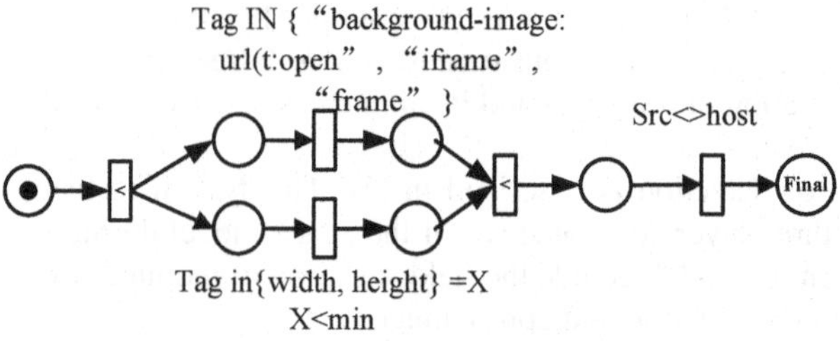

Fig. 5. The second model of the mechanism with iframe and CSS

Attack "α_3" [tag, src, width, hight]

Tag= {" background-image: url(t:open(" ,"iframe" , "frame", "window.open"}

Src= URLs

Tag= { "width", "height"}

Host=URLh <>Src

Relation =FOLLOW (AND(tag,OR(width,hight),host))

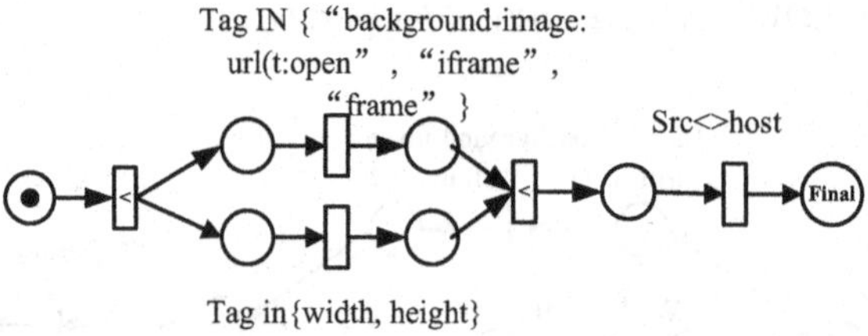

Fig. 6. The third model of the mechanism with iframe and CSS

5 Experiment

For this study, 3,478 instances of malicious and 5,000 instances of benign web pages were input into the machine learning algorithm. The generated classifier was used to classify a new sample. Of the 31,000 URLs included in the sample, 1,590 URLs were marked as malicious by the presented classification method. This amounts to a false positive rate of 5.88% and a false negative rate of 46.15% for the new classification method. The evaluation against the 5,000 URLs provided by HauteSecure resulted in similar metrics.

6 Conclusions

In order to detect and defend the drive-by downloads attacks, we propose a classification algorithm that identifies malicious web pages based on statically analyzing the initial HTTP response denoted by the URL. The attack pattern is established from an example and generalized by inductive learning, which can be used to detect unknown attacks whose behavior is similar to the example. Application of this method leads to significant performance improvements.

References

1. Seifert, C., Komisarczuk, P., Welch, I.: Application of divide-andconquer algorithm paradigm to improve the detection speed of high interaction client honeypots. In: 23rd Annual ACM Symposium on Applied Computing. ACM, Ceara (2008)
2. Anagnostakis, K.G., Sidiroglou, S., Akritidis, P., Xinidis, K., Markatos, E., Keromytis, A.D.: Detecting Targeted Attacks Using Shadow Honeypots (August 2005)
3. David, S.A., Fleizach, C., Savage, S., Voelker, G.M.: Spamscatter: Characterizing Internet Scam Hosting Infrastructure. In: Proceedings of the USENIX Security Symposium (August 2007)
4. Barford, P., Yagneswaran, V.: An Inside Look at Botnets. Advances in Information Security. Springer, Heidelberg (2007)
5. Bem, J., Harik, G., Levenberg, J., Shazeer, N., Tong, S.: Large scale machine learning and methods. US Patent: 7222127
6. Peishun, L., Dake, H.: Worm detection using CPN. In: 2004 IEEE International Conference on Systems, Man & Cybernetics, October 2004, pp. 4941–4946 (2004)
7. Kienzle, D.M., Elder, M.C.: Recent worms: a survey and trends. In: Proceedings of the 2003 ACM Workshop on Rapid Malcode, October 27, pp. 1–10. ACM Press, New York (2003)
8. Chen, S., Tang, Y.: Slowing Down InternetWorms. In: Proceedings of IEEE International Conference on Distributed Computing Systems, ICDCS (2004)
9. Christodorescu, M., Jha, S., Seshia, S.A., Song, D., Bryant, R.E.: Semantics-aware malware detection. In: Proc. IEEE Symposium on Security and Privacy, pp. 32–46 (2005)
10. Kruegel, C., Robertson, W., Vigna, G.: Detecting kernel-level rootkits through binary analysis. In: Proc. 20th Annual Computer Security Applications Conference (ACSAC 2004), pp. 91–100 (2004)
11. Kinder, J., Katzenbeisser, S., Schallhart, C., Veith, H.: Detecting malicious code by model checking. In: Julisch, K., Krügel, C. (eds.) DIMVA 2005. LNCS, vol. 3548, pp. 174–187. Springer, Heidelberg (2005)
12. Provos, N., Mavrommatis, P., Rajab, M.A., Monrose, F.: All Your iFRAMEs Point to Us, Google Technical Report provos-(2008a)

The Handicap Principle for Trust in Computer Security, the Semantic Web and Social Networking

Zhanshan (Sam) Ma[1], Axel W. Krings[2], and Chih-Cheng Hung[3]

[1] IBEST (Initiative for Bioinformatics and Evolutionary Studies) & Departments of Computer Science and Biological Sciences, University of Idaho, Moscow, ID, USA
[2] Department of Computer Science, University of Idaho, Moscow, ID, USA
[3] School of Computing, Southern Polytechnic State University, GA, USA

Abstract. *Communication* is a fundamental function of life, and it exists in almost all living things: from single-cell bacteria to human beings. *Communication,* together with *competition* and *cooperation*, are three fundamental processes in nature. Computer scientists are familiar with the study of *competition* or '*struggle for life*' through Darwin's evolutionary theory, or even evolutionary computing. They may be equally familiar with the study of *cooperation* or *altruism* through the Prisoner's Dilemma (PD) game. However, they are likely to be less familiar with the theory of *animal communication*. The objective of this article is three-fold: (*i*) To suggest that the study of animal communication, especially the *honesty* (*reliability*) of animal communication, in which some significant advances in behavioral biology have been achieved in the last three decades, should be on the verge to spawn important cross-disciplinary research similar to that generated by the study of *cooperation* with the PD game. One of the far-reaching advances in the field is marked by the publication of "*The Handicap Principle: a Missing Piece of Darwin's Puzzle*" by Zahavi (1997). The 'Handicap' principle [34][35], which states that communication signals must be costly in some *proper* way to be reliable (honest), is best elucidated with evolutionary games, e.g., Sir Philip Sidney (SPS) game [23]. Accordingly, we suggest that the Handicap principle may serve as a fundamental paradigm for trust research in computer science. (*ii*) To suggest to computer scientists that their expertise in modeling computer networks may help behavioral biologists in their study of the reliability of animal *communication networks*. This is largely due to the historical reason that, until the last decade, animal communication was studied with the dyadic paradigm (sender-receiver) rather than with the *network* paradigm. (*iii*) To pose several open questions, the answers to which may bear some refreshing insights to trust research in computer science, especially secure and resilient computing, the semantic web, and social networking. One important thread unifying the three aspects is the evolutionary game theory modeling or its extensions with survival analysis and agreement algorithms [19][20], which offer powerful game models for describing time-, space-, and covariate-dependent *frailty* (uncertainty and vulnerability) and *deception* (honesty).

Keywords: Handicap Principle; Animal Communication Network; Signal Honesty; Deception; Extended Evolutionary Game Theory; Semantic Web; Trust; Social Networking; Eavesdropping; Pervasive and Resilient Computing.

W. Liu et al. (Eds.): WISM 2009, LNCS 5854, pp. 458–468, 2009.

1 Introduction

Competition, *cooperation* and *communication* are three fundamental processes in nature. In his landmark volume "*Origin of Species*" (Darwin 1859), Darwin focused on the *competition*, or the *struggle for life*, but he was also clearly concerned with the cooperation or altruism in nature and suggested that selection can be applied upon the family—the prototype of kin selection [6]. Therefore, cooperation, rather than competition, within a family still increases individual fitness ultimately. It took near a century for scientists to formalize the idea mathematically with a simple equation, known as Hamilton's rule [6], first formalized by William Hamilton (1964) [9]. About two decades later, Hamilton's collaborative work with political scientist Robert Axelrod [Axelrod & Hamilton (1981), which implemented Robert Trivers' (1971) suggestion of using Prisoner's Dilemma (PD) game to study altruism], led to the eruption of the study of cooperation in the last three decades. Darwin was also clearly interested in communication, evidenced by his treatise titled "*The Expression of the Emotions in Man and Animals*" published in 1872. More than a century later, "*The Handicap principle: a missing piece of Darwin's Puzzle*" by Zahavi (1997) opened a new chapter in the study of animal *communication*. The Handicap principle by Zahavi [34], which states that that communication signals must be costly in some *proper* way to be reliable, did not receive wide acceptance until it was formulated as evolutionary game models by Grafen (1990a, 1990b) and Maynard-Smith (1991). Particularly, Maynard-Smith's lucid explanation with Sir Philip Sydney (SPS) game helped to establish Handicap principle as the main theory to explain honest communication in animals.

Indeed, the honesty of animal communication is one of those deep questions that bear fundamental importance in science, humanity and philosophy. For example, the question is somewhat equivalent to the question whether human beings stand out as a separate group in terms of the truthfulness of their communication; it is also relevant to the evolution of often puzzling *altruism* (cooperation) [21][30]. Despite huge research interests in biology generated by the Handicap principle, the study of animal communication and its evolutionary game models (e.g., SPS game) has not spawned significant cross-disciplinary research activities similar to those spawned by the study of *cooperation* and PD games, which in my opinion, should have been the case.

As outlined in the abstract above, this paper has three objectives: (*i*) To introduce the Handicap principle and the study of reliability of animal communication to computer scientists and further suggest that it can serve as a paradigm for trust research in computer science. (*ii*) To suggest the bidirectional research potentials: on the one direction, the potential cross-disciplinary research that the study of animal communication may spawn (e.g., utilizing the animal communication network as a model for pervasive and resilient computing); on the other direction, the theory for modeling computer networks may offer highly sought methods for studying animal communication networks. (*iii*) To propose several research topics (in the form of open questions) that are highly relevant to computer security, the semantic web, social networking, and some other broader areas in computer science.

2 The Handicap Principle

This section draws heavily on excellent monographs by Searcy & Nowicki (2005), Maynard-Smith & Harper (2003), and Zahavi (1997).

2.1 Basic Definitions

In behavioral biology, *signals* are the behavioral, physiological, or morphological characteristics that have been selected by evolutionary process to convey information to other organisms [18][21][22][30]. Searcy & Nowicki (2005) defined two conditions for a *reliable signal*: (1) There is a *consistent correlation* between some characteristics of the signal (including, its presence or absence) and some attributes of the signaler or its environment; (2) Receiving the information by receivers is beneficial to them.

One option to define *deceitful signal* is to adopt the opposite of *reliable signal*, but the resulting definition would contain more ambiguous terms. Instead, a definition by Mitchell (1986) and further extended by Searcy and Nowicki (2005) is simple yet sufficient. They suggested that *deception* happens if: (1) A receiver records (register) something *Y* from a signaler; (2) The receiver's response is appropriate if *Y* means *X*, and the receiver's response is beneficial to the signaler (3) and it is not true here that *X* is the case. The term *beneficial* means the increase of individual fitness. Hauser (1997) termed this kind of deception *functional deception*, implying that there is no need to have cognitive ability to commit the deception.

There are other types of definitions for deception and most of them assume more complex cognitive foundation. The signaler possesses *intention* to trick the receiver to *believe* a false assessment about the true situation. This kind of deception is called *intentional deception.* Whether or not animals can commit the *intentional deception* boils down to a fundamental open question, *i.e.*, the nature of animal cognition. The question is of great interests to philosophers, cognitive ethologists and general public [30][31]. However, from the perspective of how natural selection shapes animal communication to be honest or deceitful, the notions of mental states (such as intention and belief) are less relevant because the cost and benefit accrued by signalers and receivers are not affected by the existence of mental states [30]. This evolutionary approach essentially circumvents the question of the mental states.

The question around mental states may also be circumvented with a strategy adopted in engineering fault tolerance research, *i.e.*, the *hybrid fault models* or its extended versions, *dynamic hybrid fault models* (Ma 2008, Ma & Krings 2008). In the fault models, the classification of faults is based on the *consequences of* the faults, rather than on the *causes* (*processes*) or "*intentions*," e.g., malicious vs. benign, asymmetric vs. symmetric, transmissive vs. omissive.

Searcy and Nowicki (2005) particularly emphasized one issue, the withholding of signals. In other words, does failure to signal constitute deceit? For example, in cooperative social foraging, if an individual fails to alert its fellow individuals for the food source it found, would that be a deceitful action? In dynamic hybrid fault models, this will be omissive "faults" (deceit) [15][16].

2.2 The Handicap Principle for the Reliability of Animal Communication

The traditional ethological view of animal communication is that the relationship between a signaler and a receiver is honest and cooperative information exchange. With the words of Dawkins & Krebs (1978) who characterized the traditional ethological view as "*it is to the advantage of both parties that signals should be efficient, unambiguous, and informative.*" Dawkins & Krebs (1978) replaced the ethological view with *non-cooperative* or *manipulative* relationship, i.e., the signaler sends signals to manipulate the receiver's behavior to the signaler's advantage. They stated "*if information is shared at all, it is likely to be false information, but it is probably better to abandon the concept of information altogether*" [3][30].

One background that motivated Dawkins & Krebs (1978) was the growing acceptance of the dominant role of *individual selection* over the *group selection* in shaping the evolution of animal behavior. The traditional ethological view could be interpreted conveniently as the consequence of group selection. The manipulative view is more consistent with individual selection. However, there is a critical flaw in the manipulative hypothesis, i.e., if the receiver does not get any benefit, why would natural selection still select the receiver who responds to the signal at all? Natural selection would favor those that ignore the signal, and then the signal loses its value of existence and the communication system would have disappeared. It should be noted that both cooperative and manipulative hypotheses, which predict the totally honest and totally deceitful communication, respectively, are two extreme scenarios. The truth should be in the middle.

The partial answer for explaining the dilemma faced by the manipulative theory was already proposed by Zahavi (1975) with the so-termed "*Handicap Principle.*" prior to Dawkins & Krebs (1978). Zahavi (1975) argued that in a signaling system such as one used in mate selection, a superior male is able to signal with a highly developed "*handicap*" to demonstrate its quality and the handicap serves "*as a kind of (quality) test imposed on the individual*" [30][35]. Two examples of the handicap Zahavi (1975, 1997) cited are the exaggerated train of the peacock and singing in exposed positions by warblers; both are the demonstrations of good quality but they also expose a male to predators. Only males with the superior quality can afford the risk.

The handicap principle did not receive favorable support initially. It was first formulated as a genetic model of three genes. The initial modeling analysis failed to show the evolution of handicaps because it assumes that only males of high viability can survive with the handicap. This type of handicap is termed "*pure handicap*" or "*Zahavi handicap*" [30]. Later two revised versions of handicaps were proposed: *conditional handicap* and *revealing handicap*; both versions can lead to reliable signaling. The conditional handicap assumes that handicaps are expressed only if a male has both handicap and high viability genes. The revealing handicap assumes that it is possible for all males of the handicap gene to express, but the *conspicuousness* of the handicap is correlated with the viability of its possessor [30][36].

What leads to the wide acceptance of handicap principle is the evolutionary game theory (EGT) modeling. Grafen's (1990a, 1990b) two papers are considered to have completed the switch of the recognition [30]. Grafen (1990a) demonstrated that evolutionary equilibrium is reachable when the signaling level is a strictly increasing function of male quality (signal is reliable), and when females prefer males with high levels of signaling (females are responding to the signal). Grafen (1990b) further showed that evolutionary stable strategies are reachable when (*i*) signaling is expensive or signaler fitness declines with the increase of signaling level; (*ii*) receiver's judgment of the signaler's quality increases with the increase of signaler's signaling level, (*iii*) the signaler benefits from the high scores judged by the receiver, and (*iv*) the ratio of marginal cost of signaling to the marginal benefits of a higher level judgment is a decreasing function of signaler quality.

Hamilton & Brown (2001) suggested that the handicap principle may explain the autumnal change in leaf color of temperate deciduous tree, which sends an *unequivocal* signal to the herbivore insects who feed on the tree. Through the change of color, which is costly to tree, it sends insects the "keep out" signal [10].

2.3 Signal Costs

The signal cost is the central variable for quantifying the handicap model. All costs have to boil down to *fitness cost* because fitness is the "*hard currency*" to evaluate portfolio in the context of natural selection and evolution of animal communication. According to Searcy and Nowicki (2005), the cost can first be divided into two categories: *receiver-dependent* and *receiver-independent* costs. The former can further be categorized as *vulnerability costs* (e.g., exposed to predation during the signaling), and *receiver-retaliation rules* (e.g., provoke opponent to attack in an aggression conflict). The receiver-independent costs consist of three parts: *production costs* (paid at the time when the signal is displayed to the receiver, e.g., fanning of the train by peacock), *development costs* (accrued at the time when the signal is developed, e.g., growing elongated feathers to build train by peacock), and *maintenance costs* (need to bear once the signal is developed, regardless of the use, e.g., the reduced flight performance due to the exaggerated train of peacock). There is yet another classification: *efficacy costs* and *strategic costs* (Maynard-Smith and Harper 2003). Efficacy costs are the costs that produce *good enough* signals but such signals may be undetected by the receiver due to the signal attenuation or background noise. The strategic costs produce signals that cost more than efficacy costs, but such signals are ensured to be detected by the receiver [21][30]. A mathematical definition of cost was developed by Grafen (1990b), who treated a signal as costly if the *partial derivative* of fitness with respect to signaling level is less than zero, under the boundary conditions that receiver assessment and signaler quality are fixed.

2.4 Alternatives to the Handicap Principle

The opinion on Zahavi's (1975) handicap principle was switched from the initial suspicion before the middle of 1980s to near universal acceptance by 1990s. However, evolution seems always generate some exceptions; the following are several hypotheses that offer alternative or supplemental explanations to the Handicap Principle.

Johnstone (1997) argued that it is the relationship between signal intensity and benefit, rather than that between signal intensity and cost as claimed in Zahavi's handicap principle, which enforces the reliability of signaling.

Honest signaling is also possible when there are no conflicting interests between the signaler and receiver. They may reach a *consensus* for the signal level acceptable to both parties and the signaling then does not have to be costly. This is the so-termed signaling when interests overlap and is discussed in detail in the chapter 2 of [30].

A third alternative to the handicap principle is what Searcy and Nowicki (2005) termed "*individually directed skepticism*". This idea was explored by several groups of authors (e.g., [31]). The hypothesis allows the signaler and receiver interact multiple times and the receiver can "*learn a lesson*" from its past experience with the individual signaler and adjust its response accordingly. The receiver may choose to ignore the signal or retaliate if it is tricked previously. The *experience-based* costs then enforce the communication reliability [30].

A fourth explanation to reliability of animal signaling is related to the concept of *index* (of signaler's quality) proposed by Maynard Smith and Harper (2003). Index of quality refers to a signal metric that is causally correlated with the signaler's quality and cannot be faked. Because the signal cannot be faked and is performance-based, its genuineness or reliability is ensured.

A fifth explanation for the reliable signaling is the coordinated games. In some cases, signalers and receivers prefer different outcomes, but share a common interest, e.g., avoiding the escalation of conflicts, or breakup of some alliance with benefits to both parties [24].

A sixth hypothesis is based on the so-called *receiver biases*. ten-Cate & Rowe (2007) reviewed recent studies that support the *receiver-bias hypothesis*. The receiver biases refer to the preferences of sensory systems or brains of animals, which can be by products of natural selection or incidental and non-selected consequences.

2.5 Deception Revisited: Is Deception Possible?

The principle of handicap—signals can be reliable as long as they are costly in a proper manner—not only led to the recognition that animal communication can be generally reliable, but also shifted the research focus to issues such as that how the signaling could ever be deceptive, and the consequence of cheating [8][30]. Grafen (1990b) analyzed the scenarios in which cheating is tolerable. With the EGT modeling of animal communication, the evolutionary stable strategy (ESS) can be reached by balancing the costs and benefits. In the ESS, the deception may exist, not even necessarily in minority. Deception can be in majority as long as the stake (benefits of responding to an honest signal) is sufficiently high enough and the cost of committing deception is sufficient low [30].

Dawkins and Guildford (1995) argued that honest-handicap signals should be replaced by less-reliable conventional signals because of the costs imposed on receivers by the handicap. The argument seems being echoed by the conventional wisdom that a "good enough" strategy may be better off than a "perfect" one.

2.6 Reliability of Animal Communication *Networks*

Our discussion of animal communication up to this point is in the context of dyadic paradigm, the signaler-receiver paradigm. It is natural to ask the question, what are the implications of adopting a network paradigm with regard to the reliability of animal signaling?

First, we look at what may still hold after introducing the third party receivers. As argued by Searcy and Nowicki (2005), the foundation on animal signaling with the dyadic paradigm is still valid for two reasons: (*i*) dyadic paradigm has been the dominant view in animal communication literature; (*ii*) the dyadic view of the conflicts between signal and receiver captures the essence of the reliability and deceit in animal communication [30]. However, currently, there are little experiment or theoretical analysis evidence that verifies the claim. Overall, the study of reliability of animal communication networks is still an emerging field; many problems in the area are not yet defined, not to mention the answers. It is also obvious that a network perspective is crucial for seeking inspiration for engineering network reliability where a dyadic paradigm is of very limited practical utility. For example, the study of eavesdropping on signaling reliability, and the reliability of animal communication networks in large, may generate profound insights for both reliability and security of computer networks. It is not unlikely that a new security/trust protocol, with the magnitude of the influence from the invention of the *public-key protocol*, being inspired by the study of animal communication networks. Animal communication networks possess all the major features of pervasive or ubiquitous computing, and they are possibly the best natural models for pervasive communication networks.

3 Trust in Security, Semantic Web, and Social Networking

Trust can be considered as the counterpart problem of honest communication in human beings. In computer science, trust is studied extensively but may be loaded with different meanings. In semantic webs, trust is particularly important because *agents* and *reasoners* have to make trust judgment routinely based on diverse sources of information with varying quality [1]. Furthermore, trust on the Semantic Web is particularly challenging [1]: (*i*) the default of a statement is only a claim until it is verified; (*ii*) providing current trust verification is often very costly; (*iii*) context matters. Artz and Gil (2007) collected three typical definitions of trust in computer science. One of the more comprehensive is due to Olmedilla et al (2005), which states: "*Trust of a party A to a party B for a service X is the measurable belief of A in that B behaves dependably for a specified period within a specified context (in relation to service X).*" Artz and Gild (2007) emphasized that a unifying theme among the definitions of trust in computer science is that modeling of trust is only useful if there is a possibility of deception: the chance of different outcomes beyond what is expected or has been agreed upon. This argument seems to echo what is used in the study of animal communication in behavioral biology, where deception was the focus of modeling efforts. Indeed, the evolutionary game theory models such as SPS game should be sufficiently flexible to offer quantitative models for various trust concepts in computer science. In addition, the extended evolutionary game theory (EEGT) model by Ma (2009a, b, c), which extends evolutionary game theory with survival analysis and agreement algorithms to better handle time-, space-, and covariate-dependent uncertainty, vulnerability, and deception, should be particularly useful for modeling trust in computer science because the idea of introducing agreement algorithms was developed in the context of distributed computing [15].

According to Artz and Gil (2007), two common ways in computer science are adopted to determine trust: policies and reputation. Policies can set the bars (credentials) for trust and the mechanism to verify the credentials. Reputation is based on the experience from repeated interactions, and the evaluation of reputation can be based on direct interaction or indirect (third party) reporting. From a perspective of the handicap principle, credentials are similar to the handicap and the mechanisms to verify the handicap is through communication (e.g., displaying aggression behavior). Reputation and the third party reporting are similar to *audience effects* and eavesdropping modeling, which can be readily described with evolutionary games.

Once trust models are built, a further study can be to test the effects of various threats on the trust models [22]. Kovac and Trcˇek (2009) suggested that the deceitful interaction between cooperating entities in service oriented architectures is an issue rarely addressed in the existing literature and they proposed a qualitative model to deal with it. The handicap principle models formulated as evolutionary games such as Grafen (1990a,b) and Maynard-Smith's (1991) SPS game, as well as the extended evolutionary game theory modeling by Ma (2009a, b, c) may provide more general architecture to deal with the specific issues such as the effects of threats and the deceits from allies. Omar et al. (2009) presented a distributed trust model that depends on self-organization of nodes, public-key certificates, and threshold cryptography for mobile ad hoc networks. The fully distributed trust depends on the self-organization principle, which again is very similar to the nature of animal communication, where a central authority rarely exists.

Dietrich et al. (2009) presented an approach that integrates *social networking* (e.g., social bookmarking and blogs) and the Semantic Web technology to share knowledge, primarily based on existing Web 2.0 services. A particular challenge in using social networking technology is the openness of the services, which requires the mechanisms to establish and manage the trustworthiness of knowledge artifacts. Social networking itself may be formulated as evolutionary game theory models, and the integration with the Semantic Web can be considered as the interaction or overlay of two networks. The relation is similar to symbiotic relation in biology and co-evolutionary game models may provide an effective modeling tool, because trust or deception can be readily incorporated.

4 Open Questions

Finally, we summarize the six *open* questions, first presented in Ma (2009c) [21]. The term *open* is used somewhat differently from its meaning in mathematics or other subjects since the motivation of raising these questions was to stimulate the cross-disciplinary exploration [21].

(*i*) What can we learn from the *Handicap principle* (including the follow-up alternative theories) for developing secure and resilient communication protocols? Specifically, is it possible to develop a protocol for distributed trust systems? is it possible to develop an encryption or key distribution protocol that is inspired by the Handicap principle?

(*ii*) Animals apparently '*tolerate*' eavesdropping; both *interceptive* and *social* eavesdroppings have been evolved. From the evolutionary perspective, it implies that eavesdropping, at least, enhances one party's fitness and the cost is still acceptable to the parties being eavesdropped. Can this kind of *delicate balance* be used in network design? For example, in a wireless sensor network, can we use eavesdropping as a strategy to perform *network monitoring*, such as *blacklisting selfish nodes*, *detecting intrusions*, or '*spying*' on a neighbor network?

(*iii*) Handicap (including much of the seemingly '*redundant*' behavior in animal communication) can be considered as a counterpart of fault tolerant communication in animal communication. For example, there are *efficacy costs* vs. *strategic costs*: the former is '*good enough*' cost and the latter is '*guaranteed service*'. Can we develop a *quantitative 'fault tolerance' theory* for animal communication that is similar to the agreement algorithms in engineering fault tolerance theory? This *quantitative* theory should supplement the Handicap principle, rather than replacing it. Furthermore, what additional complexities will be needed for this theory to deal with animal communication *networks* beyond the simple sender-receiver dyads? Can we estimate the difference between *efficacy costs* and *strategic costs* for animal communication using such a theory?

A conjecture is: how do animals manage their redundancies? Do they '*vote*' for a consensus? How do they manage time synchronization in a highly distributed setting? The study of social insects has already revealed many interesting mechanisms, but can we develop a systematic '*distributed computing*' theory for the colony or colonies of social insects such as ants and bees?

In a *recursive* manner, what are the implications of such research on the study of engineering fault tolerance, e.g., resilience of computer networks?

(*iv*) *Deceptions* in animal communication can be considered as the counterpart of *failures* in engineering fault tolerance. Can we develop a classification of *deceptions*, similar to the *fault models* in engineering fault tolerance? for example, *malicious* vs. *benign* deception, *asymmetric* vs. *symmetric* deception; *transmissive* vs. *omissive* vs. deception. What are their relationships with tactical deception, intentional deception, and functional deception?

In a recursive manner, what are the implications of such a 'deception model' on the study of engineering fault tolerance, e.g., reliability and survivability of computer networks?

(*v*) Is there any general applicability about the 3-C processes (competition, cooperation and communication) in computing and communication systems, e.g., robot swarm control, strategic information warfare, or the study of general military warfare? How does communication *modulate* competition and cooperation?

(*vi*) Is cognition necessary for animal deception? In other words, what is the exact relationship between animal cognition and communication with regard to deceptive communication? Compared with human beings, does the limited brain cognition in 'primitive' animals such as insects makes them dumber or smarter with respect to the ability to deceive? Is social intelligence (swarm intelligence) more important to insects than to human beings and why?

In summary, *deception* is a reality and it may even become prevalent. It is one of the most intriguing and elusive elements in many social, natural, political, economic, military and engineered systems [19]-[21]. *Handicap* can make deception too costly to sustain, and therefore it may enforce honesty and *trust* mechanisms. Mathematically, deception should be treated as a dynamic game system, further complicated by dynamic frailty, EGT [12][23] and the EEGT [19][20] can be effective tools to study the system.

References

1. Artz, D., Gil, Y.: A survey of trust in computer science and the Semantic Web. J. of Web Semantics 5, 58–71 (2007)
2. Axelrod, R., Hamilton, W.D.: The evolution of cooperation. Science 211, 1390–1396 (1981)
3. Dawkins, R., Krebs, J.R.: Animal Signals: information or manipulation? In: Krebs, J.R., Davies, N.B. (eds.) Behavioral Ecology, pp. 282–309. Blackwell, Oxford (1978)
4. Dawkins, M.S., Guildford, T.: An exaggerated preference for simple neural models of signal evolution? Proc. R. Soc. Lond. B Biol. Sci. 261, 357–360 (1995)
5. Dietrich, J., Jones, N., Wright, J.: Using social networking and semantic web technology in software engineering. The Journal of Systems and Software 81, 2183–2193 (2008)
6. Dugatkin, L.A.: The Altruism Equation: Seven Scientists Search for The Origins of Goodness. Princeton University Press, Princeton (2006)
7. Grafen, A.: Sexual selection unhandicapped by the Fisher process. J. Theor. Biol. 144, 473–516 (1990a)
8. Grafen, A.: Biological signals as handicaps. J. Theor. Biol. 144, 517–546 (1990b)
9. Hamilton, W.D.: The genetic evolution of social behavior (I) & (II). J. Theor. Biol. 7, 1–52 (1964)
10. Hamilton, W.D., Brown, S.P.: Autumn tree colors as a handicap signal. Proc. R. Soc. London B Biol. Sci. 268, 1489–1493 (2001)
11. Hauser, M.D.: The Evolution of Communications. MIT Press, Cambridge (1997)
12. Hofbauer, J., Sigmund, K.: Evolutionary Games and Population Dynamics. Cambridge Univ. Press, Cambridge (1998)
13. Johnstone, R.A.: The evolution of animal signals. In: Krebs, J.R., Davies, N.B. (eds.) Behavioral Ecology, pp. 157–178. Blackwell, Malden (1997)
14. Kovac, D., Trcěk, D.: Qualitative trust modeling in SOA. J. of Syst. Arch. 55, 255–263 (2009)
15. Ma, Z.S.: New Approaches to Reliability and Survivability with Survival Analysis, Dynamic Hybrid Fault Models, and Evolutionary Game Theory. PhD Dissertation, University of Idaho (2008)
16. Ma, Z.S., Krings, A.W.: Dynamic Hybrid Fault Models and their Applications to Wireless Sensor Networks. In: The 11-th ACM/IEEE International Symposium MSWiM, Vancouver, Canada, p. 9 (2008)
17. Ma, Z.S., Krings, A.W., Sheldon, F.T.: An outline of the three-layer survivability analysis architecture for modeling strategic information warfare. In: Proc. 5th ACM CSIIRW, Oak Ridge National Lab. (2009)
18. Ma, Z.S., Krings, A.W.: Insect Sensory Systems Inspired Computing and Communications. Ad Hoc Networks 7(4), 742–755 (2009)
19. Ma, Z.S.: Towards an Extended Evolutionary Game Theory with Survival Analysis and Agreement Algorithms for Modeling Uncertainty, Vulnerability and Deception. LNCS (2009a) (accepted)
20. Ma, Z.S.: Extended Evolutionary Game Theory Approach to Strategic Information Warfare Research. Journal of Information Warfare (2009b) (accepted)

21. Ma, Z.S.: Animal Communication Inspired Computing and Communication: A Review for Computer Scientists with a Focus on Reliability of Animal Communication Networks and the Extended Evolutionary Game Theory (2009c) (submitted)
22. Mármol, F.G., Peréz, G.M.: Security threats scenarios in trust and reputation models for distributed systems. Computers and Security (2009), doi:10.1016/j.cose.2009.05.005
23. Maynard Smith, J.: Honest signaling: Sir Philip Sydney game. Animal Behavior 42, 1034–1035 (1991)
24. Maynard Smith, J., Harper, D.: Animal Signals, 164 pp. Oxford University Press, Oxford (2003)
25. McGregor, P.K. (ed.): Animal Communication Networks. Cambridge University Press, Cambridge (2005)
26. Mitchell, R.W.: A framework for discussing deception. In: Mitchell, R.W., Thompson, N.S. (eds.) Deception: perspectives on human and nonhuman deceit. State University of New York Press (1986)
27. Vincent, T.L., Brown, J.L.: Evolutionary Game Theory, Natural Selection and Darwinian Dynamics, 382 pp. Cambridge University Press, Cambridge (2005)
28. Olmedilla, D., Rana, O., Matthews, B., Nejdl, W.: Security and trust issues in semantic grids. In: Proceedings of the Dagsthul Seminar, Semantic Grid, vol. 05271 (2005)
29. Omar, M., Challal, Y., Bouabdallah, A.: Reliable and fully distributed trust model for mobile ad hoc networks. Computers & Security 28, 199–214 (2009)
30. Searcy, W.A., Nowicki, S.: The Evolution of Animal Communications. Princeton Univ., Princeton (2005)
31. Seyfarth, R.M., Cheney, D.L.: Signalers and receivers in animal communications. Annu. Rev. Psychol. 54, 145–173 (2003)
32. Silk, J.B., Kaldor, E., Boyd, R.: Cheap talk when interests conflict. Anim. Behav. 59, 423–432 (2000)
33. ten Cate, C., Rowe, C.: Biases in signal evolution: learning makes a difference. Trends in Ecology and Evolution 22(7), 380–387 (2007)
34. Trivers, R.L.: The evolution of reciprocal altruism. Quarterly Review of Biology 46, 189–226 (1971)
35. Zahavi, A.: Mate selection–a selection for a handicap. J. Theor. Biol. 53, 205–214 (1975)
36. Zahavi, A.: The Handicap Principle: A Missing Piece of Darwin's Puzzle. Oxford Univ. Press, Oxford (1997)

Formal Analysis of Fairness in E-Payment Protocol Based on Strand Space

Hong Wang[1], Jianping Ma[2], and Bo Chen[3]

[1] ShenYang Normal University
horoscope_leo@126.com
[2] Zhejiang University of Technology
majp@zjut.edu.cn
[3] Zhejiang Vocational College of Commerce
chbchyma@126.com

Abstract. Fairness as one of most important properties in E-payment protocol, it is important to verify it through informal and formal analysis approach. In this paper, we present the procedure to analyze iKP protocol in terms of strand space model, which consists of three key steps: First, we use strand space terminology to create the bundle of iKP protocol. Second, we enumerate all possible traces of each participant's strand in the run process of iKP. Third, according to the definition of fairness, determine whether iKP satisfy fairness. Then, we present the algorithm to verify fairness of E-payment protocol in pseudo code and use this algorithm to analyze ISI protocol successfully. This algorithm allows us to automatically verify fairness of E-payment protocol.

Keywords: strand space; e-Payment protocol; fairness; iKP.

1 Background

With the popularity of the Internet, E-commerce has been growing swiftly and violently in recent years. E-payment is one of many bottlenecks, which restrict the development of E-commerce. Usually E-payment is over the Internet that is an open network, so the study of E-payment has focused on secure E-payment protocol, which enables the online payment process to meet certain security requirements. Formal method is one of the most important approaches to prove the protocol satisfy security requirements.

Formal analysis of E-payment protocol stems from 1996 when Kailar [1] published a paper called Accountability in Electronic Commerce Protocols. In Kailar's paper, Kailar logic is proposed to analyze accountability of E-payment. According to [2]-[3], Kailar logic cannot prove fairness in E-commerce transaction, so that in [4], Qing Sihan et al. provided extended Kailar logic to analyze fairness as well as accountability. Bai Shuo et al. proposed logical verification with non-monotomic dynamic logic [5] and based on these, key fragments of SET protocol is verified logically. Strand space is proposed by Fabrega, Herzog and Guttma in 1998, which describes the run processes of protocols in terms of graph theory. Even though strand space mainly

W. Liu et al. (Eds.): WISM 2009, LNCS 5854, pp. 469–478, 2009.

analyzed the properties of authentication and secrecy, many researchers try to formalize other security requirements.

Therefore, this paper proposes a model to analyze fairness in E-payment protocol based on strand space. The rest of this paper is organized as follows: Section2 introduces basic idea of strand space. Section3 describes the iKP protocol and the detailed illustration on analytic process of iKP's fairness via strand space in section4. In section5, a general algorithm is provided to analyze fairness of E-payment protocol and the last section is the conclusion.

2 Strand Space

In this paper, we use the well-studied strand space model [6] to analyze iKP protocol. Basic definitions about strand space are summarized in this section. For ease of understanding strand space, we demonstrate the relationships amongst these definitions. See Fig1.

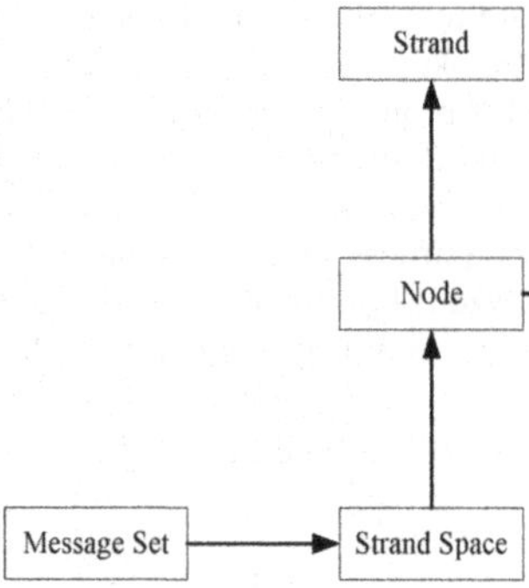

Fig. 1. Relationships among definitions of strand space

Definition 1: A message set M consists of the possible messages that can be exchange between principals in a protocol.

Definition 2: A strand space over M is a set Σtogether with a trace mapping tr:$\Sigma \rightarrow \pm M^*$, usually written in Σ.

Definition 3: A node of some strand space Σ is a pair of $\langle s,i \rangle$ with $s \in \Sigma$and i is an integer satisfying $1 \leq i \leq length(tr(s))$.

Definition 4: N denotes the set of nodes. We will say the node $\langle s,i \rangle$ belongs to the strand s. Every node belongs to a unique strand.

Definition 5: Suppose$\rightarrow_c \subset \rightarrow$; suppose $\Rightarrow_c \subset \Rightarrow$; and suppose C =$\langle N_c,(\rightarrow_c \cup \Rightarrow_c) \rangle$. C is a bundle if:

1. C is finite.
2. If $n2 \in N_c$ and term (n2) is negative, then there is a unique n1, such that $n1 \rightarrow_c n2$.

3. If n2∈Nc and n1 ⇒ n2 then n1 ⇒ c n2
4. C is acyclic.

3 iKP Protocol

In [7], Bellare M et al. proposed a new macro-payment protocol, namely iKP, to solve online transaction payment. Here, we describe iKP protocol that is analyzed below. There are three participants called M (Merchant), C (Customer) and A (Acquirer). The process flows of iKP protocol could be described as follows:

C→M: SALTC,CID, $CERT_C$
M→C: Clear, Sig_M
C→M: EncSlip,Sig_C
M→A: Clear, H(DESC, $SALT_C$), EncSlip, Sig_C, Sig_M
A→M: Sig_A
M→C: Sig_A
M→C: Goods and services
Where:Common: PRICE, ID_M, TID_M, DATE, $NONCE_M$, CID, H(DESC, $SALT_B$)
SigA: {Y/N, H(Common)} K_A^{-1}.

4 Formal Analysis of iKP Using Strand Space

We can formalize the protocol of iKP as Fig2. The strands of customer, merchant and acquirer are enumerated as followed.

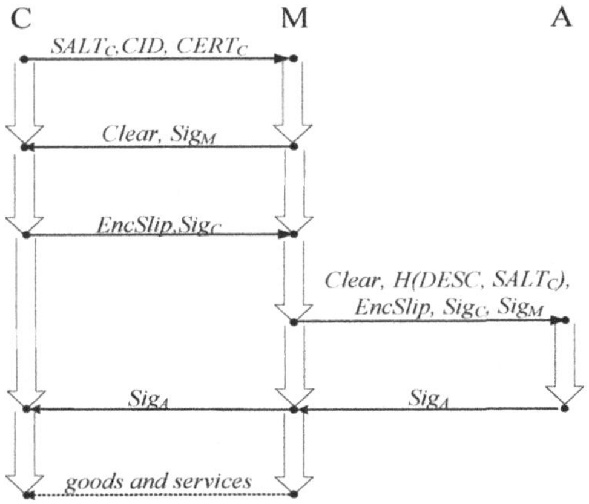

Fig. 2. Bundle of iKP Protocol

Customer strand S_1 with nodes:
<S_1, 1> +{$SALT_C$,CID,$CERT_C$}
<S_1, 2> -{Clear,Sig_M}
<S_1, 3> +{EncSlip,Sig_C}

$<S_1, 4>$ -{Sig_A}
$<S_1, 5>$ -{goods and services}
Merchant strand S_2 with nodes:
$<S_2, 1>$ -{$SALT_C$,CID,$CERT_C$}
$<S_2,2>$ +{Clear,Sig_M}
$<S_2, 3>$ -{EncSlip,Sig_C}
$<S_2, 4>$ +{Clear, H(DESC, $SALT_C$), EncSlip, Sig_C, Sig_M}
$<S_2, 5>$ -{Sig_A},+{Sig_A}
$<S_2, 6>$ +{goods and services}
Acquirer strand S_3 with nodes:
$<S_3, 1>$ -{ Clear, H(DESC, $SALT_C$), EncSlip, Sig_C, Sig_M}
$<S_3, 2>$ +{Sig_A}

Definition 6: Suppose E-payment protocol meets the fairness requirements such that:

1. At the end of protocol run, make sure that the sender receives EOR (evidence of receipt) and the receiver receives EOO (evidence of origin).

2. If the protocol is interrupted at any time, no any party has an advantage over the other party. Namely, either both of them get their expected messages respectively or no advantaged messages are obtained by either of them.

For iKP protocol, definition of fairness is described as follows:

1. If payment is made successfully then EOR=Sig_A={Y,H(Common)} K_A^{-1} and EOO=Goods and Services (which indicates that the customer gets the goods and services and the merchant gets the money), else EOR=EOO=Sig_A={N,H(Common)} K_A^{-1} (which indicates that the customer does not credit his account and the merchant does not debit his account).

2. If the protocol is interrupted by accident then the customer receives EOR iff the merchant receives EOO.

Definition 7: Suppose $n=<s,i>\in$ N then index(n) =i and strand(n) =s. Define term(n) to be $(\mathrm{tr}(s))_i$, i.e., the ith signed term in the trace of s. Similarly, uns_term(n) is the unsigned part of the ith signed term in the trace of s.

Lemma 1: Let tr(S_1) and if EOR$\in$ uns_term$<S_1,i>$(where $1\le i\le$length(tr(S_1))) and the sign of $<S_1,i>$ s negative then the sender receives message EOR. Similarly, let tr(S_2) If EOO$\in$ uns_term$<S_2,i>$(where $1\le i\le$length(tr(S_2))) and the sign of $<S_2,i>$ is positive then the receiver receives message EOO.

Proof: Given each run of the protocol, there must exist one trace of strand S_1 written tr(S_1) and one trace of strand S_2, written tr(S_2).

By Definition of strand space, we could know that if the sign of term(n) in space S_1 is negative then means that the sender receives the message of term(n).

And if EOR$\in$ uns_term$<S_1,i>$ (where $1\le i\le$length(tr(S_1))) then the sender receives message EOR.

Similarly, we could conclude that If EOO$\in$ uns_term$<S_2,i>$ (where $1\le i\le$length(tr(S_2))) and the sign of $<S_2,i>$ is positive then the receiver receives message EOO.

Lemma 2: For all possible trace pair$<$tr(S_1),tr(S_2)$>$of space S_1 and S_2 , if exists EOR$\in$ uns_term$<S_1,i>$ and EOO$\in$ uns_term$<S_2,i>$ then the sender receives EOR iff the receiver receives EOO.

Proof: Given the run of the protocol, for all possible trace pair <tr(S_1),tr(S_2)>, if EOR∈ uns_term<S_1,i> then according lemma1, the sender receives the message EOR, and EOO∈ uns_term<S_2,i> then the receive receives the message EOO.

By the above, at any time if the sender receives the message EOR then the receivers receives the message EOO.

Proposition 1: iKP protocol meets the fairness requirements.

Proof: we can analyze the fairness from two aspects:

1. If the protocol is not interrupted at the run time, there can be two trace pairs as following:

S_1^1=<+{SALT$_C$,CID,CERT$_C$},-{Clear,Sig$_M$},+{EncSlip,Sig$_C$},-{Sig$_A$},-{goods and services}>

S_2^1=<-{SALT$_C$,CID,CERT$_C$},+{Clear,Sig$_M$},-{EncSlip,Sig$_C$},+{Clear,H(DESC, SALT$_C$), EncSlip,Sig$_C$, Sig$_M$},-{Sig$_A$},+{Sig$_A$},+{goods and services}>

Because EOR∈ uns_term<S_1^1,4>and the sign of term<S_1^1,4>is negative, according to lemma1 we conclude that the customer receives EOR message.

Similarly, EOO∈ uns_term<S_2^1,7> and the sign of term <S_2^1,7> is negative, according to lemma1 we conclude that the merchant receives EOO message.

2. If the protocol is interrupted by accident, the below list all possible trace pairs:

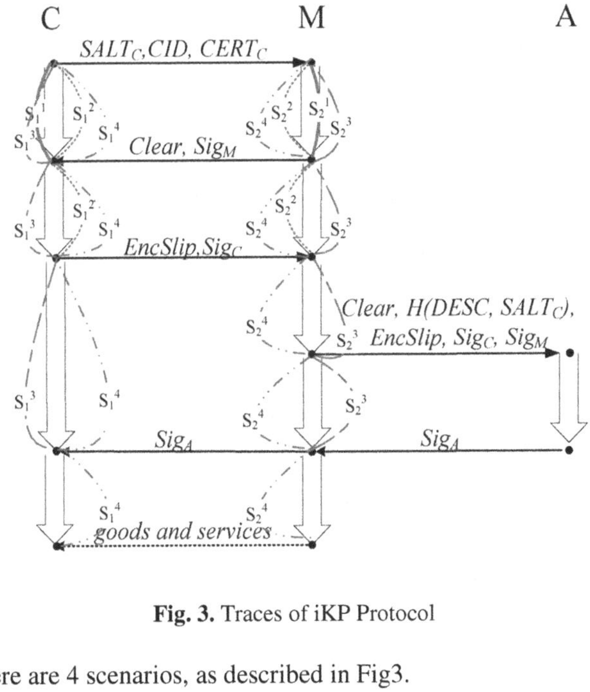

Fig. 3. Traces of iKP Protocol

Namely, there are 4 scenarios, as described in Fig3.

Scenario1:

S_1^1=<+{SALT$_C$,CID,CERT$_C$},-{Clear,Sig$_M$}>

S_2^1=<-{SALT$_C$,CID,CERT$_C$},+{Clear,Sig$_M$}>

Scenario2:

S_1^2=<+{SALT$_C$,CID,CERT$_C$},-{Clear,Sig$_M$},+{EncSlip,Sig$_C$}>

S_2^2=<-{SALT$_C$,CID,CERT$_C$},+{Clear,Sig$_M$},-{EncSlip,Sig$_C$}>

Scenario3:

S_1^3=<+{SALT$_C$,CID,CERT$_C$},-{Clear,Sig$_M$},+{EncSlip,Sig$_C$} -{Sig$_A$} >

S_2^3=<-{SALT$_C$,CID,CERT$_C$},+{Clear,Sig$_M$},-{EncSlip,Sig$_C$},+{Clear,H(DESC, SALT$_C$), EncSlip, Sig$_C$, Sig$_M$},-{Sig$_A$},+{Sig$_A$} >

Scenario4:

S_1^4=<+{SALT$_C$,CID,CERT$_C$},-{Clear,Sig$_M$},+{EncSlip,Sig$_C$},-{Sig$_A$},-{goods and services}>

S_2^4=<-{SALT$_C$,CID,CERT$_C$},+{Clear,Sig$_M$},{EncSlip,Sig$_C$},+{Clear,H(DESC, SALT$_C$), EncSlip, Sig$_C$, Sig$_M$ },-{Sig$_A$},+{Sig$_A$},+{goodsand services}>

In Scenario1, it is easy to prove that EOR$\notin S_1^1$ and EOO$\notin S_2^1$.

Likewise, in scenario2 we can prove that EOR$\notin S_1^2$ and EOO$\notin S_2^2$.

Obviously, EOR$\in S_1^3$ and the sign of Sig$_A$ is negative, but EOO$\notin S_2^3$.

According to Lemma2, the iKP protocol unsatisfied this condition: the sender receives EOR iff the receiver receives EOO and just because of this; we can conclude that proposition1 is not tenable.

5 General Analysis Algrithm of Fairness

In section3, we study fairness of iKP through strand space model. Our result shows that iKP protocol is unfair and by above reasoning idea, a general analytic model as followed is presented to judge whether E-payment protocol meets the criteria of fairness.

Typically, the analysis algorithm involves the following steps:

1. List the message sets of EOO and EOR of E-payment protocol.

Declare two one-dimension arrays to denote messages sets of EOO and EOR respectively, written as EOO[ ] and EOR[ ].

2. By definition5, construct the strand spaces and bundle of E-payment protocol.

Declare three one-dimension arrays to denote nodes of each participant and messages received and sending, written as S_1[ ],S_2[ ] and S_3[ ].

3. List all possible traces of each strand in E-payment protocol.

Where n, m stands for the number of nodes of strand S_1 and S_2 respectively and S_1_trance[ ] and S_2_trance[ ] are declared to denote all possible traces of strand S_1 and S_2.

```
ListAllPossibleTrace(S1[ ],S2[ ]){
n=Length(S1[ ]);
m=Length(S2[ ]);
p=Length(S3[ ]);
for (i=j=k=0;i<n,j<m;i++,j++,k++){
if (S1[i]→ S2[j]|| S1[i]←S2[j]) Then {
for (x=0,y=0;x<i ,y<j;x++,y++){
S1_trance[k]= S1_trance[k]+S1[i];
S2_trance[k]= S2_trance[k]+S2[j];
}}}
j=0;
```

```
while(j<m){
if !( S3[p-1]→S2[j]) then j=j+1;
else break;
}
if j>=m then exit;
token2=j;
while(i<n){
if (S1[i]→S2[token2]|| S1[i]←S2[token2]) then break;
else i++;}
if i>=n then exit;
else{
for(i=0,j=token2-1;i<m,j>0;i++,j--){
if (S1[i]→S2[j]|| S1[i]←S2[j]) then {
token1=i;
break;}
}
for(i=0;i<token1,i++){
S1_trance[k]= S1_trance[k]+S1[i];}
for(i=0;i<token2,i++){
S2_trance[k]= S2_trance[k]+S2[i];}
}}
```

4. Check if lemma1 and lemma2 are satisfied to verify fairness of E-payment protocol.

```
CheckFairness(S1_trance[ ] ,S2_trance[ ]){
i=0;
length= Length(S1_trance[ ]);
m=Length(EOR[ ]);
n=Length(EOO[ ]);
while(i<length){
j=k=0;
EOR=EOO=true;
while(j<m){
If (EOR[j]∈S1_trance[i] ) Then
j=j+1;
Else {
EOR=false;
Break;}
}
while(k<n){
if (EOO[k]∈S2_trance[i] ) then
k=k+1;
else {
EOO=false;
Break;}
}
if (EOO&&EOR) then i=i+1
else return (false);
}
return (true);
}
```

5. If E-payment protocol can go through above checks, then the protocol achieves fairness.

The return value of function CheckFairness( ) is true, then we conclude the protocol is fair.

6. Otherwise, the protocol is unfair.

The return value of function CheckFaireness( ) is false, then we conclude the protocol is unfair.

Confirming to the above algorithm, we can analyze some other E-payment protocols. As an example, we just present analytic process of ISI[8] protocol.

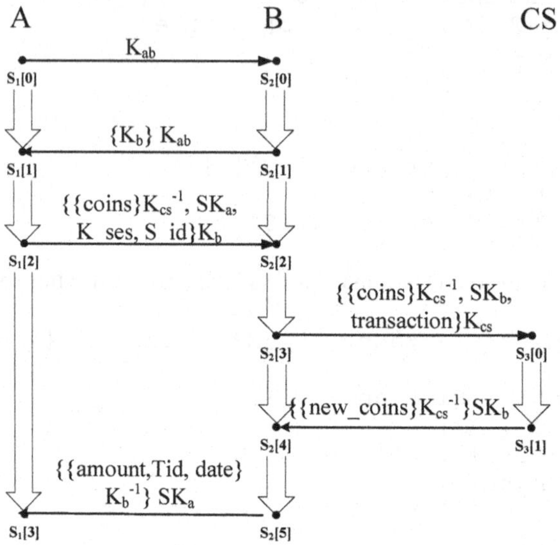

Fig. 4. Bundle of ISI Protocol

ISI protocol is described as followed, see Fig4:

1.A→B:K_{ab}

2.B→A:{K_b} K_{ab}

3.A→B:{{coins}K_{cs}^{-1}, SK_a, K_ses, S_id}K_b

4.B→CS: {{coins}K_{cs}^{-1}, SK_b, transaction}K_{cs}

5.CS→B: {{new_coins}K_{cs}^{-1}}SK_b

6.B→A:{{amount,Tid, date} K_b^{-1}} SK_a

Here:

1. List EOO and EOR.

EOO[0]= {{new_coins}K_{cs}^{-1}}SK_b

EOR[0]={{amount,Tid, date} K_b^{-1}} SK_a

2. Construct bundle of ISI.

S_1[0]=<+K_{ab}>

S_1 [1]=<-{K_b}K_{ab}>

S_1 [2]=<+{{coins}K_{cs}^{-1}, SK_a, K_ses, S_id}K_b>

S_1 [3]=<-{{amount,Tid, date} K_b^{-1}} SK_a >

$S_2[0]=<-K_{ab}>$
$S_2[1]=<+\{K_b\}K_{ab}>$
$S_2[2]=<-\{\{coins\}\ K_{cs}^{-1}, SK_a, K\_ses, S\_id\}K_b>$
$S_2[3]=<+\{\{coins\}\ K_{cs}^{-1}, SK_b, transaction\}K_{cs}>$
$S_2[4]=<-\{\{new\_coins\}K_{cs}^{-1}\}SK_b>$
$S_2[5]=<+\{\{amount,Tid, date\}\ K_b^{-1}\}\ SK_a>$
$S_3[0]=<-\{\{coins\}K_{cs}^{-1}, SK_b, transaction\}K_{cs}>$
$S_3[1]=<+\{\{new\_coins\}K_{cs}^{-1}\}SK_b>$
3. Using function ListAllPossibleTrace( ), we can get :
$S_1\_trance[0]=\{+K_{ab}\};$
$S_2\_trance[0]=\{-K_{ab}\};$
$S_1\_trance[1]=\{+K_{ab}, -\{K_b\}K_{ab}\};$
$S_2\_trance[1]=\{-K_{ab}, +\{K_b\}K_{ab}\};$
$S_1\_trance[2]=\{+K_{ab}, -\{K_b\}K_{ab}, +\{\{coins\}K_{cs}^{-1}, SK_a, K\_ses, S\_id\}K_b\};$
$S_2\_trance[2]=\{-K_{ab}, +\{K_b\}K_{ab},-\{\{coins\}\ K_{cs}^{-1}, SK_a, K\_ses, S\_id\}K_b\};$
$S_1\_trance[3]=\{+K_{ab},-\{K_b\}K_{ab},+\{\{coins\}K_{cs}^{-1},SK_a, K\_ses, S\_id\}K_b,-\{\{amount,Tid, date\}\ K_b^{-1}\}SK_a\};$
$S_2\_trance[3]=\{-K_{ab}, +\{K_b\}K_{ab},-\{\{coins\}K_{cs}^{-1}, SK_a, K\_ses, S\_id\}K_b,+\{\{coins\}\ K_{cs}^{1}, SK_b, transaction\}K_{cs}, -\{\{new\_coins\}\ K_{cs}^{-1}\}SK_b,+\{\{amount,Tid,date\}K_b^{-1}\}\ SK_a\};$
$S_1\_trance[4]=\{+K_{ab}, -\{K_b\}K_{ab}, +\{\{coins\}\ K_{cs}^{-1}, SK_a, K\_ses, S\_id\}K_b,\};$
$S_2\_trance[4]=\{-K_{ab},+\{K_b\}K_{ab},-\{\{coins\}K_{cs}^{-1}, SK_a, K\_ses, S\_id\}\ K_b,+\{\{coins\}\ K_{cs}^{1}, SK_b, transaction\}K_{cs}, -\{\{new\_coins\}\ K_{cs}^{-1}\}SK_b\};$
4.Using function CheckFair ( ) to check ISI protocol.

Because $S_1\_trance[4] \notin EOR$ and $S_2\_trance[4] \in EOO$, ISI is unfair. The result is same with [4].

6 Conclusion

Most of Formal analysis methods are focus on verifications of cryptographic and authentication protocols, while E-payment protocols have some special requirements that are different from other protocols. In this paper, we put strand space mode to analyze E-payment protocol and provide algorithm to analyze automatically fairness of E-payment protocol. In the future, we put focus on other security requirements of E-payment protocol.

References

1. Kailar, R.: Accountability in electronic commerce protocols. IEEE Transactions on Software Engineering, 313–328 (1996)
2. Zhou, D., Qing, S., Zhou, Z.: Limitations of Kailar Logic. J. Journal of Software 12, 1238–1245 (1999)
3. Gou, Y., Gu, T., Dong, R., Cai, G.: On the Limitation of Kailar Logic. J. Computer Engineering and Applications 17, 77–79 (2003)
4. Zhou, D., Qing, S., Zhou, Z.: A New Approach for the Analysis of Electronic Commerce Protocols. J. Journal of Software 19, 1318–1327 (2001)
5. Bai, S., Sui, L.: The Verification Logic for Secure Protocols. J. Journal of Software 11, 240–250 (2000)

6. Javier Thayer Fabrega, F., Herzog, J.C., Guttman, J.D.: Strand Spaces: Proving Security Protocols Correct. J. Journal of Computer Security 7, 191–230 (1999)
7. Bellare, M., Garay, J.A., Hauser, R., Herzberg, A., Krawczyk, H., Steiner, M., Tsudik, G., Waidner, M.: iKP- A Family of Secure Electronic Payment Protocols. In: First USENIX Workshop on Electronic Commerce, pp. 89–106. IEEE Press, NewYork (1995)
8. Medvinsky, G., Natcash, N.B.C.: A Design of Practical Electronic Currency on the Internet. In: ACM Conf. on Computer and Communication Security, pp. 76–82. ACM Press, NewYork (1993)

Research for Data Mining Applying in the Architecture of Web-Learning*

Chaonan Wang[1], Youtian Qu[1], Lili Zhong[1], and Tianzhou Chen[2]

[1] College of Mathematics, Physics and Information Engineering,
Zhejiang Normal University, Jinhua, Zhejiang, China, 321004
[2] College of Computer Science, Zhejiang University, Hangzhou, Zhejiang, China, 310027
quyt@zjnu.cn

Abstract. This paper lists several shortages in the construction of network educative resource database based on general model of Web-learning, improving and enriching the system framework and storage form of network educative resource after introducing the technique of Data Mining , and under this condition, it actualizes individuation education recommendation. The innovative point of this paper is proposing a model of Web-learning architecture based on Data Mining, which accords with Web learners' characteristic of humanity and intelligent.

Keywords: Data Mining, Web-learning, educative resource database, individuation recommendation.

1 Introduction

With the development and application of network technology widely, information time has already been coming recent years. Education is changing into a new way of information, network and for all life. More and more people plunge into Web-learning and they could find some knowledge which they interest and need most in the mass information. So we may say, Web-learning based on Internet is a new force suddenly rises, which favored and assented by most people. Web-learning is different from the traditional education than has so many limitations, such as time, place, teaching ways and so on. Learners can study on the network platform by themselves according to their own demands. But there is so much information in the educative resource database with different structures, how to offer learners high-quality and valuable information has been a hot topic in the field of network education.

2 General Mode of Web-Education

With the rapid development of Internet, people can enjoy the services provided by Online Schools at home. Now most of Web-Based Distance Education is still in

* Founding information: This work is partially sponsored by the Natural Science Foundation of Zhejiang Province, China (M603245, Y106469) and National high tech research and development plan (863 plan) 2007AA01Z105-05.

W. Liu et al. (Eds.): WISM 2009, LNCS 5854, pp. 479–488, 2009.

resource sharing state, only transplant traditional classroom education simply into distance education, the education system is a single model and puts the system itself as the center, so that students can only accept the same study content passively, and do not really reflect the advantages of personalized education. Despite there are various network teaching forms, Network education has several aspects in generally, and they are bulletin before classes, Web-learning courses, homework downloading and uploading and discussion between learners and teachers after classes. After analysis, the general model of network teaching can be described in detail as follows.

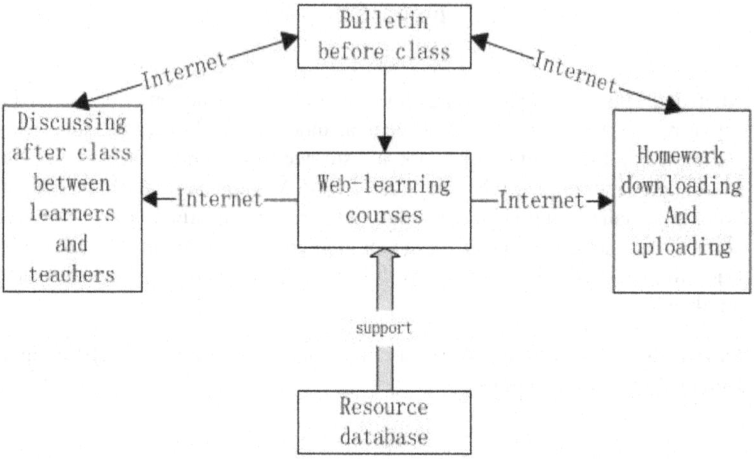

Fig. 1. General Mode of Web-education

3 Problems about Network Resource Database

From the above model, we can see that abundant resource database is the support of network education, which is a aggregate with different resources on education and all of the information is dealt, processed and made in a digital way, and can run on the network and be visited through Internet and LAN. Now, the objective of educative resource database is to provide learners some functions based on developed courseware, teaching plans, exercises, material and other educative resource, such as browsing, downloading, and uploading and so on.

The resource database mentioned above are focused on information not on learners, which is different from the real life. For example, different learners have different learning style, learning demands and learning ability. Educative resource may not match with the demand of individuation. So we summary the following problems on constructing network educative resource database.

3.1 Insufficiency of Utilizing Resource

Resource not made full use of is a main problem. It contains two complexions: one is re-construction of educative resource, and another is nonstandard sort of resource.

* Re-construction of educative resource: some network courses are similar on content and teaching manners. For instant, universities or colleges often open some hot major courses as long-distance education, such as computer, economy, English and so on. This may make learners puzzled when they are searching one subject, because they would get so much similar or even the same information. Learners have to spend much time on picking useful information in the mass, which is a waste not only on manpower but also on finance and economics.
* Nonstandard sort of resource: resource is usually classified into test questions database, teaching plan database, multimedia material database, courseware database and teaching software database. Obviously, it is sorted by tache in the activity of traditional teaching. But it falls short of the standard and mean of metadata. Sometimes, one item of information may need defining by many times, or some information contains and superposes others. All the above conditions would lead the low rate of resource used.

3.2 Long Time in Searches Resource [1, 3]

Several factors may affect the speed of searches information. One is network circumstance is not so good that makes the speed of downloading and researching lower than usual. Another factor is the difference between developer of resource database and learners on knowledge structure, expression ways and comprehension ability, which would waste learners much time in finding appropriate information in the database. For example, the cultural background of learners, language barriers, ways of thinking, age and intellectual capacity to accept, all aspects will be the reasons cause time-consuming. The third aspect is the setting of organization and query. Most resource database is constructed by the storage style of information, so it is easier to construct than to use. If learners want to query correlative knowledge points, they have to open page layout one after one, which would waste much time in researching. In addition, simple setting of researching function makes learners difficult to obtain information they needed at once.

3.3 Learning Process Inconformity with Individuation Study [2]

Inconformity individuation studying is an essential problem in network education. It mainly represents that the organization of resource lacks diversity and the content lacks pertinence and byelaw. Dynamic or static educative resource can not satisfy the demands of individuality, and learners self-study have more higher request about learning resource, which should fulfill different learners' learning goal and innovate their interest. So, it is exigent to introduce a individuation self-learning system to assistant learners to filter useful information, and discard those information with no value or impact, acquire information resources learners need accurately and rapidly to meet the demand of personalized study.

4 Data Mining Applying in the Construction of Network Educative Resource

Aiming at existent problems, we introduce Data Mining into network education to make the learning process more intelligent and humanistic and provide learners high-quality and valuable learning information. Data Mining is the process of analyzing data from different perspectives and summarizing into useful information - information that can be used to increase revenue, cuts costs, or both. To be short, Data Mining is the process of extracting hidden patterns from data. The method of Data Mining usually has two types, one is statistical and another is machine learning in artificial intelligence. The former contains probability analyse, relativity, clustering analyse and distinction analyse. The latter often gets pattern or parameter through a set of samples and exercise. Due to each one has different function and application fields; we usually use them combined to get a good result.

4.1 The Concept of Data Mining

DM(Data Mining) also is also called KDD (Knowledge Discovery in Database), refers to find the process of trend or pattern from data, it is an important information analysis method of gaining knowledge from the information collected, and it can also find relationship models which are valid, innovative with potential value and easy to understand from exist data. It is mainly based on artificial intelligence, machine learning, statistics, etc. [5]

DM is a process of distilling information and knowledge that is implicit, people do not know in advance, but potentially useful from a mass of incomplete, noisy, fuzzy and random practical application data. The "data" here refers to a collection, record and original information of the relevant facts. While "knowledge" refers to a more abstract description of information contained. The analysis process to large amounts of data includes data preparation, pattern search, knowledge evaluation and repeated modification. The requirement of mining process is extraordinary, that is to have a certain degree of intelligence and automatic. [2]

4.2 Features of Data Mining

Data Mining can make proactive and knowledge-based decision through the prediction of future trends and behavior. Data Mining's most prominent application in online education is personalized service; the following five are its functional characteristics [5, 10].

(1) Automatically Predict Trends and Behavior
(2) Correlation Analysis
(3) Clustering
(4) Concept Description
(5) Deviation Detection.

4.3 Steps of Resource Database Construction with Data Mining

There are five steps when Data Mining applying in the construction of educative resource database.

(1)Data selection and character extraction [4]
(2) Data pretreatment [2]
(3) Result analyse and evaluation
(4) Knowledge expression and resource conformity
(5) Implement individuation recommendation

It is convenient to follow the learning activity at the platform in the environment of network learning with individuation recommendation. It can help learners to improve where they are not good at. The right cue may come from the activities and mode of successful learners, which would make the logical structure of course and the step of learners more consistent. The process of individuation recommendation is showed in details as follows. [1]

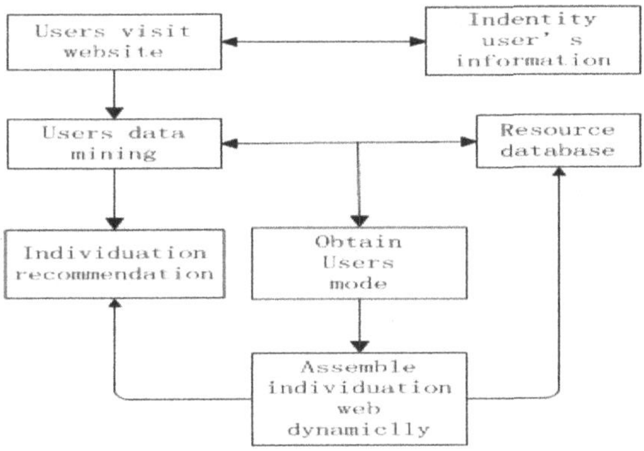

Fig. 2. The process of Implement individuation recommendation

When users visit some website, system would identify users' information through their logging in and begin to collect all the information and activities of them. Firstly, learners would look for something they interested in the mass resource database. Then, they may pay more attention to some information. At this time, system would create corresponding users patter, through which could assemble different web page together dynamically to form a temporary individuation recommendation. Different users have different individuation recommendations and all of them are stored in a given resource database. When system provides individuation recommendations to users, it could transfer them in corresponding resource database, where contains all the web pages assembled temporarily. An advantage of doing this is to save time, and also to reduce system's spending.

5 The Architecture of Network Educative Resource Based on Data Mining

5.1 Functions of Architecture

According to the above analyse, we make up a network educative resource architecture based on Data Mining with the characteristic of individuation, intelligent and high-efficiency. There are five parts in it: user interface, data research module, data management module, data analyse module and network education module. The details are explained by the following picture.

In the fig 3, user interface is a platform where system can communicate with users. Users can browse, download and searches the educative resource through the interface, and accept the recommendation of individuation educative information. During the browsing, system would pick up the users 'characters automatically and store them in the character module.

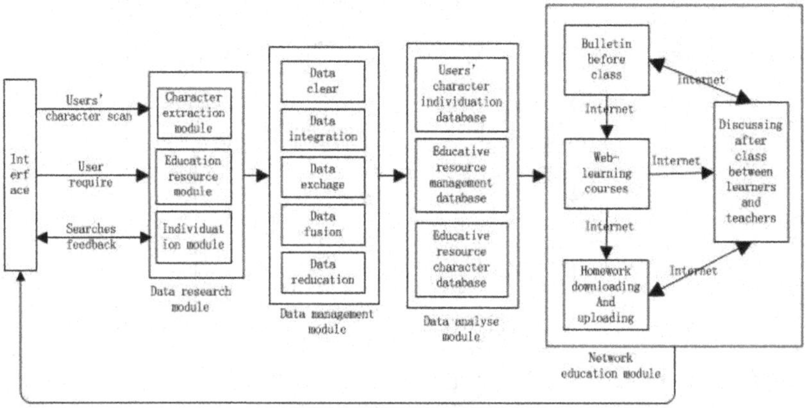

Fig. 3. The architecture and function of network educative resource based on Data Mining

Data research module mainly notes users' activity and information about network resource. At the same time, it sends the information coming from character extraction model, education resource model and individuation module to data management model, waiting to be dealt. It also collects the relative information about individuation to update and track in time, and prepares to recommend the individuation.

Data management module is an integration of the technology of data clear, data exchange, data fusion and data reducation. It extracts the key words from user individuation database to get the interested vectors, and then deals with the data of network educative resource database to send them to data analyse module to analyze and evaluate in further.

Data analyse module makes full use of managed data to mine the relationship between data according to Data Ming rules, such as clustering analyse and relevancy formula and son on. we can know from the picture that educative resource character database is exchanged from traditional educative resource, noting the key character of teaching material; educative resource management database can carry out the function of uploading, downloading and primary searches information. Users' character individuation database and educative resource character database are the support of network education module at background.

Network education module is similar to the mode which mentioned in part 2, but it has more individuality than the former one after introducing the technology of Data Mining. This module is a bridge of learners and system as a preceding service interface to improve learners' better self-study.

5.2 Capabilities of Architecture

The network educative resource mode based on Data Mining is a mostly perfect architecture to actualize individual, intelligent and efficient learning. It uses the technique of Data Mining to find out potential relationship between data, and it is the most important part in network education. Meanwhile, popularizing the mined structure to practice teaching can promote network education. The following aspects are represented its capability.

* Easy utility: this mode has overcome the shortcomings of educative resource reconstruction. It classifies and layout all the information in a uniform way to consider users' different demands quite well so that its using area is wider, its using frequency is higher and its service provided is more simple.
* Good expansibility: because of the separation of data management module and data analyse module, platform managers can add new educative resource in the relevant catalog at background without affecting learners' studying. It is easy for them to add new catalogs in the database just adding new modules.
* Convent maintenance and management: it is the key factor to guarantee system to run in a long term. This mode implements automatization management so that can reduce the working load of platform managers. It would provide multi-measures to help managers' work when they are regulate or modify the system.
* Stability: each module works independently but also has relationship with others in this Web-learning architecture. When some detail of certain module has problems, it would not affect other departments working of other modules, which implements the stability of the whole system. System managers only need very short time to locate broken-down module and service or improve it only in this module without touching others working. The objective of design is to reduce the working burden of system managers.

5.3 Performance Evaluation

To test the practicality and effectiveness of the model, performance evaluation can retrieve a certain percentage of data from the original database as test data, and set up model for testing. We can measure the accuracy of model through the compare between results gave by the model and original data information. If the accuracy rate is higher than the established standards, this model would be considered effective; if less than the established standards, it would take to find out the cause of the error, re-excavation.[6]

To analyse the quantity of learners, on the growth of the quantity analysis, we use GAM (The General Regression Model of Linear and Logistic Regression), extraction of data bases on the A-type data, and combines with the behavior records of visiting series B-type data and the satisfaction, response rate to get more scientific results. A-type data refers to lasting duplicate data, that is the learners' basic information; B-type data refers to learners' behavior information, that is, learners' clicks, downloads, visit frequency, duration, etc. C-type refers to the learners' subjective information, such as learners' study or the improvement of the model, etc. On the reduction of the quantity analysis, we adopt a classification model, because it is benefit to reduce the judge.[7]

A-type data belongs to stable information, C-type data belongs to subjective aspect, which is not conducive to quantify, therefore, we use analytic hierarchy process to analyse the B-type data, and we compare the study efficiency at different stages. First, we establish a hierarchical structure model, and abstract the actual questions into the corresponding hierarchy. The details are as follows:

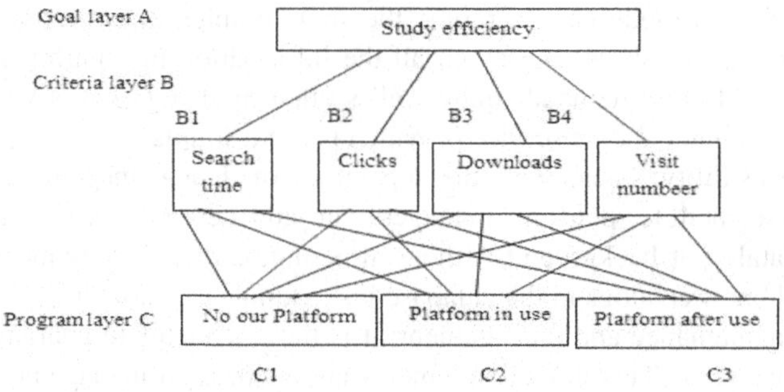

Fig. 4. A hierarchical structure model of platform using

We can see from the data on the table below obviously that the students' study efficiency is much higher than before that the platform hasn't been used. Weight of the total is actually ordered by the efficiency of learning.

A	B_1	B_2	B_3	B_4
B_1	1	1/3	1/3	1/2
B_2	3	1	1/3	1
B_3	3	3	1	3
B_4	2	1	1/3	1

Fig. 5. Criteria layer

B_1	C_1	C_2	C_3
C_1	1	1/2	1/4
C_2	2	1	1/2
C_3	4	2	1

B_2	C_1	C_2	C_3
C_1	1	1/4	1/3
C_2	4	1	1/3
C_3	3	3	1

B_3	C_1	C_2	C_3
C_1	1	1/2	1/3
C_2	2	1	1/2
C_3	3	2	1

B_4	C_1	C_2	C_3
C_1	1	1/3	1/4
C_2	3	1	1/2
C_3	4	2	1

Fig. 6. Program layer

Table 1. The total order

Criteria layer		B1	B2	B3	B4	Weight of the total order
The layer permission of criteria layer		0.1078	0.2173	0.4853	0.1896	
Program level	C_1	0.1429	0.1638	0.1279	0.1226	0.6242
Single ranking	C_2	0.2857	0.2973	0.3185	0.3202	1.2217
Weight	C_3	0.5714	0.5390	0.5536	0.5571	2.2211

6 Conclusions

Network is a new learning form different from traditional education, which can provide learners a self-study platform. But multi-data, distributing and isomerous problems are very common in network education. How to searches the most valuable information during the visiting process through Data Mining, and how to classify and unite information in a reasonable way in the resource database are concerned by most network educators. Taking into account each learner has their own different backgrounds, languages, cultures, interests, and different study methods and study path, what we advocate in the network teaching system architecture must be based on learner-centered, provide a resource-based learning, teaching resources can be adapted to a variety of learners' needs and background and carry out different combinations. The organization of teaching content is not the same and immutable, but can be adapted to individual characteristics. In the whole architecture, the best teaching method different from the ordinary network is the realization of personalized service, which thanks to a strong and rich teaching resource library as a backup to support the entire teaching system. Therefore, developers of educative resource database should be oriented by need, supported by network technology and try to process the educative resource deeply and provide best and individuality information production. Only combined with technology, resource and intelligence closely, it will bring ordered information space, optimize the network resource utility and offer Web-learners a more horizontal room.

References

1. Yuanyang, T., Erjia, H.: The Technique of Information Mining and the Organization of Network Education Resource. Electricity Education Research (6), 53–55 (2003)
2. Chenyu, Z., Ping, W., Yi, Z.: Analysis, Design and Implementation of Subject Information Navigation System Based on Data Mining. Network resource and construction (2), 57–60 (2003)
3. Xiaoping, W., Liang, Y.: The Application of the Data Mining to the Construction of Web-based Education Resources Base. Journal of Southwest China Normal University (Natural Science) (2), 163–167 (2005)
4. Zhihong, S.: Data Mining: the Necessary Skill of Information Service in Network Times. Sci-tech. Information Development and Economy (10), 17–19 (2007)
5. Chunyan, W., Shanshan, W.: The Design of Education Resources Base based on Data Mining. Fujian Computer (10), 17–19 (2007)
6. Lili, W., Aijie, X., Bo, Z., Tao, L.: Research on Network teaching Resource System Based on Data Fusion and Data Mining. Journal of Chongqing Institute of Technology (Natural Science) (6), 170–174 (2008)
7. Rui, W., Shesheng, G., Xia, Z.: The System Mode Combining with Data Mining and Data Fusion. Computer Project and applications (18), 181–183 (2006)
8. Lei, C., Huachuan, Z.: The Research of Extracting and Mining Iinformation from MEDLNE Database. Journal of Information (4), 425–433 (2003)
9. Discussion Group of Data Mining: `http://www.dmgrou.org.cn/`
10. Lin, D., Changyong, W.: The Application of Data Mining in the Individuation Service of Long-distance Education. Electricity Education Research (9), 43–46 (2002)

Design E-Learning Recommendation System Using PIRT and VPRS Model

Ayad R. Abbas and Liu Juan

School of Computer, Wuhan University, Wuhan 430079, China
ayad_cs@yahoo.com, liujuan@whu.edu.cn

Abstract. Recommendation systems depend on appropriate decision making in order to personalize information content upon an individual's user needs, preferences, interests and browsing behaviors. These systems usually neglect the user's ability and the difficulty level of the recommended item. Therefore, this study proposes a comprehensive recommendation system to provide adaptive learning, which is composed of two main agents: *PIRT_Recommendation* agent, helping the learner to personalize learning resource based on Polytomous Item Response Theory (PIRT), which considers both the difficulty of the learning resource and the ability of the learner, and *VPRS_Recommendation* agent, providing decision rules as an instrument or guide for learner's self-assessment based on improved Variable Precision Rough Set (VPRS). Experimental results show that the proposed system can exactly provide closer learning resource with appropriate feedback to the learner, resulting in increased the learning efficiency and learning performance.

1 Introduction

Personalized recommender systems improve access to relevant items and information by making personalized suggestions based on previous items of an individual user's likes and dislikes. In the last few years, recommendation systems have addressed the technology used to generate recommendations, focusing on the application of data mining techniques [1]. Others have given considerable attention to personalize e-learning system using IRT. Including, [2–4] presented a prototype of personalized Web-based instruction system based on dichotomous IRT to perform personalized curriculum sequencing through simultaneously considering courseware difficulty level and learner's ability. However, the traditional rough set based WebCT learning improves the state-of-the-art of web learning by providing virtual student/ teacher feedback and making the WebCT system much more powerful [5–7]. Thereafter, [8–9] have improved VPRS in order to handle totally ambiguous and enhance the precision of rough set, and to deal with both two and multi decision classes. This study proposes PIRT_Recommendation agent tacking into account learner's ability. For allowing more information about ability level to be extracted from a fixed set of learning resources, the estimation of this ability does not depend only on explicit feedback response (i.e., understanding response), but also depend on implicit feedback responses (i.e., learning time, learning attempt, learning score), because these responses have a direct impact on the learner's ability. Furthermore, to obtain more accurate estimation of learner's ability, these responses are classified into three

W. Liu et al. (Eds.): WISM 2009, LNCS 5854, pp. 489–498, 2009.

polytomous response categories at each item (learning resource). We have assumed that each learner has a three understanding levels, ''low understanding'', ''moderate understanding'' and ''high understanding'', and we have assigned them into 0, 1 and 2 respectively.

Lack of contact and immediate feedback between teachers and online learners is one of the main problems in Web-based learning. Therefore, we also propose VPRS_Recommendation agent to provide decision rules as an instrument or guide for learner's self-assessment when taking on-line course using VPRS.

2 The Proposed Recommended System Architecture

Figure 1 illustrates the proposed recommended system architecture, which composes of two intelligence services, learner's feedback mechanism and recommendation mechanism, the learner's feedback mechanism intends to assemble both explicit and implicit responses, update learner's ability and evaluate the difficulty parameters of learning resources. However, the recommendation mechanism composes of two recommendation agents PIRT_Recommendation agent and VPRS_Recommendation agent. PIRT_Recommendation agent aims to recommend suitable learning recourses with appropriate difficulties to the learners using polytomous IRT and decides on the next activity to be recommended to the learner. While, VPRS_Recommendation agent aims to provide general recommendation rules to the teachers and learners using improved VPRS.

The system operation procedure of the proposed system architecture is summarized as follow:

Step 1-2: The learner accesses to the learner's interface agent, then the system gets his or her learning profile from the learner profile database.

Step 3: The learner's interface agent gets the course form course database.

Step 4-5: The learner's feedback mechanism collects explicit responses from the learner's interface agent and also collects implicit responses from the learner's profile database, and then re-evaluates the learner's ability.

Step 6: When the learner's ability is estimated, the learner's feedback mechanism will send this ability to the PIRT_Recommendation mechanism.

Step 7-8: The PIRT_Recommendation mechanism calculates the information value for each item according to estimated learner's ability and the item difficulty parameter, and sends a list of ranking learning paths to the learner's interface agent.

Step 9-10: VPRS_Recommendation Extracts student grade information from learning profile to find the rules behind the information table; then compute the reduct with respect to *Fail* concept, and find dispensable attributes.

Step11-12: In order to find decision rules we need to compute positive, negative and boundary regions, then the general significant rules will send to learner's interface agent.

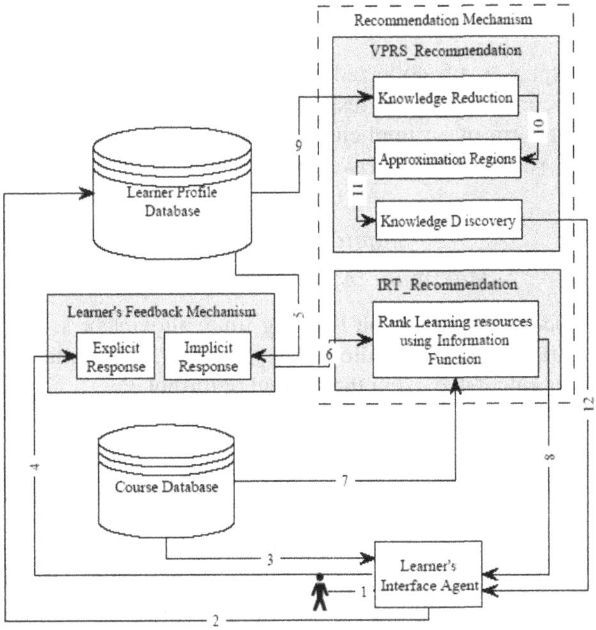

Fig. 1. The proposed system architecture

3 Learner's Feedback Mechanism

Learner's feedback mechanism collects explicit responses from the learner's interface agent, while it collects implicit responses from the learner's profile database; these responses are explained as follow:

Explicit Feedback Response: A direct learner's response (multiple choice), which contains only one question associated with understanding learning resource. This question has three understanding levels, ''low understanding'', ''moderate understanding'' and ''high understanding''.

Implicit Feedback Responses: An indirect learner's response that associated with learner's characteristic responses. Below is a description for each implicit response.

1. Learning time response: is the sum of all of the learner's session times.
2. Learning score response: is the learner's score for the specific learning resource.
3. Learning attempt: is the number of attempts on the activity.

During learner's activity (taking a lesson), the learner experiences two types of learning recourses: *Learning Material* or *Tests* (Pre-Test, Post Test). If the learner experiences the learning material, he or she will give three responses: understanding response, learning time response and learning attempt response. If the learner experiences tests, he or she will also give three responses: learning time response, learning score response and learning attempt response. The main functions of learner's feedback mechanism are explained in the following subsections.

3.1 Tuning Implicit Reponses

Once learner's responses are collected by the learner's feedback mechanism, these responses will be converted to the response categories. Therefore, this study firstly proposes a general form of an implicit response rate *IRR*, which converts actual response values into values limited between 0 and 1, as demonstrated by the following relationship:

$$IRR = \frac{Implicit\ response}{Maximum\ Information\ allowed} \tag{1}$$

where *implicit response* is either the learning time, attempt or score divided by the maximum time ,attempt or score allowed respectively. Thereafter, the implicit response category j is calculated using the following formula:

$$j = \begin{cases} 0\ (A1) & w_1 \le IRR \le w_2 \\ 1\ (A2) & w_3 \le IRR \le w_4 \\ 2\ (A3) & w_5 \le IRR \le w_6 \end{cases} \tag{2}$$

where $A1$, $A2$ and $A3$ denote verbal response categories and $w_1, w_2, \ldots, w_6$ denote the limiting values. Calculating j associated with learning score using above formula, whereas $w_1 = 0$, $w_2 = \alpha$, $w_3 = \alpha$, $w_4 = 1 - \alpha$, $w_5 = 1 - \alpha$ and $w_6 = 1$. While, calculating j associated with learning time or attempt also using above formula, whereas $w_1 = 1 - \alpha$, $w_2 = 1$, $w_3 = \alpha$, $w_4 = 1 - \alpha$, $w_5 = 0$ and $w_6 = \alpha$.
Usually, α is in the interval (0, 0.5).

3.2 Learner's Ability Estimation

A Rasch Model (RM) is applied to items that are scored polytomously (more than two responses). If the options of rating scale are successfully ordered, as in Likert scale [10]; then the Rasch family of the polytomous IRT models including Rating Scale Model (RSM) [11].

Learner's ability algorithm estimates the level of a latent trait of the learner demonstrated in an observed polytomous response pattern. The term of polytomous means that the learner can give more than two response categories in a specific item. By using RSM, we consider that the polytomous scored items are labeled $i = 1, 2, \ldots, l, \ldots, 3n$, and the response scales used to measure learner's ability M with item's difficulty D_i have $t+1$ ordinal response categories $0, 1, \ldots, j, \ldots, t$, which are separated by a set of item specific step difficulties F_k ($k = 0, 1, \ldots, j, \ldots, t$, with $F_0 = 0$ and $\sum_{k=0}^{j} F_k = 0$). The Joint Maximum Likelihood Estimation (JMLE) is used for estimating the learner's ability that maximizes the likelihood function for particular response [12]. This method involves an iterative process in which initial estimates of the learner's ability M is improved until the expected score ES agrees with the empirical sum R observed in the data. This process is controlled by the probability P_{aij} that the learner a will select the jth response category of item i, which is described as follow:

$$P_{aij} = \frac{e^{j(M - D_i) - \sum_{k=0}^{j} F_k}}{\sum_{h=0}^{t} e^{h(M - D_i) - \sum_{k=0}^{h} F_k}} \tag{3}$$

The expected score *ES* is defined as the sum of the expected value of the ratings over all individual items, thus the modeled expected rating sum over all items is.

$$ES = \sum_{i=1}^{l}\sum_{j=0}^{t} jP_{aij} \tag{4}$$

While, modeled variance *V* of the expected score *ES* at specific learner's ability *M* is given by the sum of the variances of the individual items' expected values:

$$V = \sum_{i=1}^{l}\left[\left(\sum_{j=0}^{t} j^2 P_{aij}\right) - \left(\sum_{j=0}^{t} jP_{aij}\right)^2\right] \tag{5}$$

Afterward, the initial estimate of learner's ability *M* can be any finite value, this value is obtained using formula (6).

$$M = D_{mean} + \log\left(\frac{R - R_{Min}}{R_{Max} - R}\right) \tag{6}$$

where $D_{mean} = \frac{1}{l}\sum_{i=1}^{l} D_i$

D_{Mean} denotes the average item's difficulty in *l* items, *R* is the raw score, R_{Min} is the minimum possible score and R_{Max} is the maximum possible score. Equation (7) is used for estimating better of the learner's ability.

$$\tilde{M} = M + \frac{R - ES}{V} \tag{7}$$

It must be noted that the expected score *ES* and variance *V* play a central role in the JMLE algorithm described in [13], which is summarized below.

Algorithm: Learner's ability estimation
Input: i, j, D, F, R, R_{Min}, R_{Max}
Output: M´: Learner's ability
Step 1: Collect learners' responses to the desired subset of l.
If R= R_{Min}, then set R= R_{Min} +3
If R= R_{Max} , then set R= R_{Max} -3
Step 2: Each item i has a calibration D_i, and each step j a calibration F_j.
Step 3: Compute the average item difficulty D_{mean}
Step 4: Compute the initial estimate of learner's ability M.
Step 5: Compute the probability, expected score and variance for M.
Step 6: Obtain a better estimate M´ of the measure M
Step 7: if |M´-M|>0.01 then do the following steps:
Set M= M´
Go back to step (5)
Step 8: Report final ability M´
Step 9: End.

4 PIRT_Recommendation Agent

When learner's ability is estimated through the feedback mechanism, the recommendation mechanism will compute the information functions for individual leaner, thus it recommends appropriate learning experiences to that learner.

In this study, Information function presents that each learning activity with the corresponding difficulty shows different information to the learners. In other words, learning activity with higher information value is more suitable to be recommended to the learner.

A further device is suggested for identifying the maximum information, whereas the item information is the expected variance of scoring functions based on probability along the observation ability *M*, so the expected value *E*(*M*) can be expressed as:

$$E(M) = \sum_{j=0}^{t} jP_{ij} \tag{8}$$

Let *I*(*M*) be the information function as shown in equation (8), that represents the information contributed by specific learning activity *i* across the range of learner's ability *M*.

$$I(M) = \left(\sum_{j=0}^{t} j^2 P_{ij}\right) - \left(E(M)\right)^2 \tag{9}$$

where *M* denotes new learner abilities estimated after *n* preceding learning resources, P_{ij} represents the probability of a *j*th response category for the learner with ability *M*, and *E*(*M*) denotes the expected score function.

5 VPRS_Recommendation Agent

In this paper, we adapted inductive learning algorithm to a web-based learning environment using improved VPRS Model [8]; to improve the state-of-the-art of Web-base learning by offsetting the lack of student/teacher feedback and provides both students and teachers with the insights needed to study better.

Three steps are defined to extract suitable decision rules as described in the following subsections.

5.1 Attributes Reduction

The attributes reduction is a process of omitting unnecessary condition attributes from the decision table. In other word, not all condition attributes are necessary to categorize the objects in the information system.

5.2 Approximation Regions

Improved VPRS model's ability is used to flexibly control approximation regions definitions in order to handle totally ambiguous and enhance the precision of Rough set, and to deal with both two and multi decision classes. The asymmetric IVPRS generalization is based on the certainty threshold parameters α_t, which satisfy the constraints as follows:

$$0 < \alpha_t \leq MIN\left(P(X_t), \sum_{c \neq t} P(X_c)\right) \tag{10}$$

To calculate and parameterize the remaining thresholds α_c using following formula:

$$\alpha_c = \forall_{c:c \neq t} \frac{P(X_c)}{\sum_{c:c \neq t} P(X_c)} \times \alpha_t \tag{11}$$

The improved VPRS Positive, Negative, and Boundary regions for multi decision classes are respectively defined as:

$$\begin{aligned} POS^{\alpha}(X_t) &= \cup\{E_i : \forall_{c:c \neq t}\ d(X_c, E_i) \leq -\alpha_c\} \\ NEG^{\alpha}(X_t) &= \cup\{E_i : \quad d(X_t, E_i) \leq -\alpha_t\} \\ BND^{\alpha}(X_t) &= \cup\{E_i : \quad d(X_t, E_i) > -\alpha_t \wedge \\ &\qquad \exists_{c:c \neq t}\, d(X_c, E_i) > -\alpha_c\} \end{aligned} \tag{12}$$

where the confirmation measurer *d(X, E)* is the difference between the posterior probability *p(X|E)* and prior probability *p(X)*.

5.3 Certainty Measures

Discriminant index is used to provide a measure of the degree of certainty in classifying the set of objects in E_i with respect to the concept represented by X_t, the index value is calculated first and the highest index value is determined the best attribute, the discriminant index is defined as:

$$\eta = 1 - \frac{card(BND^{\alpha}(X_t))}{card(U)} \tag{13}$$

where, η is the discriminant index, *card* is the cardinality of the set and *U* is the universal set.

6 Experimental Results and Evaluation

In order to provide the proof of feasibility study and verify the learning performance for the proposed system, we used a SCORM 2004 Photoshop examples version 1.1 released by [14], it contains a collection of learning resources about Adobe Photoshop. In this system, the range of both learner's ability and learning resource difficulty are limited from – 3, which is very weak ability or very easy material, to 3, which is very high ability or very difficult material. To evaluate the performance of the proposed system, an experience was performed with *a recommended choice mode* means learner can click on 19 recommended learning resources. Figure 2 present the relationship between the difficulty of the learning resource and the learner's ability respectively.

Paired t-test was used to investigate the statistical difference between learning resource difficulty and the learner ability at significant level (α) of 0.05.

Table 1. Paired t-Test for recommended choice

Data	Mean	SD^2	SD	SE	Minimum
difficulty	-0.222	2.361	1.536	0.352	-3.000
Ability	0.038	3.023	1.738	0.398	-2.456

t = -2.4813 and p= 0.02319 (lower significant level)= higher correlation coefficient.

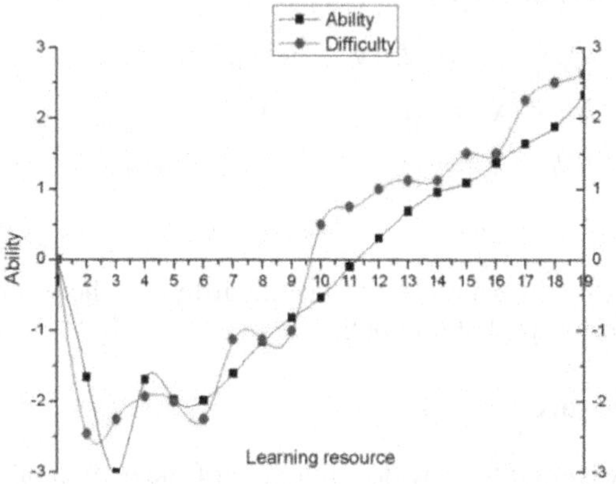

Fig. 2. Recommended choice mode

The results are shown in Table 1. It can be seen that the P-value is 0.0010. This indicates that a higher correlation was found. The results also revealed that the standard deviation between the analysis groups (i.e. difficulty & ability) is much close. In order to implement VPRS_Recommendation agent, Student Information Table (SIT) [6] was used; SIT table contains 296 students and includes some of fields; domain *U*, condition attributes *C*={*Q1*,*Q2*,*Q3*}, decision attribute *D*={*Final*} and frequency field (*Freq*), the score results of self tests should be Excellent (*A*), Very Good (*B*), Good (*C*), Fair (*D*) or Poor (*F*), Students use online self test program, whereas students receive one of the five scores (*A*, *B*, *C*, *D* and *F*) and frequency field counts the number of students have the same records. However, The target concept is "why students are poor in the final test". Therefore, X_t=X_F and X_c={X_A, X_B, X_C, X_D}; because there are no redundant attributes, SIT can not be reduced. All extracted decision rules in term of multi classes are shown below.

1. *If Q2=F then Final=F Confidence=59%*
2. *If Q1=F and Q2=D then Final=F Confidence =100%*
3. *If Q3=F and Q2=D and Q1=F then Final=F Confidence =53%*

That is, the experimental results show that the proposed system can recommend suitable learning experience to the learner with high correlation degree between learner's ability and recommended learning recourse. Moreover, the extracted rules can give appropriate feedback to the learners, resulting in increased the learning efficiency and learning performance.

7 Conclusions

We design a comprehensive recommendation system taking into account learner's ability and learner's self-assessments. First, the proposed personalized learning resource can estimate learner's ability using PIRT. Second, both non-crisp explicit feedback response and implicit feedback responses can be collected by using the proposed learner's feedback mechanism. Finally, the VPRS model can recommend a group of learners to guide them in their learning. For repeating learners, it specifies the areas they should focus on according to how they fit the rules. For new learners, it tells them which activities need extra effort in order to pass the course. A Paired t-test was successfully applied to investigate the statistical difference between leaner's abilities curve and the difficulty of learning resources curve. Experimental results show that the proposed system can exactly provide recommendations to the learners, resulting in increased the learning efficiency and learning performance.

References

1. Schafer, J.B.: The Application of Data-mining to Recommender Systems. In: Encyclopedia of Data Warehousing and Mining, pp. 44–48. Idea Group Reference, Hershey (2005)
2. Chen, C.M., Lee, H.M., Chen, Y.H.: Personalized E-learning System using Item Response Theory. Computers & Education 44(3), 237–255 (2005)
3. Chen, C.M., Liu, C.Y., Chang, M.H.: Personalized Curriculum Sequencing using Modified Item Response Theory for Web-based Instruction. Expert Systems with Applications 30(2), 378–396 (2006)
4. Chen, C.M., Duh, L.J., Liu, C.Y.: A Personalized Courseware Recommendation System based on Fuzzy Item Response Theory. In: International Conference on e-Technology, e-Commerce and e-Service 2004 (EEE 2004), pp. 305–308. IEEE Press, Los Alamitos (2004)
5. Liang, A.H., Maguire, B., Johnson, J.: Rough Set Based WebCT Learning. In: Lu, H., Zhou, A. (eds.) WAIM 2000. LNCS, vol. 1846, pp. 425–436. Springer, Heidelberg (2000)
6. Hongyan, G., Maguire, B.: A Rough Set Methodology to Support Learner Self-Assessment in Web- Based Distance Education. In: Wang, G., Liu, Q., Yao, Y., Skowron, A. (eds.) RSFDGrC 2003. LNCS (LNAI), vol. 2639, pp. 295–298. Springer, Heidelberg (2003)
7. Liang, A.H., Ziarko, W., Maguire, B.: The Application of a Distance Learning Algorithm in Web-Based Course Delivery. In: Ziarko, W.P., Yao, Y. (eds.) RSCTC 2000. LNCS (LNAI), vol. 2005, pp. 338–345. Springer, Heidelberg (2001)
8. Abbas, A.R., Juan, L., Safaa, O.M.: A New Version of Bayesian Rough Set Based on Bayesian Confirmation Measures. In: International Conference on Convergence Information Technology (ICCIT 2007), pp. 284–289. IEEE Press, Korea (2007)
9. Abbas, A.R., Juan, L., Safaa, O.M.: Improved Variable Precision Rough Set Model and its Application to Distance Learning. In: Proceedings of the 2007 International Conference on Computational Intelligence and Security, pp. 191–195. IEEE Press, China (2007)

10. Likert, R.A.: A Technique for the Measurement of Attitudes. Archives of Psychology (June 1932)
11. Andrich, D.: Application of a Psychometric Rating Model to Ordered Categories which are Scored with Successive Integers. Applied Psychological Measurement 2, 581–594 (1978)
12. Wright, B.D., Masters, G.N.: Rating scale analysis. MESA Press, Chicago (1982)
13. Linacre, J.M.: Estimating Measures with Known Polytomous Item Difficulties. Rasch Measurement Transactions 12(2), 638 (1998)
14. SCORM Photoshop Examples Version 1.0 (2004), `http://www.adlnet.gov/downloads/index.cfm`

Dynamic Content Manager – A New Conceptual Model for E-Learning

Terje Kristensen[1], Yngve Lamo[1], Kristin Ran Choi Hinna[2], and Grete Oline Hole[3]

[1] The Department of Computer Engineering, Bergen University College,
Nygårdsgaten 112, N-5020 Bergen, Norway
tkr@hib.no, yla@hib.no
[2] Faculty of Education, [3] Faculty of Health and Social Sciences,
Bergen University College, Bergen, Norway
khi@hib.no, goh@hib.no

Abstract. In this paper a conceptual model for e-learning, which uses elements from learning Objects and Concepts Maps, is introduced. The learning material is divided into atomic units and organized in graphs called Knowledge Map, Learning Map and Student Map. Such a structure provides an easy-to-use navigation interface for existing learning material. Any course content created is stored in a repository for future reference. The model is used to structure a course in geometry for post-graduate teacher students. In teacher education one has to account for the ability to transfer knowledge to students, in addition to the assimilation of knowledge. The established model is therefore discussed in a learning or didactical context.

1 Introduction

The main goal of the teaching process is to develop knowledge. A methodology to structure and model the learning process is a means to achieve this. One widely used tool for organizing, representing and building knowledge is Concept Map (CM) [10]. Since its introduction there has been extensive research regarding how to use CM to enhance teaching and learning [4].

CM is a tool well suited for representing knowledge structures. However, it does not address the dynamic process of learning. To represent the entire learning process, the Dynamic Content Manager (DCM) model for e-learning was introduced in 2007 at Bergen University College [8],[9]. It makes it possible to create knowledge elements at a finer granularity level to reuse them in various courses. Resources (R), assessments and evaluations (E) are attributed to the knowledge elements, and hence may be imported from existing learning material.

The model is used to represent structures for knowledge, learning scenarios and individual students' learning. These are implemented as Knowledge Map, Learning Map and Student Map. This approach promotes adaptive learning, flexibility in the learning process and share and reuse of learning material [1],[8].

The model provides a more flexible tool for teachers who are planning to use other learning scenarios than traditional Learning Management Systems (LMS). Within

W. Liu et al. (Eds.): WISM 2009, LNCS 5854, pp. 499–507, 2009.

DCM, the learning resources and progress may be structured and organized to maximize flexibility for both teachers and students to promote tailored learning [3].

A teacher giving a course can divide the learning material into atomic units that may be organized into the different maps. The Knowledge Map provides an overview of all the learning resources available to an instructor. The Learning Map is based on these resources, providing a functional overview of a given course and the didactical approach to the material. The Student Map monitors each student's learning progress and provides evaluation and feedback mechanisms. This structure provides an easy-to-use navigation interface for existing learning material. Any course content created is stored in the repository for future reference. As shown in [1], the DCM Learning Objects may lead to the use of these Knowledge Objects as basic elements, to construct a more elaborated version of CM.

An example on how to use these structures, when teaching geometry for teachers, is presented. In contrast to ordinary education, which is mostly skill- and fact-based, teacher education also requires understanding of didactical processes and meta-knowledge of the given subject. The DCM model is especially suited to learning scenarios, where it is important to create deep learning [11]. The evaluation perspective is especially important in teacher education. One must construct the best learning path for the students through the knowledge space and how to evaluate their learning outcome. This is the job of the teacher, and she has to do it appropriately to succeed in promoting learning. By using the DCM model one is able to create a more thoroughly didactical understanding of the topic.

2 Knowledge Modelling

All successful transfer of knowledge requires understanding of the concepts within the teaching subject. Knowledge modelling becomes a tool promoting consciousness of the subject in this meta-perspective.

2.1 Organizing Knowledge

To promote learning one has to discover the inner relationship between the learning components and convey them to the learners. According to the cognitive learning perspective, the goal is to facilitate mental processes which mediate between existing and new knowledge [10]. Within this perspective the focus is on how students understand and solve problems by symbol processing. Information is received through attention and integrated in memory. Furthermore, it is translated into knowledge and integrated into the learner's' cognitive structure for later retrieval.

In this context the notion of schema, information processing, storage and retrieval are important. Knowledge is considered as abstract symbolic representations. The teacher's primary role is to transfer knowledge by lecturing and explaining the concepts [4].

2.2 Concept Maps

CM has been used for measuring the quality of learning, as a tool for letting the students illustrate the difference between *Deep* and *Surface* learning [11]. By use of CM the teacher may be guided to adjust her teaching to different *Learning Styles* [3]. Through this, one wishes to promote meaningful learning instead of rote-learning. By considering CM as a tool for facilitating learning, it can be viewed as a mirror to the students cognitive structures within a given domain, in order to gain insight into their understanding. As Kinchin and Alias [6] would say, as a "*window into the thinking of the students*". In this way CM is an efficient meta-cognitive tool to promote understanding on how new materials interacts with existing cognitive structures. This links the way constructivism uses CM to more socio-cultural learning theories [7].

2.3 Content Units

In the DCM model knowledge is represented by Content Units (CU), which consist of Learning Resources and Evaluations (E). Constructivism views learning as a process in which the learner actively constructs or builds new ideas or concepts, based upon current and past knowledge. Knowledge is closely connected to previous experiences. Learners need to construct their own understanding. The primary role of teaching is to design situations for the learners to promote their creation of the necessary mental constructions. The learner will internalize concepts, rules and general principles which further on can be applied in a real-world context. Knowledge is considered as a constructed entity made by every learner through the learning process [10].

The DCM facilitates learning by constructivism. It has been designed to enhance the functionality of knowledge elements, courses and resources. The changes made to the underlying knowledge elements are then carefully treated by versioning and history tracking. This ensures that a specific course or aggregation of knowledge elements, which the teacher has created, can appear unchanged. She is then able to follow the various revisions made on it. Great emphasis is also placed on the design of knowledge elements to provide seamless addition of external functionalities.

2.4 Knowledge Maps

The basic requirement of the DCM is the knowledge repository, from which knowledge elements may be drawn and organised into a hierarchical structure of the course. This is represented by the Knowledge Map. The CU must be structured and organised in such a way that the teacher actually gets the information she needs to design a course. There must also be a way of adding knowledge to the actual repository.

The Knowledge Map provides the overview of the total knowledge in a given learning domain. The map is a graph where the nodes represent content units of the repository. The arrows placed between any two related CU is representing the relations between contents. When someone has created a course, there is a relation between the actual units used. The instructor who creates a new course, can use the arrows to indicate which units that can be of interest.

The DCM's "atomic" units of knowledge make it possible to construct CM-like structures as a tool for both the teachers and the students. This structure will enable DCM to provide the teacher with a graphical navigation tool to explore the knowledge repository. Such a visual presentation gives the teacher an overview of the resources of the repository and enables her to create the knowledge elements.

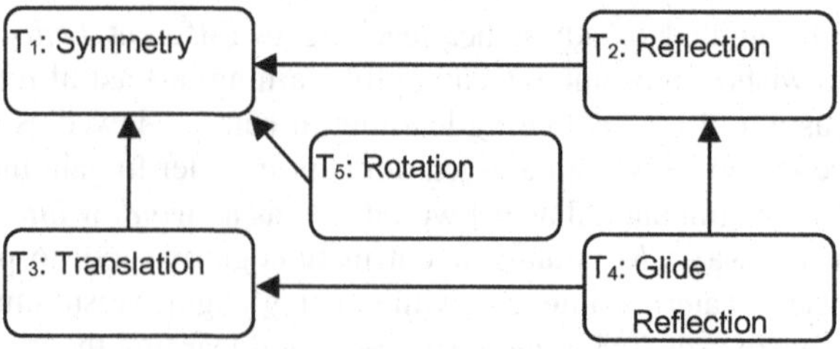

Fig. 1. An example of a Knowledge Map labeled Theme (T_1 - T_5)

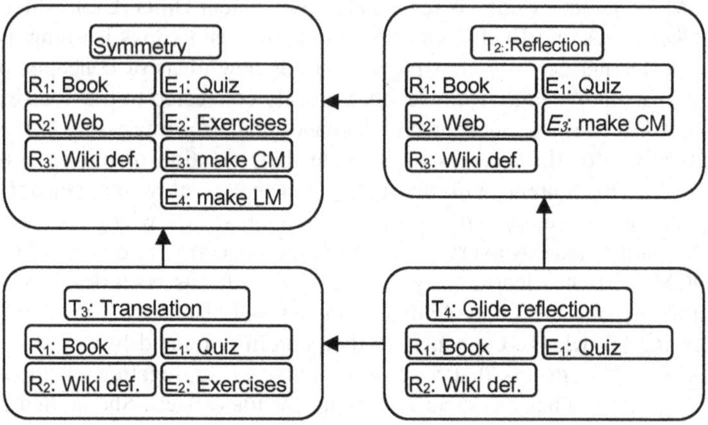

Fig. 2. Extracted Knowledge Map displaying different Resources and Evaluations of four CUs. Due to space limitations Rotation is omitted.

The interconnection between the various knowledge elements defined by many educators can also be used for data mining purposes and for creating better structuring of the knowledge repository. In a specific course the motivation of using a Knowledge Map is to model the dependencies in the learning process. The interconnection between the various knowledge elements defined by many educators can also be used for data mining purposes and for creating better structuring of the knowledge repository. In a specific course the motivation of using a Knowledge Map is to model the dependencies in the learning process.

However, these units should not be viewed as imposing requirement on the content selection. The teacher is free to select other nodes and use them as prerequisites, or to create her owns. Figure 2 displays a sample Knowledge Map consisting of four knowledge units from the geometry domain. A teacher navigating the Knowledge Map would also be interested in the inner structure of the nodes, in order to select the elements she needs when creating the course. The system therefore needs to keep track of this inner structure. The user interface should allow for easy expansion of nodes to display the resources and evaluations, as shown in figure 3.

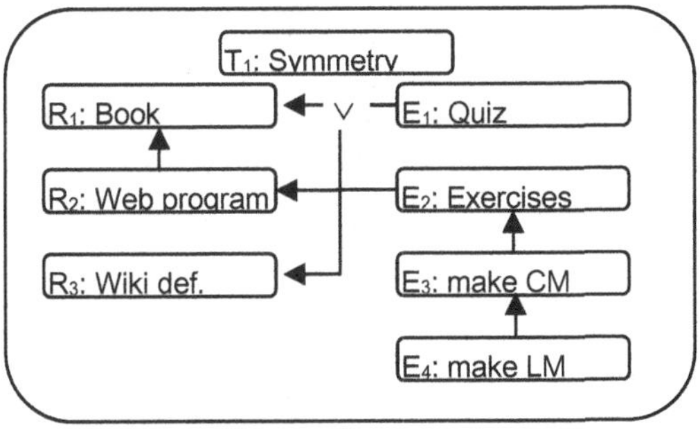

Fig. 3. A learning Map constructed from the Knowledge Map of figure 1 and 2. Notice the disjunctive relation between Evaluation and Resources 1 and 3.

3 The Learning Process

Structuring the learning process is essential to construct optimal learning scenarios from the learning material that is represented in the Knowledge Map. The actual learning scenario (Learning Map) is created by individual teachers due to their didactical understanding of the topic.

3.1 Knowledge Map

A Learning Map is a representation of the learning process. The Content Units are selected from the Knowledge Map and expanded to ensure that the Resources and Evaluations become nodes in the graph. Formally, there is a graph homomorphism from the Learning Map to the Knowledge Map.

Resources or Evaluations not needed in the course may be omitted. The Evaluations may be weighted to indicate their importance in the grading of the entire course. The Learning Map relations may occur between Content Units or between Resources and Evaluations. Such relations indicate that one element is a prerequisite of another.

4 Student Modelling

The DCM system also supports student modelling. The system defines a simple model of the learner and categorizes him, based on the actual Evaluations. The teacher must often update the category of the learner, based on assessments of the students' assignments. The DCM has a module where the learner is guided through questions that have been associated with the knowledge elements of the course he is taking. The learner's category is taken into consideration by a default optional filtering mechanism and provides the learner with questions.

4.1 Student Map

A Student Map is the model of the learning process of an individual student. It displays the Resources and the Content Units that the student has encountered. The most important aspect is probably the Evaluations which show the student's answers and results.

Figure 4 displays the maps of two different students who have taken a course based on the Learning Map of figure 3. The Learning Map has a branching point where each student may choose what resources to take, resulting in different looking Student Maps.

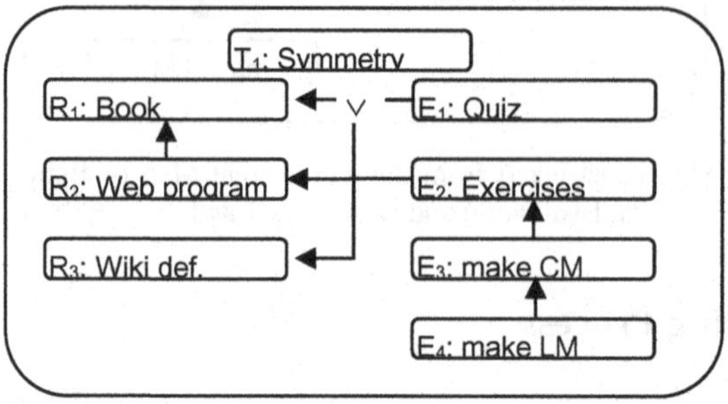

Fig. 4. Two different student maps derived from the Learning Map of figure 3

5 DCM in Geometry

Knowledge modelling is often used to represent learning of mathematics [5]. In this paper we emphasize the use of knowledge modelling, both for learning of mathematics and as a tool to develop meta-knowledge. This is of special importance for teacher students. The students have to make their own conceptual domain models represented as Knowledge and Learning Maps.

In geometry, there are many different concepts which are crucial for understanding of the topic. For instance, a Symmetry Relation may be realized by doing a Reflection, a Rotation, a Translation or a Glide Reflection. The Symmetry Concept may be interpreted at a higher abstraction level than these basic concepts.

This means that to understand the Symmetry Concept one has to first understand the basic concepts such as Reflection, Rotation, Translation and Glide Reflection. This is reflected in the Knowledge Map of Figure 1.

The activities defined by the term Evaluations in table 1 have no weights. The Evaluations document student activities, which are an important part of the learning process, in addition to grading. By Evaluation $E_1 - E_4$ one measures the different aspects of the knowledge, such as facts, skills, meta-knowledge or didactical knowledge.

The different kinds of Evaluations are connected to the outcomes stated in the syllabus and the curriculum. All these aspects are important to ensure that the student has the desired knowledge to teach mathematics, for instance in Primary School.

The Resources given by R_1–R_4 illustrate some of the learning material presented to the students. To ensure that the students achieve knowledge about facts, practical skills, concept structures and strategies the teacher needs to select adequate resources. Available resources such as text-books and online resources like animations, videos, images and interactive programs, may be selected from the Knowledge Repository. An awareness of choosing the optimal learning material can later be used by the teacher students in their own teaching practice.

Table 1. A Sample Content Unit

T1: Symmetry		
Resources	Evaluations	Aspects
R1: Book Chap 3 R2: Origami web program R3: Wiki definition R4: Video of geometrical constructions	E1:Quiz E2: Exercise, ruler/compass E3: Making Knowledge map E4: Creating a course, making Learning Map	Facts Skills Meta-knowledge Didactical knowledge

A Learning Map of the Symmetry Concept is presented in Figure 3. The teacher has selected two sources to obtain the factual knowledge which is required to answer the Quiz, given by E_1. The Student Map in Figure 4 offers an opportunity to monitor the knowledge that the teacher students have acquired. One of the students uses a textbook and the other one a wiki. The educator may evaluate the results of the students based on their learning paths. The teachers may get a clear picture of the efficiency of using different approaches to the learning of the Symmetry Concept after some repetitions of the course.

6 Discussion

One key issue of the DCM model is that different learning aspects may be emphasized. In the course designed for teacher students the learning outcome could be divided into factual, skill-based, meta- and didactical knowledge. In geometry factual knowledge is theoretical knowledge about formulae, names and symbols. Skill-based knowledge is practical use of a procedure or to design a solution of a problem. In geometry, for instance, it is crucial to have practical skills of using ruler and compass. To be a good teacher one also needs to have knowledge of the didactics of mathematics and meta- knowledge. Meta-knowledge is crucial knowledge for planning, modelling, learning and modification of domain-knowledge. This implies that the students have to be conscious about their own thinking about mathematics. It is important that teacher students have a conscious attitude to their own learning process. They must be able to reflect upon their "doings" to develop a solid basis for their own teaching.

To achieve resilient knowledge it is essential for the students to get awareness about their own possible learning outcome and their teaching practice. By resilient knowledge we mean knowledge that is resistant to cultural influences. This kind of knowledge distinguishes it from the traditional way of considering learning. For teachers it is important to develop a strong conceptual understanding of the problem domain to be able to understand the thinking of individual students. This is necessary to give the students problems which are suited to their cognitive level.

However, in this project the students learn to model their own knowledge structures by using different graphical techniques. Such graphs may be used by the students to reflect upon their knowledge acquisition. The students document their learning progress by monitoring their Student Maps. To be a good teacher it is important to be conscious about your own knowledge. If you are not able to express knowledge to yourself, how can you then be able to explain it to your students?

It is difficult to evaluate different aspects of knowledge. One often considers only the aspect of knowledge which fits to the framework of the learning outcomes. The evaluation of other aspects can be done in different ways. One possibility is a multiple-choice quiz to evaluate factual knowledge. To evaluate skills one have to make a practical test where the students carry out a construction by compass and ruler. In an e-learning scenario this may, for instance, be done by using interactive ICT-tools.

The teacher students also have to demonstrate connection between practical skills and knowledge structure by creating Knowledge Maps. However, the Evaluation of their pedagogical and didactical knowledge must be done in another way. One may let them write about their own learning process where one is reflecting upon the connection between different aspects of the subject.

The different types of assignments may be weighted differently. The students will then know what kind of knowledge that is emphasized in the course. Such kind of assessment is different from traditional practice in Norwegian schools. In the DCM model a student will be assessed in a more comprehensive way by considering his own knowledge construction. In this way the student strengthens his meta-cognitive understanding. This is one important aspect of the DCM approach. By practical experience one develops conceptual images that can be expressed in Knowledge Maps. Such a language of concepts may be used to transfer "internalized knowledge" to the learning society [12].

7 Conclusion

In this paper we have used a graph based approach, DCM, to model the learning process of teacher students in geometry. It is shown how the Knowledge Map is used to systematize the content of a course. Different kinds of resources and ways of evaluating them are demonstrated. The approach is flexible in respect to organising the content of a course. The content includes learning resources, practical tasks and evaluations. In teacher education one uses special ways of evaluation, since one has to consider different aspects of knowledge.

The Learning Map is created by the teacher to design a course. It describes a selected scenario as a path through the content. The actual Learning Map may vary between different teachers, even for courses with the same content and syllabus. This is due to the individual didactical understanding of each teacher.

The Student Map is created by the system, based on the results and weights of the Evaluations. It represents a model of the learning progress of individual students taking the course. The Student Map may be used by the educator to monitor the learning progress.

In teacher education the students also need to create their own maps to describe their conceptual understanding of the subject and to create their own teaching scenarios. In this way the students gain experience with mathematical (Meta) modelling.

By using a system such as DCM, based on Knowledge Map, Learning Map and Student Map, one adapts the e-learning system to the learning process. This is in contrast to traditional Learning Management Systems where the learning scenario has to be fitted to the system.

References

1. Eide, S., Kristensen, T., Lamo, Y.: A Model for Dynamic Content Based E-learning systems. In: ACM Proceedings EATIS, Aracaju, Brasil (2008)
2. Greeno, J.G., Collins, A., Resnick, L.B.: Cognition and Learning. In: Berliner, D.C., Robert, C.R.C. (eds.) Handbook of Educational Psychology. Macmillan Library Reference, New York (1996)
3. Gardener, H.: Frames of mind. The theory of multiple intelligences. Basis Books, New York (1993)
4. Hay, D., Kinchin, I.M.: Using Concept Mapping to Measure Learning Quality. In: Education & Training, pp. 167–182 (2000)
5. Kinchin, I.M.: Using Concept Maps to Reveal Understanding: A Two-Tier Ana-lysis. School Science Review 81(296), 41–46 (2000)
6. Kinchin, I.M., Alias, M.: Exploiting variations in concept map morphology as a lesson-planning tool for trainee teachers in higher education. Journal of In-Service Education 31(3), 569–592 (2005)
7. Kinchin, I.M., Hay, D., Adams, A.: How a qualitative approach to concept map analysis can be used to aid learning by illustrating patterns of conceptual development. Educational Research 42(1), 42–57 (2000)
8. Kristensen, T., Lamo, Y., Mughal, K., Tekle, K.M., Bottu, A.K.: Towards a dynamic, content based e-learning platform. In: Uskov, V. (ed.) Computers and Advanced Technology in Education, pp. 107–114. ACTA Press (2007)
9. Kristensen, T., Hinna, K., Hole, G.O., Lamo, Y.: Different E-learning paradigms - a Survey. In: Proceeding of MIT-LINC (Massachusets Institute of Technology Learning International Network Consortium) 4th International Conference: Technology-Enabled Education: a Catalyst for positive Change, Amman, Jordan (October 2007)
10. Novak, J., Canãs, A.: The theory underlying Concept Maps and how to construct them. Technical report, IHMC CMaps Tools, Florida Institute for Human and Machine Cognition, USA (2006/2008)
11. Marton, F., Hounsell, D.J., Entwistle, N.J.: The experience of learning. Scottish Academic Press, Edinburgh (1977)
12. Perkins, D.N., Salomon, G.: Are cognitive skills context-bound? Educational Researcher 18(1), 16–25 (1998)

A User Behavior Perception Model Based on Markov Process

Zhan Shi[1], Yuhong Li[1], Li Han[2], and Jian Ma[2]

[1] State Key Laboratory of Networking and Switching Technology, BUPT
Beijing, 100876, P.R. China
bupt2007shzh@gmail.com, hoyli@bupt.edu.cn
[2] Nokia Research Center Beijing
Beijing, 100176, P.R. China
ext-li.1.han@nokia.com, jian.j.ma@nokia.com

Abstract. Nowadays user behavior perception becomes more and more important in scientific experiments and people's daily lives, especially in mobile services. This paper presents a model which can be used to perceive user's behavior on intelligent mobile terminals based on Markov process. The model combines the syntax model and the probability transition model, and introduces also other parameters like weight. It can perceive users behavior through different kinds of contexts, such as users' reading history. According to the perceived behavior, the model can better perceive the user's behavior, and therefore allows better mobiles services experiences to users. A system architecture based on the presented model has been introduced and implemented, and a verification method of the model has also been given.

Keywords: Markov process, user behavior perception model.

1 Introduction

With the increase in computing power and the development of optimizing methods and network technologies, a variety of software and hardware can work more and more harmonically and more and more new services can be provided to users. Especially, different kinds of intelligent mobile terminals have been produced and many mobile services are provided by different operators. This situation better meets the user's day-to-day requirements, but it still not completely done in accordance with specific user (personal device) to provide specific services. Nowadays, mobile terminals and mobile services have become the necessity of people's daily lives. And from another point of view, more and more new requirements are putting forward on mobile terminals and services. However, due to the limitation of mobiles devices, such as the operation convenience, and the cost on network usages, users' need cannot be well satisfied. In order to achieve better user experience on terminals, this paper presents a new method based on the machine self-learning and context-aware, and finally provides a better result.

Some work has been done related with context-aware and user behavior perception. For example, some automatic reconfiguration methods have been implemented

W. Liu et al. (Eds.): WISM 2009, LNCS 5854, pp. 508–517, 2009.

in some web services from the perspective of servers [1][2]. Here the websites or servers can adapt themselves to the majority of users by statistics and context-aware analysis, so that it is more simple and practical for users to browse information. However, these methods cannot meet all of the users. This paper presents a solution based on the self-learning on the smart terminal which adapt to the individual user. A generalized control strategy that enhances fuzzy controllers with self-learning capability for achieving prescribed control objectives in a near-optimal manner is presented in [3]. And a neural network method is proposed in [4], where the system can learn its own accord to control a nonlinear dynamic system. "Self-learning Disk Scheduling" [5] presents four algorithms about machine-learning, and "RAP - A Basic Context Awareness Model" [6] firstly gives the general and basic model which considers mostly parameters, such as context resources, a set of actors. In the second step, we get the specific model through quantifying these parameters. It provides us the basic context-aware model. Our model has applied the self-learning idea from [3] to [6], however, we can solve user behavior perception and prediction problems that cannot be solved by simply using the above algorithms.

A mixture probabilistic user behavior model was proposed in [7], where the Markov process was used to solve the problem of predicting user's behavior. The global model for the existing users is personalized by assigning each user individual component weights for the mixture model. But the mixed model is only for the web user and can't solve the problem of mobile devices. Faten Khalil etc. also present a model based on the Markov process [8], and gave three modified Markov models which were "the all *kth* Markov model. Frequency pruned Markov model and the Accuracy pruned Markov model have been presented for predicting web page access. Dairazalia Sánchez provided us a hidden Markov model for activity recognition in ambient intelligence environments and a solution to the context recognition in [9]. Our "probability matrix transfer model" for mobile devices has borrowed some ideas from these work, however, we introduce self-learning and self-adapt on intelligent terminal, which can better meets users' need.

Teuvo Kohonen present the thought about self-learning musical grammar based on the thought of Markov process and the music produced by this method generally sounds smooth, continuous, and pleasant [10]. But the syntax regulation cannot solve other problems, while in this paper, we raise a new question about perceiving user's behavior based on the thought of the syntax model, and give the solution on the state which is a set and finally give the "syntax model".

This paper presents a novel user behavior perception model for intelligent mobile terminals. The model can perceive the users' behavior according to different kinds of context information, and can therefore decrease the users' interactive actions with mobile terminals, whereas satisfy the users' requirements through perceiving users' behavior. The model is based on the Markov process, which introduces also the idea of machine self-learning and context-awareness. It combines the syntax model and the probability transition model, and uses also other parameters like weight. The user behavior histories are used to discover user's preference [11], and an approach based on information gathered from users are described to perceive the user behaviors.

The rest of the paper is organized as follows. In section II, we introduce briefly the system architecture where the proposed model is used. Following this in section III, the perception model based on Markov process is described, and finally the implementation of the system and the verification method of the model are presented.

2 System Architecture

Figure 1 depicts the network architecture of the user behavior perception model. The architecture is used for information broadcasting to mobile users, and the client is developed on Nokia N800 [12]. In detail, the central server access and sift the information from the internet, then hand these information out to local servers. We focus on the part in the dotted line which is the part of local server and terminals. In this part, the local server is always sent messages to all devices containing the perception model in the same wireless LAN, while users could subscribe specific services based on interests. Initially, user selects his interests and hobbies concerned about the contents, then the user will receive the messages he chose. After a long term of self-learning of the device which is with the user behavior perception model, the device could perceive users' choice and become more intelligent.

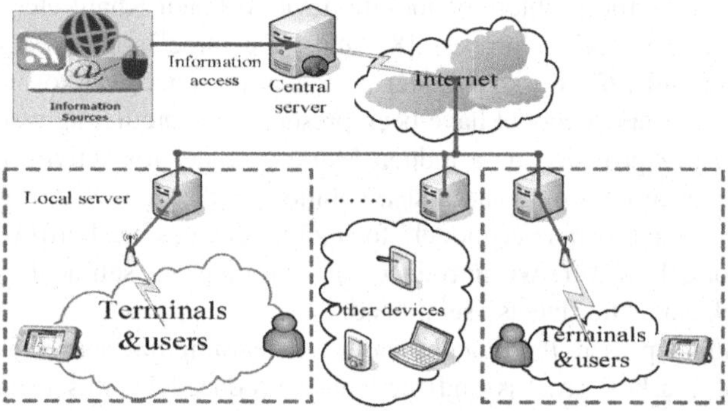

Fig. 1. Network architecture of the system

While the crux of the matter we are discussing is how to use the long-term records of the user's reading contents that recorded which categories user has read to perceive user's interests and select the categories to receive automatically next time. Here, we consider the three-level directory structure, give an top-level category example described as the Figure 2 and define the state which describes categories of users' reading as a set. For example, user chose the "CurrentAffairs", "H.K.Macao. T" from "China" which is a secondary directory and "GlobalView" from "International" and so on, then a set $A = \{CurrentAffairs, H.K.Macao.T, GlobalView, \cdots\cdots\}$ will be got. All the elements for user choosing are at the third class of the directory structure.

Figure 2 describes the directory structure of "NEWS" which is one of the top-level categories in the structure which has three levels.

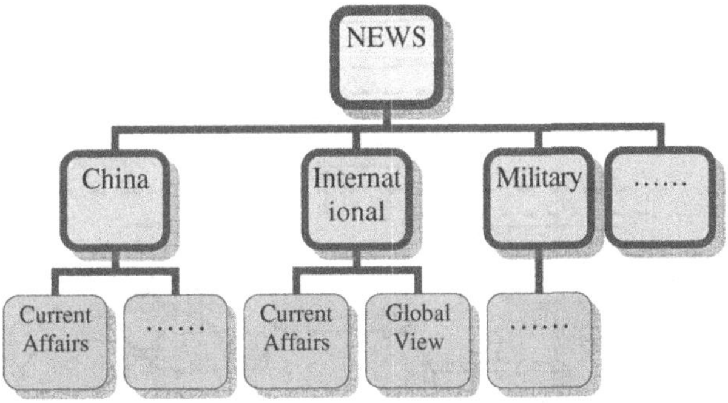

Fig. 2. Directory structure of "NEWS"

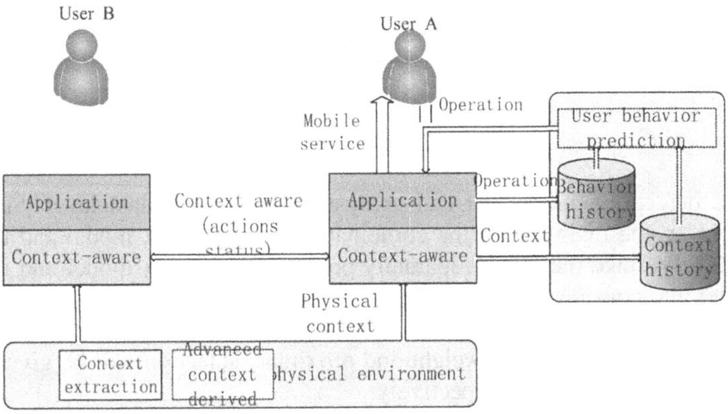

Fig. 3. Context information collection

User behavior perception model is based on collecting the useful context information. Figure 3 which is a highly abstract description of context-aware in mobile computing illustrates the relationship between the context information and the user behavior perception. Here, the context-aware refers to systems able to detect and make effective use of contextual information (such as user location, time, environmental parameters, adjacent equipment and personnel, user activity, etc.). We could extract context and derive the advanced context from the physical environment and use these parameters to perceive user's behavior considering behavior history and context history. The mobile terminals with the perception model are at the center of the network topology and the surrounding environment is implicit input. In this way, with the changing environment, mobile service will be automatically adjusted.

3 User Behavior Perception Model Based on Markov Process

The perception model is described in detail by Figure 4, it includes two prediction models, user list and user log. Here we use the syntax model and transition matrix model together, and consider other factors, finally give the proximate selection.

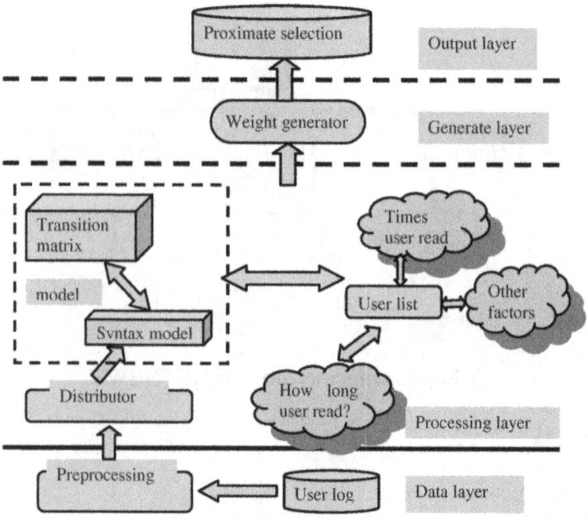

Fig. 4. The perception model

First, the perception model get and analysis user's log in data layer. Second, in processing layer, the main part is the combination of the syntax model and transition matrix model. We take the states separately predicted by syntax model and transition matrix model into consideration and integrate other factors, such as "How many times user read a category?", "How long user read a category?" etc. to get more accuracy results in processing layer. Then weight and proximate selection will be given in generate layer and the output layer respectively.

The main method solving problems is based on the Markov process which can be used in many areas to predict the probability of next case happened, the emergence and the probability of a certain status. Based-on the Markov process, we introduce three ways to achieve the purpose of the perception. The first is that there exists a grammar or a syntax formed after recording user's long-term reading habits, and through the syntax, we can forecast the next state, that is to say, to perceive which categories are users most interest on. Obviously, the most import thing is the forming process of the grammar. Second, we use the one step probability transition matrix which demonstrates the state switching from one to another to do the perception. But the second method is only on the basis of the last condition, so the coverage isn't broad enough. For this reason, we use the first method in chief and auxiliary use the second method, then the third model which uses the first way to predict the next state and uses the second method auxiliary emendation the prediction is got. Simultaneously, as the context is always changed, we should consider other factors which are influence the prediction results, such as "how long the user read this category" and "how many times the user opened the web pages" in this category, this information indeed reflects users' interests indirectly. We expound these models respectively in the following description.

3.1 A Prediction Syntax Model

The deduction is on the basis of records which are stored in the database as a log file keeping track of user's operation over a long period of time and the "no after-effect" property of the Markov process. We firstly give the typical example and summarize the algorithm about this grammar. As described in the previous, a state on behalf of each type of content user read and it's a set.

Simple context example:
$B, A, S, L, D, J, F, L, S, W, J, F, L, W, E, U, R, U, G,$
$H, K, A, S, F, H, S, F, H, G, H, L, W, S, J, F, E$

Consider the state U, the next state would be R or G, so which state will happen is not clear, either R or G, but we see EU could decide R and RU decide J, this solves the problem. So we can obtain the syntax shown in the table below for each state.

$B: B \to A$

$A: A \to S$

$S: S \to L, S \to W, S \to F, S \to J \Rightarrow BAS \to L,$

$LS \to W, KAS \to F, WS \to J$

$L: L \to D, L \to S, L \to W \Rightarrow SL \to D, FL \to W,$

$FL \to S, HL \to W \Rightarrow WJFL \to W, DJFL \to S, HL \to W, SL \to D$

$D: D \to J$

$J: J \to F$

$F: F \to L, F \to H, F \to E \Rightarrow JF \to L, SF \to H,$

$JF \to E \Rightarrow WJF \to L, SF \to H, SJF \to E$

$W: W \to J, W \to E, W \to S \Rightarrow SW \to J, LW \to E,$

$LW \to S \Rightarrow SW \to J, FLW \to E, HLW \to S$

$E: E \to U$

$U: U \to R, U \to G \Rightarrow ER \to R, RU \to G$

$R: R \to U$

$G: G \to H$

$H: H \to K, H \to S, H \to G, H \to L \Rightarrow GH \to K, FH \to S,$

$FH \to G, GH \to H \Rightarrow UGH \to K, HGH \to L, ASFH \to S,$

$HSFH \to G$

$K: K \to A$

If we don't track user's data enough, coverage of the rule will be narrow and small. On the contrary, more records, more accuracy of the rule, but with the coverage scope expanding, the complexity of the algorithm is increasing. For this reason, a regulation to cease the deduction of the rule appears very important, otherwise, if let the device learn according with this rule unrestricted, the efficiency of the algorithm will certainly be descending. In this algorithm, we will deduce until the level up to 8. In case of the device use the rule up to level 8 and can't determine the next state, we will consider select the front categories ordered by the probability. This algorithm not only considers the coverage of the data sets, but also the efficiency.

Figure 5 illustrates the forming process of the decision tree when "L" as the root. Here only 3 levels form the rule that could predict the next state.

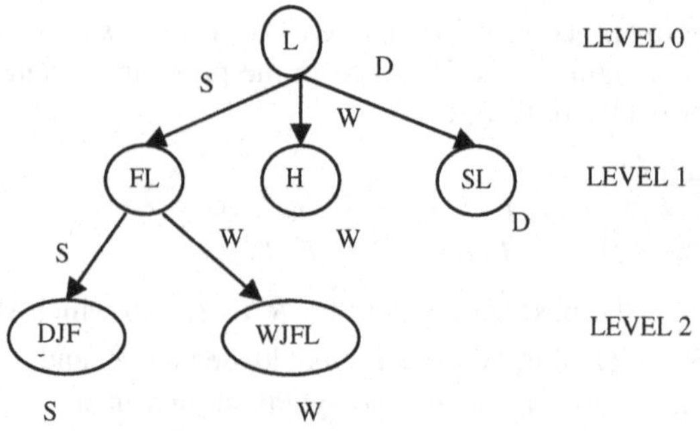

Fig. 5. Illustration of a simple example

Algorithm

```
int main()
{
int level=0;
for ( int i=1, i<count (state); i++,)
  { E_i =root;
        While ( count (E_i →next) !=1 && level<=8)
        {  E_i = E_i →pre+ E_i ; level++; }
  }
Return 0;
}
```

3.2 The One Step Probability Transition Matrix Model

Utilizing above algorithm, terminal could perceive the most probably categories user will read next time, but the result may be not absolutely accurate, there is a need to amend the algorithm to get more accuracy result. We use the one step probability transition matrix to achieve this objective.

The transition matrix model also accesses and analyzes user's operation records firstly. We consider the categories which user read one time as a state. Because of the categories of our information broadcasting system are many, such as military, education, sports and so on, the combination will be more. Assuming the state space recorded in database is $\{ A, B, C, D, E, F \cdots\cdots \}$, each element represents a set defined above. Then we will get the step probability transition matrix as Figure 6.

$$\begin{matrix} & A & B & C & \cdots \\ \begin{matrix} A \\ B \\ C \\ \vdots \end{matrix} & \multicolumn{4}{c}{\left\{ \begin{matrix} P_{AA} & P_{AB} & P_{AC} & \cdot\cdot \\ P_{BA} & P_{BB} & P_{BC} & \cdot\cdot \\ P_{CA} & P_{CB} & P_{CC} & \cdot\cdot \\ \cdots & \cdots & \cdots & \cdot\cdot \end{matrix} \right\}} \end{matrix}$$

Fig. 6. One step probability transition matrix

From the matrix described in Figure 6, we can get the probability from A to A is P_{AA} and from A to B is P_{AB}, and so on.

3.3 Other Factors

Now we come to consider other factors of the third class categories in which the pages are how many times opened and how long read are accessible by user's log. We define the time of user reading is B_{ij}, (i represents the ith category and j indicates the jth page in the ith category) and times of user reading is T_{ij}, page number of the ith category is N_i, the number of all categories is S, Then the "$AllTime$" parameter is defined as follows:

$$AllTime_i = \sum_{j=1}^{N_i} \sum_{j=1}^{T_{ij}} B_{ij} \quad (0 \le i \le S, 0 \le j \le N_i)$$

Then weight could be deduced,

$$weight_i = \frac{AllTime_i}{\sum_i^S AllTime_i} \quad (0 \le i \le S)$$

Thus, a sort will be obtained. The following chart is given by investigating operation of one user. It represents weight of each category.

Another problem we have is that we should decide how many categories will be chose in the predicting model. Solution to the problem is still to analyze user data in database. We could get the number of user choose most frequently and define this number as T in above model. And we also define the element number of the combination of syntax model and transition model as C. There would be three cases:

- $C = T$, we only choose the combination;
- $C < T$, we should choose the combination and $T - C$ categories in the remaining categories considering the "weight" factor;
- $C > T$, the first T categories would be chose in accordance with the order of the "weight".

Above three steps set up the whole model for user behavior perception. It predicts the next state through the syntax model and the transition matrix model, then determines T, gives the final result based on other factors, such as "weight".

4 Implementation and Verification

The perception model has been implemented on Nokia N800 with the operating system 2008 developing on "Maemo" platform [13]. User should use the device containing the perception model for a long term, so the machine could form a set of rules which would be more and more accuracy as the rules becoming more and more stable through context-aware and self-learning. Figure 7 illustrates perception result by user behavior perception model.

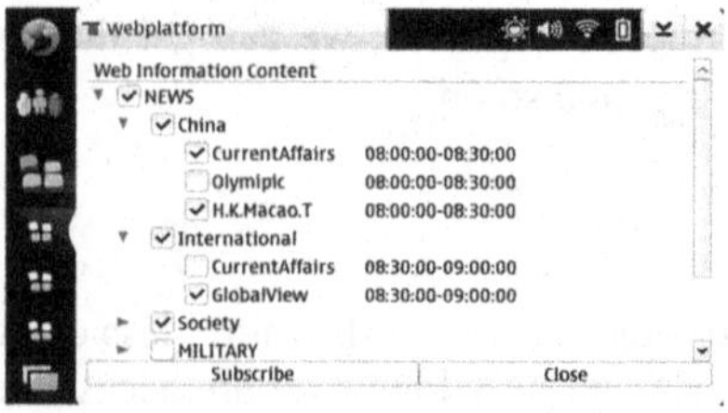

Fig. 7. Graphical interface for user selection

To verify the effect of the presented model, we could investigate a mount of typical users, each of whom uses the device containing the perception model for a long term. Then, we compare the differences between the state which is got from the perception model and the state which is the user's choice. Assuming the next state we get from the perception model is $A=\{a_1,a_2,a_3,\cdots\}$, but user's choice is $B=\{b_1,b_2,b_3,\cdots\}$, we define the ratio of the element number in the intersection between A and B to the total number of B as the accuracy of perception model. In the following formula, $\#A$ represents the element number in the set A.

$$accuracy = \frac{\#(A \cap B)}{\# B}$$

We investigate the questionnaire to compare the degree of the satisfaction on accuracy before and after using the perception model.

The 100 typical users mentioned above participate in the questionnaire, and figure 15 and 16 give the comparison of the satisfaction degree between the system without the perception model and the system with the perception model.

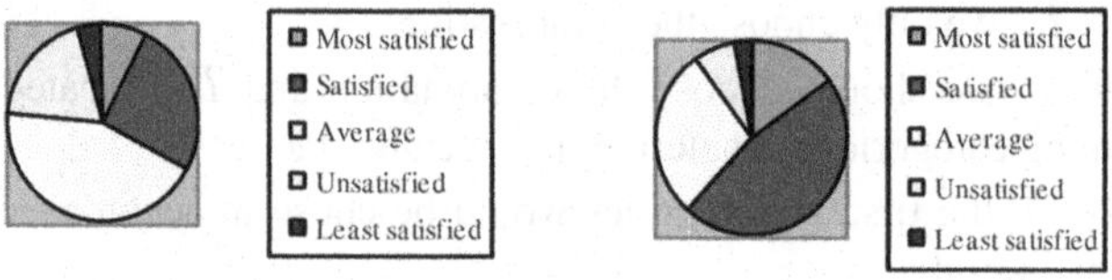

Fig. 8. The accuracy comparison

From the comparison in figure 8 between the left figure which is the system without the prediction model and the right with model, we can clearly see the satisfaction degree raised by using perception model, which verifies the assumption and achieves the objective of our perception model.

5 Conclusion and Future Work

This paper proposes a new perception model, which utilizes the syntax model and transition matrix model, and considers other factors, like "AllTime", "weight", gives the proximate decision finally. The model focuses on the description of the mechanism of machine self-learning, give a better perception of user's behavior.

References

[1] Schilit, B.N., Adams, N., Want, R.: Context-Aware Computing Applications. In: IEEE Workshop on Mobile Computing Systems and Applications, December 8-9 (1994)

[2] Chen, G., Kotz, D.: A Survey of Context-Aware Mobile Computing Research, Dartmouth Computer Science Technical Report TR2000-381

[3] Jang, J.-S.R.: Self-learning fuzzy controllers based on temporal backpropagation. IEEE Transactions on Neural Networks 3(5), 714–723 (1992)

[4] Nguyen, D.H., Widrow, B.: Neural networks for self-learning control systems. IEEE Control Systems Magazine 10(3), 18–23 (1990)

[5] Zhang, Y., Bhargava, B., Fellow: Self-Learning Disk Scheduling. IEEE Transactions on Knowledge and Data Engineering 21(1) (January 2009)

[6] Salomie, I., Cioara, T., Anghel, I., Dinsoreanu, M.: RAP – A Basic Context Awareness Model. In: 4th International Conference on Intelligent Computer Communication and Processing, ICCP 2008 (2008)

[7] Manavoglu, E., Pavlov, D., Lee Giles, C.: Probabilistic User Behavior Models. In: Proceedings of the Third IEEE International Conference on Data Mining, ICDM 2003 (2003)

[8] Khalil, F., Li, J., Wang, H.: A Framework of Combining Markov Model with Association Rules for Predicting Web Page Accesses. In: Proc. Fifth Australasian Data Mining Conference, AusDM 2006 (2006)

[9] Sánchez, D., Tentori, M., Favela, J.: Hidden Markov Models for Activity Recognition in Ambient Intelligence Environments. In: Eighth Mexican International Conference on Current Trends in Computer Science. IEEE, Los Alamitos (2007)

[10] Kohonen, T.: A Self-Learning Musical Grammar, or "Associative Memory of the Second Kind". In: Proceedings of the 1989 International Joint Conference on Neural Networks (1989)

[11] Chen, Y., Yu, Y., Zhang, W., Shen, J.: Analyzing User Behavior History for Constructing User Profile. In: Proceedings of 2008 IEEE International Symposium on IT in Medicine and Education (2008)

[12] `http://en.wikipedia.org/wiki/Nokia_N800`

[13] `http://www.maemo.org`

The Knowledge Sharing Based on PLIB Ontology and XML for Collaborative Product Commerce

Jun Ma, Guofu Luo, Hao Li, and Yanqiu Xiao

Department of Mechanical-Electronic Engineering, Zhengzhou University of Light Industry, Zhengzhou 450002, China
{Majunfirst,guofuluo,haoLI,Yanqiuxiao}@sohu.com

Abstract. Collaborative Product Commerce (CPC) has become a brand-new commerce mode for manufacture. In order to promote information communication with each other more efficiently in CPC, a knowledge-sharing framework based on PLIB (ISO 13584) ontology and XML was presented, and its implementation method was studied. At first, according to the methodology of PLIB (ISO 13584), a common ontology—PLIB ontology was put forward which provide a coherent conceptual meaning within the context of CPC domain. Meanwhile, for the sake of knowledge intercommunion via internet, the PLIB ontology formalization description by EXPRESS mode was converted into XML Schema, and two mapping methods were presented: direct mapping approach and meta-levels mapping approach, while the latter was adopted. Based on above work, a parts resource knowledge-sharing framework (CPC-KSF) was put forward and realized, which has been applied in the process of automotive component manufacturing collaborative product commerce.

Keywords: Collaborative Product Commerce, XML, Knowledge sharing, PLIB, Ontology.

1 Introduction

With various customer demand and intense market competition in modern commercial environment, enterprises are compelled to communicate and share the crucial information with their customers, manufacturers, sale agents and suppliers all around the world quickly and effectively.

The idea of Collaborative Product Commerce (CPC) is to put the separated enterprise product activities (such as design, analysis, sale, market, customer service and so on) together for forming a information communication net based on internet/intranet, by which the enterprises distributing in different place could take on the different task in the product commercializing process using different application system [1]. Usually, the enterprise implementing CPC have some common traits, which are [2][3]: Its supply chain reaches to a long range comparatively. The suppliers act as an important role for its product market success, moreover, its collaborator including suppliers would be changed during the CPC. At last, the enterprise could provide customized product quickly at favorable price.

The key during CPC is the knowledge sharing based on mutual and coherent meaning. At present, most researches have pay attention to data integration inside single

W. Liu et al. (Eds.): WISM 2009, LNCS 5854, pp. 518–526, 2009.

organization, which is short of a mechanism of information reusing and processing in knowledge level [4][5].

This paper discussed what kind of role ontology, PLIB (ISO 13584) standard ontology especially, would play in knowledge sharing for CPC. Moreover, corresponding knowledge representation method based on XML was presented for network environment. At last, a knowledge-sharing system for CPC was put forward, and the knowledge-sharing process for automobile component manufacture collaborative product commerce was presented.

2 The Knowledge Sharing Method Based on PLIB Ontology and XML

2.1 PLIB Ontology and Its Methodology

Since the term ontology was borrowed from philosophy by John Mc Cathy in the 70's and introduced in the computer science vocabulary, many definitions have been offered. The most commonly cited definition is one by T. Gruber "An ontology is a formal explicit specification of a shared conceptualization" [6].

In order to share and reuse the knowledge in CPC, it is necessary that the different application systems could comprehend the knowledge in common domain without ambiguity. So, the standard and general technology term aggregation, that is so called domain ontology, should be built to provide expression specification for the definition of concept and relation in CPC. The PLIB standard (ISO 13584) published by ISO/WC3 is just one of the basal specification for building component resource standard ontology in manufacture domain.

ISO13584 is a series of International Standards for the computer-sensible representation and exchange of parts library data. The objective is to provide a mechanism capable of transferring parts library data, independent of any application system that is using a parts library data [7]. The nature of this description makes it suitable not only for the exchange of files containing parts data, but also as a common standard ontology in industrial automation domain for representing knowledge and information of product and component. At present, there have been developed many kinds of domain ontology model based on PLIB ontology such as IEC 61360-4, JEMIMA, CNIS 13584-511 and so on [8].

The figure 1 shows the so-called three levels description mechanism in PLIB ontology for defining conceptual object.

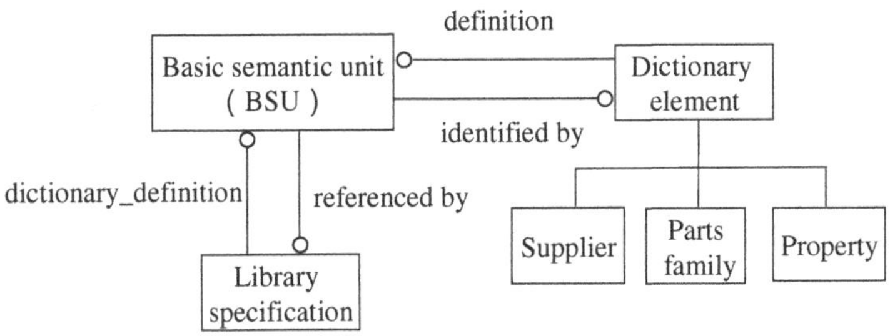

Fig. 1. The three layers description mechanism for parts resource concept in PLIB ontology

The concept involved in parts resource domain consists of a set of conceptual units including supplier, parts family and properties. PLIB ontology describes these concepts in three levels that are basic semantic unit (BSU), dictionary element and library specification [9]. The BSU identifies a concept consistently. A dictionary element provides a computer-sensible and human-readable definition for the concept. The library specification specifies the instantiated-description for these conceptual objects. e.g. geometric, schematics, procurement data etc[10].

2.2 The Transformation between EXPRESS and XML Schema for Knowledge Representation

The PLIB ontology could resolve the interoperable problem in semantics levels during the knowledge sharing. However, the knowledge modeling formalism with EXPRESS in PLIB ontology could not meet the open, readable and extensible information expression requirement for CPC in networked environment. So, it has restrained the application of PLIB ontology in certain extent. XML emerged as the most popular format for exchanging over the internet. Thus, it became desirable to allow knowledge modeling based on PLIB ontology to be exchanged using XML.

Establishing a mapping rule between XML schema and EXPRESS mode correctly is the key. There are two approaches: direct mapping and meta-levels mapping.Direct mapping consists in representing each entity instance in PLIB ontology, by the tag objects in XML document directly [11]. The meta levels mapping approach consists in redefining the PLIB ontology meta-data, such as it is defined in the EXPRESS mode, including the data types, entity declaration, constraints, schema declaration and so on, by corresponding XML Schema meta-model.

Direct mapping is a kind of explicit method that may induce information redundancy. Comparatively, The meta-levels mapping approach is easy to ensure the correctness and integrality of knowledge model. At first, a mapping rule between XML schema and EXPRESS data source should be established. Thereafter, according to this rule, XML schema would be created to achieve the data structure definition of the

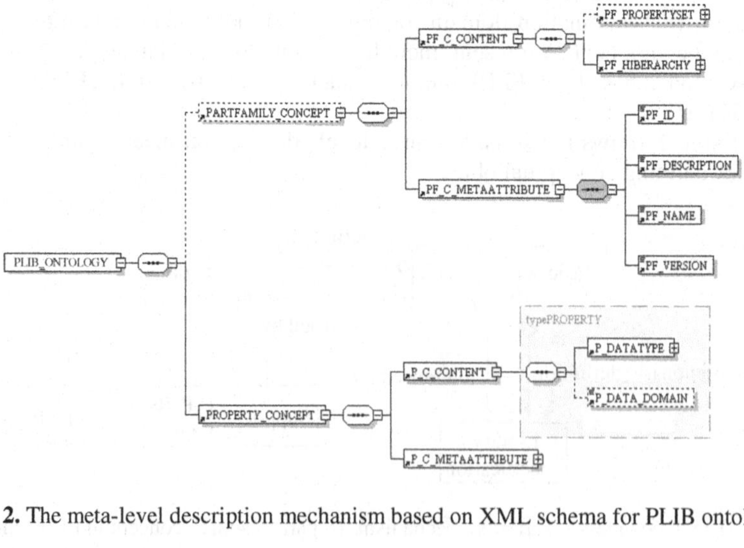

Fig. 2. The meta-level description mechanism based on XML schema for PLIB ontology

information integration. At last, XML documents could be obtained based on the corresponding schema and it would be expressed in diverse presentation form according to XSLT and transferred to application system via SOAP/HTTP protocol.

The result of conversion from the EXPRESS mode describing parts family dictionary element in PLIB ontology (as shown in figure 1) into XML schema adopting meta-levels mapping approach as shown in figure 2.

3 The Knowledge Sharing Framework for CPC

3.1 The Knowledge Sharing Framework for CPC

In order to fulfill the demands of knowledge sharing in CPC, a web-based knowledge-sharing framework (CPC-KSF) according to the methodology of PLIB (ISO 13584) ontology and XML was put forward. As illustrated in figure 3, there are four layers in the framework that are knowledge storage layer, middleware layer, business logic layer and end user layer. The detail functions of each layer as follows.

1) **Storage layer:** Storage layer provide enterprise in CPC with a platform independent way to built knowledge repository severally. These repositories were independent of one another and distributed in network. So enterprises repository could be integrated easily, which fulfill the dynamic change requirement of leaguers in CPC.

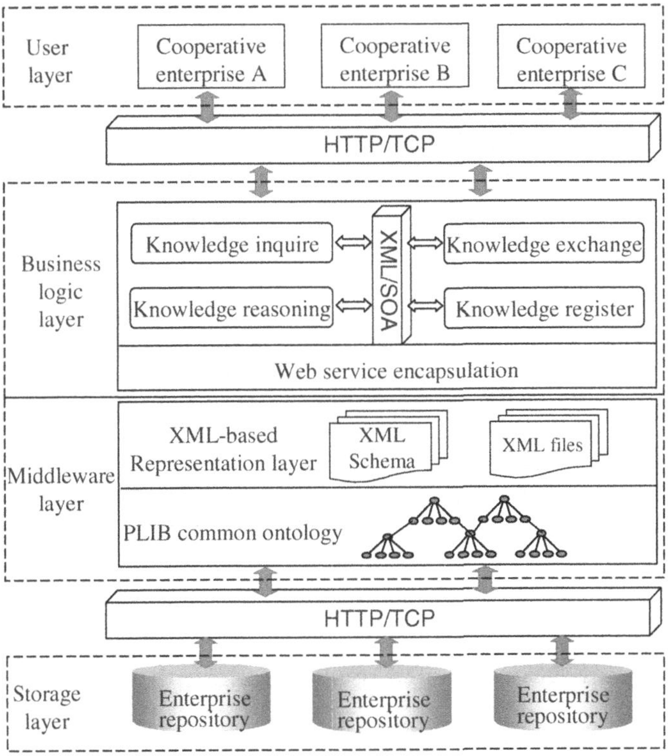

Fig. 3. Knowledge sharing framework for CPC

2) **Middleware layer:** The middleware is the key component of KSF, and it is comprised of common ontology layer and XML-based representation layer.The common ontology based on PLIB meta-model provides a shared and common glossary aggregation for the integration of heterogeneous repositories stored in knowledge storage layer. The glossary aggregations include the basic terms especially within the context of products and component resource domain in CPC, which could resolve the problem of ambiguity of terms and implement synchronizing of local knowledge. Based on PLIB common ontology in semantics, the knowledge representation layer introduces XML in syntax to represent the knowledge instance uniformly.

3) **Business logic layer:** Business logic layer, provide pivotal knowledge sharing function for end user by using encapsulating mechanism of WEB Service. For example, knowledge discovery and reasoning function could complement and renew repository content constantly. The knowledge conversion function could realize the semantic interoperation between heterogeneous knowledge resources and make it possible for knowledge sharing among CPC leaguers.

4) **Use layer:** Use layer provide CPC leaguers with a convenient and safe visiting gateway, by which the enterprises could access to the specific knowledge view in light of their roles in CPC.

3.2 The Process of Knowledge Sharing for CPC

The CPC-KSF provide strong knowledge processing ability by virtual of the support of PLIB ontology and XML for knowledge defining, representing and exchanging, by which knowledge increment could be implemented finally. The process of the knowledge sharing is illustrated in figure 4.

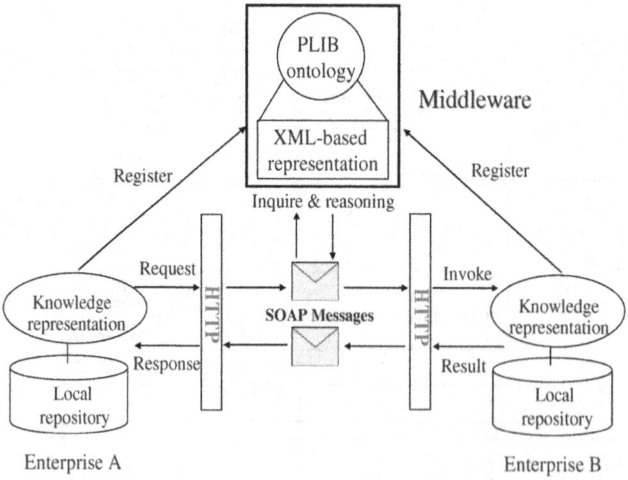

Fig. 4. Knowledge sharing process in CPC-KSF

1) At first, interior knowledge in each enterprise would be represented using certain expression specification and stored in enterprise repository respectively.

2) The common ontology for knowledge sharing would be built by third party (such as trade association or the leading manufacturer in enterprise alliance). For example, as an acceptable standard ontology within component resource application domain, PLIB could provide a common representation for component knowledge in automobile industry.

3) Enterprises register their repositories in middleware and notify CPC enterprise alliance which of knowledge and information in their repository could be shared and what purview accessing to these knowledge could be granted to other enterprise user.

4) According to the assignment in CPC, enterprise A sent a request to the middleware for inquiring information of enterprise B. As a response of the request, enterprise B sent the knowledge information described by its local term and semantics to the middleware. In the middleware, the knowledge would be redefined using standard terms and their semantics based on PLIB ontology, and then transmitted to enterprise A using XML.

5) Enterprise A could attain new knowledge by reasoning and recombining old knowledge for the purpose of knowledge increment and innovation.

4 Application in Automobile Component Manaufacture CPC

As showed in figure 5, automobile manufacture is a kind of highly collaborative and specialized industry, and an automobile usually comprised of hundreds and thousands of components. With the intense market competition gradually, the development of demand for components in automobile industry make for global purchase and modularized supplying. In order to meet this tide, the component manufacturer must work together effectively with the customers, manufacturer, sale agents and suppliers all around the world under a uniform knowledge sharing framework for CPC.

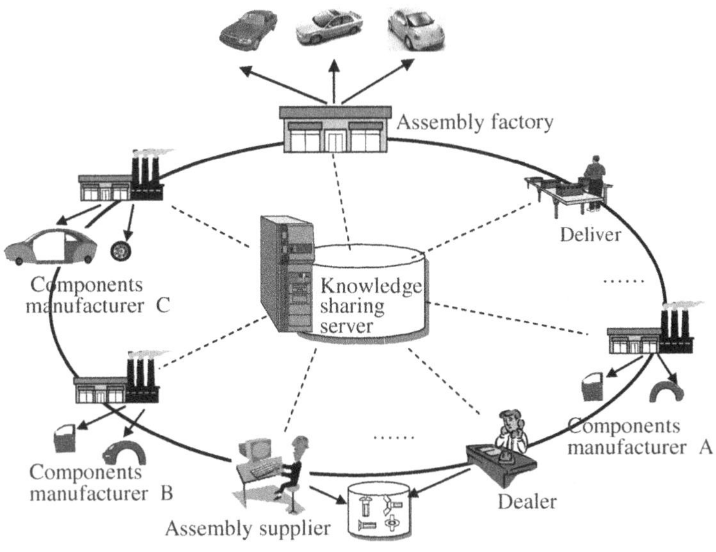

Fig. 5. CPC pattern for automobile components manufacture

Figure 6 illustrate the master application interface of CPC-KSF in the automobile component enterprise, by which the automobile component manufacture, as developer of the local repository, release knowledge information about its product and other participator of CPC could share these knowledge.

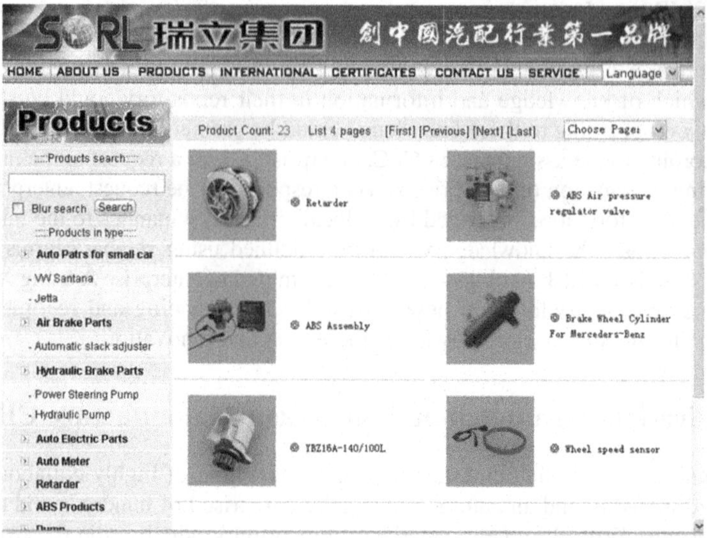

Fig. 6. The application of CPC-KSF in automobile component enterprise

As shown in figure 7, EXPRESS formalization description for PLIB ontology is provided and then converted into XML based on meta-level mapping mechanism.

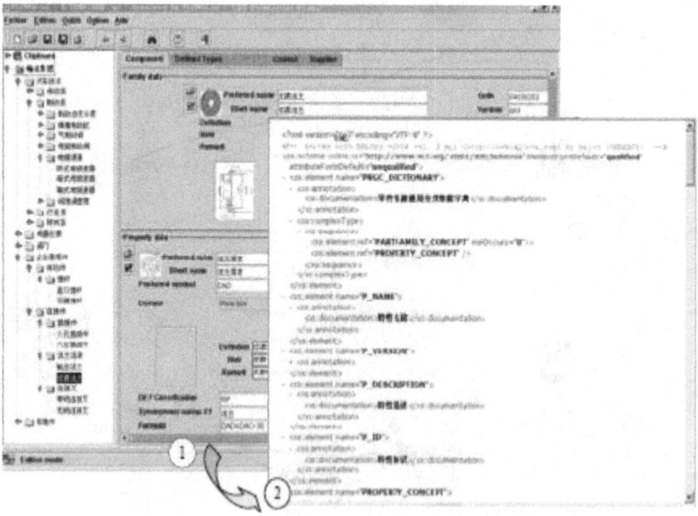

Fig. 7. Knowledge conversion interface of CPC-KSF

As shown in figure 8, with the help of tools for product collaborative design integrated in CPC-KSF, designer could share the parts resource and communicate with each other.

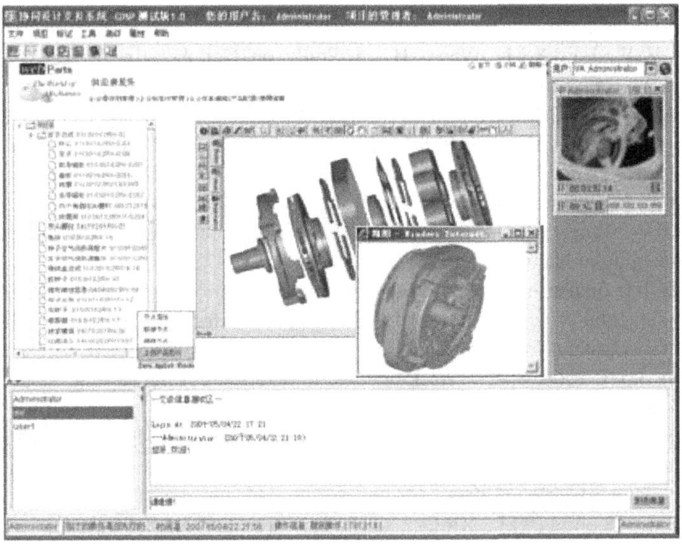

Fig. 8. Sharing the parts resource with the help tools for collaborative design integrated in CPC-KSF

5 Conclusion

In order to realize the information share in knowledge level in CPC, this paper provide a knowledge-sharing framework for CPC based on PLIB common ontology and XML, and the method of implementing knowledge sharing in this framework was studied. By the practical application case in automobile components manufacture CPC, it has been proved that the CPC-KSF could support the inter-organizational knowledge communication and accelerates the knowledge innovation effectively. At the same time, with the development of the CPC theory, more function and technology in the CPC-KSF would be perfected.

Acknowledgement

The research work is obtained financial support from the"Innovation Scientists and Technicians Troop Construction Projects of Henan Province (NO. 084100510047)"and "Doctor Scientific Research Fund of Zhengzhou University of Light Industry".

References

1. Aberdeen, G.: Collaborative Product Commerce: Delivering Product Innovation at Internet Speed. Aberdeen Group, Market Viewpoint 12, 1–9 (1999)
2. Tong, B.S., Tian, L.: Agile Design Supporting System for Collaborative Product Commerce. Journal of Computer-Aided Design & Computer Graphics 15, 800–804 (2003)
3. Zhang, W.F., Lai, X.M., Wang, H.: A Study on Collaborative Product Commerce in Automotive Enterprise. Automotive Engineering 26, 246–249 (2004)
4. Sun, G.Z., Jiang, C.Y., Wang, N.S.: Virtual Product Data Management in Collaborative Product Commerce. Mechanical Science and Technology 22, 134–137 (2003)
5. Zhang, X.F., Deng, J.T.: Data and Data Management in Collaborative Enterprises Under CPC. Aeronautical Manufacturing Technology 3, 79–82 (2004)
6. Deng, Z.H., Tang, S.W., Zhang, M.Y.: Overview of Ontology. Universitatis Pekinensis, Acta Scientiamm Naturalium 38, 730–737 (2002)
7. ISO13584-1: Industrial automation systems and integration-Parts Library-Part 1: Overview and fundamental principles. International Organization for Standardization (1999)
8. Pierra, G.: Context-Explication in Conceptual ontologies: The PLIB Approach. In: Proceedings of CE 2003, Special track on Data Integration in Engineering, Madeira, Portugal, pp. 253–254 (2003)
9. ISO13584-42: Industrial automation systems and integration-Parts Library-Part 42: Description methodology: Methodology for structuring parts families. International Organization for Standardization (1998)
10. Jun, M.A., Guoning, Q.I., Xinjian, G.U.: Research and Implementation of Product Resource Common Model Based on PLIB Ontology. Computer Integrated Manufacturing Systems 13, 631–637 (2007)
11. Xie, L.Y., Hao, Y.: Application and Expression of Product Information Model Based on XML. Computer Integrated Manufacturing Systems 4, 263–268 (2002)

Reinforcement Learning Based Web Service Compositions for Mobile Business

Juan Zhou and Shouming Chen

School of Economics and Management, Tongji University, Shanghai, 200092, China
wendyzhou7@hotmail.com, schen@tongji.edu.cn

Abstract. In this paper, we propose a new solution to Reactive Web Service Composition, via molding with Reinforcement Learning, and introducing modified (alterable) QoS variables into the model as elements in the Markov Decision Process tuple. Moreover, we give an example of Reactive-WSC-based mobile banking, to demonstrate the intrinsic capability of the solution in question of obtaining the optimized service composition, characterized by (alterable) target QoS variable sets with optimized values. Consequently, we come to the conclusion that the solution has decent potentials in boosting customer experiences and qualities of services in Web Services, and those in applications in the whole electronic commerce and business sector.

1 Introduction

In the E-Commerce sector nowadays, Web Service technologies characterized by their being obturate, loosely coupled, and highly integrated, are much preferred. However, single Web Services have inevitable limitations in functionality, thus hard to fulfill requirements raised by realistic applications. Therefore, Web Service Composition (WSC) is introduced to further dig potentials of shared Web Services via their combination and uniting. WSC triumphs in that it functions in such a manner that makes use of small, simple and easily-executable services to construct various user-specific complex services. And as an increasingly important subdivision of E-Commerce, mobile business also sees a prosperous perspective in applying WSC, especially as Wi-Fi and WAP based wireless data (rather than conventional voice) services boost.

Reinforcement Learning (RL) is a form of machine learning for interaction problems [1] [2] [3]: An agent employs a trial-and-feedback strategy to explore the environment, receives (positive or negative) rewards for certain actions under certain environment states (collaterally changes the environment conditions), and modifies its action policy accordingly. Such an action sequence with feedbacks eventually leads to an optimized decision making policy that helps acquire the highest reward, and the route towards it. Thus, RL algorithms can be effectively applied in WSC to realize a true dynamic methodology.

In this paper, we shall first briefly cast a summary on reactive WSC and RL respectively, and then look into how RL could be used in WSC to raise its dynamic capability, finally taking QoS into consideration in the RL-Based WSC model to make it realistically useful in the real mobile business.

W. Liu et al. (Eds.): WISM 2009, LNCS 5854, pp. 527–534, 2009.

2 Web Service Composition

Literally, Web Service Composition refers to the technique of composing arbitrarily complex services from relatively simpler services available over the Internet [4]. WSC technologies are usually classified in generation methodologies of the composition solutions. Namely, proactive WSC means that the composition procedure is carried out during the design or installation phase, well before even the very first user makes a service request to the system. This composition service may integrate heterogeneous services into a single one on a single platform, or communicates with the various services on distinct hosts. Such a composition technique provides high stability yet merely none flexibility, thus suitable for composite services with invariable data and control flow forms such as e-ticketing.

On the contrary, reactive WSC has its composition procedure executed on-the-fly (simultaneously with the actual services), thus, of course, bringing along more flexibility. In other words, service compositions are only created at customers' requests. This is particularly useful in situations where service requests are occasional and component services are unstable also, and is therefore the ultimate trend of WSC evolutions.

A typical realization of a reactive WSC system consists of 2 user roles (a service requester, and a service provider) and 5 parts (a Translator, a Composition Manager, an Executive Engine, a Service Matcher, and a Service Library), and can be depicted as Fig. 1.

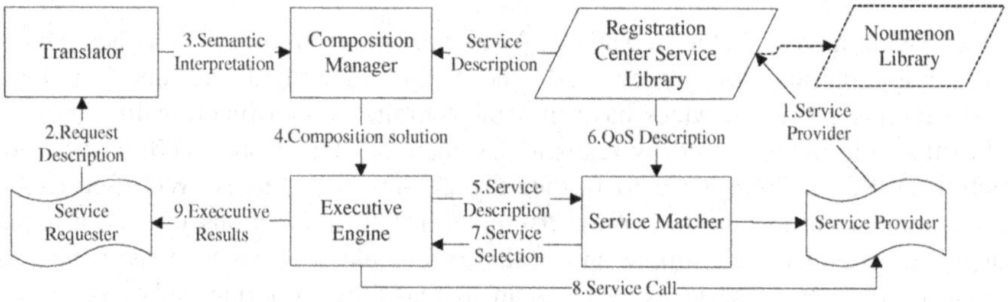

Fig. 1. A typical realization of a reactive WSC system

And procedures of WSC can be described as follows:

The service provider releases service information to Registration Center Service Library.

Steps 2 and 3: A service request raised by a requester is translated from natural language into meaningful specification description by the Translator, and transmitted to the Composition Manager.

Step 4: The Composition Manager generates a composition solution that meets the requests according to the specification description and service descriptions from service libraries, and transmits it to the Executive Engine.

Steps 5 to 7: The Executive Engine transmits the composition solution to the Service Matcher. The Service Matcher selects the most suitable services accordingly, and returns their handles to the Executive Engine.

Step 8: The Executive Engine calls and supervises executions of the Web Services according to the composition solution and the Web Service handles.

Step 9: The system returns the execution result to the requester.

3 Reinforcement-Learning-Based WSC

LI et al. [5] and TANG et al. [6] have reported applications of Reinforcement Learning in WSC respectively. However, Rewards and Transition Probabilities in the RL model were not made full use of for simplicity of modeling. And Quality of Service (QoS), for which considerations of service requesters and providers are growing rapidly currently, was also excluded. These have significantly limited applications of the models. Thus, we hereby propose a new approach to model WSC with RL, which takes QoS variables as a decisive factor of the composition solution.

3.1 Modeling Basics

The RL-Based WSC model is set up as follows.

Web Services can be abstracted as externally accessible functions with interfaces with others. Let S be the set of states (services available at the Service Library), i.e. the environment, and $S_{1..N} \in S$ be states (services) within the former; $A_{1..M} \in A$ be transitions from one service to another (thus the two are decently interfaced); as Fig. 2 presents.

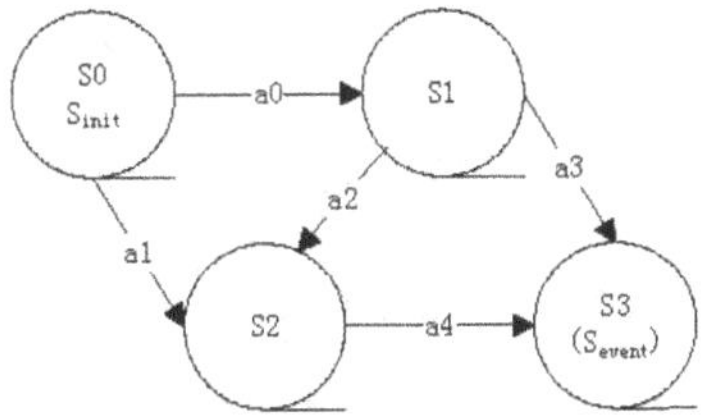

Fig. 2. Abstractions of web services and intermediate transitions

Here, state transitions are quantified with 2-tuples <R, T>, with the same definitions as those in the RL Model. Note that states without intermediate transitions mean services without interfaces. And in the transition probability matrix, the correspondent element is simply 0. In such a situation, rewards become meaningless. Hence, only states (services) with available transitions (interfaces) are considered in this model. Relative concepts with RL are explained in the coming chapter.

Thus, Web Service Compositions are denoted as paths among states (services) via actions (transitions among services). Suppose an initial state S_{init} and an eventual state S_{event} are given by the Translator as it generates the specification description. Consequently, WSC is transferred into search for paths from S_{init} to S_{event} with an "optimal value", which we shall explain soon in the next chapter. And Reinforcement Learning resultantly provides a sound approach to its solution.

3.2 Reinforcement Learning and Markov Decision Process

Reinforcement learning (RL) is a broad class of optimal control methods based on estimating value functions from experience, stimulation, or search [7]. RL has been initially used in artificial intelligence domain, yet also has a sound application in various domains including but not limited to robotics, cell-phone network routing, efficient web-page indexing and stochastic planning problems that are important for industries and governments.

This system reinforces the agent to choose the best actions through trial and error experience under specific rules, environments, and sometimes uncertain information. Markov decision processes provide the basic formalism of RL. A task is defined by four basic factors, namely, a set of states: $S, s \in S$, a set of actions: $A, a \in A$, state transition probabilities p (for each s and a, p is a distribution over the state space S), and a reward function of the agent $R: S \times A \rightarrow R$. We also define $\gamma \in [0, 1]$ as the discount factor, which devalues rewards received in the future.

Accordingly, can define the value of the state as followings:

$$R(S_0) + \gamma R(S_1) + \gamma^2 R(S_2) + \cdots \text{ or } \sum_{t=0}^{\infty} \gamma^t R^t \tag{1}$$

where R^t is the reward of time t.

The goal of RL is to learn a policy $\pi: S \rightarrow A$ mapping from the states to the actions, through which we take action series $a = \pi(s)$ to hop among states to maximize the expected sum of discounted reward.

$$V(s, \pi) = \sum_{t=0}^{\infty} \gamma^t E(R_t | \pi, s_0 = s) \tag{2}$$

$V(s, \pi)$ is the expected sum of discounted rewards starting in state s_0 and taking actions under policy π.

Given fixed π, the value function $V(s, \pi)$ satisfies the Bellman equations:

$$V(s, \pi) = R(s, a_\pi) + \gamma \sum_{s' \in S} p(s'|s, a) V(s', \pi) \tag{3}$$

a_π is the action under policy π. There is an optimal policy π^*, that can help reach the optimal value $V(s, \pi^*)$ for the state set S.

$$V(s, \pi^*) = \max V(s, \pi) \tag{4}$$

Substituting $V(s, \pi)$ with (3) yields:

$$V(s, \pi^*) = \max \left\{ R(s, a) + \gamma \sum_{s' \in S} p(s'|s, a) V(s', \pi^*) \right\} \tag{5}$$

We also define the optimal policy $\pi^*: S \rightarrow A$ as:

$$\pi^*(s) = \arg\max \sum_{s' \in S} p(s'|s, a) V(s', \pi^*) \tag{6}$$

Thus, we can find that π^* as defined above is the optimal policy for all $s \in S$.

4 QoS Considerations for Mobile Business

As stated above, it is of high necessity that QoS be considered decisive factor of WSC. In RL, rewards, transition probabilities, and discounts are 3 variable sets deciding the value function, thus the strategy solution for given environment states and transition availabilities (whether transition probabilities are 0. See Fig. 2).

As a result, modeling QoS variables as rewards, transition probabilities, and discounts in RL enables significant rise in service efficiencies and customer experiences of WSC. For mobile businesses, sensitivities of QoS variables differ considerably for different business models, access methods, network types and topologies, and many other factors. And such QoS-sensitive modeling has the further advantage of flexibility allowing rewards, transition probabilities, and discounts to be changed along with RL-based WSC application situations.

4.1 AgFlow's 5-Dimensional QoS Model and Modifications

ZENG et al. [8] proposed Quality Criteria for Composite Services, as Table 1 shows.

Table 1. Aggregation function for computing the QoS of execution plans

Criteria	Aggregation function
Price	$Q_{pr}(p) = \sum_{i=1}^{N} Q_{pr}(S_i, op(t_i))$
Duration	$Q_{du}(p) = CPA(p, Q_{du})$
Reputation	$Q_{rep}(p) = \frac{1}{N}\sum_{i=1}^{N} Q_{rep}(S_i)$
Success rate	$Q_{rat}(p) = \prod_{i=1}^{N} (Q_{rat}(S_i^{Z_i}))$
Availability	$Q_{av}(p) = \prod_{i=1}^{N} (Q_{av}(S_i^{Z_i}))$

AgFlow's model contains five dimensions as above, which are used to evaluate the quality of web service. The quality vector of composite service's execution plan is:

$$Q(p) = (Q_{pr}(p), Q_{du}(p), Q_{rep}(p), Q_{rat}(p), Q_{av}(p)) \tag{7}$$

Here we give a brief introduction for it:

1. P is the plan.P = {$<t_1, s_1>$,$<t_2, s_2>$...,$< t_n, s_n>$ }, i.e. a set of 2-tuples, where the tuple is denoted as <t, s>. t is the time and s is the service.
2. Execution Price: $Q_{pr}(p)$ is a sum of prices of invoked operations in plan P, and $op(t_i)$ denotes task t_i.
3. Execution Duration: $Q_{du}(p)$ is calculated by using CPA (critical path algorithm). The critical path of a project diagraph is a path from the initial state to the final state which has the longest total weights labeling its nodes.

4. Reputation: $Q_{rep}(p)$ is the average of reputations of services ($Q_{rep}(s_i)$) in plan P.
5. Successful execution rate: we assume the re-execution of failed task is allowed, therefore $Q_{rat}(p)$ is the product of $Q_{rat}(S_i^{Z_i})$. Z_i is equal to 0 ($Q_{rat}(S_i^{Z_i})=1$) if S_i is not on the critical path, otherwise Z_i is equal to 1.
6. Availability: $Q_{av}(p)$ is the product of $Q_{av}(S_i^{Z_i})$, where $Z_i=0$ if S_i is not on the critical path and $Z_i=1$ if S_i is a critical service. Also, the service can be reselected when noncritical service is unavailable. So the whole $Q_{av}(p)$ will not be affected.

The model stated above is meant to evaluate QoS of composite services whose topology is an engineering network with multiple paths from S_{init} to S_{event}, thus critical paths are introduced to evaluate the overall execution of the service. However, here in RL-based WSC, the task itself is to find a single optimal path from S_{init} to S_{event}, excluding concerns for critical paths. Thus some modifications have to be made. Rewrite the aggregation functions for duration, success rate and availability as follows.

Table 2. Modified aggregation functions

Criteria	Aggregation function
Duration	$Q_{du}(p) = \sum_{i=1}^{N}(Q_{du}(S_i))$
Success Rate	$Q_{rat}(p) = \prod_{i=1}^{N}(Q_{rat}(S_i))$
Availability	$Q_{av}(p) = \prod_{i=1}^{N}(Q_{av}(S_i))$

4.2 High-Reliability B2C Service

Mobile web banking (mostly via cell phones) may be the most popular mobile business type nowadays. A mobile subscriber visits the web banking host over WAP or public WLAN accesses, which host provides various basic transaction services (or component services) such as inquiries and transfers. And on a lot of occasions like B2C trading, service compositions are asked for. Such applications of WSC require high reliability (for safety's concern) and low web communication flux (seen as cost in the QoS model above) between the host and the subscriber (for wireless data services are not usually cheap). Service duration is also usually taken into consideration by subscribers, yet not so sensitive to. Therefore, in solving such WSC problems, modeling is carried out as followings.

Suppose a web banking host has N basic services (or component services). Each service has a reliability (Rel_i) gained through theoretical calculation or statistics of testing. Thus, let:

$$Q_{rat}(S_i) = Rel_i, 1 \le i \le N \tag{8}$$

Each service also has a data communication flux with subscriber (Flu_i), and at interfaces of each two services, another communication flux ($iFlu_{i,j}$) is recorded. Thus, define:

$$Q_{pr}(S_i, op(t_i)) = Flu_i + iFlu_{i,j}, 1 \le i, j \le N \tag{9}$$

where $op(t_i)$ can be interpreted as the transition from S_i to S_j (or the state transition $A_k, 1 \leq k \leq M$ in the RL-Based WSC model above).

And for variables in the RL-Based WSC model, we let:

$$R(S_i, A_k) = \frac{Rel_i}{Flu_i + iFlu_{i,j}}, 1 \leq i, j \leq N, 1 \leq k \leq M \tag{10}$$

where $A_k, 1 \leq k \leq M$ is the transition from S_i to S_j.

Transition probabilities $p(s'|s, a)$, as the RL model defines, are relatively difficult to define analytically, yet can be acquired according to statistics of service composition forms from previous static WSC solutions. Or taking further factors into consideration, one can re-define the variable set. Yet we exclude such discussions due to limit of length of the paper.

The discount factor γ for this model can be used to adjust typical length of the composition solution. A compromise may be needed in decision of it, as a large discount factor near 1 discourages efficiency, and a small one near zero may cause the solution not achievable.

With these definitions and assumptions, the RL-Based WSC model can be such worked out (with various proposed algorithms, including but not limited to Sutton's TD(λ) [9], Watkins' Q-Learning [10], and Real-Time Dynamic Programming [11] [12]) that as the maximum value of the value function and the optimal strategy achieved, theoretical minimum price and maximum reliability are also realized.

5 Conclusion

QoS-Sensitive Reinforcement-Learning-Based Web Service Composition can be very useful in the mobile business sector which experiences fast changes in business models. Various service providers can apply this tool to enable reactive WSC at the same time of optimizing QoS. However, it is note-worthy that its utility is not limited to mobile business. Any online service concerning WSC can make decent use of it. Proposed further work of this topic include development and testing of a prototype system, and more importantly, more in-depth analysis of integration of this and real mobile business.

References

1. Bertsekas, D.P., Tsitsiklis, J.N.: Neuro-Dynamic Programming. Athena Scientific, Belmont (September 1996)
2. Kaelbling, L.P., Littman, M.L., Moore, A.W.: Reinforcement learning: A Survey. Journal of Artificial Intelligence Research 237, 285 (1996)
3. Barto, A.G.: Reinforcement learning in the real world. In: Proceedings, 2004 IEEE International Joint Conference on Neural Networks, Budapest, Hungary (2004)
4. Chakraborty, D., Joshi, A.: Dynamic service composition: State-of-the-art and research directions, Report No.: TR-CS-01-19, Department of Computer Science and Electrical Engineering, University of Maryland, Baltimore County, MD, USA (2001)
5. Li, C., Ding-yuan, N., Dong, L.: Applications of Markov Decision Process in Web Service Composition. Journal of Higher Correspondence Education (Natural Sciences) 20(2) (April 2007)

6. Tang, P.-p., Wang, H.-b.: Web Service Composition Based on Reinforcement – Learning. Computer Technology and Development 18(3) (March 2008)
7. Barto, A.G., Bradtke, S.J., Singh, S.P.: Real-Time Learning and Control Using Asynchronous Dynamic Programming, Technical report (TR-91-57), Amherst, Massachusetts (1991)
8. Liangzhao, Z., Benatallah, B., Ngu, A., et al.: QoS-Aware Middleware for Web Services Composition. IEEE Trans. on Software Engineering 30(5), 311–327 (2004)
9. Sutton, R.S.: Learning to Predict by the Method of Temporal Differences. Machine Learning 3, 9–44 (1988)
10. Watkins, C.J.C.H.: Learning from Delayed Rewards, Ph.D. Thesis, Cambridge University, Cambridge, England (1989)
11. Barto, A.G., Bradtke, S.J., Singh, S.P.: Learning to Act Using Real-Time Dynamic Programming. Artificial Intelligence 72(1-2) (January 1995)
12. Bradtke, S.J.: Incremental Dynamic Programming for On-Line Adaptive Optimal Control, Ph.D. Thesis, University of Massachusetts (1994)

Will Commodity Properties Affect Seller's Creditworthy: Evidence in C2C E-commerce Market in China

Hui Peng and Min Ling

School of Economics & Management
Beijing University of Posts and Telecommunications
Beijing 100876, P.R. China
penghuigrace@126.com, lingmina@gmail.com

Abstract. This paper finds out that the credit rating level shows significant difference among different sub-commodity markets in E-commerce, which provides room for sellers to get higher credit rating by entering businesses with higher average credit level before fraud. In order to study the influence of commodity properties on credit rating, this paper analyzes how commodity properties affect average crediting rating through the degree of information asymmetry, returns and costs of fraud, credibility perception and fraud tolerance. Empirical study shows that Delivery, average trading volume, average price and complaint possibility have decisive impacts on credit performance; brand market share, the degree of standardization and the degree of imitation also have a relatively less significant effect on credit rating. Finally, this paper suggests that important commodity properties should be introduced to modify reputation system, for preventing credit rating arbitrage behavior where sellers move into low-rating commodity after being assigned high credit rating.

Keywords: E-commerce; credit rating; commodity property; trustworthy; credit distribution; credit cluster; rating arbitrage.

1 Introduction

Credit environment is the foundation of development of C2C E-commerce. Many scholars study on the credit-related factors of C2C E-commerce. Friedman, E., and Resnick(2001) analyzed the seller behavior in view of revenue and cost of fraud. Bernard Conein, Richard Arena(2008) described the game process in credit in open e-business community. Wang YF, Hori Y, and Sakurai K (2008) discussed characterizing properties of trust and reputation in peer to peer environment.

Some scholars began to research on the credit factors in specified commodity market. Wei'an Li, Desheng Wu and Hao Xu (2008) examined the relationship of reputation and trading volume in online game prepaid card C2C market. But it only analyzed the prepaid card sub- market. This paper tries to explore whether and how the properties of commodities have significant impact on credit rating in some C2C E-commerce sub-market, not only focus on one sub-market.

W. Liu et al. (Eds.): WISM 2009, LNCS 5854, pp. 535–544, 2009.

Is credit reputation really different among C2C E-commerce commodity sub-market? From the Descriptive Statistics of Average Credit Rating ,We found that online game good has the highest average credit rating of 3.72, while home furnish has the lowest of 0.91. The range of between the highest and the lowest level goes to 2.81. The standard deviation of average credit rating is 0.68. This shows that the average credit rating of commodities is indeed significantly different.

Table 1. Descriptive Statistics of Average Credit Rating

Range	Minimum	Maximum	Mean	Std. Deviation	Variance
2.81	0.91	3.72	2.5394	0.68138	0.464

In the view of different types of commodities how the commodities properties (price, volume, the extent of imitation, etc.) affect the behavior and credit rating of sellers in e-commerce market can be explored in depth.

2 Whether Credit Is Dependent on Commodity Properties: Theory Discussion

2.1 Literature Review

Whether the credit behavior is dependent on the internal personal factors or on external environment factors is controversial. The psychologists consider that both the internal personal factors and external environment factors affect the human behavior, while no consensus has been reached on which factor, internal factors or external factors, is more important in credit behavior. There are mainly two arguments concerning the determinants of creditworthy:

(1) Both Internal Factors and External factors have Same Impact on Behaviors
Heider F. (1958) wrote in his book "The psychology of interpersonal relations" that factors of behavior includes internal factors and external factors. Internal factors are personality, motivation, emotion, attitude, ability, mood, effort and other personal factors, and external factors are incentives, luck, job difficulty and other environmental factors. Behavior is the result of both. From this point of view, credit behavior is determined by the human nature and external environment. External environment has an impact on credit behavior.

(2) External factors are more important to individual behavior
Weiner's findings (1986) places external factors in a more important position and illustrates that the individual behaviors may not reflect the nature of personality, attitude, etc. He considers that when individuals obey the social environment, individual behavior may not meet internal personality. Even many scholars have shown that the C2C e-commerce credit behavior is closely related to the extent of information asymmetry (Jagdip Singl and Deepak Sirdeshmukh, 2006), the return and cost of fraud(Friedman and Resnick, 2001), price (Lee, Z., Im, I., and Lee, SJ, 2000), consumer trust (Javenpaa.SL and N. Tractinsky, 1999), trading volume (Wei'an Li, Desheng Wu and Hao Xu, 2008).

(3) Related research in commodity properties and credit rating
Melnik, M. I. and J. Alm(2002) observed that different reputation will affect the buyers' behavior in E-bay auction. Paul A. Pavlou, Angelika Dimoka(2006) thought that the feedback text comments about product attribute can influence buyer's decision. Dewan, S. and V. Hsu(2004) found out the buyer's adverse selection in online stamp auctions. Wei'an Li, Desheng Wu and Hao Xu (2008) find out that the positive relationship of reputation and trading volume in online game prepaid card C2C market in China.

Based on previous research, this paper tries to explore whether and how the properties of commodities have significant impact on credit rating in some C2C E-commerce sub-market, not only focus on one sub-market. And it also want to examine both buyer's and seller's adverse selection in E-commerce.

2.2 Commodity Properties and Credit Rating: Propositions

In C2C E-commerce reputation system, the average credit rating of a commodity market reflects the overall credit behaviors of sellers in the market. In different commodities' markets, the different degree of credit rating does exist. The aspects in which the commodity properties affect the credit rating are as follows:

(1) Commodity properties determine the degree of information asymmetry.
Different commodities do not have same degree of information asymmetry. Variance in the degree of information asymmetry makes the sellers different in the probability of fraud punishment. It results in different fraud possibility of sellers in different commodities' markets.

(2) Commodity properties determine the return and costs of fraud.
Commodities differ in the return and cost of fraud. For example, expensive goods (high return from fraud) with little possibility of consumer complaints (low cost from fraud) are more prone to fraud.

(3) Commodity properties determine credibility perception and fraud tolerance.
Credibility perception is the extent of the satisfaction the buyers get from the products and seller's behavior. Fraud tolerance is the possibility of no complaint when fraud occurs. For different merchandise, the consumer's criteria for integrity are not the same. So for same credit behavior in different commodities' markets, the same buyer's satisfaction will be different. Even if the buyer has same satisfaction, fraud tolerance may differ. For example, the buyer will have higher possibility of complaint of an expensive product than that of a cheap one.

3 Methodology and Data Preparation

To study the impact of different commodity properties on average credit rating (reflection of credit behavior in reputation system) and verify the assumptions in section II, this paper selects 8 indicators closely related to commodity property.

Table 2. Credit Rating and Commodities Properties

Commodity Classification	Average Credit rating	Brand rate	Delivery	Average Price	Average trading Volume	Standardization degree	Imitation degree	Complaint possibility
Cell phone recharge service	3.71	100	1	96	669	5	2.55	3.7
Online game goods	3.72	100	1	114	535	4.35	3.25	2.7
QQ goods	2.62	100	1	26	144	4.05	3.25	2.7
Internet phone service	2.55	96	1	288	192	4.35	2.95	3.75
Cell phone	2.51	74	2	1348	381	3.1	3.7	4.1
Camera	2.7	95	2	1149	215	3.75	2.7	4.7
Office equipment	2.18	95	2	1175	222	1.65	4.7	2.7
Flash memory card	2	91	2	106	157	2.95	4.4	2.7
Lady shoes	2.83	22	2	236	151	1.35	4.35	2
Lady clothes	2.27	6	2	228	201	1	4.7	2
Watch	2.76	38	2	1244	220	1.65	4.7	2.7
Accessory	1.81	13	2	180	259	1.3	4.7	1.65
Men's wear	3.12	46	2	263	174	1.35	4.7	2
Men's shoes	2.41	44	2	259	113	1.35	4.7	2.35
Health care	2.53	41	2	217	34	1	4.4	1.65
Home furnishing	0.91	0.46	2	1074	243	1.35	4.4	1.65

(1) The brand rate

The brand rate is equal to the number of well-known brand products divided by the number of all products in this category. Well-known brands usually are under strict management. Buyer can check the imitation through the inspection hotline of company who owns the brand. The companies also have high requirements on credibility of distribution sellers and proxy sellers. Thus the brand rate can reflect the degree of the information asymmetry.

(2) Delivery

In C2C E-commerce market, there are 2 kinds of deliveries: virtual delivery and physical delivery. Virtual delivery refers to delivering the product on-line, usually the virtually-delivered products are virtual products, such as mobile phone recharge service. Physical delivery refers to delivering by express or by post.

(3) Average price
Average price is calculated on the basis of all the products in this submarket. This indicator relates to the return from fraud. The higher the price is, the greater profits the seller may get from cheating.

(4) Average trading volume
The average trading volume is the total number of history transaction in this submarket divided by the number of sellers, which refers to the average trading volume per seller. The low average trading volume means that the consumer has low credibility perception in the products markets.

(5) The degree of standardization
The products with higher degree of standardization will be told from imitation by buyers, which means has less degree of information asymmetry.

(6) The degree of imitation
The products can be imitated more easily, the sellers has less cost of fraud.

(7) Complaint possibility
The indicator refers to the possibility of complaints to Taobao.com when buyers face frauds. It relates to credibility perception and fraud tolerance. Low complaint possibility may imply that buyer could not identify the fraud even if it happens, or may tolerate with the fraud even if being cheated or dishonestly treated.

We will validate the relationships between average credit rating and these indicators. Indicator A-D is collected from Taobao.com on Jan. 12, 2008. Indicator E-G is collected from 31 buyers in Taobao.com. In delivery, '1' represents virtual delivery and '2' represents physical delivery. In degree of standardization, the degree of imitation and complaint possibility, the value ranges from 1 to 5, 1 represents the lowest degree of imitation or complaint possibility (see Table 2)

4 Commodity Property and Credit Rating: Empirical Analysis

4.1 Correlation Analysis

Pearson correlation analysis is conducted in SPSS.

Table 3. Correlations with Average Credit rating

	Brand rate	Trading Volume	Standardization	Imitation degree	Integrated factor of average Price, delivery and complaint possibility
Pearson Correlation	0.516 (0.04)	0.890 (0.00)	0.53 (0.04)	-0.59 (0.02)	-0.613 (0.01)

Through the correlation analysis (Table 3), we can know that the brand rate, the degree of standardization, the degree of imitation and average trading volume have significant linear correlation with average credit rating.

An Integrated factor is constructed (D represents delivery, P represents average price, and Cp represents complaint possibility):

$$F=D\times \ln P/ \ln Cp \quad (1)$$

The new indicator is a comprehensive reflection of delivery, average price and complaint possibility. The Pearson correlation coefficient of new variable and average credit rating is -0.613 and significance test value is less than 0.05. It indicates that the new variable and the average credit rating have a significant linear correlation.

4.2 The Regression Model

Through correlation analysis, the brand rate, the degree of standardization, the degree of imitation, and integrated factor of delivery, average price and average trading volume are selected to do recession analysis with average credit rating in Stepwise method in SPSS.

Table 4. R^2, F test and T test of Regression

Model	R Square	Sig. of F test	T test		
			Constant	average trading volume	integrated factor of average Price, delivery and complaint possibility
a[1]	0.792	0.000	1.612 (0.00)	0.04 (0.00)	
b[2]	0.848	0.000	2.08 (0.00)	0.003 (0.00)	-0.02 (0.048)

We make Cr represents average credit rating, Tv represents average trading volume, D is delivery, P is average price and Cp is complaint possibility. Regression equation based on model 2 is

$$Cr = 2.08+0.003\times Tv-0.02(D\times \ln P / \ln Cp) \quad (2)$$

4.3 Mechanism of Commodity Properties Affecting Credit Rating

Through data analysis above, we found that delivery, average trading volume, average price, and complaint possibility have important impact on the average credit; the brand rate, the degree of standardization and the degree of imitation also have relative impact on the average credit.

(1) The effect of average trading volume
Regression model shows that average trading volume has positive impact on average credit rating in commodity transactions. If the average trading volume increases one unit, credit rating will increase by 0.003. Average trading volume, the total number of transactions divided by seller quantity, reflects the consumer's credibility perception.

[1] Model a. Predictors: (Constant), average trading volume.

[2] Model b. Predictors: (Constant), average trading volume, integrated factor of average Price, delivery and complaint possibility.

Buyers will buy more products if they feel satisfied and trust the sellers in this commodity in C2C e-commerce market.

(2) The effect of delivery
Credit rating of virtual delivery merchandise is higher than that of physical delivery one, the reason, as regression shows, is that the quality of virtual delivery merchandise is easily checked but physical delivery merchandise needs a period of time to be tested. In Taobao.com, buyers need to give comments as soon as they receive the merchandise. Buyers need a period of time to tell whether the physically delivered merchandise is good or not. Before buyers can tell, they have already given their comments.

(3) The effect of average price
In the regression model, average price and average credit rating in commodity classification has a negative correlation. However, the negative effect is diminishing in margin. The high price means that the return of fraud is high. So the motivation of expensive merchandise is strong.

(4) The effect of complaint possibility
In the regression model, complaint possibility has a positive effect on average credit rating in commodity classification. However, the positive effect is diminishing in margin. Buyer complaints will force the sellers to integrity, that is, consumers with lower fraud tolerance will force sellers to integrity.

(5) The integrated effect of average Price and complaint possibility
At the same level of average trading volume, the possible combinations of average price and complaint possibility will affect the cost-return from fraud (shown in the table 5).

Table 5. Average credit rating and combinations of average price and complaint possibility

Combination		Effect	Example
Price	Complaint possibility		
high	high	So the possibility of fraud is less and negative impact on average credit rating is smaller	camera
high	low	the possibility of fraud is more and negative impact on average credit rating is greater	furnishing
low	high	the possibility of fraud is least and negative impact on average credit rating is smallest	cell phone recharge service
low	low	the possibility of fraud is less and negative impact on average credit rating is smaller	accessory

(6) The effect of other factors on average credit rating in commodity classification
Brand rate, the degree of standardization and the degree of imitation are removed from the regression equation. But in the correlation analysis, coefficients of correlation of them are more than 0.51 to illustrate the brand rate, the degree of standardization and the degree of imitation also has impact on the average credit.

5 Credit Rating Difference Affecting Main Business Choices of Sellers

5.1 Credit Cluster: Low-Credit Seller Expelling High-Credit Seller

In C2C e-commerce, adverse selection does exist (Dewan, S. and V. Hsu,2004). In sub-market with low credit level or products with high degree of imitation, sellers who sell low value or quality products can get high profit. However, the buyers will be unsatisfied and lose trust in this sub-market, they tend to cut down their online demand for this category of commodity or lower the price they are willing to pay, which is the adverse selection under information asymmetry. The sellers who sell good products cannot earn profit. So the high-credibility sellers begin to migrate into other sub-market with higher average credit level. In the end, the creditable sellers will be stay in higher-credit-level sub-market.

5.2 Arbitrage in Different SUB-markets

In order to investigate whether sellers will immigrate among sub-market, we collect the trading record of 9 sellers in Taobao (Table 6). We can find out that The No.1-6 sellers got a high credit level through cheap goods, then cheat on expensive goods.

The No.7 and No.8 sellers also got a certain credit level through QQ goods, and then began to sell other products. They haven't get any negative rating until now.

The No.9 sellers sold the sports shoes, and then moved to the laptop market. The average credit level in the laptop market is 2.21, which is higher than that of sports shoes' market, 2.17. It means, the vicious competition makes the creditable seller move into other markets.

Table 6. History Trading Statue of Sellers

ID	Before			After		
	Sub-market	Price	Good Rate	Sub-market	Price	Good Rate
Repeat trade	Cell phone	921	Be closed by Taobao	lady clothes	138	100%
hzking2008	cell phone charge	10	99.46%	accessory	28	98.67%
cn403	cell phone charge	7	100%	lady clothes	178	95.59%
YaominVs Jianlian Yi	QQ goods	1	100%	lady clothes	35	99.01%
JIangshan Buyi	lady clothes	80	96.59%	lady clothes	246	95.38%
Boya trade limited	watch	700	99.71%	watch	1700	98.62%
ccm117	QQ goods	24	99.57%	laptop	3200	100%
liulishang	QQ goods	0.2	100%	men's shoes	258	100%
wangmaoan88	sports shoes	118	100%	laptop	3155	100%

6 The Commodity-Dependent Credit Rating Model

Based on the empirical and theoretical analysis above, we hereby propose the Commodity-Dependent Credit Rating Model: Credit rating and creditworthy behavior depends on the commodity type.

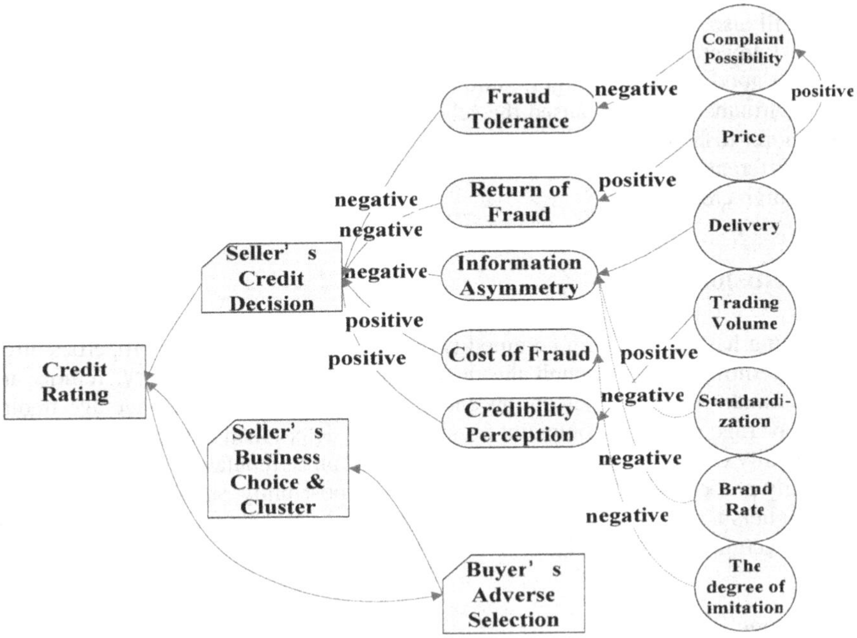

Physical delivery has higher degree of information asymmetry.
Virtual delivery has higher degree of information asymmetry.

Fig. 1. The way of commodity properties to affect on credit rating

6.1 Seller's Credit Decision in Different Sub-market

(1) Complaint possibility, trading volume, standardization and brand rate are positive factor to credit rating

Complaint possibility has negative effect on fraud tolerance and fraud tolerance has negative effect on credit rating, so the compound effect on credit rating is positive. Standardization and brand rate have negative effect on information asymmetry, while information asymmetry has negative effect on credit rating, so the compound effects on credit rating are positive. Trading volume has a positive effect on creditability perception and creditability perception has positive effect on credit rating, so the compound effect on credit rating is positive.

(2) Price and imitation degree are negative to credit rating

Price has a two-way effect on credit rating. It has a positive effect on return of fraud and return of fraud has negative effect on credit rating, so the compound effect on credit rating is positive. Price also has a positive effect on complaint possibility and complaint possibility is positive factor on credit rating. However, the negative effect is larger than positive effect, so the total effect is negative. And this effect is diminishing in margin. The degree of imitation has positive effect on cost of fraud and cost of fraud has negative effect on credit rating, so the compound effect on credit rating is negative.

6.2 Sellers Cluster in Same Credit Level Sub-market

Buyers will also react with voting with feet, they may leave this online sub-market, or tend to ask lower price for the commodities they buy. This adverse selection will drive seller with good credibility migrant into sub-market with higher credit level and better credit environment, which caused the sellers with similar credit level cluster together. Furthermore, dishonesty sellers may also take advantage of the credit difference existed in different commodities.The move into other commodity sub-market after being assigned high credit rating.

7 Conclusion

Credit rating has difference in commodity classification. Commodity properties affect average crediting rating through the degree of information asymmetry, returns and costs of fraud, credibility perception and fraud tolerance. Delivery, average trading volume, average price and complaint possibility has determined impact on credit rating in commodity classification. In the future modification of reputation system, important commodity properties, such as price and compliant possibility, should be introduced to prevent sellers transferring their main business in order to obtain a higher credit rating and then becomes dishonesty later.

References

1. Friedman, E., Resnick, P.: The Social Cost of Cheap Pseudonyms. J. Journal of Economics and Management Strategy 10(2), 173–199 (2001)
2. Conein, B., Arena, R.: On virtual communities: individual motivations, reciprocity and we-rationality. J. Bibliographic Citation, International Review of Economics 55(1), 185–208 (2008)
3. Wang, Y.F., Hori, Y., Sakurai, K.: Characterizing economic and social properties of trust and reputation systems in P2P environment. J. Journal of Computer Science and Technology 230, 129–140 (2008)
4. Li, W., Wu, D., Xu, H.: Reputation in China's online auction market: Evidence from Taobao.com. J. Frontiers of Business Research in China 2(3), 323–338 (2008)
5. Singl, J., Sirdeshmukh, D.: Agency and mechanisms in consumer satisfaction and loyalty judgments. J. Academy of Marketing Science Journal 28, 150–167 (2000)
6. Melnik, M.I., Alm, J.: Does a Seller's Ecommerce Reputation Matter? Evidence From eBay Auctions. J. The Journal of Industrial Economics 50(3), 337–349 (2002)
7. Pavlou, P.A., Dimoka, n.: The Nature and Role of Feedback Text Comments in Online Marketplaces: Implications for Trust Building, Price Premiums, and Seller Differentiation. J. Information Systems Research 17(4), 392–414 (2006)
8. Dewan, S., Hsu, V.: Adverse Selection in Electronic Markets: Evidence from Online Stamp Auctions. J. Journal of Industrial Economics 52(4), 497–516 (2004)

Trust Model Based on M-CRGs in Emergency Response

Shasha Deng [1,2], Pengzhu Zhang[1], and Zhaoqing Jia[1]

[1] Department of Management Information Systems, Shanghai Jiaotong University, Shanghai 200052, China
[2] Department of Computer Science, Shanghai University of Electric Power, Shanghai, 200090, China
dengshasha1108@gmail.com, {pzzhang,jiazhaoqing}@sjtu.edu.cn

Abstract. Many research results demonstrate that government itself cannot handle all the requests from residents in emergency response. Some scholars proposed that building community response grids which utilized pre-existing communities to support citizen request. Unfortunately, little attention has been given to achieve effective and trustworthy collaboration between professional emergency responders and residents. In this paper, the authors modify the architecture of CRGs to provide a valid organizational pattern in emergency response. Based on the modified CRGs (M-CRGs), the trust modeling framework is discussed in detail. Through recording the total behaviors and evaluation of all agents in the systems, the society network is built and the global trustworthiness which reflects the agents' true synthetical ability is gained in the model. An application of this model to Snow Disasters in Southern China is illustrated. Analysis shows that the model contributes to developing efficiency in emergency response.

Keywords: Trust, Emergency response, Community response grids, E-government, Society network.

1 Introduction

When various disasters especially major catastrophe such as Wenchuan earthquake and snow disasters in southern China occur, government itself cannot handle all the requests from residents [1]. "Such disasters may require massive coordination of public and private agencies, plus cooperation from millions of citizens." [1, 2] But effective collaboration of all kinds of organization and citizens assistance activities in disaster remains elusive [2].

During emergency preparedness and response, Palen considers local citizens are the true first responders [3]. Shneiderman and Preece put forward to build community response grids (CRGs) [1]. CRGs focuses on cooperation among millions of citizens, in contrast to Department of Homeland Security (DHS) information network, the later is designed to build network for professional emergency responders. If these communities in CRGs are well-ordered and used in regular period, community members can know each other and develop closer community trustworthy contacts [1]. The CRGs would exert favorable influence on emergency response [4].

W. Liu et al. (Eds.): WISM 2009, LNCS 5854, pp. 545–553, 2009.

Unfortunately, little attention has been given to impact of e-government systems in CRGs. Cherrie and Dickson consider e-government platform is infrastructure which provides information integration from various governmental departments and public services [5]. It is obvious that influence of e-government systems cannot be neglected in emergency response.

Although in CRGs the residents are divided into communities, such network is loosely structured [6]. If every active object such as community member in CRGs is viewed as an agent, it is difficult to select suitable agents to complete the collaborative activities and also cannot assure the information provided is trustworthy. In pear-to-pear environment, it is effective to differentially treat agents according to agent's trustworthiness [7]. Jeager and his research group point out effective trust mechanism is key factor to guarantee CRGs systems operate normally [4, 8].

Kini and Choobineh define trust from perspective of social psychology and economics, saying "Trust in a system is defined as an individual's belief in the competence, dependability, and security of the system under conditions of risk." [9] Generally, trust model refers to creating measurable evaluation system, using trustworthiness to measure agent's trust degree. Through interaction history among these agents, the trustworthiness essentially is mapping of synthetical ability between agent's actual ability and its participant strategy. The research of trust model concentrates on the question of malicious behavior. The trust model is based on society trust network utilizing sociology research results. The hypothesis of this method is higher trustworthiness recommender is more trustworthy. But, it is not always true [10].

Compared with P2P environments, there are many different features in CRGs. For communities are divided by geography, the collaboration among them become more frequent. As agent usually belongs to one or more leagues, their interaction and cooperation are conducted in form of group. Hereby, ***the kernel question of trust model based on M-CRGs in emergency response is to achieve effective and trustworthy collaboration between professional emergency responders and citizens. And its main challenge is how to calculate trustworthiness which truly represents agent's synthetical ability***.

In this paper, we develop a trust model based on proposed M-CRGs. The details of the architecture of modified community response grids are given in Section 2. The trust model framework is presented in Section 3. The model and its contributions are discussed in Section 4. Finally, the conclusion for future research is provided.

2 Modified Community Response Grids (M-CRGs)

Shneiderman and Preece consider CRGs can "support registering of households, reporting of incidents, requesting assistance, and responding to requests" [1]. But, in CRGs, e-government systems are neglected. We think that is not suitable, because e-government systems are indispensable factors in emergency response [5]. In addition, we consider residents, professional emergency responders and e-government systems should be treated equally. The sole difference among them is their function. Thus, we give the architecture of modified CRGs (M-CRGs).

2.1 M-CRGs Architecture

As illustrated in Fig. 1, the architecture of M-CRGs is designed to hierarchical structure divided into three levels.

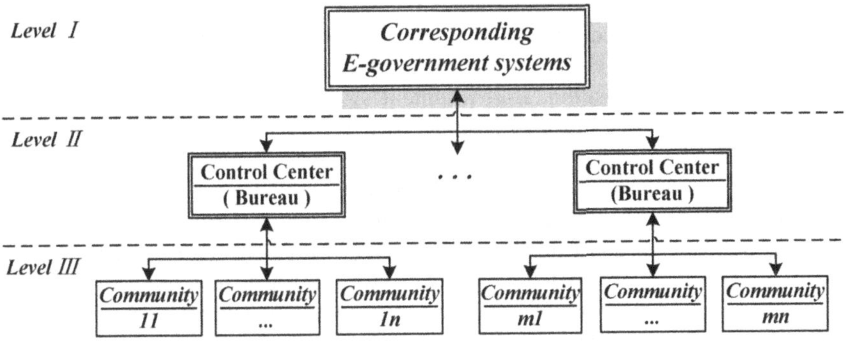

Fig. 1. Architecture of Modified Community Respond Grids

In Level III, we define the grid based on geographical position. Each community includes one or more grids on street basis. According to the registered authentic address information, the register affiliates to corresponding community. Considering temporary movement of population such as business trip or journey, the register can enter any community belonged on M-CRGs conveniently. Governors preside over the control center in Level II. They directly serve limited communities by region. Control center staff can separate out suspicious report and trustworthy report in light of the resident's trustworthiness. According to requesting information, control center can help community members to organize self-rescue group and assign appropriate governmental resources to them. Every control center has corresponding E-government systems in Level I, they exchange information and instruction each other.

2.2 E-government Systems in M-CRGs

During emergency management, efficient e-government systems are crucial to exchange the information collected from different area and make collaboration among different organizations. Especially today, many government departments have built their e-government systems that produce plenty of useful resource. However, E-government systems may be not suitable if directly applied in emergency response. Among the architecture of M-CRGs, the community provides a channel to gather incident reports, quickly sending needs of resident to professional emergency responder through E-government systems. Meanwhile, the results of resident-to-resident assistance feed back to e-government systems via control center. In this way, the emergency response organizations are familiar with the disaster site promptly, so as to rationally assign resources. Especially, when emergency response organizations cannot immediately arrive at the disaster site, through M-CRGs, local citizens can rescue themselves under the instruction of professional emergency responders to reduce loss.

2.3 Functions of M-CRGs

The most significant value of M-CRGs is that it can build perfect relieving system which includes professional emergency responders and resident before, during and after disasters.

In M-CRGs, control center and e-government systems act as information processor and supervisor. After receiving community members reports, control center process and send information to E-government to identify an emergency. After synthesizing and analyzing the information, E-government systems deliver instruction to control center. And then, these control centers transmit them to corresponding one or more communities to help to organize self-rescue.

Community is the basis of whole architecture, without which, the control center and corresponding E-government system would not operate. Residents in community can report incidents, request help, and help each other. Once request arrives, control center would help to organize self-rescue group and push request information and instructions to group members via email, mobile communication device.

When centralized services such as control center or father are overwhelmed by catastrophe, these communities in level III will play vital role in whole M-CRGs. Residents can also provide the local disaster information and request each other. When residents request assistance, relevant information is directly pushed to neighborhood, who will organize group to rescue them.

3 Trust Modeling Framework

Shneiderman and his research group consider that in CRGs information authentication is guaranteed by professional staff. The staff is responsible for monitoring the CRGs and picking out suspicious information [1, 8]. We think it improper to take professional staff as monopolistic judge. Firstly, professional staff is not in the site, he is difficult to give fair judgment. Secondly, if all the authentication work is assigned to one department, cheat become easily in CRGs. Finally, such trust model is not fit for application in large scale. Effective trust mechanism depending not on one organization is necessary.

Because emergency time differs from ordinary period, inaccuracy would lead to loss of life and wealth. M-CRGs should promote a real-name system, which would force register to notice their words and actions to refrain from slander or diffuse harmful information. But real-name system is not almighty. Different from trust model in e-commerce, M-CRGs emphasizes on accuracy and ability of completing task. Due to personal knowledge and experience are diverse, intended cheating is not ultimate reason of some mistakes. In this section, we describe the details of the proposed trust model based on M-CRGs, and give the calculation equation of trustworthiness.

3.1 Definition and Representation

Although there are two categories of objects in above M-CRGs, all objects are treated equally. These objects not only community members but also administrator such as control center or corresponding e-government systems are viewed as agent with some features. These features are defined as a serial of attributes. In our model, the agent is described as a tuple: $agent_i$ = <*flag, address, field, job, age*>, where *flag* is used to discriminate the type of agents. *address* records the agent authentic geographic position. *field* is the agent's professional field. In M-CRGs, we define some ordinary field in advance to supply the register to choose. *job* refers to the agent's career. *age* indicates agent's age, which can calculate automatically from identification card. If the agent is administration, age is its runtime.

As reporting incident and coordinating responses are two main activities in M-CRGs, the local trustworthiness denoted by LT_i which consists of $reportT_i$ and $actionT_i$. $reportT_i$ is trustworthy degree of his report which is adjusted dynamically by other agents' comments on his report. $actionT_i$ measures collaborative work ability of $agent_i$ in the action, which will be altered after action according the others' evaluation.

In our model, when report occur in community, we will record some information use a tuple: $report^{(L)}$ = <*reporter, CommMember*>, where *reporter* is agent who publishes report. *CommMember* is set of agents who comment on this report. *L* is serial number of this report and is unique identification in the system.

Once requesting information appeared in community, these agents would organize a group to accomplish the task. We define $Action^{(K)}$ to record the activity information. $Action^{(K)}$ is a tuple: $Action^{(K)}$ = <*Tasktype, Region,* $\psi^{(K)}$, *Success, GroupMember*>, where *Tasktype* is the property of the action, which is defined in advance. *Region* refers to address of task. $\psi^{(K)}$ is difficulty coefficient of this task . *Success* describes the completing degree of this task. *GroupMember* is set of agents who participate in this task. When an agent is selected to participate in the action, the agent would be inserted in the set. *K* is serial number whose identification is unique in the systems.

3.2 Local Trustworthiness

As LT_i consists of $reportT_i$ and $actionT_i$, we calculate them respectively.

(1) $reportT_i^{(L)}$

$$reportT_i^{(L)} = \sum_{j \in A} (reportE_{ij}^{(L)} * LT_j) \Big/ CommN \tag{1}$$

When a report has been inspected, we can obtain the *i* from the attribute of $report^{(L)}$. $reportE_{ij}^{(L)}$ is the report evaluation of $agent_i$ given by $agent_j$, where *L* is serial number of this report. The evaluation value is limited in the set {-1, 0, 1}. The set A comes from *CommMember,* the attribute of $report^{(L)}$. *CommN* is the number of effective evaluation. If $agent_j$ completely agree with this report he would choose $reportE_{ij}^{(L)} = 1$, otherwise $reportE_{ij}^{(L)} = -1$. "0" means the evaluation is invalid, and then *j* would be deleted from set A. $reportT_i^{(L)}$ is trustworthy degree of the report released by $agent_i$.

(2) $actionT_i^{(K)}$

$$actionT_i^{(K)} = \left[\sum (actionE_{ij}^{(K)} * LT_j) \Big/ GroupN \right] * \psi^{(K)} \tag{2}$$

After an action completed, the participators are asked to evaluate others' performance. $actionE_{ij}^{(K)}$ represents collaborative ability evaluation of $agent_i$ given by $agent_j$. If $agent_j$ consider $agent_i$ spared no effect in this action, he can choose $actionE_{ij}^{(K)} = 1$. When $agent_j$ chooses $actionE_{ij}^{(K)} = 0.5$, it imply he is basically satisfied with $agent_i$. If $agent_i$ behaved badly during entire action, $actionE_{ij}^{(K)}$ is equal to "-1". Please note that to an estimator, the possibility of $actionE_{ij}^{(K)} = 1$ or $actionE_{ij}^{(K)} = -0.5$ is restricted in certain range. $\psi^{(K)}$ refers to the difficult coefficient of this action, whose value comes from $Action^{(K)}$. B is participator's set, which also comes from $Action^{(K)}$. *GroupN* is the number of participator based on set B.

(3) LT_i

$$LT_i = \alpha * \left(\sum_{L \in RT} reportT_i^{(L)} \Big/ rt \right) + \beta * \left(\sum_{K \in AT} actionT_i^{(K)} \Big/ at \right) \tag{3}$$

LT_i is local trustworthiness of $agent_i$ and its initial value is zero. The proportion of $reportT_i$ and $actionT_i$ in LT_i is α and β respectively. Set *RT* is collection of report's serial number which post by $agent_i$, similarly, set AT is collection of collaborative activity which the $agent_i$ participated. And *rt* and *at* is sum of set *RT* and *AT* respectively.

From equation (1) and (2), it is possible that the value of $reportT_i$ and $actionT_i$ are negative. In result, the value of LT_i also may be negative from equation (3). Conversely, if the LT_i is negative number, it would lead to equation (1) and (2) produce wrong calculation. So we need to adjust the value of LT_i. If the LT_i is negative number, we consider the $agent_i$ is un-trustworthy, and his evaluation to others can be ignored, and then we assigned zero to LT_i.

3.3 Society Network

In M-CRGs, when an agent reported incident, some effective evaluation would be produced and reserved. At the same time when an agent released a request in community, $Action^{(K)}$ would be instantiated. After selection of group member and completing real action, these participators evaluate each other. Through the two kinds of evaluation to these agents, society network would be created gradually.

$$S = \begin{bmatrix} w_{11} & w_{12} & \cdots & w_{1\mu} \\ w_{21} & w_{22} & \cdots & w_{2\mu} \\ \cdots & \cdots & w_{ij} & \cdots \\ w_{\mu 1} & w_{\mu 2} & \cdots & w_{\mu\mu} \end{bmatrix} \tag{4}$$

The society network would be viewed as a graph S with vertex set *V(S)* and edge *E(G)*. These agents in M-CRGs constitute set *V(S)*. Edge *E(G)* denotes the relationship among set *V(S)*. If the sum of agents is μ, we can represent the society network

S as a μ-by-μ matrix, as it is denoted in equation (4). The entry w_{ij} is element that lies in the *i-th* row and the *j-th* column of the matrix. It represents weighted average evaluations of $agent_i$ given by $agent_j$ during all mutual behaviors including reporting and collaborative activities. Its initial value assigned infinity ∞ which means there is no edge from $agent_j$ to $agent_i$. That means no inter-activity between them.

$$w_{ij} = \alpha * \left(\sum_{L \in M} reportE_{ij}^{(L)} \Big/ m \right) + \beta * \left(\sum_{K \in N} actionE_{ij}^{(K)} * \psi^{(K)} \Big/ n \right) \tag{5}$$

Equation (5) is concrete calculation expression of w_{ij}, where α and β is weighted value came from equation (3). The element of set *M* refers to report serial number of $agent_i$'s effective evaluation given by $agent_j$, so have the set *N*. The m and n refers to the sum of set *M* and set *N* respectively.

3.4 Global Trustworthiness

$$T_i = \sum_{j \in T} \left(w_{ij} * LT_j \right) \Big/ t + T_0 \tag{6}$$

Global trustworthiness vector is $T = (T_1 \ T_2 \ \dots \ T_\mu)'$. From equation (6), the global trustworthiness T_i not only related to evaluation from other agents but also has impact on initial value denoted by T_0. Note that because agents in M-CRGs are real name and their authentic information is recorded in tuple $agent_i$, T_0 is obtained according to the proportion calculation of attributes in the tuple.

The other part of the global trustworthiness T_i is the average of weighted local trustworthiness of all agents that collaborate with $agent_i$. Those agents constitute the set of *T*. *t* is the number of those agents.

4 Discussion and Contributions

From above equations, we can find local trustworthiness comes from evaluation of every action. If we use only local trustworthiness to measure one agent's trust degree, then the trust system is easily to be attacked by some malicious agents. Although global trustworthiness in this model relates with local trustworthiness, in fact it is accumulation of evaluation from all participators. And cheating becomes difficult in such situation. When an agent releases request information, the system would select high global trustworthiness agents according to the attribute information of $Action^{(K)}$. When an agent reports incident, we can also estimate his report according to his global trustworthiness.

Take Snow Disasters in Southern China for instance. As hitting by the worst winter storm in five decades, thousands of passages were trapped in iced road, and residents met difficulty in daily life. Government cannot respond all resident requests. For example, if a citizen finds the nearby road become too dangerous to pass through for vehicles, he can report the incident. The control center can deal with this report according to his trustworthiness. When a lot of reports rush into community, the trustworthiness of the reporter is prone to discrimination. If those reports with high trustworthiness are paid high attention, then the residents' participative passion will be stimulated to report more true incidents.

Sometimes the report can be changed to request information. In the above example, if citizen want to call on his neighbor to clean out the frosted ice, $Action^{(K)}$ is created in M-CRGs. According to requirement of action, under the help of systems, the control center would select high trustworthiness residents and push request and instruction information to their communication. At the same time, control center send the information to corresponding e-government system to remind the governmental department. If his neighbors agree to participate in this action, he would answer the message. After completing this road ice cleaning, every participator can evaluate each other. And so their trustworthiness would be altered. In special situation, action would demand some professional agents. The procedure of selection will be adjusted slightly. According to professional requirement, the system would choose eligible agents and rank them on the basis of their trustworthiness, and then front agents would be selected.

M-CRGs utilize the pre-existing communities to deal with the daily emergency, build the trustworthy relationship among agents. When major catastrophe occurred, M-CRGs can easily respond residents' requests. Effective trust mechanism can maintain its operation. Existing literatures show little research concerns on such resident-to-resident assistance. Our work modifies CRGs by adding the e-government systems and creating hierarchical structure to develop relationship among professional emergency responders and residents.

Based on M-CRGs, we build the trust model, which is another contribution. Through massive literature search, we find there is hardly any study on trust mechanism in emergency response from microcosmic perspective. This trust model can restrain the attack of malicious agents in certain degree. This trust mechanism can assure the operation of M-CRGs.

5 Conclusion

In this paper, we modify the architecture of CRGs, which will serve better for emergency response. Based on the M-CRGs, we are the first attempt to develop the trust model in emergency response. Through recording the total behaviors and evaluation of all agents in the systems, the society network is built and the global trustworthiness which reflects the agents' true synthetical ability is gained in the model. By analyzing case of Snow Disasters in Southern China, our model contributes to resident-to resident assistance, enhancing collaboration between government and citizen, ultimately developing efficiency in emergency response. Future research would focus on simulating such systems.

Acknowledgement

This research was supported by the National Natural Science Foundation of China under Grant 70533030.

References

1. Shneiderman, B., Preece, J.: Public Health 911.gov. Science 315, 944 (2007)
2. Stephenson, M.: Making Humanitarian Relief Networks More Effective: Operational Coordination, Trust and Sense Making. Disasters 29, 337–350 (2005)
3. Palen, L., Hiltz, S.R., Sophia, B.L.: Online Forums Supporting Grassroots Participation in Emergency Preparedness and Response. Communications of the ACM 50, 54–58 (2005)
4. Jaeger, P.T., Shneiderman, B., Fleischmann, K.R., et al.: Community Response Grids: E-government, Social Networks, and Effective Emergency Management. Telecommunications Policy 8, 592–604 (2008)
5. Cherrie, W.W., Dickson, K.W.: E-government Integration with Web Services and Alerts: A Case Study on an Emergency Route Advisory System in Hong Kong. In: 39th Hawaii International Conference on System Sciences, pp. 1–10. IEEE Computer Society, New York (2006)
6. Waugh, W.L., Sylves, R.T.: Organizing the war on terrorism. Public Administration Review 26, 145–153 (2002)
7. Mäntymäki, M.: Does E-government Trust in E-Commerce When Investigating Trust? A Review of Trust Literature in E-Commerce and Government Domains. In: Oya, M., Uda, R., Yasunobu, C. (eds.) Towards Sustainable Society on Ubiquitous Networks. IFIP International Federation for Information Processing, vol. 286, pp. 253–264. Springer, Boston (2008)
8. Jaeger, P.T., Fleischmann, K.R., Preece, J., et al.: Community Response Grids: Using Information Technology to Help Communites Respond to Bioterror Emergencies. Biosecurity and Bioterrorism: Biodefense Strategy, Practice, and Science 5, 335–345 (2007)
9. Kini, A., Choobineh, J.: Trust in Electronic Commerce: Definition and Theoretical Considerations. In: 31st Annual Hawaii Int'l. Conference on System Sciences, pp. 51–61. IEEE Computer Society, New York (1998)
10. Kamvar, S.D., Schlosser, M.T., Garcia-Molina, H.: The Eigentrust Algorithm for Reputation Management in P2P Network. In: 12th Int'l. World Wide Web Conference, pp. 640–651. ACM Press, Budapest (2003)

Research on a Queue Scheduling Algorithm in Wireless Communications Network

Wenchuan Yang[1], Yuanmei Hu[1], and Qiancai Zhou[2]

[1] No. 95 Box, Beijing University of Posts and Telecommunications
yangwenchuan@bupt.edu.cn,
huymxie@gmail.com

[3] Room 1501, Beijing Gehua, No. 1 Building of Qinglong Hutong, Dongcheng District, Beijing
zhouqiancai@cnpc.com.cn

Abstract. This paper proposes a protocol QS-CT, Queue Scheduling Mechanism based on Multiple Access in Ad hoc net work, which adds queue scheduling mechanism to RTS-CTS-DATA using multiple access protocol. By endowing different queues different scheduling mechanisms, it makes networks access to the channel much more fairly and effectively, and greatly enhances the performance. In order to observe the final performance of the network with QS-CT protocol, we simulate it and compare it with MACA/C-T without QS-CT protocol. Contrast to MACA/C-T, the simulation result shows that QS-CT has greatly improved the throughput, delay, rate of packets' loss and other key indicators.

Keywords: QS-CT, MACA/C-T, Ad hoc, Queue Scheduling Mechanism.

1 Introduction

Modern networks are syncretic, in which the Ad hoc network (called multi-hop network or self-organizing network) is a new type of communications network being independent of the existing framework. In Ad hoc network hosts for communications are generally mobile terminal equipment such as portable computers, personal digital assistant (PDA) and so on. Different with the traditional wireless networks, the Ad hoc network, without the support of cable infrastructure, does not have specialized routers, base stations and other fixed infrastructure. It realizes communication through transponder of the adjacent nodes. In the network, mobile terminals have not only the functions of host, but also the functions of router. The emergence of Ad Hoc network promotes the process of achieving freedom communication in any circumstances, at the same time it also provides an effective solution for military communications, disaster relief and temporary communication.

At any time, in Ad hoc network, a topology structure with random shape can be formed by connecting the mobile nodes with wireless transceiver device through wireless channel. Nodes can be mobile, which may lead to the topology structure being changed. In this environment, as a result of the limited coverage of terminal's wireless communication, the two terminals which can't be directly linked with each other must achieve data communications through other terminals' packet forwarding.

W. Liu et al. (Eds.): WISM 2009, LNCS 5854, pp. 554–562, 2009.

Because of the following characteristics, the design and utility of Ad hoc network protocol are faced with great challenges. First of all, wireless Ad Hoc network does not have a central control node, and a node can access or leave at any time, which makes the topology network become highly dynamic. Secondly, the wireless channel environment is very poor, due to such as multi-path fading, shadow effects, inter-user interference. The channel access protocol, as a sub-layer protocol stack of Ad hoc network, controls the sending and receiving of datagram on wireless channel. Ad hoc networks' bandwidth resources are limited. It will take a crucial part in network performance to use of limited bandwidth effectively. So, proposing a better Ad hoc channel access protocol is of great significance.

2 Sub-layer of MAC(Media Access Control)

2.1 Exiting Problems

Due to the own characteristics, Ad hoc network's MAC sub-layer is different from traditional MAC. These differences mainly manifest on channel-sharing approach of multi-hop, hidden terminals, exposed terminals, invasion of node and so on.

Multi-hop channel-sharing method. In Ad hoc network, a node only sends datagram to the nodes that are in the range of its communications, and other nodes can't detect the existence of that communication. This makes Ad hoc networks have more than one simultaneous communication at the same time. This kind of channel is called multi-hop sharing radio channel. It increases utilization of network resources, but makes the datagram collision involved with the node's geographical location. In Ad Hoc network, the datagram collisions are local incidents, and the channel statuses perceived by the sending nodes and the receiving nodes are not always the same, which will bring hidden terminal and exposed terminal problems.

Hidden terminal problem[2]. Hidden terminal node is a node which is within the scope of receiving nodes' communication, and beyond the scope of sending nodes' communication. In Figure 1, node H is in the range of D's communications, and outside the scope of the node S's communication. When sending data from S to D, H can't perceive the ongoing transmission, and then sends transmission to D at the same time, which must lead to datagram collision. This is so-called Ad hoc of hidden terminals problem.

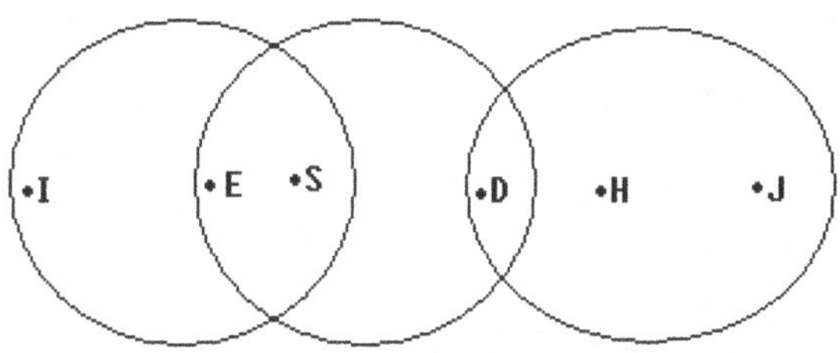

Fig. 1. Hidden terminal and exposed terminal

Exposed terminal problem. Exposed terminal node is a node which is within the scope of sending nodes' communication, and beyond the scope of receiving nodes' communication. In Figure 1, node D and node E are all in the range of S's communications. When sending data from S to D, E perceives the ongoing transmission, and then delays its own transmission. But in fact, transmission between E and I is feasible. The exposed terminal problem greatly reduces the utilization of bandwidth resources.

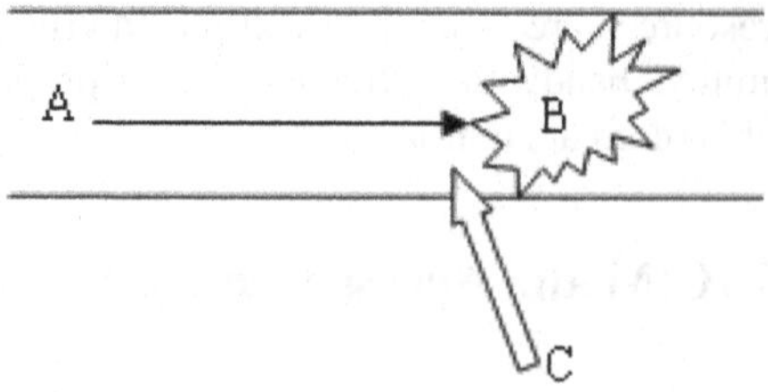

Fig. 2. The issue of invasion of terminals

The issue of terminals invasion[1]. The issue of terminals invasion is caused by nodes' mobility. In Figure 2, the transmission between A and B is going on, mean while node C walks into the scope of B's communication and sends data to B. Because all the nodes are sharing the same channel, data collision at node B will take place.

2.2 Description of Existing Protocol Queue Scheduling Mechanism

When speaking based on the types of channel, the existing Ad hoc MAC layer protocol can be divided into single-channel protocol, dual-channel protocol and multi-channel protocol. Single-channel protocol is used in the Ad Hoc network which has only one sharing channel. One of the main objectives of this channel access protocol is to design appropriate strategies to avoid conflict, typical examples are MACA, 802.11 DCF, etc.; Dual-channel protocol is used in the Ad Hoc network with two sharing channels. Using the two different channels, control messages and data messages will not conflict, typical example is DBTMA; Multi-channel protocol makes the access control more flexible. This protocol can use one channel as a public control channel, also allows control messages and data packets mixed in the same transmission channel. This channel access protocol concerns the major two issues: channel allocation and access control, typical examples are PAMAS, MACA / C-T and so on.

3 Scheduling Mechanism

All paragraphs must be indented. All paragraphs must be justified, i.e. both left-justified and right-justified.

3.1 Description of Scheduling Mechanism

Figure 3 shows a model of queue scheduling. As is shown in the Figure, we add two queues to MAC layer, in which queue Q2 is a FIFO (First In First Out) queue, and plays the role of storing packets temporarily. Queue Q1 is a sender queue, supplying packets that are to be sent to MAC layer. Different from Q2, Q1 is not a FIFO queue, and its selection of sending-packets depends on packets' destination addresses, retreat time and NAV time. The different characteristics of Q1 and Q2 are respectively manifested by rows and columns in Figure 3. When working, the upper package will enter the queue Q1 firstly. But if Q1 is full, the upper package will have to wait for access to Q2, until Q1 is available. If Q2 has been filled, the packet will be discarded. The scheduling mechanism selects packet from Q1 and send it, if the sending task fails, then starts the retreat mechanism, in which the failed package is placed again in Q1. Meanwhile the Q1 finds packet of which back-off time is 0, and NAV time is 0, and with a different destination node from the pre-packet to continue to send.

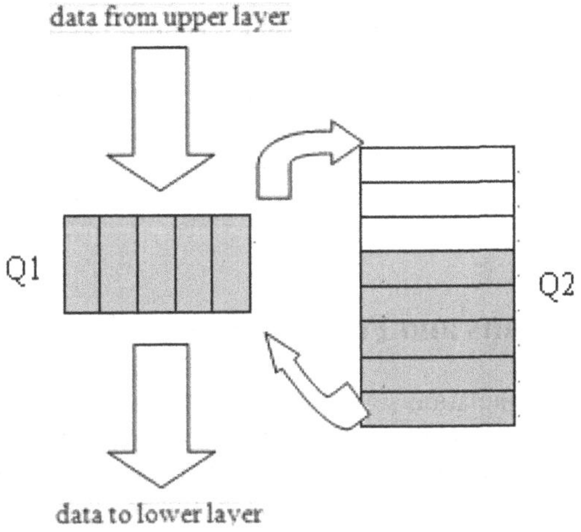

Fig. 3. The Model of scheduling Mechanism

When the sending task failed ,there are three principle of packets' choice: (1) retreat time is 0; (2) NAV time is 0; (3) with a different destination node from the pre-packet. The retreat is not a behavior of the whole MAC layer, but the packets which have been fail to send. MAC layer will continue to choose a packet that has different purpose node to send. In the same way, setting of NAV is only the behavior of packages in the Q1. So, the queue scheduling mechanism changes the ways from a passive sending into an active choice. And then, the pre-packet will participate in the competition of seizing channel together with other packets, which makes the access to channel more fairly. In addition, we know that the reason of packages' failed sent are largely that the different terminal nodes send data to the same node at the same time, leading to a conflict in the purpose node. Introducing the principle (3), re-sent packets have different destination nodes, reducing the probability of a conflict at the same purpose node. It will certainly bring a good performance.

3.2 QS-CT Protocol

QS-CT protocol come into being after adding the queue scheduling mechanism to the RTS-CTS-DATA. This article will compare QS-CT protocol with MACA / CT which is also based on RTS-CTS-DATA multiple access protocol. MACA / CT protocol adopts the method of competitive access. It utilizes the RTS-CTS-DATA handshake mechanism to solve the hidden terminal problem of the Ad hoc network; It utilizes the method of spread-spectrum to approach channel into public channel and data channel, and sends RTS / CTS in the public channel, sends data packets in the data channel, then solves the exposed terminal problem and the invasion problem. MACA / CT protocol works well when the network business is small, but when the network business is relatively larger, it does not work so well. This is because the competition mechanism makes excessive retreat when access business is large, which will increase the number of nodes fit to send RTS at the next moment , and that will increase the probability of RTS packets' collisions in public channel. In order to improve the performance of MACA / CT, we introduced the MAC layer queue scheduling mechanism (Figure 3), so that MAC layer can continue to select packets with other different destinations to send when the pre-packet failed to be sent and was retreat. So that it not only improves the utilization of channel, but also reduces the probability of packet collision engendered when sending data to the same node. It is bound to improve the network performance.

4 Simulation Results and Performance Analysis

We use NS2 network simulation software to simulate the queue scheduling multiple access protocol QS-CT and multiple access protocol MACA / CT which bases on the sending code. Since what we are simulating is MAC layer protocol, when evaluating its performance, the impact of router protocols should be excluded from, so, we simulate the performance of QS-CT and MACA / CT under single-hop environment firstly. However, Ad hoc networks' actual environments are all multi-hop and dynamic. Single-hop network can't simulate the actual environment, thus, we introduce the DSR (Dynamic Source Routing Algorithm) router protocol to simulate the performance of QS-CT and MACA / CT under the multi-hop environment. As a result, the simulation work contains two parts: the simulation under the single-hop environment and the simulation under the multi-hop environment.

4.1 Simulation Results and Performance Analysis under Single-Hop Environment

In the single-hop environment, the simulation time is 400 seconds, the topology square area is of 170 meters × 170 meters, the longest distance between two nodes in the area is $170 \times \sqrt{2} = 240$, less than the spread of the node range (250 meters), which ensures the communication between any two nodes does not need to relay. This is called single-hop action. Nodes are randomly distributed in the topology structure, sending Poisson distribution data, each packet will randomly select destination node. The maximum number of retransmission is 6, the minimum and maximum of contention window CW is respectively 7 and 127, the length of retreat step is 20us. The sum of control packet RTS's size and CTS's size is 83 bytes, and the size of data packet is 800 bytes. In our algorithm the queue is very important, MACA / C-T choose queue of witch the length of is 10 packets, and the length of Q1 and Q2 in QS-CT respectively are 3 packets and 7 packets. In the figure, the abscissa shows the mean of Poisson business λ, and the ordinate shows the number of packets sent the average node per second. Figure 4 shows how the throughput volume changes with business volume. We can see that the number of nodes in the network at 10,15,20, respectively, QS-CT's performance is all better than MACA / C-T's. Now, we can say that the queue Q1 in QS-CT plays a very significant role to the improvement of performance. It can reduce the possibility of conflict by selecting different data packets in Q1 to send.

Figure 5 is a curve that indicates how the amount of end-to-end delay changes with the business volume. It is obvious that QS-CT reduces the delay comparing with the corresponding MACA / CT significantly, especially when the business is comparatively larger. This is because when nodes conflict or random retreat, QS-CT changes the MACA / CT's ways from negative waiting into positive sending, takes full advantage of the retreat time, reduces delay. The time delay in simulation comes mainly from queuing and transmission delay, of which the queuing time is major.

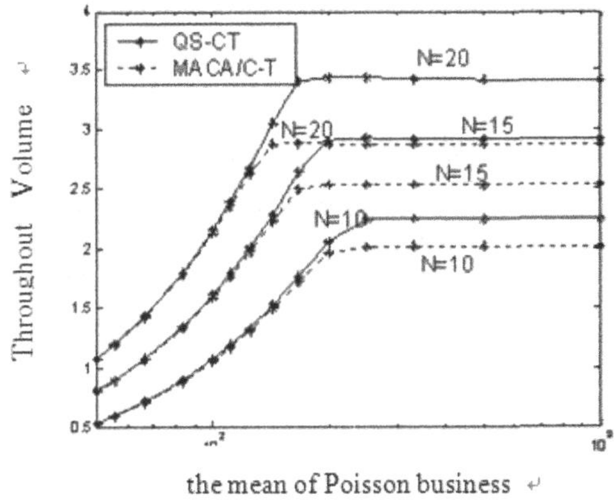

Fig. 4. Single-hop environment volume of throughout v volume of business

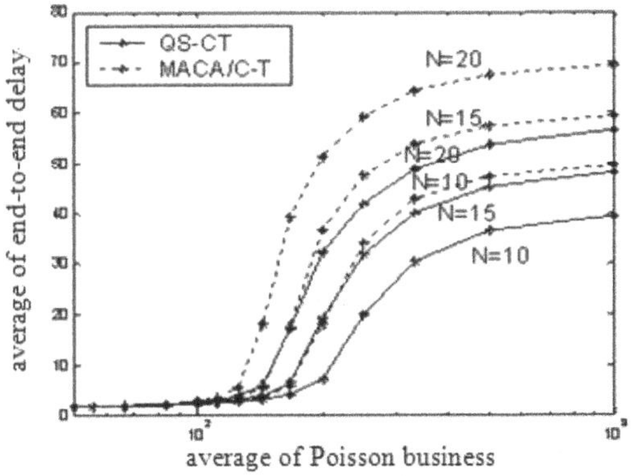

Fig. 5. Single-hop environment delay v volume of business

It can be seen from the map that the finally delay stabilizes with the increasing volume of business. The reason is that the queue is full at that time, and each packet's average queuing time stabilizes.

Following is a brief analysis of the maximum queuing delay. We do the following assumptions: the length of queue is lq data packets, the maximum number of retransmission is 6, the minimum and maximum of contention window CW is 23 and 27 respectively. So, the largest delay can be got from the following simple formula:

$$DelayTime \approx \left(\sum_{i=0}^{4} 2^{i+3} + 2^{7} \times 2 \right) \times StepTime \times Lq \quad (1)$$

Put the parameters of the simulation into the formula, we can calculate the maximum delay to 100.8 ms .In fact, It is impossible that each retreat selects the current largest CW, because it is chosen in accordance with the uniform distribution, therefore, it can be know the average size of the delay is about 50.4 ms, which is same as the result of Figure 5.

4.2 The Simulation Results and Performance Analysis under Multi-hop Environment

In the Multi -hop environment, the simulation time is 400 seconds, the topology square area is of 600 meters × 600 meters, the longest distance between two nodes in the region is $600 \times \sqrt{2} = 848$, which is equal to nearly 4 single-hops. MACA / C-T protocol chooses two queues of witch the length of is 10 packets, and the length of Q1 and Q2 in QS-CT respectively are 2 packets and 8 packets. Other conditions remain the same with single-hop.

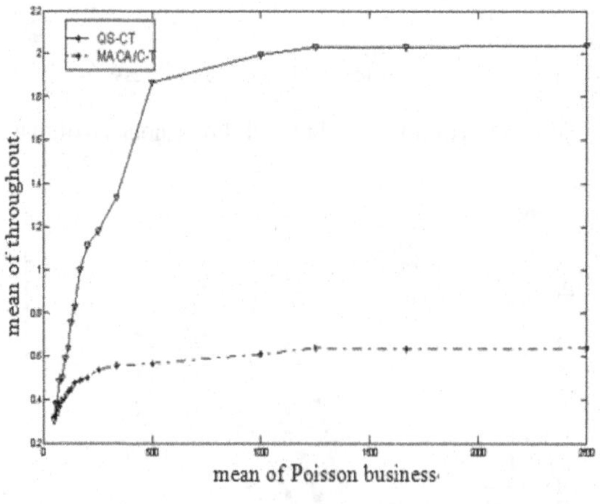

Fig. 6. Multi-hop environment volume of throughout v volume of business

Figure 6 shows how the throughput volume changes with business volume under multi-hop environment. Under the multi-hop environment, because of the transmission's limitations, the data transmission between two nodes needs to go through a transit node, together with the dynamic changes of topology in network, makes the throughput volume reduce a lot than single-hop. However, comparing with MACA / CT, QS-CT greatly improves the performance of the network. From Figure 6 can be seen that maximum throughput of QS-CT and MACA / CT are respectively 2.04 and 0.64, increased by 210 percent.

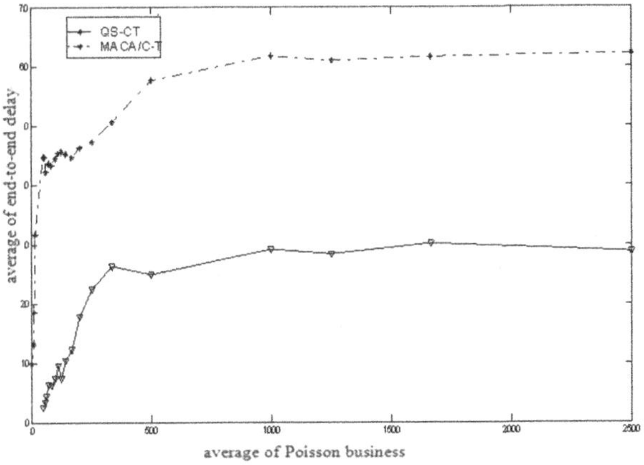

Fig. 7. Multi-hop environment delay v volume of business

Figure 7 is a curve that indicates how the amount of end-to-end delay changes with the business volume under multi-hop environment. It is obvious that QS-CT reduces the delay comparing with the corresponding MACA / CT significantly, especially when the business is comparatively larger. This is because when nodes conflict or random retreat, QS-CT changes the MACA / CT's ways from negative waiting into positive sending, takes full advantage of retreat time, and reduces delay. In multi-hop environment, the delay of simulation comes not only from the process of queuing and transmission, but also from the router-selecting. It can be seen from the map that the finally delay stabilizes with the increasing volume of business. The reason is that the queue is full at that time, and each packet's average queuing time stabilizes.

5 Conclusion

Media access control layer (MAC) has a coordination role in the allocation of wireless resources. Nowadays, the wireless networks resources have been limited, but the requirements of performance of network are increasing. The performance of MAC layer is very important for the development of Ad hoc networks without central control. In this paper, introduces a queue scheduling multiple access protocol QS-CT, which joins the queue scheduling mechanism to the MAC layer, uses queues packet buffer, does different packets scheduling in accordance with the characteristics of each packet, as well as network characteristics, thereby reduces the probability of conflict and improves utilization of resources. Through the analysis of the simulation results under single-hop and multi-hop environment, we can see that the QS-CT protocol clearly improves access performance in delay, throughput.

References

1. Joa-Ng, M., Lu, I.: Spread Spectrum Medium Access Protocol with Collision Avoidance in Mobile Ad-hoc Wireless Network. In: Proc. of IEEE INFOCOM 1999, March 21-25, vol. 2, pp. 776–783 (1999)
2. Ware, C., Wysocki, T., Chicharo, J.: Hidden terminal jamming problems in IEEE 802.11 mobile ad hoc networks. In: IEEE International Conference on Communications, ICC 2001 (2001)
3. Haas, Z.J., Deng, J.: Collision-free medium access control scheme for ad-hoc networks. In: Proceedings of IEEE Military Communications Conference. MILCOM 1999 (1999)
4. Inmon, W.H.: The operational data store. PRISM Tech. Topic 1(17) (1993)
5. Valides-Perez, P.: Principles of human-computer collaboration for knowledge-discovery in science. Artificial Intelligence 107, 335–346 (1999)
6. Widrow, B., Rumelhart, D.E., Lehr, M.A.: Neural networks: Application in industry, business and science. Communication of ACM 37, 93–105 (1994)
7. Wang, R., Storey, V., Firth, C.: A framework for analysis of data quality research. IEEE Trans. Knowledge and Data Engineering 7, 623–640 (1995)
8. Knorr, E., Ng, R.: Algorithms for mining distance-based outliers in large datasets. In: Proc.1998 Int. Conf. Very Large Data Bases (VLDB 1998), August 1998, pp. 392–403 (1998)
9. Jagadish, H.V., Koudas, N., Muthukrishnan, S.: Mining deviants in a time series database. In: Proc.1999 Int. Conf. Very Large Databases (VLDB 1999), Edinburgh, UK, September 1999, pp. 102–113 (1999)

Mixed H_2/H_∞ Control for Networked Control Systems (NCSs) with Markovian Packet-Loss

Jie Fu and Yaping Dai

Dept. of Control Science and Engineering, Beijing Institute of Technology
egarimocean@gmail.com,
daiyaping@bit.edu.cn

Abstract. This paper addresses the mixed H_2/H_∞ control design problem of networked control systems(NCSs) with Markovian packet loss. Under the assumption that the packet loss is bounded and governed by Markov chains, a packet-loss dependent stabilizing controller is found by minimizing an upper bound of H_2 norm, under the restriction of a prespecified H_∞ norm bound. By transforming the stability criterion into an optimization problem, the mean-square stabilizing packet-loss dependent controller is thus obtained. The linear matrix inequality (LMI) approach is employed to calculate the robust controller for packet loss, with special respect to the uncertainty in the transition Markov probability matrix. Illustrative numerical examples are presented to demonstrate the effectiveness of derived methods.

Keywords: Networked control systems, packet loss, Markov Jump linear systems, Lyapunov functions, robust control, LMIs, convex optimization.

1 Introduction

Networked control systems (NCSs) are feedback control systems with control loops closed via a real-time network. Its primary advantages are cost efficiency, reduced system wiring, ease of systems maintenance and increased agility of systems [1], because of which, NCS have many potential applications in industrial world.

Different to traditional control systems, in a NCS, both network protocol and the topology have influences on network communication and stability of control systems. A top issue in NCS is packet loss problem. Up to now, two approaches have been applied to deal with packet dropout. One is delayed systems approach[2] and the other is switched system approach[3,4,5]. In paper[3], with modeling the bounded packet loss as a Markov Chain, sufficient and necessary conditions for the means square stability of NCSs with random packet loss is presented and a general controller is established via Lyapunov approach. In paper[5], a packet-loss dependent Lyapunov function is adopted to design packet-loss dependent stabilizing output feedback controllers by resolving some linear matrix inequalities.

W. Liu et al. (Eds.): WISM 2009, LNCS 5854, pp. 563–575, 2009.

As it is noticed, previous studies in the switched system approach are based on two assumptions: a). environmental disturbance is ignored; b). Markov chain governing the packet loss process is preciously known. To our best knowledge, in reality, the disturbance from the environment may inevitablely influence the stability and performance of systems. Besides, it is difficult to know the precious packet loss rate or the precious Markov chain underlying the networks. The real situation is that either the packet loss rate or the Markov chain may be changing responding to varying network congestion situations.

In this paper, both of the limitations mentioned above will be studied. Under the disturbance consideration, with previous results in Markov Jump Linear Systems(MJLS)[6,7,8], the mixed H_2/H_∞ dynamic state feedback control design problem of NCS with Markovian packet loss is studied. The aim is to find a robust mean-square stabilizing packet loss dependent controller that minimize the upper bound for the H_2-norm of NCS, in the meantime, the H_∞-norm is less than a pre-specified value δ. Moreover, in order to describe the uncertainty in packet loss process, it is assumed that the Markovian packet loss is subjected to polytopic-type parameter uncertainty in transition probability matrix. In this paper, for either certain or uncertain Markov packet loss, the convex approach is derived, which enables us to find robust packet loss dependent controller in only one shot, avoiding the iterative process and convergence difficulties.

The paper is organized in the following way: Section 2 gives the description of the systems studied and introduces some auxiliary results about H_2 and H_∞ control in MJLSs. Section 3 studies the mixed H_2/H_∞ control problem for NCS with packet loss. By taking the environmental disturbance as white noise, a sufficient condition is presented for a mean square stabilizing controller that makes the H_∞-norm of NCS with bounded packet loss less than a specified value δ. The condition is written as a Lyapunov-like equation. A LMI optimization problem that leads to an approximation for mixed H_2/H_∞ control problem is derived. For both certain and uncertain Markov chains that governing the packet loss process, the LMIs approaches for finding robust packet-loss dependent controller are presented. Numerical examples are presented in section 4 and section 5 conclude the paper.

Notation: In this paper, if not explicitly stated, matrices are assumed to have compatible dimensions. $\mathbb{N}$ and $\mathbb{Z}^+$ denote the set of natural numbers and the set of nonnegative integer numbers. $\mathbb{R}^n$, $\mathbb{R}^m \times n$ denote the n-dimensional Euclidean space and the set of $m \times n$ real matrices. $\mathbb{M}(\mathbb{R}^{m\times n})$ denotes the normed linear space of all $m \times n$ real matrices. For simplicity, $\mathbb{M}(\mathbb{R}^{n\times n}) = \mathbb{M}(\mathbb{R}^n)$. $L > (\geq, <, \leq)\, 0$ is used to denote a symmetric positive-definite (positive-semidefinite, negative, negative semidefinite) matrix. The superscript "T"denotes the transpose for vectors or matrices. I is identity matrix. $E(.)$ denotes the mathematical expectation operator.

2 Problem Formulation and Auxiliary Results

A discrete-time linear networked control systems is described as follows,

$$\begin{aligned} x(k+1) &= Ax(k) + Bu(k) + Jw(k) \\ z(k) &= C_z x(k) + D_z u(k) + E_z w_z(k) \\ u(k) &= Kx(k). \end{aligned} \tag{1}$$

Where $x(k) \in R^n$ is plant state vector, $u(k) \in R^m$ is control input, $w(k) \in R^p$ is the environmental disturbance, $z(k) \in R^r$ is the controlled output. Matrices in the equations are assumed to have appropriate dimensions.

Set sequence $\mathcal{I} = \{i_1, i_2, i_3, \ldots\}$, where i_k stands for the package that arrived at the actuator $Kx(i_k)$. If $i_k \in \mathcal{I}$, then the package $Kx(i_k)$ is not lost.

Definition 1. *[3]. Packet-loss process is defined as*

$$\left\{\eta(i_k) \triangleq i_{k+1} - i_k; \quad i_k \in \mathcal{I}\right\} \tag{2}$$

where $\eta(i_k) \in \mathcal{N} = \{1, 2, ...N\}$ *and* $N = \max_{i_k \in I}(i_{k+1} - i_k)$.

Definition 2. *[3] Packet-loss process (2) is said to be Markovian if it is governed a discrete-time homogeneous Markov chain on a probability space* $(\Omega, \mathcal{F}, \mathcal{P})$ *and takes values within a transition probability matrix, which follows the description of transition matrix* $\Pi = (\pi_{ij})_{i,j\in\mathcal{N}}$ *in MJLS [9], where the transition probabilities of* $\{\eta(k), k \in Z^+\}$ *are given by*

$$\Pr\{\eta(k+1) = j | \eta(k) = i\} = \pi_{ij}, \quad for\ i, j \in \mathcal{N} \tag{3}$$

with $\pi_{ij} \geq 0$ *for* $i, j \in \mathcal{N}$, *and* $\sum_{j=1}^{N} \pi_{ij} = 1$ *for* $i \in \mathcal{N}$.

In order to model the uncertainty in Markovian packet loss process, here we give the definition of uncertain Markovian packet loss that is studied in this paper.

Definition 3. *For uncertain Markovian packet loss, the transition probability matrix that governing the Markov chain* $\Pi = [\pi_{ij}]$ *is unknown, but belongs to a given polytope*

$$\Lambda_\Pi := \left\{\Pi : \Pi = \sum_{t=1}^{v} a_t \Pi^{(t)}, a_t \geq 0, \sum_{t=1}^{v} a_t = 1\right\} \tag{4}$$

where $\Pi^{(t)} = \left[\pi_{ij}^{(t)}\right]$, *for* $i, j = 1, \ldots, N$, $t = 1, \ldots, v$ *are known transition probability matrices. Note the convex hull of transition probability, each* $\Pi^{(t)}$ *has a probability* $a_t \geq 0, \sum_{t=1}^{v} a_t = 1$.

Packet-loss dependent feedback controller is expected to change the state feedback gain $K = \{K_1, K_2 \ldots, K_N\} \in \mathbb{M}(\mathbb{R}^{m\times n})$ according to the different packet loss situation, correspondingly. At the i_k sampling time, the state feedback gain at next sampling period is calculated according to the lost packet in the previous sampling time $\eta(i_{k-1})$ and thus $u(k) = K_{\eta(i_k-1)}x(i_k)$. Without lost generality[10], it is assumed that $w(k)$ is piecewise constant, that is, $w(i_k) = w(i_k+1) = \cdots = w(i_k+\eta(i_k)-1)$. The evolution of the plant states can be described as follows.

$$
\begin{aligned}
x(i_k+1) &= \left(A + BK_{\eta(i_k-1)}\right)x(i_k) + Jw(i_k) \\
x(i_k+2) &= Ax(i_k+1) + \left(BK_{\eta(i_k-1)}\right)x(i_k) + Jw(i_k) \\
&= \left(A^2 + \left(ABK_{\eta(i_k-1)} + BK_{\eta(i_k-1)}\right)\right)x(i_k) + (AJ+J)\,w(i_k) \\
&\vdots
\end{aligned}
$$

$$
x(i_k+\eta(i_k)) = \left(A^{\eta(i_k)} + \sum_{i=0}^{\eta(i_k)-1} A^i BK_{\eta(i_{k-1})}\right)x(i_k) + \sum_{i=0}^{\eta(i_k)-1} A^i Jw(i_k) \tag{5}
$$

The closed-loop system is expressed as $\mathcal{G}_{clNCS}$:

$$
\begin{aligned}
&x(i_{k+1}) = \left(A^{\eta(i_k)} + B_{\eta(i_k)}K_{\eta(i_{k-1})}\right)x(i_k) + J_{\eta(i_k)}w(i_k), \\
&z(i_k) = \left(C_z + D_zK_{\eta(i_{k-1})}\right)x(i_k), \quad i_k \in \mathcal{I} \\
&\text{where } B_{\eta(i_k)} = \sum_{i=0}^{\eta(i_k)-1} A^i B, \quad J_{\eta(i_k)} = \sum_{i=0}^{\eta(i_k)-1} A^i J, \\
&u(k) = K_{\eta(i_k-1)}x(i_k), \quad z(i_k) \text{ is the controlled output.}
\end{aligned} \tag{6}
$$

Definition 4. *Refer to the definition of H_2-norm in [9], assuming that $\mathcal{G}_{clNCS}$ is stable, the H_2-norm of systems $\mathcal{G}_{clNCS}$ is defined as*

$$
\begin{aligned}
&\|\mathcal{G}_{clNCS}\|_2^2 = \sum_{i=1}^{N} Tr\left(J_i^T E_i(S)\,J_i\right), \text{ where } S_i = S_i^T > 0 \text{ solves} \\
&S_i = E_i\left(\left(A^j + B_jK_i\right)^T S_j\left(A^j + B_jK_i\right)\right) + (C_z + D_zK_i)^T(C_z + D_zK_i)
\end{aligned} \tag{7}
$$

Where $x(k) \in R^n$ is plant state vector, $u(k) \in R^m$ is control input, $w(k) \in R^p$ is the environmental disturbance, $z(k) \in R^r$ is the controlled output.

Definition 5. *The H_∞-norm of a stable system $\mathcal{G}_{clNCS}$ from the input w to the output z is given by*

$$
\|\mathcal{G}_{clNCS}\|_\infty^2 = \sup \frac{\|z\|_2^2}{\|w\|_2^2} \tag{8}
$$

3 Main Results

Theorem 1. *If for $\delta > 0$ a fixed real number, there exists $P = (P_1, \ldots, P_N) \geq 0$, $P = \mathbb{M}(\mathbb{R}^n)$ and $K = \{K_1, K_2 \ldots, K_N\} \in \mathbb{M}(\mathbb{R}^{n\times m})$ such that $\forall\ i$, the* *following inequalities could be satisfied, then NCS with Markovian packet-loss process $\mathcal{G}_{clNCS}$ (6) is mean-square stable.*

$$
\begin{aligned}
&-P_i + E_i\left(\left(A^j + B_j K_i\right)^T P_j \left(A^j + B_j K_i\right)\right) + \\
&\left(C_z + D_z K_i\right)^T \left(C_z + D_z K_i\right) + \frac{1}{\delta^2} P_i E_i\left(J_j J_j^T\right) P_i \leq 0
\end{aligned}
\tag{9}
$$

where $K = \{K_1, K_2 \ldots, K_N\} \in \mathbb{K}$ satisfying the following conditions:

1. The H_∞-norm of $\mathcal{G}_{clNCS}$

$$
\|G_{clNCS}\|_\infty^2 \leq \delta^2 (1 - v) \leq \delta^2, \text{ where } v \in \left(0, \left(1/\delta^2\right) \sum_{i=1}^{N} tr\left(J_i^T P_i J_i\right)\right) \tag{10}
$$

2. The H_2-norm of $\mathcal{G}_{clNCS}$

$$
\|\mathcal{G}_{clNCS}\|_2^2 \leq \sum_{i=1}^{N} tr\left(J_i^T E_i(P) J_i\right). \tag{11}
$$

Proof. For simplicity's sake, $\hat{A}_{\eta(i_k)} = A^{\eta(i_k)} + B_{\eta(i_k)} K_{\eta(i_{k-1})}$, $\hat{C}_{\eta(i_k)} = C_z + D_z K_{\eta(i_{k-1})}$, $\eta(i_{k-1}) = i$, $\eta(i_k) = j$. Then

$$
\begin{aligned}
&E_{\eta(i_{k-1})}\left(x(i_{k+1})^T P_{\eta(i_k)} x(i_{k+1})\right) \\
=&E_{\eta(i_{k-1})}\left(\left(\hat{A}_{\eta(i_k)} x(i_k) + J_{\eta(i_k)} w(i_k)\right)^T P_{\eta(i_k)} \left(\hat{A}_{\eta(i_k)} x(i_k) + J_{\eta(i_k)} w(i_k)\right)\right) \\
=&E_i\left(\left(\hat{A}_j x(i_k)\right)^T P_j \left(\hat{A}_j x(i_k)\right)\right) + E_i\left(\left(J_j w(i_k)\right)^T P_j \left(J_j w(i_k)\right)\right. \\
&\left. + \left(\hat{A}_j x(i_k)\right)^T P_j \left(J_j w(i_k)\right) + \left(\hat{A}_j x(i_k)\right) P_j \left(J_j w(i_k)\right)^T\right)
\end{aligned}
\tag{12}
$$

From e.q. (9), we have

$$
\begin{aligned}
&E_i\left(\left(\hat{A}_j x(i_k)\right)^T P_j \left(\hat{A}_j x(i_k)\right)\right) \\
<&x(i_k)^T \left(P_i - \left(C_z + D_z K_i\right)^T \left(C_z + D_z K_i\right) - \delta^{-2} P_i E_i\left(J_j J_j^T\right) P_i\right) x(i_k)
\end{aligned}
\tag{13}
$$

Using the result in e.q. (13) to e.q. (12), and also notice $z(i_k) = (C_z + D_z K_i) x(i_k)$,

$$
\begin{aligned}
&\left\|P_j^{1/2} x(i_{k+1})\right\|_2^2 - \left\|P_i^{1/2} x(i_k)\right\|_2^2 + \|z(i_k)\|_2^2 \\
\leq& -\delta^{-2}\left\|J_j^T P_i x(i_k)\right\|_2^2 + \|E_i(P_j) J_j w(i_k)\|_2^2 + E_i\left(\left(\hat{A}_j x(i_k)\right)^T P_j \left(J_j w(i_k)\right) + \left(\hat{A}_j x(i_k)\right) P_j \left(J_j w(i_k)\right)^T\right) \\
=& -\delta^{-2}\left\|J_j^T P_i x(i_k)\right\|_2^2 + \delta^{-2}\left\|J_j^T P_j x(i_{k+1})\right\|_2^2 - \delta^{-2}\left\|J_j^T P_j x(i_{k+1})\right\|_2^2 - \\
&E_i\left(\left(J_j w(i_k)\right)^T P_j \left(J_j w(i_k)\right)\right) + 2E_i\left(w(i_k)^T J_j^T P_j \left(\hat{A}_j x(i_k) + J_j w(i_k)\right)\right)
\end{aligned}
$$

Thus

$$\left\|P_j^{1/2}x(i_{k+1})\right\|_2^2-\left\|P_i^{1/2}x(i_k)\right\|_2^2+\|z(i_k)\|_2^2-\delta^{-2}\left\|J_j^TP_jx(i_{k+1})\right\|_2^2+\delta^{-2}\left\|J_j^TP_ix(i_k)\right\|_2^2$$
$$\leq-\delta^{-2}\left\|J_j^TP_jx(i_{k+1})\right\|_2^2+2E_i\left(w(i_k)^TJ_j^TP_jx(i_{k+1})\right)-\delta^2\|w(i_k)\|_2^2+$$
$$E_i\left(w(i_k)^T\left(\delta^2I-J_j^TP_jJ_j\right)w(i_k)\right)$$
$$=-\left\|\delta^{-1}J_j^TP_jx(i_{k+1})-\delta w(i_k)\right\|_2^2+E_i\left(w(i_k)^T\left(\delta^2I-J_j^TP_jJ_j\right)w(i_k)\right)$$
$$\leq E_i\left(w(i_k)^T\left(\delta^2I-J_j^TP_jJ_j\right)w(i_k)\right)$$

The initial state is $x(0)=0$. While $k\to\infty$, $\|x(k)\|_2\to 0$. Then we get

$$\|z(i_k)\|_2^2\leq\delta^2\sum_{k=0}^{\infty}E_i\left(w(i_k)^T\left(I-\delta^{-2}J_j^TP_jJ_j\right)w(i_k)\right)$$

Since $E_i\left(J_j^TP_jJ_j\right)=\sum_{j=1}^{N}tr\left(\pi_{ij}J_j^TP_jJ_j\right)\leq\sum_{j=1}^{N}tr\left(J_j^TP_jJ_j\right)$

Set $v\in\left(0,(1/\delta^2)\sum_{j=1}^{N}tr\left(J_j^TP_jJ_j\right)\right)$, then $\|z(i_k)\|_2^2\leq\delta^2(1-v)\|w\|_2^2$.

$$\text{So } \|\mathcal{G}_{clNCS}\|_\infty=\sup\frac{\|z\|_2}{\|w\|_2}\leq\delta(1-v)^{1/2}<\delta$$

For $V_i\geq 0$, $\quad i=(1,\cdots,N)$, e.q. (9) could be written as

$$\begin{aligned}P_i=\;&E_i\left(\left(A^j+B_jK_i\right)^TP_j\left(A^j+B_jK_i\right)\right)+\\&\left(C_z+D_zK_i\right)^T\left(C_z+D_zK_i\right)+\frac{1}{\delta^2}P_iE_i\left(J_jJ_j^T\right)P_i+V_i\end{aligned}$$

From Definition 4,

$$S_i=E_i\left(\left(A^j+B_jK_i\right)^TS_j\left(A^j+B_jK_i\right)\right)+\left(C_z+D_zK_i\right)^T\left(C_z+D_zK_i\right)\leq P_i$$

Therefore, $\|\mathcal{G}_{clNCS}\|_2^2=\sum_{i=1}^{N}Tr\left(J_i^TE_i(S)J_i\right)\leq\sum_{i=1}^{N}Tr\left(J_i^TE_i(P)J_i\right)$.

The proof is completed.

Corollary 1. *Given $\delta>0$, find $\mathbb{K}^{\mathrm{H}}=\left(K_1^H,K_2^H,\ldots,K_N^H\right)\in\mathbb{M}\left(\mathbb{R}^{m\times n}\right)$ which minimizes ξ subject to $\|\mathcal{G}_{clNCS}\|_2\leq\xi^{1/2}$, $\|\mathcal{G}_{clNCS}\|_\infty\leq\delta$, under the governance of transition matrix Π. The solution is obtained via an LMI optimization problem, which is defined as*
Find $\mathrm{P}=(P_1,P_2,\ldots,P_N)\in\mathbb{R}^{n\times n}>0$, symmetry matrices $\mathrm{Q}=(Q_1,Q_2,\ldots,Q_N)\in\mathbb{R}^{n\times n}>0$, and $\mathrm{Y}=(Y_1,Y_2,\ldots,Y_N)\in\mathbb{R}^{m\times n}$. Set $K_i^H=Y_iQ_i^{-1}$ and $\xi=\sum_{i=1}^{N}tr\left(J_i^TE(P)J_i\right)$, Then $K_i^H\in\mathbb{K}^H$ and $\|\mathcal{G}_{clNCS}\|_2\leq\xi^{1/2}$, $\|\mathcal{G}_{clNCS}\|_\infty\leq\delta$, such that

$$\xi = \min tr\left(\sum_{i=1}^{N}\left(J_i^T E_i\left(P\right)J_i\right)^{1/2}\right) \tag{14}$$

subject to: For all $i, j = \{1, 2, \ldots, N\}$, *have*

$$\begin{bmatrix} P_i & 0 \\ 0 & Q_i \end{bmatrix} \geq 0 \tag{15a}$$

$$\begin{bmatrix} Q_i & M_i^T & Q_iC_z^T & Y_i^TD_z^T & F_i \\ M_i & W & 0 & 0 & 0 \\ C_zQ_i & 0 & I & 0 & 0 \\ D_zY_i & 0 & 0 & I & 0 \\ F_i^T & 0 & 0 & 0 & \delta^2 I \end{bmatrix} \geq 0 \tag{15b}$$

where

$$M_i = \left[\left(\pi_{i1}\right)^{\frac{1}{2}}\left(AQ_i + BY_i\right)\left(\pi_{i2}\right)^{\frac{1}{2}}\left(A^2Q_i + B_2Y_i\right)\cdots\left(\pi_{iN}\right)^{\frac{1}{2}}\left(A^NQ_i + B_NY_i\right)\right] \tag{15c}$$

$$F_i = \sum_{j=1}^{N}\sqrt{\pi_{ij}}J_j \tag{15d}$$

$$W = diag\left[Q_1, Q_2, \cdots, Q_N\right] \text{ and } B_j = \sum_{i=0}^{j-1}A^iB, \quad J_j = \sum_{i=0}^{j-1}A^iJ, \tag{15e}$$

Proof. According to Corollary 1, in order to minimize the H_2-norm, which is denoted as ξ^2, it is assumed that there exist $\mathrm{Q} = (Q_1, Q_2, \ldots, Q_N) \in \mathbb{M}(\mathbb{R}^n) > 0$ satisfying $P_i \geq Q_i^{-1}$, for all $i = 1, 2, \ldots, N$, $j = 1, 2, \ldots, N$. Then $Q_iP_iQ_i \geq Q_iQ_i^{-1}Q_i = Q_i$, according to e.q. (9), we have

$$\begin{aligned}
Q_iQ_i^{-1}Q_i \geq\ & Q_iE\left(\left(A^j + B_jK_i^H\right)^T P_j\left(A^j + B_jK_i^H\right)\right)Q_i + \\
& Q_i\left(C_z + D_zK_i^H\right)^T\left(C_z + D_zK_i^H\right)_i Q_i + Q_iP_iE\left(\delta^{-2}J_jJ_j^T\right)P_iQ_i \\
\geq\ & Q_iE\left(\left(A^j + B_jK_i^H\right)^T Q_j^{-1}\left(A^j + B_jK_i^H\right)\right)Q_i + \\
& Q_i\left(C_z + D_zK_i^H\right)^T\left(C_z + D_zK_i^H\right)Q_i + Q_iQ_i^{-1}E\left(\delta^{-2}J_jJ_j^T\right)Q_i^{-1}Q_i \\
=\ & E\left(\left(Q_i\left(A^j\right)^T + Y_i^TB_j^T\right)Q_j^{-1}\left(A^jQ_i + B_jY_i\right)\right) + \\
& \left(Q_iC_z^T + Y_i^TD_z^T\right)\left(C_zQ_i + D_zY_i\right) + \delta^{-2}E\left(J_jJ_j^T\right)
\end{aligned} \tag{16}$$

Applying Schur complementary to e.q. (16), e.q. (15b) is obtained. The robust packet-loss dependent feedback control law is $K_i^H = Y_iQ_i^{-1}, \forall\, i$.

Remark 1. In order to obtain the H_∞ controller, we have to minimize δ in e.q. (9) in Theorem 1. The solution is also obtained via an LMI optimization problem and the following result is derived.

Corollary 2. *Find* $\mathrm{P}=(P_1,P_2,\ldots,P_N)\in\mathbb{R}^{n\times n}>0$*, symmetry matrices* $\mathrm{Q}=(Q_1,Q_2,\ldots,Q_N)\in\mathbb{R}^{n\times n}>0$*, and* $\mathrm{Y}=(Y_1,Y_2,\ldots,Y_N)\in\mathbb{R}^{m\times n}$ *to solve the optimization problem, which is described as follows,*

minimize $u(u^{1/2}=\delta=\|\mathcal{G}_{clNCS}\|_\infty)$*, subject to:*
For all $i,j=\{1,2,\ldots,N\}$*, have*

$$\begin{bmatrix} Q_i & M_i^T & Q_iC_z^T & Y_i^TD_z^T \\ M_i & W & 0 & 0 \\ C_zQ_i & 0 & I^p & 0 \\ D_zY_i & 0 & 0 & I^m \end{bmatrix} \geq \begin{bmatrix} 0 & F_i \\ F_i^T & -u \end{bmatrix} \tag{17}$$

where M_i*,* F_i*,* W *is the same as defined in Corollary 1.*

Proof. Applying Schur Complimentary to e.q. (15b), e.q. (17) is obtained.

Remark 2. If the transition probability matrix in the Markovian packet loss is not preciously known, but belong to a given polytope (4) in Definition 3, the robust packet-loss dependent feedback controller could also be obtained via LMI optimization problem.

Corollary 3. *Given* $\delta>0$*, find* $\mathbb{K}=(K_1,K_2,\ldots,K_N)\in\mathbb{M}(\mathbb{R}^{m\times n})$ *which minimizes* ξ *subject to* $\|\mathcal{G}_{clNCS}\|_2\leq\xi^{1/2}$*,* $\|\mathcal{G}_{clNCS}\|_\infty\leq\delta$*, under the governance of transition matrix* $\Pi^t\in\Lambda_\Pi$*. The solution is obtained via an LMI optimization problem, which is defined as*
Find $P^{(t)}=\left(P_1^{(t)},P_2^{(t)},\ldots P_N^{(t)}\right)\in\mathbb{R}^{n\times n}>0$*,* $\mathrm{Q}=(Q_1,Q_2,\ldots,Q_N)\in\mathbb{R}^{n\times n}$*, and* $\mathrm{Y}=(Y_1,Y_2,\ldots,Y_N)\in\mathbb{R}^{m\times n}$*.*
Set $K_i=Y_iQ_i^{-1}$ *and* $\xi=\sum_{a_t=1}^{v}\sum_{i=1}^{N}a_t tr\left(J_i^TE_i\left(P^{(t)}\right)J_i\right)$*, Then* $K\in\mathbb{K}$ *and* $\|\mathcal{G}_{clNCS}\|_2\leq\xi^{1/2}$*,* $\|\mathcal{G}_{clNCS}\|_\infty\leq\delta$*, such that*

$$\xi=\min\sum_{a_t=1}^{v}\sum_{i=1}^{N}a_t tr\left(J_i^TE_i\left(P^{(t)}\right)J_i\right) \tag{18}$$

where a_t *is the probability of occurrence of network condition with transition matrix* $\Pi^{(t)}$*.*
For all $i,j\in\{1,2,\ldots,N\},t\in\{1,2,\ldots,v\}$*, have*

$$\begin{bmatrix} -Q_i-Q_i^T+P_i^{(t)} & {M^{(t)}}^T & Q_iC_z^T & Y_i^TD_z^T & F^{(t)} \\ M^{(t)} & W^{(t)} & 0 & 0 & 0 \\ C_zQ_i & 0 & -I & 0 & 0 \\ D_zY_i & 0 & 0 & -I & 0 \\ {F^{(t)}}^T & 0 & 0 & 0 & -\delta^2 I \end{bmatrix}<0 \tag{19}$$

where

$$M^{(t)} = \left[\left(\pi_{i1}^{(t)}\right)^{\frac{1}{2}} (AQ_i + BY_i) \left(\pi_{i2}^{(t)}\right)^{\frac{1}{2}} (A^2 Q_i + B_2 Y_i) \cdots \left(\pi_{iN}^{(t)}\right)^{\frac{1}{2}} (A^N Q_i + B_N Y_i) \right] \tag{20a}$$

$$F^{(t)} = \sum_{j=1}^{N} \sqrt{\pi_{ij}^{(t)}} J_j \tag{20b}$$

$$W^{(t)} = diag\left[-P_1^{(t)}, -P_2^{(t)}, \cdots, -P_N^{(t)}\right], \text{ and } B_j = \sum_{i=0}^{j-1} A^i B, \quad J_j = \sum_{i=0}^{j-1} A^i J, \tag{20c}$$

Proof. It should be noticed that the matrices $P_i, i = 1, \ldots, N$ in e.q. (15b) are independent of the system uncertainty. If the P_i is constant for all transition probability matrices in the convex set, the robust stability condition can be conservative. Therefore, here $P_i^{(t)}$ varies for each transition probability matrix $\Pi^{(t)}$. The rest of the proof is easy to obtain. The total number of linear matrix inequality for condition e.q. (19) is $N \times v$.

Remark 3. As for NCSs, it is often the case that the packet loss might change unexpectedly to the network condition and thus the transition probability matrix of packet loss process will change accordingly. Therefore, Corollary 3 may have more practical usage while it comes to the real-world situation, which means the robust packet loss dependent controller can stabilize the system under different network conditions.

4 Numerical Examples

Example 1. Considered the six-dimension system - flexible inverted pendulum in Fig. 1.

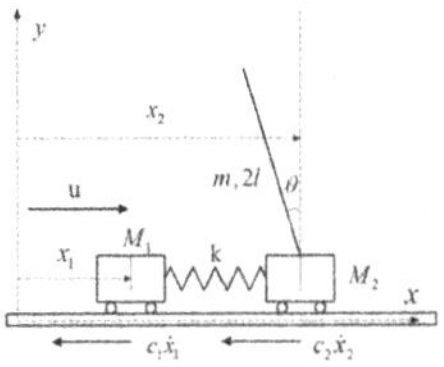

Fig. 1. One-stage Flexible Inverted Pendulum

Consider the discrete-time system model for flexible inverted pendulum :

$$x(k+1)=\begin{bmatrix} 0.9719 & 0.005 & 0.0281 & 0 & 0 & 0 \\ -11.1525 & 0.9718 & 11.1525 & 0.0281 & 0 & 0 \\ 0.0273 & 0 & 0.9727 & 0.005 & 0 & 0 \\ 10.8279 & 0.0273 & -10.8279 & 0.9718 & 0.0042 & 0 \\ 0.082 & 0.0001 & -0.082 & -0.0001 & 1.0004 & 0.005 \\ 32.4878 & 0.082 & -32.4878 & -0.0846 & 0.1596 & 1.0004 \end{bmatrix} x(k)+\begin{bmatrix} 0 \\ 0.0045 \\ 0 \\ 0 \\ 0 \\ 0.0001 \end{bmatrix} u(k) \tag{21}$$

The eigenvalue of A is $\left[0.9443+0.3283i \ 0.9443-0.3283i \ 1.0281 \ 0.9727 \ 0.9996 \ 1\right]^T$. The open loop system is unstable. The packet-loss upper bound is set as $N=4$ per 5 packets, which means that the maximum lost packet is 80% of the whole.

Case 1: The Markovian packet loss process transition matrix is preciously known as $\Pi=\begin{bmatrix} 0.5 & 0.2 & 0.1 & 0.1 & 0.1 \\ 0.2 & 0.5 & 0.3 & 0 & 0 \\ 0 & 0.2 & 0.5 & 0.3 & 0 \\ 0 & 0 & 0.2 & 0.5 & 0.3 \\ 0.1 & 0.1 & 0.1 & 0.2 & 0.5 \end{bmatrix}$. The initial state of this system is $x(0)=\left[5\ 0\ 0\ 0\ -5\ 0\right]$ and the initial distribution of packet loss is $Init=\left[0\ 1\ 0\ 0\ 0\right]$. The velocity of cart 1 ($x(2)$), cart 2 ($x(4)$) and the angular velocity ($x(6)$) is observed.

By Theorem 11 in paper [3], general controller $K=\left[-1281.5\ -72.1\ 1290.7\ 91.9\ -136\ -39.9\right]$.

According to Theorem 1 and Corollary 1, we use *mincx* in Matlab LMI Toolbox to solve the H_2/H_∞-control problem with $\delta=80$. The controller $\mathbb{K}^H$ is obtained as follows:

$$\begin{aligned} K_1^H &= \left[-4556.0\ -94.9\ 7575.9\ 1417.8\ -2860.3\ -508.6\right] \\ K_2^H &= \left[9000\ -141\ 13533\ 2512\ -4795\ -927\right] \\ K_3^H &= \left[-5545.5\ -107\ 9159\ 1731.2\ -3301.8\ -625.5\right] \\ K_4^H &= \left[-3681.8\ -88\ 6256\ 1205.9\ -2288.7\ -433\right] \\ K_5^H &= \left[-3810.3\ -86.2\ 6420.5\ 1237.7\ -2289.8\ -449.8\right] \end{aligned}$$

In simulation, we add white noise as the disturbance of this system and use robust packet-loss dependent controller $\mathbb{K}^H$. The simulation result is presented in Fig. 2c and corresponding Markovian packet loss process is presented in Fig. 2a. For comparison, we also present the simulation result of the closed-loop systems with general controller in Fig. 2b.

Moreover, by convex approach in Corollary 2, the H_∞ controller $\mathbb{K}^{hinf}$ can be obtained using *gevp* in Matlab:

$$\begin{aligned} K_1^{hinf} &= \left[-1623.9\ -79.5\ 1697.5\ 117.9\ -181\ -47.6\right] \\ K_2^{hinf} &= \left[-3079\ -110.1\ 3188.4\ 164.5\ -258.7\ -69.4\right] \\ K_3^{hinf} &= \left[-1882.8\ -84.6\ 1959.5\ 128.8\ -207.5\ -53.2\right] \\ K_4^{hinf} &= \left[-1199.2\ -69.9\ 1259.7\ 105.7\ -167.3\ -425\right] \\ K_5^{hinf} &= \left[-1052.3\ -66.5\ 1109.7\ 99.9\ -155.7\ -39.8\right] \end{aligned}$$

Table 1. System Performance Comparison (according to pendulum angular velocity: $x(6)$)

Comparison Results	Maximum Error	Steady-state error	Regulation Time($\pm$ 5%)
NCS with K	-882.2	-7.38	110
NCS with $\mathbb{K}^H$	-352.9	-2.879	55

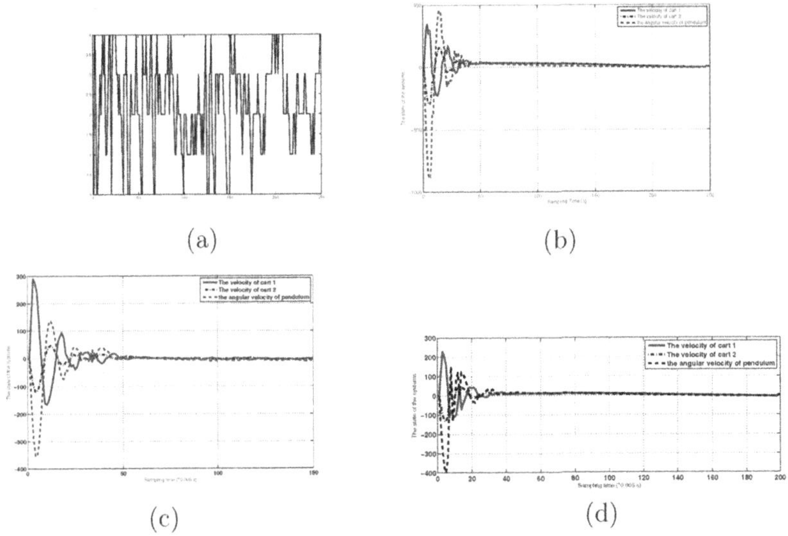

Fig. 2. (a) is Markov chain (250 sampling times); (b) is the state response with the general controller; (c) is the state response with mixed H_2/H_∞ controller.; (d) is the state response with H_∞ controller.

The calculated minimum of H_∞-norm of pendulum system is 316.2. The simulation result of pendulum controlled by H_∞ controller is shown in Fig. 2d.

As shown in Table 1, the pendulum with robust packet loss dependent controller has improved performance than it with a general controller.

Case 2: The Markovian packet loss process is not preciously known but belong to a convex set Λ_Π with 3 vertices:

$$\Pi_1 = \begin{bmatrix} 0.8 & 0.2 & 0 & 0 & 0 \\ 0.4 & 0.5 & 0.1 & 0 & 0 \\ 0.1 & 0.3 & 0.5 & 0.2 & 0 \\ 0.1 & 0.1 & 0.2 & 0.5 & 0.1 \\ 0 & 0.1 & 0.2 & 0.3 & 0.4 \end{bmatrix}, \Pi_2 = \begin{bmatrix} 0.5 & 0.2 & 0.1 & 0.1 & 0.1 \\ 0.2 & 0.5 & 0.3 & 0 & 0 \\ 0 & 0.2 & 0.5 & 0.3 & 0 \\ 0 & 0 & 0.2 & 0.5 & 0.3 \\ 0.1 & 0.1 & 0.1 & 0.2 & 0.5 \end{bmatrix}, \Pi_3 = \begin{bmatrix} 0 & 0.8 & 0.2 & 0 & 0 \\ 0 & 0.5 & 0.5 & 0 & 0 \\ 0 & 0.2 & 0.5 & 0 & 0.3 \\ 0 & 0 & 0 & 0.8 & 0.2 \\ 0 & 0 & 0 & 0.2 & 0.8 \end{bmatrix}$$

Network condition: Π_1 - good; Π_2 - normal; Π_3 - bad.

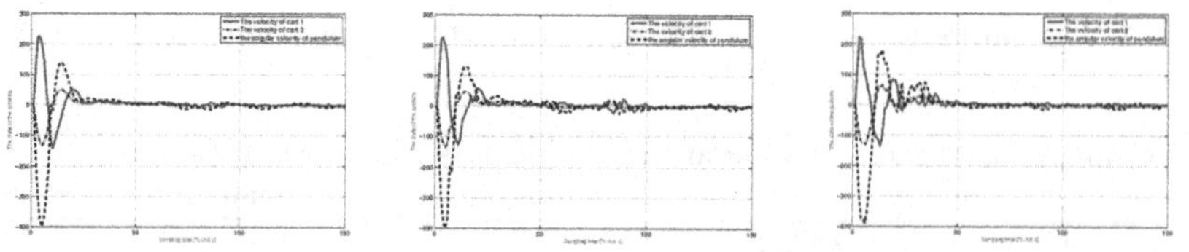

(a) (b) (c)

Fig. 3. State response under different network conditions: (a) is the state response under good network condition; (b) is the state response under normal network condition; (c) is the state response under bad network condition

For good, normal and bad conditions, the loss packets took up 10%, 45% and 67% of the total number of transferred packets. $a_{(t)}$ for the distribution of $\Pi^{(t)}$ is 30%, 40%, 30%, respectively. The initial state of system and Markovian packet loss distribution is the same as in *Case.1*.

The robust packet-loss dependent controller for NCS with uncertain Markov packet loss $\mathbb{K}$ is obtained by Corollary 3. The simulation results are shown in Fig. 3a, 3b, 3c.

$\mathbb{K}$:

$$K_1 = \begin{bmatrix} -5961.4 & -107.8 & 9834.6 & 1815.5 & -3740.9 & -653.1 \end{bmatrix}$$
$$K_2 = \begin{bmatrix} -8673 & -140 & 16108 & 3119 & -6010 & -1110 \end{bmatrix}$$
$$K_3 = \begin{bmatrix} -4951 & -98 & 10821 & 2336 & -4140 & -819 \end{bmatrix}$$
$$K_4 = \begin{bmatrix} -3942.1 & -90.5 & 8622.2 & 1888.9 & -3285.6 & -663.7 \end{bmatrix}$$
$$K_5 = \begin{bmatrix} -3505.1 & -84.9 & 9182.3 & 2153.8 & -3521.4 & -749.5 \end{bmatrix}$$

According to the simulation results of robust packet loss dependent controller for NCS with uncertain Markovian packet loss, it is noticed that the set of state feedback gains - $\mathbb{K}$ stabilize the system in mean square stable sense for all transition probability matrices in the convex set Λ_{Π}.

5 Conclusion

In this paper, the problems of robust packet-loss dependent state feedback control design of NCSs with Markovian packet loss are studied. The definition and theories from mean square stability in MJLSs are employed and implemented for robust stabilization problem of NCSs with Markovian packet loss. Besides studying NCS with certain Markovian packet loss, it is assumed that the transition probability matrix may belong to an appropriate convex set so that the uncertainty in network condition can be well described. The convex approaches of solving the LMIs and thus obtaining the robust packet-loss dependent feedback gain are given. The control method presented in this paper minimizes an upper bound for the H_2 norm, under the restriction that the H_∞- norm is less than a pre-specified value δ. Testified in numerical simulation, robust packet-loss dependent controller ensures the mean square stability of NCS with Markovian packet loss and has improved the performance of the same system controlled by a general controller. The potential of robust packet-loss dependent controller for NCS with uncertain Markov packet loss was also illustrated in example and simulation.

References

1. Zhang, W., Branicky, M.S., Phillips, S.M.: Stability of networked control systems. IEEE Control Systems Magazine 21, 84–99 (2001)
2. Nilsson, J., Bernhardsson, B.: Analysis of real-time control systems with time delays. In: Proc. of the 35th Conf. on Decision and Contr., pp. 3173–3178 (1996)
3. Xiong, J., Lam, J.: Stabilization of linear systems over networks with bounded packet loss. Automatica 43, 80–87 (2007)
4. Mei, Y., Long, W., Guangming, X., Tianguang, C.: Stabilization of networked control systems with data packet dropout via switched system approach. In: IEEE International Symposium on Computer Aided Control Systems Design, pp. 362–367 (2004)
5. Junyan, Y., Long, W., Mei, Y., Yingmin, J., Jie, C.: Stabilizability of networked control systems via packet-loss dependent output feedback controllers. In: American Control Conference, pp. 3620–3625 (2008)
6. Costa, O.L.V.: Stability result for discrete-time linear systems with markovian jumping parameters. Journal of Mathematical Analysis and Applications, 157–178 (1993)
7. Junlin, X., James, L.: Robust h2 control of markovian jump systems with uncertain switching probabilities. Intern. J. Syst. Sci. 40, 255–265 (2009)
8. Wu-Hua, C., Jian-Xin, X., Zhi-Hong, G.: Guaranteed cost control for uncertain markovian jump systems with mode-dependent time-delays. IEEE Transactions on Automatic Control 48, 2270–2277 (2003)
9. Costa, O.L.V., Fragoso, M.D., Marques, R.P.: Discrete-Time Markov Jump Linear Systems. Probability and its Applications. Springer, Heidelberg (2005)
10. Wang, Y.L., Yang, G.H.: H[infinity] controller design for networked control systems via active-varying sampling period method. Acta Automatica Sinica 34, 814–818 (2008)

Research on the Trust-Adaptive Scheduling for Data-Intensive Applications on Data Grids

Wei Liu[1,2,3] and Wei Du[1,4]

[1] College of Computer Science and Technology, Wuhan University of Technology, Wuhan 430063, China
[2] State Key Laboratory for Novel Software Technology, Nanjing University, Nanjing 210093, P.R. China
[3] State Key Lab of Software Engineering, Wuhan University, Wuhan 430063, China
[4] College of Computer Science and Technology, Huazhong University of Science and Technology, Wuhan 430074, China

wliu@whut.edu.cn, smilekitty@smail.hust.edu.cn

Abstract. Flexible trust requirements are increasingly become major concern for applications running on data grids. However, existing scheduling strategies are unable to satisfy the trust needs of complex data-intensive applications. To address the issue, we introduce the trust heterogeneity concept for our scheduling model in the context of data grid. Based on the concept, we propose the trust-adaptive scheduling algorithm, which strives to maximize the probability that all tasks are executed with the flexible trust requirements. Extensive experimental studies indicate that the scheduling performance is affected by trust heterogeneities and other performance metrics. Additionally, empirical results demonstrate that with respect to trust and performance, the proposed scheduling algorithm achieves a good trade-off between trust heterogeneity and responsiveness.

Keywords: Trust Heterogeneity, Trust-Adaptive Scheduling Algorithm, Data-intensive Applications, Resource Management, Data Grid.

1 Introduction

Data grid is defined as an infrastructure that manages large scale files and provides computational resources across widely distributed computing [1, 2, 3]. Storage systems including hard disks and any storage media are the most commonly used resources in data grid. Our study is intended to introduce a concept of trust heterogeneity, which provides a means of measuring trust overhead incurred by trust services in the context of data grid.

There have been some efforts to develop data-intensive applications like bioinformatics and high energy physics in data grid [4, 5]. In recent years, many studies addressed the issue of scheduling for data grid [6]. Although existing scheduling algorithms provide high performance for data-intensive applications, these algorithms are not adequate for data-intensive applications with trust-sensitive constraints.

W. Liu et al. (Eds.): WISM 2009, LNCS 5854, pp. 576–585, 2009.

It is imperative for computer systems to embrace a wide range of trust services to support applications. Importantly, trust requirements of next generation data-intensive applications are flexible, meaning that it is highly desirable for storage resources in data grid to dynamically adjust trust service to meet the flexible trust needs of complex applications. Providing flexible trust service for storage resources in data grids is challenging, because requests in some data-intensive applications need to be completed within desired response times. In this study, guaranteed response times and flexible trust service are two performance goals to be achieved in data grids. Automatic adjusting trust service is of a critical issue in development of storage systems for data grids.

In this paper, we propose a framework for adaptable trust service in data grid based on the trust heterogeneity. The framework is applicable to other distributed computing environments and thus, we can integrate the trust-adaptive window into resource scheduling strategy to make the resource management adapt to changing trust requirements for data-intensive applications. We implement the trust-adaptive window controller to demonstrate the effective adaptability of trust service. Experimental results show that the proposed framework achieves high quality of trust while guaranteeing desired response times.

The rest of this paper is organized as follows. Section 2 includes a summary of related work in this area. Section 3 describes the system model. Section 4 models the tasks with trust requirements. Section 5 introduces the scheduling strategy. In Section 6 analyses the simulation parameters and results, and evaluates the effectiveness of the proposed framework. Section 7 concludes the paper with summary and future research directions.

2 Related Work

Trust service is important to various data-intensive applications in grid computing. This is mainly because data stored in storage resources of data grid requires special protection against any untrusted or disreputable resource access.

Gui proposed a novel security framework for securely handling privacy sensitive information on the grid [7]. Hassan designed a large-scale, self-managing trust management framework that makes efficient use of apparently invisible evidences that are scattered across potentially global networks [8]. Now, the issue of trust has been paid more attention both experimentally and theoretically in distributed computing systems. Azzedin suggested associating trust-index only to a domain, then to a broker within a domain [9]. Kwei-Jay Lin suggested to having a single broker for each domain, leading to a single-point failure [10]. Jon B. Weissman proposed adaptive reputation-based scheduling on unreliable distributed infrastructures [11, 12]. Song proposed security-driven scheduling algorithms for grids [13, 14]. Although the above works address trust requirements in distributed systems, none of them can provide adaptable trust services for data-intensive applications in grid environments.

Consequently, we need to find a solution which can quantitatively evaluate the trust overhead and adjust the trust service, because flexible trust requirements will be one of the main characteristics in the next-generation data grid. However, to the best of our knowledge, the way of trust-adaptive resource management has received little attention. Jon B. Weissman analyzed several existing reputation-based trust algorithms and adapted them to the problem of service selection in grid environment, thus performed a quantitative comparison of both the accuracy and overhead associated with these techniques [15]. Weisong Shi compared different trust recommendation algorithms in open network environments focusing on several performance metrics [16]. Their works provided us an insightful view of trust overhead measurement and resource management.

In this paper, we propose a trust-adaptive window task management for data grids. Our scheme can be integrated into existing scheduling algorithms to substantially improve adaptability of trust service of the systems without the loss of conventional performance metrics.

3 System Model of Data Grid

In this study we use the data grid model, which can be envisioned as a collection of storage subsystems. We model data grid as a network of n storage subsystems with various system resources. Each storage subsystem consists of data stores and computational facilities. The trust-adaptive task manager in each storage subsystem accommodates data-intensive jobs submitted to the data grid. The trust-adaptive task manger aims at tracking load information by periodically changing its load status in a data grid. The incoming data-intensive jobs are placed into a waiting queue managed by a job acceptor in each subsystem.

4 Modeling Tasks and the Trust Requirements in Data Grid

4.1 System Model of Trust-Adaptive Task Manager

We consider a queuing model of the n sites system in a data grid are connected via network to process independent tasks submitted by m users. M = {M1, M2, …, Mn} denote the set of data grid sites. The system model is composed of a schedule queue, TATM (Trust-adaptive task manager) and n local task queues. The function of TATM is aimed to make a good task allocation decision for each arrival task to satisfy its trust requirements and maintain the ideal performance.

TATM scheduler then processes all arrival tasks in First-Come First-Served (FCFS) manner. After being processed TATM, the tasks are dispatched to one of the designated site Mi for execution. The major component of the system model above is TATM, which is composed of four modules: (1) Execution Time Estimator, (2) Trust Overhead Manager, (3) Trust Shortage Degree (TSD) Calculator, (4) Trust Window Adapter. Since execution time of each task can be estimated by statistical prediction [17], it considers that the execution time of the arrival task for each site is a prior and this information is managed in the execution time estimator. Similarly, we assume that the trust overhead for each arrival task on each site is a prior, and the information is maintained in the trust overhead manager. The TSD calculator is used to calculate discrepancies between the arrival task's trust level and the trust service level that each site offers.

The function of trust window adapter is to adjust the size of window to discover a perfect site for the current task so that its trust demands can be well satisfied, then the total execution time will be as small as possible. To illustrate the work principle of trust window adapter, we give an example as below.

After searching site information (such as execution time, trust overhead, trust shortage degree and the size of trust window adapter), the trust window adapter will choose the appropriate site to be assigned to the task. Each site above is inherently heterogeneous in trust. While each task has an array of trust service requests, each site offers the different trust services. The level of the trust service provided by a site is changed in the range from 0.1 to 1.0.

4.2 Modeling Tasks with Trust Requirements

The paper considers a data grid where an application is comprised of an aggregate of tasks performed to accomplish an entire mission. It is assumed that tasks are independent of one another. Each task requires a set of trust services with various trust levels specified by a user. Values of trust levels are diversified in the range from 0.1 to 1.0 as well. Suppose there is a task Ki submitted by a user, Ki is expressed as a set of rational parameters, e.g., Ki= (Si, Ei, Fi, Di, Ti), where Si and Fi are the start and finish time, and Di denotes the amount of data to be protected. Ei is a vector of execution times for task Ki on each site in M, and Ei = (1ie, 2ie, ..., nie). Suppose Ki requires q trust services, Ti = (1is, 2is,..., qis), a vector of trust levels, characterizes the trust requirements of the task.

The TATM has to make use of a effective function to measure the trust shortage degree gained by each arrival task. In particular, the trust shortage degree of task Ki is quantitatively modeled as a function of the discrepancy between trust levels requested and the trust levels offered. The trust shortage degree function for task Ki on site Mj is denoted by TSD: (Ti , Rj) : (where $0 \le Z_i^k \le 1, \sum_{k=1}^{q} Z_i^k = 1$)

$$\mathrm{TSD}(\mathrm{T_i}) = \sum_{k=1}^{q} Z_i^k H(T_i^k, R_j^k{}_{avg}) \tag{1}$$

Notice that Z_i^k is the weight of the kth trust service for task i. Users express the weights to reflect relative priorities given to the required trust services. Trust shortage degree is defined to be a weighted sum of discrepancy values between trust levels requested by a task and the trust levels offered by a site. For each task, a small TSD value means a high satisfaction degree. 0 TSD value implies that a task's trust requirements can be perfectly met. Thus, there exists at least one site Mj that can satisfy the following condition:

$$\forall k \in [1,q], T_i^k \le R_i^j{}_{avg}$$

Let Xi be all possible schedule for task Ki, xi∈X be a scheduling decision of Ki. Given a task, the Trust Shortage Degree of Ti is expected to be minimized:

$$\mathrm{TSD}(\mathrm{X_i}) = \min_{xi \in Xi} \{TSD(X_i)\} = \min_{xi \in Xi} \{ \sum_{k=1}^{q} Z_i^k H(T_i^k, R_j^k{}_{avg}) \} \tag{2}$$

TATM strives to minimize the system's overall TSD value defined as the sum of the trust shortage degree of submitted tasks (See Equation 1). Thus, the following TSD function needs to be minimized:

$$\mathrm{TSD(X)} = \min_{xi \in X} \{\ \mathrm{TSD}(X_i)\} \tag{3}$$

Thus, our proposed TATM scheduling algorithm makes an effort to schedule tasks in a way to minimize it. Now that the TSD value for Ki expresses the trust service satisfaction degree experienced by the task, it is effective to measure quality of trust for Ki during its execution. Therefore, we derive in this section the probability Dpsr (Ki, Mj) that Ti remains task processing success rate during the course of its execution. The trust requirements of a task Ki with respect to the kth trust service is calculated as note that this model assumes that task processing success rate is a function of trust levels, and the distribution of task success rate for any fixed time interval is approximated using a Poisson probability distribution. The task processing success rate model is just for illustration purpose only.

4.3 Trust Overhead Model

We introduce in this section the trust overhead based the above model for data-intensive jobs running on data grids.

At first, we consider a case where input data is locally available and processing is performed on the local site. Let TC_i^k and $T_j^k\,(TC_i^k)$ be a trust level and the overhead of the kth trust level for job Ji . Likewise, let TD_i^k and $T_j^k\,(TD_i^k)$ denote a trust level and the overhead of the kth trust service for the job's data set. The trust overhead Tij experienced by job Ji on local site Mi where the data set is available can be computed as a sum of times spent in trusting the application code and data set. Thus, we can obtain the following Equation (4), where Tck ∈TC and Tdk ∈TD.

$$T_{ij} = \sum_{k=1}^{q} T_j^k\ (TC_i^i\ ) + \sum_{k=1}^{q} T_j^k\ (TD_i^k\ ) = \sum_{k=1}^{q} [T_j^k\,(TC_i^k) + T_j^k (TD_i^k)] \tag{4}$$

Secondly, we derive an expression to calculate the trust overhead of a locally executed job Ji accessing its remote data set. In this scenario, job's data set needs to be fetched from the remote site through networks. Suppose there are p number of trust services provided for a link between site v and site j. The trust overhead of the kth trust service for the data set delivered from site v to site j is expressed as $T_{vj}^k\,(TD_i^k\,)$. The total trust overhead is the sum of the trust overhead caused by data transfer and dataset protections. Therefore, the trust overhead in this case can be written as Equation (5):

$$T_{ij} = \sum_{k=1}^{p} T_{vj}^k (TD_i^k\ ) + \sum_{k=1}^{q} [T_j^k\,(TC_i^k) + T_j^k (TD_i^k)] \tag{5}$$

Third, Expression (5) is used to compute the trust overhead of a remotely executed job Ji that accesses its data set on a local site. Thus, the application code needs to be transmitted to the remote site where the data is stored. The trust overhead of the kth trust service for transmitting the application code from remote site v to local site j is denoted by T_{vj}^{k} (TC_i^k). Finally, the total trust overhead in this case can be calculated as Equation (6):

$$T_{ij} = \sum_{k=1}^{p} T_{vj}^{k}(TC_i^k) + \sum_{k=1}^{q} [T_j^k(TC_i^k) + T_j^k(TD_i^k)] \tag{6}$$

5 The Scheduling Strategy

The quality of trust of a task Ki in data grids with respect to the kth trust service is calculated as $\exp(-\lambda_i^k (e_i^j + \sum_{l=1}^{q} T_{ij}^l (T_i^l)))$ where λ_i^k is the task's processing success rate of the kth trust service, and $T_{ij}^l(s_i^l)$ is the trust overhead experienced by the task on site j. The success rate is expressed as Equation (7) as below:

$$\lambda_i^k = 1 - \exp(-\alpha(1 - T_i^k)) \tag{7}$$

Note that this model assumes that success rate is a function of trust levels, and the distribution of task process success rate for any fixed time interval is approximated using a Poisson probability distribution. The task process success rate model is just for illustration purpose only. Thus, the model can be replaced by any task process success rate model with a reasonable parameter α.

The quality of trust of task Ki on site Mj can be obtained below by considering all trust services provided to the task. Using equation 6, we obtain the overall quality of trust of task Ki in the system as follows, where pij is the probability that Ki is allocated to site Mj. Given a task set K, the probability that all tasks are free from being attacked during their executions is computed.

By substituting the task processing success rate model into Equation 13 and 4, we finally obtain Drf(T) as Equation (8) as shown below:

$$D_{rf}(x) = \prod_{k_i \in K} \{\sum_{j=1}^{n} \{P_{ij} \exp(-(e_i^j + \sum_{l=1}^{q} T_{ij}^l (T_i^l)))\}\} \tag{8}$$

In summary, TSD values show us trust service satisfaction degrees experienced by tasks, while task process success rate probability measured by Equation 14 defines quality of trust provided by data grids.

Given a data grid M and a sequence of jobs J, our scheduling strategy is intended to generate an allocation X minimizing the trust shortage degree computed by (2). Finally, we can obtain the following non-linear optimization problem formulation as Equation (9):

$$\text{Minimize TSD (M,J,X)} = \sum_{j=1}^{n} [x_{ij} TD_{ij}] \tag{9}$$

Now we present the trust-adaptive task management (TATM). The earliest start time of job Ji on site Mj can be approximated as Equation (10) below:

$$\text{es}_j(\text{Ji})= \sum_{J_i \in M_j} (e_l^j + \sum_{k=1}^{q} [T_{lj}^k (TC_l^k) + T_{lj}^k (TD_l^k)]) \tag{10}$$

where rj represents the finish time of a job currently running on the j th site, and the second term on the right-hand side is the overall execution time (trust overhead is factored in) of waiting jobs assigned to site Mj prior to the arrival of Ji .

6 Experimental Results

We use simulation experiments to evaluate benefits of the TATM strategy. To demonstrate the efficiency of TATM, we compared it with two well-known data grid scheduling algorithms: JobRandom and JobDataPresent, among which JobDataPresent is the best algorithm based on results reported in [4].

6.1 Simulation Parameters

Dataset popularities are randomly generated with a regional uniform distribution. All submitted jobs in the trace fall into three categories based on their execution times. The categories include short jobs, medium jobs and long jobs. Accordingly, we assign each category a dataset size range.

6.2 Overall Performance Comparisons

The goal of this experiment is to compare the proposed TATM algorithm against the two heuristics in two dataset replication methods, DataDoNothing and DataRandom.

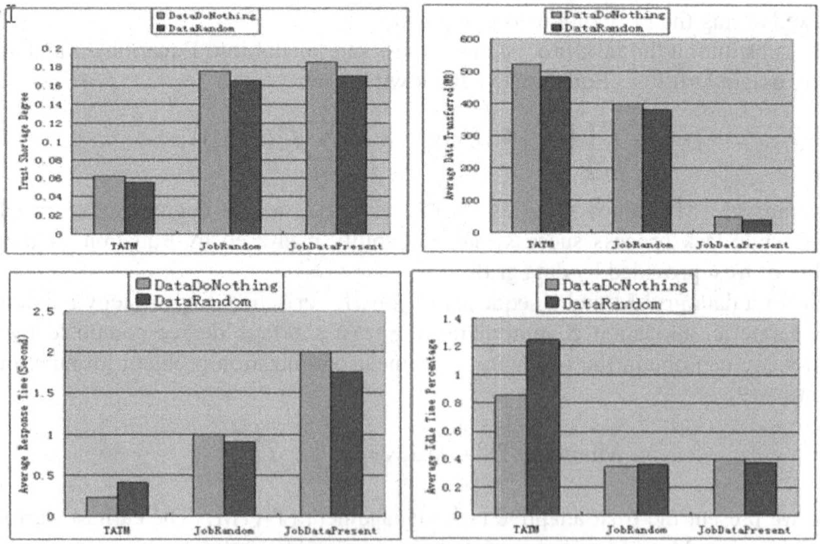

Fig. 1. Overall Performance Comparisons

Figure 1 shows the simulation results for the three scheduling algorithms on data grid with 32 sites. We observe from Figure 1 that TATM significantly depress the trust shortage degree, which improve the trust service satisfaction degree. We attribute the performance improvement to the fact that TATM is a trust-adaptive task manager and judiciously assigns a job to a site not only considering its computational time but also its demands. In addition, TATM is greatly superior to the two alternatives in conventional performance metrics such as average response time, which are mostly concerned by users. The performance improvements of TATM are at the cost of a higher volume of data transferred. The higher idle time percentage suggests that the data grid has potential to accommodate more jobs when the TATM scheduling algorithm is in place.

6.3 Characteristics of Datasets

The popularity of a given dataset is defined as the total number of requests issued by the entire job set on the dataset. Since the distribution of dataset popularity influences the system performance, we tested three distributions of dataset popularity, i.e., uniform distribution, normal distribution, and geometric distribution in this experiment. All other experiments use a local uniform distribution of dataset popularity. The uniform distribution resembles an ideal case where each job randomly selects a dataset to access. The normal distribution reflects the fact that a majority of jobs only request a subset of datasets. The geometric distribution models the scenario in which a group of users interests on some datasets more than others.

There are several important observations based on Figure 2. Firstly, TATM significantly outperforms the other two scheduling algorithms in trust metrics and average response time in the distribution cases. Moreover, the distribution of dataset popularity has no impact on trust performance. Secondly, the normal distribution leads to the

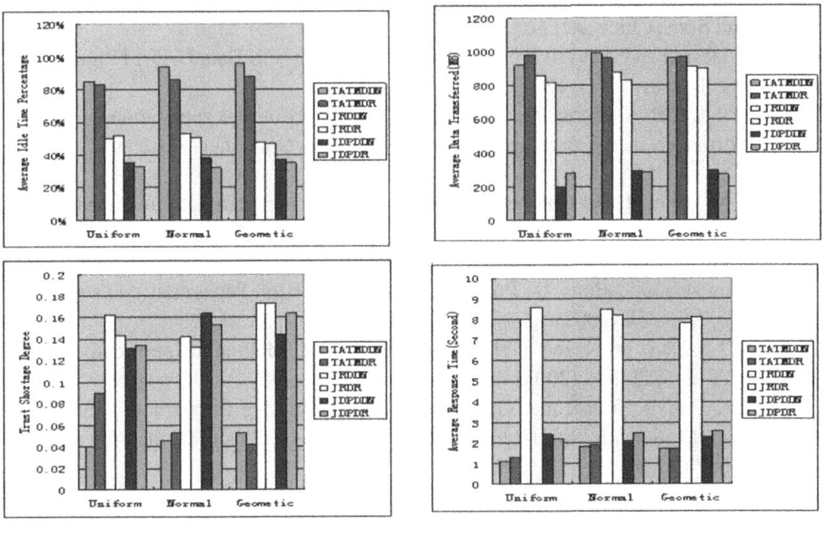

Fig. 2. Performance Impact of Distribution of data set

shortest average response time. This is because the requests on the datasets are evenly distributed, and thus, the workload is well balanced. Thirdly, the geometric distribution results in the lowest volume of average data transferred. This can be explained by the fact that more dataset replicas of a particular dataset are created during the course of job scheduling for the dataset is frequently requested, which is a characteristic decided by geometric distribution. Thus, the subsequent jobs that issue requests on the dataset are most likely to be assigned to a site where a replica of the dataset was already there.

7 Summary and Future Work

In this paper, we considered the flexible trust requirements of applications in the context of task scheduling in data grid. This is important because increasing number of applications running on data grid requires not only descent scheduling performance but also adaptable trust service. To solve this problem, we proposed a trust-adaptive heuristic scheduling that is based on the concept of trust heterogeneity. Experimental results demonstrate that our strategy outperforms existing approaches in both trust and performance on simulated data grid.

In future research, the heuristic will be extended to schedule parallel applications. This work can be accomplished by factoring in precedence constraints among tasks. Further research will be needed to address the issue of trust-adaptive in other distributed computing environments.

References

1. Qin, X., Jiang, H.: Data Grids: Supporting Data-Intensive Applications in Wide Area Networks. In: High Performance Computing: Paradigm and Infrastructure, pp. 481–494. John Wiley and Sons, Chichester (2005)
2. Qin, X.: Design and analysis of a load balancing strategy in Data Grids. Future Generation Computer Systems 23(1), 132–137 (2007)
3. Tang, M., Lee, B.-S., Tang, X., Yeo, C.-K.: The impact of data replication on job scheduling performance in the data grid. Future Generation Computer Systems 22(3), 254–268 (2006)
4. Park, S.-M., Kim, J.-H.: Chameleon: a resource scheduler in a data grid environment. In: Proc. Int'l. Symp. Cluster Computing and the Grid, pp. 258–265 (2003)
5. Ranganathan, K., Foster, I.: Decoupling computation and data scheduling in distributed data-intensive applications. In: Proc. IEEE Int. Symp. High Performance Distributed Computing, pp. 352–358 (2002)
6. Winton, L.: Data grids and high energy physics. A Melbourne perspective.Space Sci. Rev. 107(1–2), 523–540 (2003)
7. Guido, J., Van 't Noordende, G.J.: A trusted data storage infrastructure for grid-based medical applications. In: 8th International Symposium on Cluster Computing and the Grid, pp. 627–632 (2008)
8. Hassan, M.W., McClatchey, R., Willers, I.W.: A Scalable Evidence Based Self-Managing Framework for Trust Management. Electronic Notes in Theoretical Computer Science 179, 59–73 (2007)

9. Azzedin, F., Maheswaran, M.: A Trust Brokering System and Its Application to Resource Management in Public Resource Grid. In: Intl. Parallel and Distributed Computing Symposium, pp. 289–298 (2004)
10. Lin, K.-J., Lu, H., Yu, T., Tai, C.-e., Hsu, J.Y.-j.: A Reputation and Trust Management Broker Framework for Web Applications. In: Proceedings of the 2005 IEEE International Conference on E-Technology, E-Commerce, and E-Service, pp. 262–269 (2005)
11. Sonnek, J.D., Chandra, A., Weissman, J.B.: Adaptive Reputation-Based Scheduling on Unreliable Distributed Infrastructures. IEEE Trans. Parallel Distributed System 18(11), 1551–1564 (2007)
12. Sonnek, J.D., Nathan, M., Chandra, A., Weissman, J.B.: Reputation-Based Scheduling on Unreliable Distributed Infrastructures. In: IEEE International Conference on Distributed Computing Systems, pp. 30–37 (2006)
13. Song, S., Kwok, Y.-K., Hwang, K.: Trusted job scheduling in open computational grids: security-driven heuristics and a fast genetic algorithms. In: Proc. Int'l. Symp. Parallel and Distributed Processing, pp. 33–40 (2005)
14. Song, S., Hwang, K., Kwok, Y.-K.: Risk-Resilient Heuristics and Genetic Algorithms for Security-Assured Grid Job Scheduling. IEEE Trans. Computers 55(6), 703–719 (2006)
15. Sonnek, J.D., Weissman, J.B.: A Quantitative Comparison of Reputation Systems in the Grid. In: Proceedings of the 6th IEEE/ACM International Workshop on Grid Computing, pp. 242–249 (2005)
16. Liang, Z., Shi, W.: Analysis of ratings on trust inference in open environments. Performance Evaluation 65(2), 99–128 (2008)
17. Braun, T.D., et al.: A Comparison Study of Static Mapping Heuristics for a Class of Metatasks on Heterogeneous Computing Systems. In: Proc. Workshop Heterogeneous Computing, pp. 15–29 (1999)

Robust Stability of Multi-variable Networked Control Systems with Random Time Delay

Li-sheng Wei[1], Ming Jiang[1], and Min-rui Fei[2]

[1] Anhui University of Technology and Science, Wuhu, Anhui 241000, China
lshwei_11@163.com, my_kjjm@163.com
[2] Shanghai Key Laboratory of Power Station Automation Technology,
Shanghai University, Shanghai 200072, China
mrfei888@x263.net

Abstract. The issue of delay-dependent robust stability for uncertain continuous-time multi-variable networked control systems with state feedback controller is researched, where its network transmission is connected with network-induced delay. The complete mathematical model is derived. And the sufficient condition for asymptotical stability is analyzed and the maximum allowable delay conditions for systems are derived by using Lyapunov stability theory combined with free-weighting matrices techniques. The efficacy and feasibility of the proposed methods is shown by presenting simulation results from multi-variable networked control example with uncertainty.

Keywords: Networked Control Systems(NCSs), Robust Stability, Delay, Free-Weighting Matrix, LMI.

1 Introduction

The concept of Networked Control Systems (NCSs) was proposed in 1980s, which was called Integrated Communication and Control Systems (ICCS) at that moment [1]. NCSs are the feedback control loops closed through a real time network. That is, in NCSs, communication networks are employed to exchange the information and control signals between control system components [2]. The aim of such systems is to ensure data transmission and coordinating manipulation among spatially distributed components. Compared with conventional point-to-point control systems, the main advantages of NCSs are low cost, reduced weight, simple installation and maintenance, and high flexibility and reliability. It is well known that NCSs have been widely applied to many complicated control systems, such as, manufacturing plants, vehicles, aircraft, and spacecraft [3].

Despite of the great advantages and wide applications, communication networks in the control loops make the analysis and design of NCSs complicated. Recently, the modeling and analysis of NCSs have been a hot research topic, and have attracted much research interest [4-7]. Many new problems such as network-induced delays and packet dropouts have emerged and become a topic of significant interest to the control community [8]. Among these problems, network-induced delays are of prior importance, and are usually the major causes of the deterioration of the dynamic

W. Liu et al. (Eds.): WISM 2009, LNCS 5854, pp. 586–595, 2009.

performance of the system and potential instability of the system. Du et al. consider the delay-dependent stability and stabilization for a class of NCSs with independent sensors and actuators [9]. Liu et al. discussed stability of NCSs with sensors, controllers and actuators clock-driven [10]. Huang et al. analyze the integrity design of NCSs with actuator failure by using delay-dependent approach and integral-inequality method. And they get the sufficient stabilizability condition of system with memoryless networked state feedback controller based on linear matrix inequalities [11]. However, in these papers, controller design of Multi-variable NCSs has not been considered. And the system uncertainties led by modeling errors, network environment and the aging of components are not involved.

Aiming at this instance, the delay-dependent robust stability of Continuous-time NCSs with uncertainty and network-induced delay is presented based on Lyapunov stability theory and free-weighting matrix approach.

The remainder of this paper is organized as follows. In Section 2, the sufficient condition for convergence of the proposed robust control algorithm is derived. And the maximum allowable transfer interval is used in its place to ensure absolute stability of an NCSs. In Section 3, simulation results with the proposed control scheme are obtained. Finally conclusion remarks are presented in Section 4.

2 Robust Stability of Multi-variable NCSs

2.1 System Modeling

Consider a class of multi-input and multi-output linear uncertain system described by the following equations

$$\begin{aligned} \dot{x}(t) &= (A+\Delta A)x(t)+(B+\Delta B)u(t) \\ y(t) &= Cx(t) \\ x(t) &= \phi(t) = 0, \qquad t\in[-\tau,0] \end{aligned} \tag{1}$$

where $x(t)\in R^n$ is the state vector, $u(t)\in R^n$ is the control input vector, $y_p(t)\in R^r$ is the measurable output vector, A , B , C are known constant real matrices with proper dimensions, ΔA and ΔB are matrix-valued functions of appropriate dimension parameter uncertainty in the system model. $\phi(t)$ is a given continuous vector valued initial function and subsection differential on $[-\tau,0]$. The parameter uncertainties considered are assumed to be norm bounded and satisfy

$$[\Delta A \quad \Delta B] = DF(t)[E_a \quad E_b] \tag{2}$$

where D , E_a , E_b are known constant real matrices of appropriate dimensions that represent the structure of uncertainty, and $F(t)$ is an uncertain matrix function with Lebesgue measurable elements and satisfies

$$F^T(t)F(t) \le I \tag{3}$$

where I denotes the identity matrix of appropriate dimension.

For the convenience of investigation, without loss of generality we make the following assumptions [12-13]:

1) All the sensors are time-driven, the sample period is T, and the time gas among the sensors' nodes can be neglected, the controllers and the actuators are event-driven.

2) The data is transmitted with one data packet through the communication network. That is, a single packet is enough to transmit the plant output at every sampling period.

Then the real memory-less state feedback control input $u(t)$ realized through zeroth-order hold in Equation (1) is a piecewise constant function.

$$u(t) = Kx(t) \tag{4}$$

If we consider the effect of the network-induced delay and network packet dropout on the NCSs, then the real control system (1) with Equation (4) can be rewritten as [14-15]

$$\begin{aligned} \dot{x}(t) &= (A + \Delta A)x(t) + (B + \Delta B)Kx(t-\tau_1), \\ y(t) &= Cx(t) \\ x(t) &= \phi(t), \qquad t \in [-\tau, 0] \end{aligned} \tag{5}$$

where τ_1 is the time delay, which denotes the time from the instant when the sensor nodes sample sensor data from a plant to the instant when actuators transfer data to the plant. Then we assume that $u(t) = 0$ before the first control signal reaches the plant and a constant $\tau > 0$ exists such that $0 \le \tau_1 \le \tau < \infty$.

According to Equation (2), system (5) can be rewritten as the following equivalent form

$$\begin{aligned} \dot{x}(t) &= (A + DF(t)E_a)x(t) + (B + DF(t)E_b)Kx(t-\tau_1) \\ y(t) &= Cx(t) \\ x(t) &= \Phi(t), \qquad t \in [-\tau, 0] \end{aligned} \tag{6}$$

For brevity of our discussion, rearranging Equation (6), we have

$$\begin{aligned} \dot{x}(t) &= (A + DF(t)E_a)x(t) + (A_1 + DF(t)E_1)x(t-\tau_1) \\ y(t) &= Cx(t) \\ x(t) &= \Phi(t), \qquad t \in [-\tau, 0] \end{aligned} \tag{7}$$

This model takes network-induced delay of the uncertainty multi-variable NCSs into consideration. With this model, the robust stability sufficient condition is proposed in the next section.

2.2 Robust Stability of NCSs

Before the development of the main result, the following lemmas will be used.

Lemma1[16](Schur complement theorem): Given any symmetric matrix $X = X^T$, the matrix inequality (8) holds.

$$S = \begin{bmatrix} S_{11} & S_{12} \\ S_{21} & S_{22} \end{bmatrix} \tag{8}$$

where $S_{11} \in R^{r \times r}$, S_{12}, S_{21}, S_{22} are known real matrices with proper dimensions. The following three conditions are equivalent

(1) $S < 0$

(2) $S_{11} - S_{12}S_{22}^{-1}S_{12}^T < 0$ and $S_{22} < 0$

(3) $S_{22} - S_{12}^T S_{11}^{-1} S_{12} < 0$ and $S_{11} < 0$

Lemma2[16]: Given any symmetric matrix $Q = Q^T$, and constant matrix H、E with appropriate dimension, if $F(t)$ is an uncertain matrix function with Lebesgue measurable elements and satisfies $F^T(t)F(t) \le I$, in which I denotes the identity matrix of appropriate dimension. Then

$$Q + HF(t)E + E^T F^T(t)H^T < 0 \tag{9}$$

If and only if there exists positive constant $\varepsilon > 0$ such that

$$Q + \varepsilon^{-1}HH^T + \varepsilon E^T E < 0 \tag{10}$$

In the following, the sufficient condition for convergence is given.

Theorem1: considering the closed-loop system (7), given positive constant $0 \le \tau$, if there exist positive matrix P、Q、Z, matrix Y、N_1、N_2 with proper dimensions and positive constant $\varepsilon > 0$ such that the LMI (11) holds for all uncertain matrix function $F(t)$.

$$\begin{bmatrix} \Phi_{11} + \varepsilon E_a^T E_a & \Phi_{12} + \varepsilon E_a^T E_1 & \tau A^T Z & PD \\ * & \Phi_{22} + \varepsilon E_1^T E_1 & \tau A_1^T Z & 0 \\ * & * & -\tau Z & \tau ZD \\ * & * & * & -\tau I \end{bmatrix} < 0 \tag{11a}$$

$$\Pi = \begin{bmatrix} X_{11} & X_{12} & N_1 \\ * & X_{22} & N_2 \\ * & * & Z \end{bmatrix} \ge 0 \tag{11b}$$

Then system (7) is asymptotically stable when all network-induced delay satisfies $0 \le \tau_1 \le \tau$.

where $*$ denotes the symmetric terms in a symmetric matrix, and

$$\Phi_{11} = PA + A^T P + N_1 + N_1^T + Q + \tau X_{11}$$
$$\Phi_{12} = \Phi_{21}^T = PA_1 - N_1 + N_2^T + \tau X_{12}$$
$$\Phi_{22} = -N_2 - N_2^T - Q + \tau X_{22}$$
$$X = \begin{bmatrix} X_{11} & X_{12} \\ X_{21} & X_{22} \end{bmatrix} \geq 0$$

Proof: Construct a Lyapunov-Krasovskii functional candidate is presented as follows:

$$V(t,x) = V_1(t,x) + V_2(t,x) + V_3(t,x) \tag{12}$$

where P 、 Q and R are symmetric positive matrices, and

$$V_1(t,x) = x^T(t)Px(t), V_2(t,x) = \int_{-\tau_1}^{0}\int_{t+\beta}^{t} \dot{x}^T(\alpha)Z\dot{x}(\alpha)d\alpha d\beta, V_3(t,x) = \int_{t-\tau_1}^{t} x^T(\alpha)Qx(\alpha)d\alpha$$

On the one hand, from the Leibniz-Newton formula, we have

$$x(t) - x(t-\tau_1) = \int_{t-\tau_1}^{t} \dot{x}(\alpha)d\alpha \tag{13}$$

Then the following equation is true for any free-weighting matrices N_1 and N_2 with appropriate dimensions.

$$2\left[x^T(t)N_1 + x^T(t-\tau_1)N_2\right] \times \left[x(t) - x(t-\tau_1) - \int_{t-\tau_1}^{t} \dot{x}(\alpha)d\alpha\right] = 0 \tag{14}$$

On the other hand, give the symmetric positive matrices X as

$$X = \begin{bmatrix} X_{11} & X_{12} \\ * & X_{22} \end{bmatrix} \geq 0 \tag{15}$$

Then we have

$$\tau\overline{x}^T(t)X\overline{x}(t) - \int_{t-\tau_1}^{t} \overline{x}^T(t)X\overline{x}(t)d\alpha \geq 0 \tag{16}$$

where $\overline{x}(t) = \left[x^T(t) \quad x^T(t-\tau_1)\right]^T$.

Calculating the derivative of $V(t)$ along the solutions of system (7) and adding the left side of equation (14) and inequality (16) into it, we have

$$\begin{aligned}
\dot{V}(t,x) &= \dot{V}_1(t,x)+\dot{V}_2(t,x)+\dot{V}_3(t,x) \\
&= x^T(t)\left(P(A+DF(t)E_a)+(A+DF(t)E_a)^T P\right)x(t) \\
&\quad +2x^T(t)P(A_1+DF(t)E_1)x(t-\tau_1)+x^T(t)Qx(t) \\
&\quad +x^T(t-\tau_1)Qx(t-\tau_1)+\tau_1\dot{x}^T(t)Z\dot{x}(t)-\int_{t-\tau_1}^{t}\dot{x}^T(\alpha)Z\dot{x}(\alpha)d\alpha \\
&\le x^T(t)\left(P(A+DF(t)E_a)+(A+DF(t)E_a)^T P\right)x(t) \\
&\quad +2x^T(t)P(A_1+DF(t)E_1)x(t-\tau_1)+x^T(t)Qx(t) \\
&\quad +x^T(t-\tau_1)Qx(t-\tau_1)+\tau\dot{x}^T(t)Z\dot{x}(t)-\int_{t-\tau_1}^{t}\dot{x}^T(\alpha)Z\dot{x}(\alpha)d\alpha \\
&\quad +2\left[x^T(t)N_1+x^T(t-\tau_1)N_2\right]\times\left[x(t)-x(t-\tau_1)-\int_{t-\tau_1}^{t}\dot{x}(\alpha)d\alpha\right] \\
&\quad +\tau\overline{x}^T(t)X\overline{x}(t)-\int_{t-\tau_1}^{t}\overline{x}^T(t)X\overline{x}(t)d\alpha \\
&= \overline{x}^T(t)\Omega\overline{x}(t)-\int_{t-\tau_1}^{t}\tilde{x}^T(t)\Pi\tilde{x}(t)d\alpha
\end{aligned} \tag{17}$$

where

$$\Omega=\begin{bmatrix}\Omega_{11}+\tau M_{11} & \Omega_{12}+\tau M_{12}\\ * & \Omega_{22}+\tau M_{22}\end{bmatrix},\Pi=\begin{bmatrix}X_{11} & X_{12} & N_1\\ * & X_{22} & N_2\\ * & * & Z\end{bmatrix}\ge 0$$

$$\tilde{x}^T(t)=\left[x^T(t)\quad x^T(t-\tau_1)\quad \dot{x}^T(t)\right]^T$$

$$\Omega_{11}=P(A+DF(t)E_a)+(A+DF(t)E_a)^T P+N_1+N_1^T+Q+\tau X_{11}$$

$$\Omega_{12}=\Omega_{21}^T=P(A_1+DF(t)E_1)-N_1+N_2^T+\tau X_{12}$$

$$\Omega_{22}=-N_2-N_2^T-Q+\tau X_{22}$$

$$M_{11}=A+DF(t)E_a)^T Z(A+DF(t)E_a$$

$$M_{12}=A+DF(t)E_a)^T Z(A_1+DF(t)E_1$$

$$M_{22}=A_1+DF(t)E_1)^T Z(A_1+DF(t)E_1$$

It can be proved that if the following matrix inequality (18) holds, then the closed-loop NCSs (7) with uncertainty is asymptotically stable by the Lyapunov-Krasovskii functional theorem.

$$\begin{aligned} &\Omega < 0 \\ &\Pi \geq 0 \end{aligned} \tag{18}$$

By the Schur complement theorem, and according to inequality (18), we have

$$\Omega = \begin{bmatrix} \Omega_{11} + \tau M_{11} & \Omega_{12} + \tau M_{12} \\ * & \Omega_{22} + \tau M_{22} \end{bmatrix}$$

$$= \begin{bmatrix} \Omega_{11} & \Omega_{12} \\ * & \Omega_{22} \end{bmatrix} - \begin{bmatrix} \tau(A + DF(t)E_a)^T Z \\ \tau(A_1 + DF(t)E_1)Z \end{bmatrix} (-\tau Z)^{-1} \begin{bmatrix} \tau A + (DF(t)E_a)^T Z \\ \tau(A_1 + DF(t)E_1)^T Z \end{bmatrix}^T$$

$$= \begin{bmatrix} \Omega_{11} & \Omega_{12} & \tau(A + DF(t)E_a)^T Z \\ * & \Omega_{22} & \tau(A_1 + DF(t)E_1)^T Z \\ * & * & -\tau Z \end{bmatrix}$$

Then rearranging inequality (18), we have

$$\begin{bmatrix} \Phi_{11} + PDF(t)E_a + E_a^T F^T(t) D^T P & \Phi_{12} + PDF(t)E_1 & \tau A^T Z + \tau E_a^T F^T(t) D^T Z \\ * & \Phi_{22} & \tau A_1^T Z + E_1^T F^T(t) D^T Z \\ * & * & -\tau Z \end{bmatrix} < 0 \tag{19}$$

where $\Phi_{11} = PA + A^T P + N_1 + N_1^T + Q + \tau X_{11}$

$\Phi_{12} = \Phi_{21}^T = PA_1 - N_1 + N_2^T + \tau X_{12}$

$\Phi_{22} = -N_2 - N_2^T - Q + \tau X_{22}$

That is

$$\Phi + \begin{bmatrix} PD \\ 0 \\ \tau ZD \end{bmatrix} F(t) \begin{bmatrix} E_a & E_1 & 0 \end{bmatrix} + \begin{bmatrix} E_a^T \\ E_1^T \\ 0 \end{bmatrix} F^T(t) \begin{bmatrix} D^T P & 0 & \tau D^T Z \end{bmatrix} < 0 \tag{20}$$

where

$$\Phi = \begin{bmatrix} \Phi_{11} & \Phi_{12} & \tau A^T Z \\ * & \Phi_{22} & \tau A_1^T Z \\ * & * & -\tau Z \end{bmatrix}.$$

Using Lemma 2, the following inequality (21) can be obtained from inequality (20).

$$\Phi + \varepsilon^{-1} \begin{bmatrix} PD \\ 0 \\ \tau ZD \end{bmatrix} \begin{bmatrix} D^T P & 0 & \tau D^T Z \end{bmatrix} + \varepsilon \begin{bmatrix} E_a^T \\ E_1^T \\ 0 \end{bmatrix} \begin{bmatrix} E_a & E_1 & 0 \end{bmatrix} < 0 \tag{21}$$

Rearranging inequality (21), we have

$$\begin{bmatrix} \Phi_{11}+\varepsilon E_a^T E_a & \Phi_{12}+\varepsilon E_a^T E_1 & \tau A^T Z \\ * & \Phi_{22}+\varepsilon E_1^T E_1 & \tau A_1^T Z \\ * & * & -\tau Z \end{bmatrix} - \begin{bmatrix} PD \\ 0 \\ \tau ZD \end{bmatrix} (-\varepsilon I)^{-1} \begin{bmatrix} D^T P & 0 & \tau D^T Z \end{bmatrix} < 0 \tag{22}$$

Now, using Schur complement theorem to simplify and rearrange (22), we have

$$\begin{bmatrix} \Phi_{11}+\varepsilon E_a^T E_a & \Phi_{12}+\varepsilon E_a^T E_1 & \tau A^T Z & PD \\ * & \Phi_{22}+\varepsilon E_1^T E_1 & \tau A_1^T Z & 0 \\ * & * & -\tau Z & \tau ZD \\ * & * & * & -\tau I \end{bmatrix} < 0 \tag{23}$$

So we can see inequality (23) is the same as inequality (11a) in Theorem 1.

The proof of Theorem 1 is completed. ■

3 Numerical Example

In this section, the effectiveness of the proposed delay-dependent robust stability of multi-variable NCSs is demonstrated by simulations.

Consider the following multi-variable system with uncertainty

$$A = \begin{bmatrix} -10.38 & -0.2077 & 6.715 & -5.676 \\ -0.5814 & -4.29 & -10 & 0.675 \\ 1.067 & 4.273 & -6.653 & 5.893 \\ 0.048 & 4.273 & 1.343 & -2.104 \end{bmatrix}, \quad B = \begin{bmatrix} 0 & 0 \\ 5.679 & 0 \\ 1.136 & -3.146 \\ 1.136 & 0 \end{bmatrix}$$

$$C = \begin{bmatrix} 1 & 0 & 1 & -1 \\ 0 & 1 & 0 & 0 \end{bmatrix}, \quad k = \begin{bmatrix} -0.0907 & 0.4222 & -1.7898 & 0.5445 \\ -0.2190 & -0.9225 & -0.1876 & -1.8060 \end{bmatrix}$$

$$E_a = diag\{0.3, 1.2, 0.06, 0.7\}, \; E_1 = diag\{0.1, 0.05, 0.8, 0.3\}, \; D = I$$

By Theorem 1 in the paper, the maximum allowable transfer interval (MATI) that guarantees the stability of system with uncertainty is 135.5ms. However, using the Theorem 2 in reference [17], the maximum value of τ is 77.4ms. Therefore, the free-weighting matrix introduced in Theorem 1 improves the result of reference [17] and has less conservation. The simulation results are shown in the flowing figure.

By simulating a multi-variable plant with uncertainty, we successfully reduce the system conservation and robustness. The results are better than the ones that the theorem 2 in reference [17] can provide.

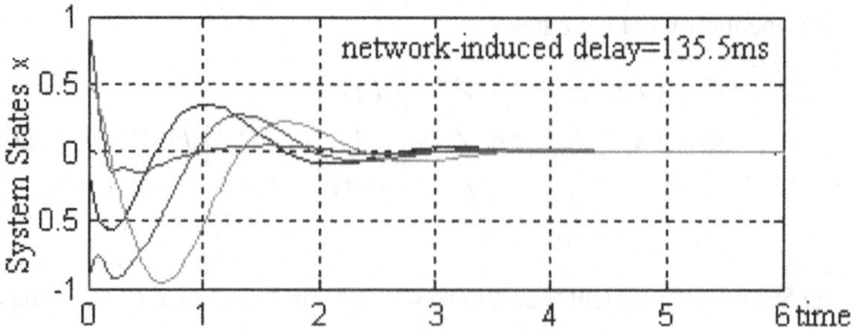

Fig. 1. Simulation results (network-induced delay=135.5ms)

4 Conclusions

In this paper, a method of analysis the robust stability sufficient condition for an uncertain continuous-time multi-variable NCSs with random communication network-induced delays has been proposed. Based on Lyapunov stability theory combined with LMIs techniques, a sufficient convergence condition is derived by introducing some free-weighting matrices which can be selected properly to lead to much less conservative results. Simulation results indicate the effectiveness and robustness of the proposed methods.

Acknowledgments. This work was supported by National Natural Science Foundation of China under grant 60774059 and 60843003, Anhui Provincial Natural Science Foundation under grant 090412071, unlight Plan Following Project of Shanghai Municipal Education Commission, Shanghai Leading Academic Disciplines under grant T0103.

References

1. Halevi, Y., Ray, A.: Integrated communication and control Systems: part I-analysis. ASME J. Dyn. Syst., Meas., and Control 110(4), 367–373 (1988)
2. Zhang, W., Branicky, M.S., Phillips, S.M.: Stability of networked control system. IEEE Control Systems Magazine 21(2), 84–99 (2001)
3. Walsh, G.C., Ye, H., Bushnell, L.G.: Stability analysis of networked control systems. IEEE Transactions on Control Systems Technology 10(3), 438–446 (2002)
4. Schenato, L.: Optimal Estimation in Networked Control Systems Subject to Random Delay and Packet Drop. IEEE Transactions on Automatic Control 53(5), 1311–1317 (2008)
5. Yang, T.C.: Networked control system: a brief survey. IEE Proceedings of Control Theory Applications 153(4), 403–412 (2006)
6. Gao, H., Meng, X., Chen, T.: Stabilization of Networked Control Systems with a New Delay Characterization. IEEE Transactions on Automatic Control 53(9), 2142–2148 (2008)
7. Wei, L.S., Fei, M.R.: A Real-time Optimization Grey Prediction Method for Delay Estimation in NCS. In: The 6th IEEE International Conference on Control and Automation, Guangzhou, China, May 30-June 1, pp. 514–517 (2007)

8. Hespanha, J.P., Naghshtabrizi, P., Xu, Y.: A Survey of Recent Results in Networked Control Systems. Proceedings of IEEE 95(1), 138–162 (2007)
9. Zhaoping, D., Qingling, Z., Lili, L.: Elay-dependent stabilization of MIMO networked control systems. In: 27th Chinese Control Conference, Yunnan, China, July 16-18, pp. 81–85 (2008)
10. Liu, L.Y., Lv, W.J., Chen, Y.Z.: Stability analysis of MIMO networked control system. Information and Control 35, 393–397 (2006)
11. Huang, H., Han, X., Ji, X., Wang, Z.: Fault-tolerant Control of Networked Control System with Packet Dropout and Transmission delays. In: IEEE International Conference on Networking, Sensing and Control, Sanya, China, April 6-8, pp. 325–329 (2008)
12. Wei, L., Fei, M., Hu, H.: Modeling and stability analysis of grey–fuzzy predictive control. Neurocomputing 72(1-3), 197–202 (2008)
13. Liu, G.P., Chai, S.C., Rees, D.: Networked Predictive Control of Internet/Intranet Based Systems. In: Proceedings of the 25th Chinese Control Conference, Harbin, Heilongjiang, August 7-11, 2006, pp. 2024–2029 (2006)
14. Schenato, L.: Optimal Estimation in Networked Control Systems Subject to Random Delay and Packet Drop. IEEE Transactions on Automatic Control 53(5), 1311–1317 (2008)
15. Yue, D., Han, Q.-L., Peng, C.: State feedback controller design of networked control systems. IEEE Trans. on Circuits and Systems-II 51(11), 640–644 (2004)
16. Peterson, I.R., Hollot, C.V.: A Riccati equation approach to the stabilization of certain linear systems. Automatica 22(4), 397–411 (1986)
17. Huaicheng, Y., Xinhan, H., Min, W.: Delay-dependent stability of networked control systems with uncertainties and multiple time-varying delays. Journal of Control Theory and Applications 25(02), 303–306 (2008) (In Chinese)

Author Index